THE AMATEUR ENTOMOLOGISTS' SOCIETY

PRACTICAL HINTS

FOR THE

FIELD LEPIDOPTERIST

BY

J. W. TUTT

A FACSIMILE REPRINT

WITH A FOREWORD BY

PAUL WARING

AND INTRODUCTION BY

BRIAN O. C. GARDINER, FRES, FLS

THE AMATEUR ENTOMOLOGIST
VOL 23
1994

This facsimile edition © 1994 by
The Amateur Entomologists' Society
(Registered Charity No. 267430)
4 Steep Close, Orpington, Kent BR6 6DS.

ISBN 09000 54 59X

Printed by Henry Ling Ltd, The Dorset Press, Dorchester DT1 1HD

FOREWORD
by
Paul Waring

Practical Hints contains a wealth of information on the behaviour and ecology of the Lepidoptera and it collects together much information on moths which is not readily available elsewhere. The book stems from the era before mercury vapour light traps, when fieldcraft was the only way to find many species and the knowledge of the immature stages of the moths was accumulating rapidly. As founder and editor of *The Entomologist's Record & Journal of Variation*, Tutt was in a position to intercept and include many scraps of useful information which came his way concerning the requirements, seasonality and behaviour of moths and butterflies. This is precisely the sort of information needed by modern day field entomologists, biological surveyors and site managers.

Being a facsimile, this work must be taken in the context of its time. Some species have disappeared from the localities mentioned and some of the sites are greatly changed or have been completely destroyed. Before adopting any of the collecting methods described, readers are reminded of the Code for Insect Collecting (Appendix 4 in Vol. 13 of *The Amateur Entomologist*).

The majority of the behavioural observations in *Practical Hints* remain valid and extremely useful today. As one who has copied out and consulted large chunks from library copies over the years and found them a great help in fieldwork, I am most grateful to the AES for making this classic widely available again at an affordable price. The accompanying cross-referencing index to modern scientific and vernacular names is an added bonus which will save a great deal of time and confusion when using this book.

INTRODUCTION
by
Brian O. C. Gardiner, FRES, FLS

Produced at the turn of this century by J. W. Tutt, *Practical Hints for the Field Lepidopterist* was a compilation of information both from his own experience and information supplied to him by his many correspondents. It was published between 1901 and 1905 in three parts, with a reprint of Part I in 1908. It has been useful to entomologists ever since. Most copies appear to have been issued interleaved for the user to make his or her own observations. The three parts were often later bound as one volume and there are a number of copies around where only any two of the parts have been bound together. In view of the intensive use to which copies were often put, with the less well-bound paperback issues in particular becoming disbound and so being discarded, extant copies have become fewer and fewer as the years have rolled by, but the demand has continued and second-hand (by now third- and fourth-hand) copies have become not only very expensive but very hard to find in the first place.

After due consideration the Council of The Amateur Entomologists' Society decided that it would be a great service to entomologists to produce a facsimile reprint and so make it available again at a reasonable price.

To those who have not read it, it might be thought that *Practical Hints* is a book that advocates collecting. This is not so, as Tutt himself stated quite clearly in his Prefaces and on page 1 of Part III. It is worth quoting part of one Preface:–

"We would again reiterate what we insisted on in the publication of Part I, *viz.*, that these have been put together in the hope that the young scientist will be enabled rapidly to make a sufficiently large number of observations to aid him to make the necessary generalizations for his work to have a scientific value. For the lepidopterist who uses these notes only to enlarge his collection and to help to exterminate our fauna, and whose work has no scientific basis whatsoever, we can only express a sorrow bordering on contempt."

Part II of the work contains an extensive 40 page biological account of the early stages and how to keep, rear, photograph and describe them. In the 1908 reprint of Part I Tutt added an extensive section on the collection and conservation of lepidoptera, together with information on pinning, setting and labelling. Since this facsimile edition is taken from the first edition, this additional material is not included, but the information it contains, updated to modern usage, is available in *A Lepidopterist's Handbook* by Richard Dickson; *Insect Photography for the Amateur* by Peter E. Lindsley; *Practical Hints for Collecting and Studying the Microlepidoptera* by Paul Sokoloff and *Breeding the British Butterflies* by Peter W. Cribb (Volumes 13, 14, 16, 18 of *The Amateur Entomologist* respectively).

During the 90 years that have passed since the first publication of *Practical Hints* there have been many changes in the names of the lepidoptera; a few indeed have not only changed completely in the intervening years but have eventually returned to the original name! Others have not been so lucky with both specific and generic names changed. It was therefore decided to prepare a separate cross-referencing index giving both the modern name of the lepidopteran and its vernacular name, although these latter were not used by Tutt. To have included this index in the book would not only have added to its bulk but made it difficult to consult both the original and the new at the same time. This cross-referencing index has therefore been published separately as Volume 23A of *The Amateur Entomologist* and will be supplied to every purchaser of this edition of Tutt's *Practical Hints*.

It has been researched and prepared by me and I would like to thank Col. Maitland Emmet in particular who pointed out a number of serious errors; Paul Waring both for writing the foreword and checking the index; Colin Hart and Bernard Skinner for also checking it. Thanks are also due to the unknown previous owner of a copy who had very thoughtfully annotated and written the English vernacular names into an interleaved copy. In quite a few instances it proved far easier to trace a modern scientific name through these than by other means. The arrangement followed is to list the name as in Tutt's index, followed by the modern name given in *The Moths & Butterflies of Great Britain & Ireland* Vol 7(2) Edited by A. Maitland Emmet and John Heath (1992). Where given in that work the vernacular name then follows, but for the majority of the Microlepidoptera the vernacular name is that used by I. R. P. Heslop in his *Revised Indexed Check-list of the British Lepidoptera* published in *The Entomologist's Gazette* (1959–1962) **10**:177–187; **11**:55–66; 169–178; 227–234; **12**:97–108; 205–230; **13**: 7–11; 86–89; 185–204. Conversely the second part of the index gives the modern name followed by Tutt's, but not the vernacular. The page references to *Practical Hints* are not given anew as these are to be found in the facsimile reprint.

The names of the foodplants quoted in *Practical Hints* were not indexed therein, nor are they here. There has been less change in the nomenclature of the flora than there has been of the lepidoptera. To those who might find an unfamiliar plant name I would refer them to any edition of Bentham & Hooker's *Handbook of the British Flora* published in the early part of this century.

I also felt it appropriate to include the portrait of Mr Tutt which is printed overleaf and was originally Plate VI in volume 23 (1911) of *The Entomologist's Record & Journal of Variation* where a full appreciation of him and his works, together with a full bibliography of his publications will be found, and I thank the present editor of that journal for permission to reproduce it here.

PRACTICAL HINTS

FOR THE

FIELD LEPIDOPTERIST

BY

J. W. TUTT, F.E.S.

Editor of *The Entomologist's Record and Journal of Variation ;*
Author of *British Lepidoptera, British Noctuae and their Varieties,*
British Butterflies, British Moths, &c.

I

Price 5s. 6d. net (interleaved 6s.).

April 1901.

LONDON:
ELLIOT STOCK, 62, PATERNOSTER ROW, E.C.

BERLIN:
R. FRIEDLÄNDER & SOHN, 11, CARLSTRASSE, N.W.

PREFACE.

In offering this little brochure to field lepidopterists the author is fully aware of its elementary nature and painfully conscious of its incompleteness. It was simply undertaken as a reprint of a series of "Practical Hints" that have been appearing for some time in *The Entomologist's Record and Journal of Variation*, and in response to the repeated solicitations of a certain number of lepidopterists who had found the original notes useful, but who required to have them in a better arranged and more concentrated form, the inconvenience of searching through the back volumes for unclassified material being very great. It is, however, much more than a reprint, about one-half of the contents being matter that has not yet been published.

It is the knowledge that only a man (or woman) who has collected his (or her) own material, who has studied the life-histories and habits of some one or many species in detail, can ever become a really useful scientific naturalist, coupled with the feeling that much preliminary labour may be saved by knowing how, when, and where, to look for certain species, that the following hints are given. One recognises that this may have a directly contrary effect to that wished, in the loss of initiative and self-reliance in young lepidopterists, and may lead to the rapid making of collections by some without advantaging the science of entomology, but such individuals as would thus abuse the information herein contained never would probably become naturalists and one can only hope that their purposeless efforts will not be inimical to the progress of science. At the same time, if these notes help the young scientist to make a sufficiently large number of observations in a short time, and thus enable him to rapidly accumulate the preliminary facts on which he can base future generalisations, they will have served a good purpose.

It may be suggested that the hints refer to a comparatively small portion of our lepidopterous fauna. This is so of necessity. A very large proportion of the species inhabiting Britain is unknown, except in the imaginal state, the habits of many more are as yet unpublished, and the actual observations of any one individual naturally only cover a small area of the whole field. Besides, there has been no intention to make the hints exhaustive, nor would they have been printed, at least, for some considerable time, had this been required, the author being far too occupied at present to undertake such a task. It is in the hope that, when the methods of working for a few species have been learned by the lepidopterist, he will extend his field of observation independently in other directions and discover the habits and ways of the unknown species that these hints are offered.

The responsibility of publication is shared with the author by the following ladies and gentlemen, to whom he offers his sincerest thanks for their kind help and generosity. He has no doubt that they, like himself, are quite prepared to bear that responsibility. Messrs. Watkins and Doncaster (12 copies), H. W. Marsden (6 copies), Mr. S. Mason (5), Rt. Hon. Earl Waldegrave (2), Captain S. G. Reid (2), F. W. Hanbury (2), A. Harrison (2), H. W. Head (2), K. König (2), G. L. W. Newman (2), A. Sich (2). The following will take one copy each:—Miss Alderson; Sir D. L. Broughton, Bart.; Revs. C. D. Ash, E. N. Bloomfield, C. R. N. Burrows, C. Chichester, F. B. Cowl, S. G. Hodge, F. E. Lowe, A. M. Moss, G. H. Raynor, J. E. Tarbat; Drs. R. Freer, H. C. Phillips, T. Steck; Major F. D. Bland; Messrs. B. W. Adkin, H. C. Arbuthnott, R. H. Archer, F. R. Atkinson, H. E. Barrett, C. Bartlett, E. D. Bostock, J. W. Boult, C. Bourne, Brazenor Bros., T. F. Brook, H. Rowland Brown, E. R. Bush, W. D. Cansdale, T. Carlyon, R. Chapman, W. G. Clutten, C. W. Colthrup, F. Cotton, B. H. Crabtree, G. O. Day, W. Daws, J. C. Dollman, H. M. Edelsten, A. H. Fisher, T. B. Fletcher, S. W. Gadge, J. E. Gardner, P. T. Gardner, A. T. Goodson, G. C. Griffiths, J. C. Haggart, W. Hewett, H. A. Hill, S. G. Hills, C. M. Holt, M. F. Hopson, Selwyn Image, W. J. Kaye, F. J. Lambillion, A. R. Leivers, T. A. Lofthouse, F. McIntyre, H. Main, J. G. Le Marchant, G. W. Mason, G. F. Mathew, C. Mayor, J. Musham, M. A. Pitman, D. H. Pearson, J. Peed, C. P. Pickett, J. R. Pickin, J. Porter, E. Ransom, C. Rea, P. C. Reid, P. W. Ridley, E. A. Rogers, D. Rosie, G. B. Routledge, A. Russell, C. Russell, S. G. C. Russell, A. H. Rydon, V. E. Shaw, G. B. Smith, H. D. Stockwell, J. H. Stott, A. M. Swain, E. H. Thornhill, T. Tunstall, H. J. Turner, J. W. Vaughan, S. Walker, F. Wallace, C. J. Watkins, H. J. Webb, G. W. Wynn, W. T. Kerr, W. Thornthwaite, Sir J. T. D. Llewelyn, Bart., M.P. (2)

PRACTICAL HINTS

FOR THE FIELD LEPIDOPTERIST.

JANUARY, FEBRUARY, AND EARLY MARCH.

During the winter months, little out-of-door work can be done by lepidopterists. Searching for eggs that are laid on twigs in autumn and do not hatch until spring, the completion of any work already begun in pupa-digging, much better done, however, in autumn, before the pupæ have been exposed to the thousand and one chances of destruction during winter, and before the pangs of hunger have made their vertebrate enemies search every available nook and cranny for food, the collection of leaves for mines of Lithocolletids, &c., searching beneath trees for cocoons spun up among the leaves in autumn and now fallen, stripping moss from tree-trunks for pupæ, collecting roots of various *Compositae* for various Tortricid larvæ that feed throughout the winter, are all occupations that help to fill up the lepidopterist's leisure. As, however, winter passes and spring slowly takes its place, other work is possible. The lamps give male *Phigalia pedaria*, *Hybernia marginaria*, &c., whilst beating tree-trunks produces the apterous females of the former, and it is worth while searching the still leafless hawthorn hedges with a lantern after dark for *Hybernia rupicapraria*, *H. marginaria*, and both hedges and tree-trunks for *Anisopteryx aescularia*, the wingless ♀ s of the Hyberniids being best beaten into an open umbrella, as they get well into the bushes to lay their eggs. The eggs of *A. aescularia* are laid round a twig, and completely covered with hairs from the anal tuft of the ♀, a characteristic which at once distinguishes it from its allies. At the foot of the hedges, Noctuid larvæ feed all through the winter when the weather is at all mild, and one can get a supply of excellent specimens of common species by collecting the larvæ in the slack season, rearing them, and breeding the imagines later in the year. As the end of March approaches, spring advances rapidly, and with increased temperature, both vegetable and animal life make vast strides, and, one may truly say that often, in some three weeks, one has passed from lepidopterological stagnation to the full bustle of active collecting—larva-beating, sallowing, obtaining eggs from hybernated ♀ s, and such like work adapted to the accumulation of specimens and knowledge. It seems unnecessary to mention that eggs of *Thecla w-album* on elm, *T. pruni* on sloe, *Zephyrus quercûs* on oak, are to be obtained throughout the winter on twigs near the base of leaf-buds ; those of *Lymantria monacha* on oak, beech and birch, of *Poecilocampa*

A

populi, Malacosoma neustria on hawthorn, oak, birch, sloe, &c., and so
on through all the species that pass the winter in the egg-stage, yet each
has its own particular mode of oviposition, and it must not be forgotten
that they are better found now than in autumn because the trees are leafless.   Similarly with regard to pupæ, one might repeat the names of a
large number of species which are much better dealt with in the
autumn notes, *e.g.*, *Smerinthus ocellatus* at roots of willow, *S. populi* at
roots of poplar, *S. tiliae* at roots of lime, and so on, although a few
may be mentioned with the general assumption that many others are
to be similarly obtained.

TINEINA (unclassified).—During January and February collect mined
oak and other leaves for the pupæ of Nepticulids, Lithocolletids, and
Incurvarias.
    Beat thatch—whether straw-thatch or chip-thatch—stacks of
bracken and hop-haulm. Hold the net directly under the place beaten
and catch everything that falls. Stacks of bracken, hop-haulm, reed, and
thatch, shelter *Coriscium bronyniartellum*, *C. cuculipennellum*, *Laverna
decorella*, *Teleia humeralis*, *Gracilaria stigmatella*, *Depressaria ciliella*,
D. chaerophyllella, *D. albipunctella*, *D. applana*, *D. arenella*, *D. carduella*,
D. subpropinquella, *D. propinquella*, *D. heracliella*, *D. purpurella*,
Cidaria miata, *Orneodes hexadactyla*, and many other species, and these
may be sometimes beaten from such places in great abundance during
the months of February and March.

TINEIDES.—The larvæ of *Teichobia verhuella* should be gathered during
the winter and spring months.   They mine in the leaves of *Asplenium
ruta-muraria* and *Scolopendrium vulgare* (Hart's Tongue) and later feed
on the indusia.

COLEOPHORIDES.—Collect the stems of sea-lavender (*Statice limonium*)
for the larvæ of *Goniodoma limoniella* (*atriplicivorella*).   These larvæ feed
on the flowers, eating out one of the petals and using it as a case, in
which the larva moves about till full-fed in December, then, crawling
down the stem, it eats its way inside, covering up the small hole with a
slight web, soon after which the case drops off (Elisha).
    Larvæ of *Coleophora lineolea* are blotching the leaves of *Ballota nigra*
and *Stachys sylvatica* in March ; they are also found in gardens on a
species of *Stachys* commonly called "Lamb's Ear."

GLYPHIPTERYGIDES.—Old stems of viper's bugloss (*Echium vulgare*)
should be collected in the winter ; *Douglasia ocnerostomella* will often
emerge from these in abundance in June.
    During the winter gather the seed-heads of cotton-grass ;
Glyphipteryx haworthana may be bred from them in June.

LYONETIIDES.—Larvæ of *Bucculatrix cristatella* are to be found on
leaves of *Achillea millefolium*.   Care must be taken not to shake the
plants or the larvæ will fall to the ground, and then are not easily to be
detected.   Later on the most beautiful white cocoons may be found
spun on the leaves and stem of the food-plant.

ELACHISTIDES.—The larvæ of *Elachista stabilella* mine the leaves of

Aira cespitosa on chalk, and may be found in February in the brown withered tips of the leaves, from which they make pale yellow mines to the stem, pupating in April at the base of the blade, under an oval-shaped web (Warren).

TORTRICIDES.—The larvæ of *Tortrix forsterana* can be found in March, between two leaves of ivy, spun flatly upon one another.

During the last week in February the larvæ and pupæ of *Ephippiphora scutulana* are to be obtained in thistle heads and stems.

Oak-galls collected in the winter give *Ephippiphora gallicolana* and *Heusimene fimbriana*.

During the winter, stems of *Stachys sylvatica* should be collected for larvæ of *Ephippiphora nigricostana*.

The roots of *Artemisia* should be collected in January and February for larvæ of *Ephippiphora foeneana* and *Dichrorampha simpliciana*.

Roots of ragwort (*Senecio jacobaea*) dug during the winter will give larvæ of *Ephippiphora trigeminana*, *Eupoecilia atricapitana*, &c.

Collect the seedheads of teasel (*Dipsaceus sylvestris*) and keep in a bandbox ; you will breed plenty of *Penthina gentiana* and *Eupoecilia roseana*.

The roots of corn woundwort (*Stachys arvensis*) should be collected in February for larvæ of *Orthotaenia antiquana*.

In March, with a fern scoop, dig up roots of thistles, plantain, ragwort, knapweed, dandelion, &c., for root-feeding larvæ. Place in flower-pots, &c. In this way, *Orthotaenia striana* (from dandelion) and many other species may be bred (Barrett).

A quantity of the common round oak-galls should be collected in March. They will contain the pupæ of *Coccyx splendidulana*, &c., the larvæ of many species spinning up in them.

Roots of tansy (*Tanacetum vulgare*), dug during the winter, give *Dichrorampha alpinana* and *D. saturnana*.

During February, pull gently the last year's flower-stalks of ragwort, and you will find that when the root contains a larva of *Argyrolepia aeneana*, the stem breaks off readily, leaving a piece about two inches high, which is slightly webbed over the hole that leads down to the root. Dig up the roots, plant in pots or boxes, and bring indoors in May.

Roots of *Daucus carota* should be collected in the winter months for larvæ of *Argyrolepia zephyrana*.

The larvæ of *Conchylis dipoltana* feed in a web among the seeds of yarrow in the autumn and winter.

In winter and early spring collect the stems of wild parsnip (*Pastinaca sativa*) for the larvæ of *Conchylis dilucidana*.

Small holes in the stems of *Daucus carota* indicate the presence of larvæ of *Conchylis francillonana*. Collect in January.

In January, collect a bundle of flower stems of *Alisma plantago*, stand them out of doors till May, then put into a cage, and *Eupoecilia udana* should emerge from June to August.

PYRALOIDES.—The roots of *Artemisia* should be collected in February for larvæ of *Exaeretia allisella*, the latter mine up the new shoots in early spring, causing them to die.

A 2

Larvæ of *Lita tricolorella* should be sought in spun-together tops of *Stellaria holostea*.

I observed the larva of *Butalis grandipennis* in the greatest profusion on the furze-bushes on the steep hill-sides between Torrington and the river. The webs were, in February, quite a feature in the landscape (Stainton).

During March *Diurnaea fagella* can be obtained in plenty in parks and woods on tree-trunks; the dark aberrations want searching for closely; the ♀ s are best obtained with a lantern after dark when they are often found on the move up the tree-trunks.

PYRALIDES.—During February and March the larvæ (some almost full-fed) of *Scoparia angustea* are to be found by pulling off the moss on the surface of rocks and old walls.

CRAMBIDES.—Full-fed larvæ of *Myelois cribrella* may now be found in dead thistle stems ; they are most partial to those of *Cnicus lanceolatus*.

Euzophera cinerosella (*artemisiella*) feeds throughout the winter inside the root stalks of *Artemisia absynthium*.

The larvæ of *Crambus falsellus* feed on and among the moss growing on walls, rocks, &c.

BREPHIDES.—*Brephos parthenias* flies freely in bright sunshine round the tops of birch trees, in which position it is difficult to obtain, but by standing in an open space between the trees, it is readily secured, as in passing from tree to tree it comes considerably nearer the earth ; out by mid-March only in early seasons.

GEOMETRIDES.—Eggs of *Ennomos autumnaria* on alder, birch and· sallow ; of *E. tiliaria* on oak, sallow, birch and alder ; of *E. juscantaria* on ash and privet ; of *E. erosaria* on oak, birch, beech, &c. Usually in small batches a few laid side by side along a twig.

Pupæ of various species may be obtained throughout January and February under moss on trees—*Eurymene dolobraria* on oak ; *Odontopera bidentata*, various trees ; *Tephrosia consonaria* near roots of beech, oak and birch ; *T. bistortata* on alder, birch, willow, &c.; *T. punctulata* on birch ; *Eupisteria obliterata* on alder ; *Eupithecia fraxinata* on ash, &c. Many of these are almost certain finds.

Other pupæ are to be found among the *débris* at the foot of trees—*Zonosoma porata* at oak, *Z. punctaria* at oak and birch, *Z. trilinearia* at beech, *Z. pendularia* at birch, *Numeria pulveraria* at sallow, &c.

Others are to be dug at the foot of trees, having pupated underground—*Nyssia hispidaria* at oak, *Selenia bilunaria* at willow, oak, &c., *S. lunaria* at sloe, oak, nut, &c., *Biston hirtaria* at lime, *Amphidasys strataria* and *A. betularia* at oak, elm, &c.

I used to have much difficulty when breeding lichen-feeding larvæ in pots, owing to the contents becoming mouldy. I have since adopted the following method with success :—In a damp and somewhat shaded part of the garden I suspend by means of a wire, between two branches of a tree, a glass cylinder (such as those used on candlesticks), or a small garden pot, and cover it with a large muslin sleeve tied to the wire towards both ends ; in this way the food is kept healthy by exposure to air and moisture, and the larvæ can always get shelter; by this means I have bred, *ab ovo*, *Cleora lichenaria*, &c. (Riding). The larva of *Cleora lichenaria* is always worth looking for in February and March (as well as later) on the lichen-covered fences and trees it loves to haunt. It is rather fond of sunning itself on bright days.

You can also beat lichen-covered thorn bushes and the low branches of oak, apple and beech trees for larvæ of *Cleora lichenaria*, and search the long hoary lichens on old beeches for larvæ of *C. glabraria*. From the thorns will also be beaten larvæ of *Miselia oxyacanthae* (Moberly).

A mild evening in January or February will be sure to give *Hybernia rupicapraria* if hawthorn hedges be searched soon after dark with a lantern (this species is usually going over when *H. marginaria* commences to appear). *Phigalia pedaria* is also sometimes abundant, the males being attracted to the lantern.

When searching for *Hybernia leucophaearia* in February, it is well to remember that the imagines appear usually to emerge between 10 a.m. and noon, and can then be found drying their wings on tree-trunks, fences, &c.

The imagines of *Hybernia marginaria* are to be found after dark from late January onwards, sitting on the bare twigs of hedges and the bushes in woods—hazel and hawthorn appear to be preferred.

Aleucis pictaria appears with the first fine days of spring, flying in the evening about the sloe-bushes in flower (Guénée) ; rarely appears so early as March in Britain except in exceedingly forward seasons.

In mid-March the imago of *Nyssia zonaria* may, in mild seasons, be already found resting on the bare sand on the sandhills of the Lancashire coasts. It emerges from the pupa about 3.0 p.m. (Birchall).

A female *Amphidasys strataria* enclosed in a small muslin bag (3in. diameter) and hung in a sheltered position on the outskirts of a wood will attract males in suitable weather from 10 p.m. to 12 p.m. (Alderson).

By the middle of March those who want *Tephrosia bistortata* should commence to search the tree-trunks regularly, especially if the weather be mild.

From February onwards the pale coloured larva of *Boarmia repandata* is very conspicuous at night on sloe and bramble twigs.

From mid-March beat Scotch fir, which will yield larvæ of *Thera variata*, *Ellopia prosapiaria* (*fasciaria*), and, in some localities, *Boarmia abietaria* and *Aventia flexula*.

The larvæ of *Gnophos obscurata* are to be found in January on *Geranium lucidum*, &c., by searching at night with a lantern, and may be so found until the beginning of April, when they are nearly full-fed. They then prefer flowers to leaves.

The pupæ of *Hypsipetes ruberata* should be searched for on willow trees. The angles made where the boughs branch from the trunk are the favourite places for pupation.

SESIIDES.—The twigs of currant bushes (cut the previous year) should be overhauled in February and March for the larvæ of *Sesia tipuliformis*. The winter cocoons will be found not far from the end of the shoot ; keep the cut twigs in damp sand.

The larva of *Sesia myopiformis* feeds just under the bark in trees that are suffering from canker, or that have been injured by the rough removal of branches (Fenn).

In February, where one knows osiers or sallows are being cut down, look over the cut sticks for the burrows of *Trochilium crabroniforme*; the larvæ rarely get more than a foot or so from the root; keep the cut sticks in sand with holes downward; the moths emerge the first fortnight in June.

PSYCHIDES.—Cases containing larvæ of *Narycia monilifera* (*melanella*) are to be found by diligently searching the lichen-covered trunks of various trees; also on lichen-covered rocks.

Larvæ of *Luffia ferchaultella* are on the move in early March if the weather be mild, on fences, tree-trunks, &c., covered with suitable lichens.

The cases of *Solenobia inconspicuella* and *S. lichenella* on similar lichen-covered tree-trunks, fences, &c., should be collected as soon as they are ready for pupation.

LACHNEIDES.—Late imagines of *Poecilocampa populi* are still to be obtained at light, if mild, during the first fortnight of January.

Eggs of *Trichiura crataegi* are laid in rows, side by side, on branches of sloe, hawthorn, sallow, birch, &c.

Eggs of *Malacosoma castrensis* are to be found on the shores of the Thames and Medway marshes, directly under the sea-wall, where the egg-rings have been washed up by the winter floods.

Collect larvæ of *Macrothylacia rubi* on sunny days in the early spring. Put each one separately into a small box (large enough, however, for the moth to emerge and expand its wings), put the boxes into a warm kitchen; the larvæ will spin up and the imagines emerge in due course without further trouble.

The imago of *Lachneis lanestris* emerges in February and March, and is to be found resting on hawthorn, the wings wrapped round the twig, and thus exactly resembling a dead leaf.

The eggs of *Lachneis lanestris* are laid on hawthorn twigs, &c., in March, the bulky, roughly spindle-shaped mass, covered thickly with scales from the abdominal tuft, is usually not difficult to spot, as it often measures an inch and a half or two inches from end to end.

SPHINGIDES.—Pupæ of *Smerinthus ocellatus* and *S. populi* may still be dug at the roots of willows, poplars, &c., those of *S. tiliae* at roots of elm (a large hollow in an old elm trunk is generally sure to harbour several pupæ of the last-named species).

Some Sphingid pupæ can readily be forced through January and February. Place them in a large flower-pot on sand (well-baked before placing in pot) and cover with moss, cover with muslin and place the whole on the kitchen mantel-piece, keep fairly damp.

NOTODONTIDES.—Cocoons of *Lophopteryx camelina* and *L. carmelita* are subterranean at foot of birch, *L. cucullina* at maple and sycamore, *Pterostoma palpina* at willows and poplars; of *Leiocampa dictaea*, at poplars and willows, and *L. dictaeoides* at birch; of *Notodonta ziczac* at poplars, sallows, and willows, of *N. trepida, Drymonia chaonia,* and *D. dodonea* at oak.

Dead wood on or about the roots of sallow-trees, should be searched for the cocoons of *Cerura furcula ;* they are often also at the base of a branch, or in a hollow where a branch has been pulled off. Those of *Cerura bicuspis* on alder and birch, of *C. bifida* on poplar.

Eggs of *Ptilophora plumigera* on maple, chiefly on the twigs, in hedges not shaded by trees ; often singly but sometimes in clusters of two, three, or many more ; circular, smooth, brown above, whitish beneath (Merrin).

Early hatching (January and February) eggs of *Asteroscopus sphinx* can be reared on birch catkins by cutting the latter open, the larvæ will not touch the catkins unless split (Greene).

ARCTIIDES.—The fluffy spider-web-like cocoons of *Gnophria rubricollis* are spun under moss on oak-trees.

Cocoons of *Spilosoma mendica* are to be found under moss on trees bordering ditches (Merrin).

LYMANTRIIDES.—Cocoons of *Dasychira pudibunda* are to be found spun up among dead leaves, &c., under various kinds of trees.

NOCTUIDES.—A small supply of calico bags is one of the most important items in the lepidopterist's outfit. They are infinitely better than tins for many larvæ, especially those feeding in flower- or seed-heads, rolled leaves, &c.

Beating and searching for larvæ by night—both in spring and autumn—are much more profitable than beating by day.

Cocoons of the Acronyctids are spun up in hollows and crevices on the bark of the various plants on which they feed :—*Triaena tridens,* on hawthorn, &c. : *T. psi,* on fruit trees, &c. ; *Acronicta leporina,* on birch and alder ; *Apatela aceris,* on horse-chestnut, maple, and sycamore ; *Cuspidia megacephala,* on poplar, &c.

Pupæ of *Moma orion* are enclosed in cocoons of rotten wood on bark of birch and oak.

Cocoons of *Craniophora ligustri* are to be found spun up under moss on ash trees and under top stones of walls or other suitable place near privet bushes. The cocoons are very hard, and feel lumpy when spun under moss on ash-trunks, or, if ivy be on the trees, they will be found adhering to the rootlets ; the moss should not be pulled off, as it spoils the trees for another year.

Search the stone dykes which fence the fields in Scotland (Aberdeen, Moray, &c.) for pupæ of *Arctomyscis* var. *myricae.* The larvæ spin up on the stones, being easily seen.

In March the larvæ of *Heliophobus hispidus* are to be found by night. The larvæ of *Stilbia anomala, Agrotis lucernea, Epunda nigra, E. lichenea,* and other species are also to be found at the same time in suitable places. Torquay is a well-known locality for all these species.

The larvæ of *Heliophobus hispidus* are nearly full-fed by the middle of February, and may be obtained resting on the grass in the evening in their known haunts ; they cling tightly when disturbed, and do not drop to the ground ; they feed only at night, burying themselves in the earth by day; in confinement they enter a sod of grass if one be available, and, as there is risk of injuring them by breaking the sod, fresh grass every day cut close to the roots and scattered in the box, the accumulated waste being cleared out every few days, proves satisfactory, and they pupate well (Brown).

The eggs of *Gortyna ochracea*, laid in heaps, are to be found on marsh-thistles, burdock, &c. (Merrin).

The larvæ of *Apamea unanimis* are to be found from January to March under the bark of old trees on the borders of marshes, &c. They are also to be found in March and April among grass at the roots of willows in similar places.

January and February (if the weather be mild and damp) is the season to take the larvæ of *Tryphaena subsequa* (*orbona*). I took it first, early in January, 1874, by sweeping tufts of *Dactylis glomerata*, and afterwards at night, feeding on the common *Triticum repens*, and it was seen still feeding on grass as late as April. It is nearly half-grown when *T. orbona* (*comes*) is very small (Williams).

Mild evenings in February give larvæ of *Triphaena orbona*, *T. ianthina*, *Leucania lithargyria*, *Phlogophora meticulosa* at the base of hedges; they are easier to find now than later, the vegetation being less thick (Mathew); larvæ of *Leucania lithargyria* are sometimes especially abundant in these situations.

Agrotis lucernea larvæ hide in daytime in isolated tufts of grass growing in clefts on bare cliff faces. On February 12th they are large and not far from full-fed. Look for them at night with a lantern (Kane).

To rear larvæ of *Agrotis agathina* plant in pots some shoots of *Erica cinerea*, surround the shoots with moss to serve as a hiding-place for the larvæ during the day, place 12-15 larvæ in each pot, cover the whole with a carefully-fastened gauze bag, place the pot in a saucer full of water, and the whole out in the open air, a most important factor for success; leave the pots in the sun, but they must be protected against heavy rain. After about a fortnight they want attending to, the moss lifted and searched with care so that no hidden larvæ be thrown away, the grass cleared, fresh moss added, and the larvæ placed back. If the heath be too much eaten a fresh piece must be planted. When full-fed the larvæ prefer to spin up between the stems of the heath almost on the ground; once they have buried themselves the pot need not be watered regularly, it is sufficient to place it in a saucer full of water and replenish the latter about once a fortnight to prevent the earth drying up entirely. The pots should now be kept in the shade, and, after July 20th, carefully watched for the imagines, which emerge from this date till about September 8th (Léon de Joannis).

Dasypolia templi hibernated females laid ova March 1st and 6th-13th. The larvæ emerged from April 5th, transparent yellow with black heads. Some burrowed into the leaf stem of *Heracleum sphondylium* (cow-parsnip), two others spun a slight silk protection and devoured the cuticle of a leaf. By April 13th nearly all were hidden in the leaf stems. On June 18th found a full-fed larva (stung, however, and full of a mass of small maggots) under a stone near Athlone. Howth is also a locality (Kane).

Many subterranean pupæ are to be still found at the foot of

various trees, *e.g.*, the Tæniocampids, Hadenids, &c., the Dianthœcias near *Lychnis*, *Silene*, &c., and the Cucullias near *Verbascum*, &c.

The purplish-brown eggs of *Polia chi* are to be found on dock, hawthorn, sallow, and probably lettuce, sow-thistle, tea-tree, &c. (Merrin).

The eggs of *Polia flavicincta* are to be found on dock, chickweed, groundsel, mint, &c., and in gardens on everlasting pea, plum, &c. (Merrin).

In Portland the green larvæ of *Epunda lichenea* are to be found from the middle of February to the end of April, at night, on the top of grass stems, dock, twigs, &c., the smallest in a sphinx-like attitude, generally in little groups of five or six, in sheltered spots at the foot of banks, beside large boulders, or in the crevices of the rock ; those that were feeding ate dock, sorrel, chickweed, and a common sort of grass. When older they become olive-coloured, and hide among the roots of low plants (Brown) ; also at Torquay and other localities.

The larvæ of *Stilbia anomala* are best found at the end of February by the aid of a lantern, for they feed quite exposed at night on grass (Norman).

Eggs of *Tethea subtusa*, on poplar twigs near leaf buds, those of *T. retusa* on sallow and willow, near the base of the old leaves or base of a leaf-bud.

Those well-placed for sallowing should commence work as soon as the earliest sallows are in blossom. The Tæniocampids and hybernated *Dasycampa rubiginea*, *Hoporina croceago*, *Xylina semibrunnea*, &c., are on the move as soon as the earliest catkins show. The pistillate catkins are just as attractive as the staminate ones.

Always sugar in the spring, when sallowing, for *Dasycampa rubiginea*.

Never forget in March and April to fill several large linen bags with sallow catkins (not too old). Turn into band-box, or well-covered tubs, or flower-pots. Many Noctuids and Tortricids will be bred. Collect from as many localities as possible and the result will be much more varied and, hence, far more interesting.

PAPILIONIDES.—Pupæ of *Papilio machaon* are to be collected on stems of various plants, dried sticks, &c., in the fens which it still frequents.

Eggs of *Thecla w-album* on elm, of *T. pruni* on sloe, of *Zephyrus quercûs* on oak ; all placed as a rule near leaf-buds on twigs.

LATE MARCH AND APRIL.

The condition of the leafage is the best test of what is likely to be an early or late season. When shall I commence to sugar ? When is the time to sallow ? are questions that the young lepidopterist frequently asks. As soon as the ash-bloom is well out, whether it be in the middle of March or well into April, then is the time for larvæ of *Cirrhoedia xerampelina* ; as soon as the sallow-catkins, both sexes, throw out their maturing anthers and pistils, then the moths will come to sallow and, if the evening be mild, to sugar. The same atmospheric conditions that have brought out the sallow-blooms, and opened the early leaf-buds, will waken up the hybernating larvæ into new life, will cause the over-wintering eggs to hatch, will bring the spring imagines from the pupal stage, and re-awaken the hybernators into activity. *Tephrosia bistortata* and *Eupithecia abbreviata* come to sallows with the newly-emerged Tæniocampids and Pachnobiids, and these are joined by hybernating *Dasycampa rubiginea*, *Hoporina croceago*, *Xylina semibrunnea*, &c., nor must it be forgotten that a long spell of mild weather is not at all necessary to produce rapidly-changed lepidopterological conditions, for the Tæniocampids, and many other early-appearing moths, mature the preceding autumn, and the imagines are already developed in the pupa, a few suitable days being ample to tempt them to emerge. At sallow, a clear sky is always accompanied by a dearth of insects, and it has never yet been explained why, on some apparently suitable nights, sugar will attract more successfully than sallow blossoms, and *vice versâ*. By hedgesides in the daytime one can capture *Dasysoma salicella*, *Choreutis vibrana*, &c., at dusk on heaths and moors *Larentia multistrigaria* is on the wing, by hedgerows and on the outskirts of woods, *Anticlea derivata*, *A. badiata*, *Cidaria suffumata*, *Aleucis pictaria*, &c., may be dusked for, or searched for later with a lantern, many of them coming freely to flowers of *Berberis*, whilst trunk-hunting now becomes a serious part of one's business, especially in Scotland, where such quarry as *Petasia nubeculosa* is to be expected. Rearing insects, however, is the all-important part of lepidopterological work at this time of the year. Larva-beating by day is much less successful until the buds are expanded and leaves rather fully out, than larva-beating and larva-searching by night, the latter on favourable evenings and in good seasons sometimes giving marvellous results, and one is sure to get a good supply of common Noctuid larvæ, as well as a fair supply of rarer ones, if they are systematically worked for by wood-ridings, hedges, and on the budding birches, hawthorns, sallows, &c., whilst sweeping on heaths is already fairly productive, many good species being now obtainable. Hunting *Brephos parthenias* with a long-handled (twelve feet at least) net gives exhilarating sport on a sunny afternoon, and the muscles of one's leg should be in training, in order to kick vigorously the tall aspens, and shake *Brephos notha* therefrom, when the sun is setting. One wants no hints as to how to catch the early butterflies, except to suggest that a web or two of larvæ *Melitaea aurinia* may now be obtained, that the earliest sheltered patch of bluebells will be sure to prove an attraction to *Cyaniris argiolus*, and that the hybernating Vanessids are already egg-laying and that large batches of eggs may readily be obtained of many of the species. The sportsman may semble for *Dimorpha* (*Endromis*) *versicolora* and *Saturnia pavonia* in the afternoon, in suitable localities, whilst quite at the end of April, and on into May our midland and northern moorlands want working for *Gastropacha ilicifolia*. It may not be out of place here to note that, in rearing insects, a want of knowledge of the natural food-plant or inability to get it, may lead to the loss of a brood of young larvæ, and that knotgrass (*Polygonum aviculare*) will be eaten by almost all species in confinement, particularly Geometrids, whilst sallow, birch, plantain, dock and dandelion are general pabula for a large percentage of the Noctuids.

ERIOCRANIIDES.—If the birch bushes were systematically worked in

April many rare species of the sun-loving genus *Eriocrania*, would doubtless be found to have a more general distribution than is at present supposed. They fly gently above the bushes in the sun, and are very easily captured.

TINEIDES.—In April, the larvæ of *Psychoides verhuella* are to be found on the underside of fronds of ferns in lanes near Alkham ; mining among the sori of *Asplenium ruta-muraria* and the hart's tongue fern (Elisha).

The larvæ of *Lampronia praelatella*, should be collected in early April from wild strawberry ; generally under the leaves.

Young currant shoots with withering leaves should be collected in April for *Incurvaria capitella*, and young raspberry shoots, showing a similar tendency, should be collected for larvæ of *Lampronia rubiella*.

ADELIDES.—Towards the end of April, the larvæ of *Nemotois schiffer-milleriella* feed on the radical leaves of *Ballota nigra*. They live in cases, and drop as soon as the plant is touched, so that the best way is to search the ground around the food-plant for the flat, oblong, figure-of-eight or fiddle-shaped cases in which they live.

The imagines of *Adela cuprella* fly round the sallow blossoms in late April and early May. Wimbledon Common is, we believe, still a good locality for this species.

GLYPHIPTERYGIDES.—At the end of March the stems of *Luzula campestris* bearing brown leaves should be pulled up for larvæ of *Glyphipteryx fuscoviridella*.

The larvæ of *Douglasia ocnerostomella* abound in dried stems of *Echium* at St. Margaret's Bay (Elisha).

The larva of *Acrolepia perlepidella* mines the lower leaves of young plants of *Inula conyza* in April and the beginning of May, completely hollowing them out until they have become mere brown bladders mottled with scattered excrement.

COLEOPHORIDES.—In April the plants of *Stellaria holostea*, give plenty of almost full-fed larvæ of *Coleophora solitariella* (Machin).

During April, beat heather, cranberry, *Myrica*, &c., into an umbrella by night ; besides Geometrid and Noctuid larvæ, you will obtain those of *Coleophora pyrrhulipennella*, *C. juncicolella* and others in abundance. It is best to tumble the beatings into a bag, and carefully overhaul them at home, or the *débris* can be put into a hat-box with close muslin lid, when the Coleophorids will crawl to the top.

During the first week in April sweeping *Calluna vulgaris* on heaths and moors will also give larvæ and cases of *Coleophora juncicolella*.

The time to get the nearly full-fed larvæ of *Coleophora anatipennella* is at the end of April, when the leaves of the sloe are beginning to shoot. They are then most easily seen.

ELACHISTIDES.—The larvæ of *Laverna raschkiella* make long irregular yellowish mines in the leaves of *Epilobium angustifolium*. When looking for them I always find myself singing—

> White mines they never hold larvæ,
> But yellow ones always contain them, I see,

and this couplet is quite true (Corbett).

The larva of *Laverna miscella* is to be found mining the leaves of *Helianthemum* towards the end of April.

ARGYRESTHIIDES.—The larva of *Argyresthia aurulentella* mines the leaves of juniper at the end of April ; it never enters the stem.

At the end of April, and during May, pick the rolled up leaves of sallow. *Argyresthia pygmaeella* will be bred.

LYONETIIDES.—The larvæ of *Bucculatrix cristatella* nibble the leaves of yarrow in March and April, causing the bitten edges to turn brown, which betrays their presence. The beautiful ribbed whitish cocoon is spun irregularly across the front of the leaf and may be found in May.

TORTRICIDES.—At the end of April, and during May, pick the rolled-up leaves of sallow for *Hypermoecia angustana*, *Penthina capreana*, *Tortrix crataegana*, *Ptycholoma lecheana*, *Ephippiphora populana*. Quite a number of species will be obtained if the leaves be gathered continuously for several weeks.

At the end of March and on through April, so long as they are available, collect the catkins of birch, and tie up tightly in linen bags for a week or two, then turn them into a band-box with a close-fitting lid ; *Paedisca bilunana* and many other species will be bred.

Aspen leaves that are rolled up like cigarettes will yield *Tortrix branderiana*.

The larva of *Tortrix teucriana* is to be found in March and April in rolled-up leaves of *Teucrium scorodonia* in Folkestone Warren.

The imagines of *Amphysa prodromana* fly during the late morning and early afternoon sunshine amongst *Calluna vulgaris* and *Vaccinium myrtillus* in April. Towards 3 p.m. or 4 p.m. they settle on the tops of the twigs, and at this time, too, the ♀ s may frequently be found *in cop*.

In March and April the larva of *Paedisca oppressana* feeds within the buds of *Populus nigra*.

Collect an abundance of sallow catkins, keep them in a large tub or bandbox, and you will breed numbers of *Grapholitha nisana*, &c. (Different localities will give a variety of species).

The larvæ of *Grapholitha geminana*, *Peronea caledoniana*, and *Penthina sauciana* feed in spring in the tops of bilberry (*Vaccinium myrtillus*) the plants being sometimes quite blighted by the larvæ.

The imago of *Steganoptycha subsequana* should be looked for amongst silver fir or spruce at the end of April. In the daytime it seems most inclined to fly in the early afternoon, when it may be beaten out of the trees, and it also flies naturally at dusk.

The sunny side of spruce hedges should be beaten during the first week of April for *Steganoptycha pygmaeana* ; it can only be disturbed from about 12.30 p.m. to 4.30 p.m. ; during this period the male flies out if disturbed, but the female drops to the ground.

The larvæ of *Sciaphila sinuana* are to be found in the flowers of hyacinth (*Scilla nutans*) just before the flowers are over. A distorted head with some silk spun among the flowers, or the corolla closed by silk threads, is sure to contain the larva.

The larva of *Retinia pinivorana* occurs in young trees of Scotch fir (*Pinus sylvestris*) during the last fortnight of April, preferring, however, the side shoots to the central ones.

About the second week in April the central shoots at the tops of branches of young trees of Scotch fir (*Pinus sylvestris*) should be collected for larvæ of *Retinia turionana*; the larva clears out the centre of the shoot. The pupæ may be collected towards the end of the month; a shoot containing a pupa does not commence to grow in the spring, and can thus readily be distinguished.

The larva of *Retinia sylvestrana* feeds in the shoots of stone-pine (*Pinus pinea*) and other pines in March and April. It may be detected by the pellet of frass which is ejected from the tunnel by the larva, and which remains attached outside.

In the months of April and May the larvæ of *Retinia buoliana* are plentiful in the leaf-buds and young shoots of different species of *Pinus*, eating out the entire centre of the shoot. The presence of a larva can be assumed by there being an opaque and hollow resinous exudation on a bud. The boring larva mounts upwards with the shoot and pupates near the top of it.

The larva of *Ephippiphora nigricostana* passes the winter in the stems of *Stachys sylvatica*, which should be collected in March and April.

The larva of *Ephippiphora cirsiana* may be found through the winter in the old flowering stems of *Inula dysenterica* just below the surface of the ground, as well as in those of other plants.

In April the larvæ of *Ephippiphora pflugiana* may be obtained from thistle-stems.

Those who neglected to do so earlier may still collect roots of *Artemisia vulgaris* for larvæ of *Ephippiphora foeneana* and *Dichrorampha simpliciana*.

The swollen roots of *Stachys palustris* should be collected at the end of April for larvæ of *Orthotaenia antiquana*.

Spruce fir cones collected in March and April should be over-hauled for larvæ of *Coccyx strobilana*, which feed up and pupate therein; keep in closed box.

The larvæ of *Coccyx hyrciniana* are plentiful on spruce firs about the middle of April.

Stems of wild cabbage, collected in April, showing little heaps of brown frass sticking out of the new shoots and leaf-stalks, contain larvæ of *Stigmonota leplastriana* (common on coast near South Foreland).

In early April you can still collect the dead teasel-heads of the previous year. Keep in a bandbox, and you will breed *Penthina gentiana*.

The larvæ of *Dichrorampha plumbana* and *D. plumbagana* are to be obtained in March and April by digging up plants of *Chrysanthemum leucanthemum*. The latter tunnels in the centre of the root-stocks, whilst the former occurs deeper down in the roots, grooving them deeply under cover of a web. The larva of *D. plumbana* also mines in stems of *Achillea millefolium*, as also does that of *D. politana*.

By the end of April the twisted ox-eye daisies should be collected for *Dichrorampha acuminatana* and *D. consortana*.

Roots of *Centaurea nigra* collected in April will produce *Xanthosetia zoegana* in July.

The larva of *Chrosis rutilana* feeds in a slight web on the shoots of juniper in April and May.

By collecting teasel-heads in April and May, and putting them in a bandbox, *Eupoecilia roseana* will probably be bred in plenty.

PYRALOIDES.—The larvæ of *Depressaria assimilella* are to be found feeding between united broom-twigs in early April.

At the end of April, and during May, pick the rolled-up leaves of sallow. *Tachyptilia populella* will be bred.

During April and May the larvæ of *Lita viscariella* are to be found in the tops of *Lychnis*.

The larva of *Lita instabilella* makes a greenish-white mine in the fleshy leaves of *Atriplex portulacoides* and is full-fed about the middle of April. It leaves its mine to pupate.

The larva of *Lita plantaginella* burrows in the root of *Plantago coronopus* and changes therein to a pupa. Its presence is indicated by a small heap of pale brown frass on the crown of the plant, sometimes partly hidden by the leaves; full-fed at end of April or early in May.

The shoots of *Anthyllis* should be examined in April; their bleached appearance betokens the presence of *Anacampsis anthyllidella*.

Seedheads of *Arctium lappa* should be collected in April for pupæ of *Parasia lappella* : the larvæ pupate in the heads and the imagines emerge in July.

Amphisbatis incongruella is much overlooked in the south of England owing to its early appearance, we have repeatedly taken it in great abundance in late April and early May on the chalk hills of Kent.

Larvæ of *Œcophora unitella* (*fuscoaurella*) have been found in April, under the decaying bark of elm.

The larvæ of *Hypercallia citrinalis* (*chrstiernana*) were found on *Polygala vulgaris* between April 27th and May 22nd, 1868, near Shoreham (Walsingham).

PYRALIDES.—The roots of sorrel collected at Folkestone in April give larvæ of *Scoparia* var. *ingratella*.

CRAMBIDES.—Search the sand about the roots of various grasses and sedges on the sandhills near the seaside for the silken tubes in which the larvæ of *Anerastia lotella* reside.

The larva of *Rhodophaea advenella* spins together the hawthorn buds in April or May, or if there be no buds it spins up the young leaves ; it is of a bright green colour with red subdorsal lines. The larva can be beaten but is best obtained by searching.

At the end of April and beginning of May the larva of *Dioryctria* (*Nephopteryx*) *abietella* sometimes feeds in a central shoot of Scotch fir. It is, however, more often found in dead shoots of the previous year's growth, eating out the pith. The stem or shoot which contains a larva may be known by its decayed or sickly appearance, the needles being shrivelled and brown.

BREPHIDES.—In late March and April *Brephos parthenias* sometimes flies in great abundance among birches, more especially in the afternoon sunshine ; the species appears to fly lower about 4 p.m. than in the earlier part of the day, and is then more easily captured ; in morning and early afternoon the moths fly higher and want a long net for their capture.

In open parts of woods, high up around the tops of young aspens, the imagines of *Brephos notha* are sometimes to be seen in March and April, on hot sunny days, flying in considerable numbers. Towards evening kick sharply the stems of the trees and look carefully at everything that falls; the imagines are thus frequently dislodged, often when *in copula*.

The full-fed larvæ of *Brephos parthenias* and *B. notha* must have a piece of old cork in which to pupate.

CYMATOPHORIDES.—In the first week of April search birches for *Asphalia flavicornis*, they are generally found on the stems and branches, but often resting on pieces of twigs on the ground.

The imagines of *Cymatophora ridens* are best found about April 15th upon the trunks of oak, generally from one to four feet up the tree ; they are difficult to see, being of the same colour as the bark (Mawson).

GEOMETRIDES.—Trunk-hunting is one of the more successful modes of work this month—*Phigalia pedaria, Nyssia hispidaria, Biston hirtaria, Amphidasys strataria, Tephrosia bistortata, Eupithecia abbreviata*, &c., may be thus captured.

One finds the species that simulate dead leaves hanging on bushes—*Selenia bilunaria, S. lunaria, S. tetralunaria, Odontopera bidentata*, &c. These are much less rarely taken on the trunks.

Scotch fir will still yield larvæ of *Thera variata, Ellopia prosapiaria (fasciaria)* and, in some localities, *Boarmia abietaria* ; from stunted alders and birches may be beaten larvæ of *Geometra papilionaria* : oaks will yield larvæ of *Boarmia roboraria, Phorodesma pustulata*.

To rear larvæ use wide-mouthed glass bottles (such as those in which the anchovies of the Compagnie de la Mediterranée are sold) about six and a half inches high, and three and a quarter inches in diameter ; at the bottom place baked sand, damp enough for larvæ but not so as to cause mould, on this baked moss, plant food in the damp sand ; lids should be constructed of wire rings (ordinary fencing wire) covered with muslin, &c., and made so as to fit easily over the mouth of the jar (Wilson).

In April the males of *Nyssia zonaria* are to be found sitting on the bare twigs of dwarf sallow or on grass tufts on the Wallasey sand-hills.

The females of *Nyssia zonaria* often outnumber the males on the Wallasey sand hills, and occur in some years literally in hundreds ; the males are often very conspicuous resting flat on the short herbage, others low down on the stems of the coarse grass. April 20th-30th, are fair dates, much, however, depends on the season.

In early April the limes which grow so freely in London streets and gardens should be searched for *Biston hirtaria*, which rest usually on the main stems, frequently within reach, and are not difficult to spot. They stick close to the trees, and are found almost everywhere in London, and are equally abundant in the afternoon as in the morning.

About March 16th is the date for *Nyssia hispidaria* in Richmond Park. In some parts of the park in some seasons they are abundant, whilst in other parts not a specimen is to be seen.

Where *Boarmia repandata* var. *conversaria* occurs the larvæ should be searched for at night about the end of April along the sides of woods and hedges; the pale larva is then easily found, and may be beaten into an umbrella, but it is of little use searching for it in the daytime; almost polyphagous in confinement, but found wild most commonly on hawthorn, blackthorn and birch in the order named (Fenn).

In gardens the end of April will give *Hemerophila abruptaria* on palings, fences, &c., near lilac and privet bushes; emergence is spread over three or four weeks.

The first warm days at the end of March and early April bring out imagines of *Tephrosia bistortata*, which should be searched for in woods on tree-trunks; it is best obtained at the bottom of the larch-trunks, &c., when drying its wings. *T. crepuscularia* does not usually appear until May and early June, when it occupies somewhat similar situations.

Near Perth *Tephrosia bistortata* is abundant in the larch woods during April, whilst *Cidaria suffumata* and the ab. *piceata* is almost everywhere at the same time (Wylie).

Search the trunks of trees in woods towards the end of April for *Tephrosia consonaria* (Moberly).

The trunks of oaks should be searched during the afternoon in March and April for newly emerged imagines of *Amphidasys strataria*.

Towards the end of March and throughout April the almost full-fed larvæ of *Gnophos obscurata* are to be found feeding by night on *Geranium lucidum*, preferably on the flowers, but the larvæ are also to be obtained in January and February.

During April the small larvæ of *Pericallia syringaria* are to be found hanging at night from their food-plant (honeysuckle).

The larvæ of *Ellopia prosapiaria* must be beaten at the end of March and early April with those of *Thera variata* and *T. firmata* from pines (the former species does not pass the winter as a pupa as Newman says). The larvæ of *Thera variata* will pupate towards the end of the month; those of *Ellopia prosapiaria* obtained with them will pupate during the first fortnight in May.

The larva of *Scodiona belgiaria* is to be obtained by sweeping heath in April; the imagines rest on the ground in June. Common on the Greetland Moors (Porritt).

By dusking along mixed hedges containing blackthorn, in April, a few *Aleucis pictaria* may be captured, but by searching low blackthorn hedges about an hour after dusk large numbers may be obtained (Raynor).

The imagines of *Aleucis pictaria* appear to prefer sitting on twigs or flowers of sloe-bushes at night after flight, whilst in the larval stage the stunted bushes growing in its special localities are those most affected.

Anisopteryx aescularia is sometimes to be found in numbers by searching tree-trunks in the late afternoon. It is often also abundant on the framework of roadside lamps in the early morning.

Larentia multistrigaria usually rests at the foot of palings, &c., in its haunts, where it is almost hidden by the herbage.

Hybernated larvæ of *Phorodesma smaragdaria*, very much like a little bunch of withered leaves, may be obtained by searching *Artemisia maritima* in April and May.

The larva of *Geometra papilionaria* must be searched for on birch-trees in April ; it sticks out from a twig and exactly resembles a birch catkin at this time of the year, so that all suspicious-looking catkins should be carefully examined.

The ova of *Lobophora viretata* are to be found during the last week in April on the terminal shoots of holly, particularly those at the tops of the trees bearing flowers ; the larvæ hatch in from 7-12 days, feed first on the flowers, next on the green berries, and, lastly, on the young leaves ; in confinement they take readily to privet, and eat bark as well as leaves. To beat the larvæ, spread a sheet on the ground and use a ladder (Tunaley).

During the latter part of April *Lobophora carpinata* can generally be found on the trunks of trees among its food-plants.

Collect plenty of sallow catkins in March and April and keep in bandboxes if you wish to breed *Eupithecia tenuiata*.

The larvæ of *Eupithecia rectangulata* feed in flowers of apple (crab and cultivated) in April and May, the blossoms being spun together by a silken web ; the pupæ are to be found under loose bark and moss on the trunks in May and June.

Fences, palings, &c., near clematis bushes, should be searched for examples of the early brood of *Eupithecia coronata*.

Eupithecia dodoneata, usually said to be attached to oak, is much more frequently met with among hawthorn, being disturbed during the day ; it is said to fly naturally in early morning sunshine as well as at dusk.

Search the trunks of apple-trees during April for *Eupithecia consignata*.

The first brood of the local *Mesotype virgata* (*lineolata*) is often well out on the coast sandhills (Deal, Wallasey), &c., during April and early May ; the second and third broods are usually still more abundant in June and August.

Much depends upon the season, but usually late April will give *Bapta taminata* (cherry and sloe), *B. temerata* (sloe), *Zonosoma pendularia* (birch), *Numeria pulveraria* (sallow), *Eupithecia pumilata* (furze), *E. vulgata* (hawthorn), *E. irriguata* (birch), *E. abbreviata* (oak), &c.

PTEROPHORIDES.—In April and May young leaves of burdock (*Arctium lappa*) will often show the little round holes made by the feeding larvæ of *Aciptilia galactodactyla*, which hide on the underside of the leaves and are well protected by their pale greenish-white colours ; the old leaves show most conspicuous traces of the feeding, but the greater number of larvæ will be found on the young and as yet not fully expanded leaves.

In late April collect the flowerheads of coltsfoot (*Tussilago farfara*), tie up in linen bags at first, and then turn out in a few days into a bandbox ; you will breed *Platyptilia gonodactyla*.

The larva of *Mimaeseoptilus plagiodactylus* eats down into the heart of *Scabiosa columbaria* in April and May, before the flowering stem is thrown up.

B

The larvæ of *Agdistis bennettii* feed on the leaves of sea-lavender (*Statice limonium*) in April and May (and again in July). The large pieces taken out of the leaves are a good guide as to the whereabouts of the larvæ.

SESIIDES.—The sickly-looking plants of dock and sorrel found along the slopes of Folkestone Warren, on the sea-face of the cliffs, give larvæ of *Sesia chrysidiformis* in early April. The presence of a larva may be readily discovered by the mines and frass in the root-stocks. Plant again the roots dug up which do not produce the desired larvæ, as such disturbed roots are always productive the next year. Plant the affected roots in a fern case ; water well, and keep in sun.

Sesia tipuliformis larvæ can still be obtained, feeding in the stems of old currant-bushes, or in the pruned branches of younger ones; the branches generally contain larvæ the year following that in which they are pruned, the females in June laying their eggs in the branches pruned a few months before.

Sesia myopiformis larvæ are now full-fed in stems of apple and plum, usually in pruned or old trees ; the sawing off of branches containing the larvæ appears to be the only means of making sure of the imagines ; sleeving an affected branch is sometimes successful.

The larva of *Sesia cynipiformis* is to be found in April (and May) between the bark and solid wood of oak-stumps, and spins a small cocoon before pupating, which may be found by pulling off the bark.

The pupæ of *Sesia formiciformis* should be obtained by cutting off the tops of osier stumps during April and May.

The larvæ of *Sesia culiciformis* are to be found in two-year-old birch stumps, e.g., those cut down in the winter of 1900-'01 will be found to contain larvæ in April, 1902. They feed just under the bark, in the wood, but the large quantities of frass thrown out between the wood and bark leave no room for doubt where this species is to be found.

Sesia sphegiformis lives in young suckers of alder, where the old tree has been cut down, the larva eats a gallery up the centre of the shoot for nine or ten inches, then turns aside, eats through the bark, and assumes the pupal stage within the burrow (Merrin) ; to find whether an alder contains a larva bend over the branches carefully to see whether the bark cracks, then gently raise the bark. If there be a round hole it is almost sure to contain a *Sesia* larva or pupa ; the branch must then be cut low down, and the sticks kept in damp sand. The month of April is the best time to look for larvæ.

In March and April look over the willow sticks that have been cut in woods (or elsewhere) during the previous winter. In them a number of orifices will often be observed ; the burrows contain the full-fed larvæ of *Trochilium bembeciforme*. [In Hereford the sallow stems cut in winter for hop-poles are often seen to be tunnelled by these larvæ, often two or three in a single pole (Hodges)].

PSYCHIDES.—The larvæ of *Diplodoma herminata* (*marginepunctella*) commence, in late April, to crawl up on fences, tree-trunks, &c., for the purpose of pupation ; its large case makes this species quite unmistakable.

About the last week in March, the larvæ, pupæ and imagines of *Solenobia inconspicuella* occur abundantly on palings and old elm trunks on Clapham Common (Coverdale) and elsewhere.

The larvæ of *Taleporia tubulosa* (*pseudobombycella*) begin to leave the hedgesides, &c., where they feed, and climb fences, walls, tree-trunks, &c., for pupation, in late April. The long, slender, cylindrical case is characteristic.

The cone-shaped cases of *Luffia lapidella* are actively on the move, on lichen-covered walls towards the end of April, the larvæ being now nearly full-fed ; the conical *Luffia* cases are immediately distinguishable from the triangular-sectioned cases of *Solenobia*.

The cases of *Acanthopsyche opacella*, should be sought for in late April and May, as the larvæ leave the ground and seek some higher position for pupation.

To obtain the cases of *Pachythelia villosella*, I search, during the months of March and April, the trunks of fir trees. The cases are generally to be found from one to two feet from the ground. This appears to be the favourite position taken up for pupation. They are also to be obtained on the ends of twigs of *Ulex europaea* (Fowler).

The cases of *Sterrhopterix hirsutella*, formerly common in the woods around London and other localities, but now rarely found, should be searched for on tree-trunks, the larvæ being almost full-fed towards the end of April and through May.

LACHNEIDES.—Beating sallows will yield young larvæ of *Trichiura crataegi* and *Poecilocampa populi* ; the larvæ of both species may also be beaten from many other trees.

The cage in which the larvæ of *Trichiura crataegi* are kept should be placed where the morning sun can shine upon it, as they love to bask in the sunshine.

DIMORPHIDES (ENDROMIDES).—*Dimorpha* (*Endromis*) *versicolora* is a very uncertain species in its appearance, in some years being quite abundant in its favourite haunts, the males fly swiftly in the morning sun, the females hanging from bare twigs or resting on the heather in April.

The females of *Dimorpha versicolora* are rarely to be found in any abundance until the males have been well out for a week or more (Andrews).

The eggs of *Dimorpha versicolora* are laid in little batches on the twigs of birch and alder ; soon resemble closely the colour of the bark on which laid.

SATURNIIDES.—The last week in April and the earlier part of May is the time to assemble *Saturnia pavonia*.

NOLIDES.—Search the trunks of trees in woods towards the end of April and in early May for *Nola cristulalis* ; it often sits head downwards on the stems of privet, beech, hornbeam, birch and oak ; rather conspicuous on beech and hornbeam.

NYCTEOLIDES (CHLOEPHORIDES).—At the end of April beating oaks will give larvæ of *Hylophila bicolorana* (*quercana*).

NOTODONTIDES.—In March and April the alder trunks should be carefully scanned for the cocoons of *Cerura bicuspis*. The earliest imagines appear towards the end of April on birch and alder trunks.

From April 10th-21st is the best time to search for *Lophoptery.r* *carmelita*. It is generally found upon the trunk of a birch or oak about mid-day, about four or five feet up the stem (Mawson), the last week in April in the Reading district on birch trunks for imagines (Holland), on birch trunks and palings near; flies in sunshine ; may be taken at sallows, and is attracted by puddles (Merrin).

Petasia nubeculosa is to be sought during April on birch trunks in the Rannoch district.

LYMANTRIIDES.—The larvæ of *Dasychira fascelina* are to be obtained on dwarf sallows on the coast sandhills of Lancashire on the northern moorlands (and elsewhere), in April and early May.

NOCTUIDES.—Sugar should always be applied to the trees before sunset, and when practicable, it should be put on warm, as the scent is much more powerful than when it is applied cold. The greatest number of moths will generally be found about three quarters of an hour after sunset. Many species visit the sugar in the morning about an hour before sunrise (Doubleday).

I find that, as a rule, evenings following fine or showery days with a west or south wind, and some sort of moisture on the grass, either rain or dew, and no, or at least a very young, moon, are good for moths if the wind is not too strong, whereas a dry night, or when the wind is north or east, is usually bad (Keyworth).

Searching for larvæ by night in April and May is sometimes exceedingly profitable. Large numbers of larvæ of *Triphaena ianthina*, *T. fimbria*, *Xylophasia scolopacina*, *Noctua baia*, *N. brunnea*, *N. triangulum*, *N. ditrapezium*, *Aplecta nebulosa* may still be captured on hawthorn, sloe, or the low plants that carpet our woods in the neighbourhood of London. Many woods in which the Rhopalocera and day-flying Geometrids have become practically exterminated, still produce many Noctuids in abundance.

Sallowing will be in full force at the end of May and early April— (for details see *Ent. Rec.*, vol. vii., pp. 241-243).

Plum blossom should be worked (as well as sallows) for the early emerging moths at dusk in April.

Larvæ of *Leucania littoralis* are to be obtained in April on the coast sand-hills by raking the sand at the roots of marram grass, (*Ammophila arundinacea*). I have generally obtained the larvæ of *Leucania lithargyria* by cutting open old stems of thistles and umbelliferous plants to which they resort for concealment during the day (Eales).

The larva of *Nonagria neurica* lives in stems of common reed ; a small hole is visible in the stem if tenanted by a larva (Merrin).

Sweeping or searching low herbage by night in favourable places still gives good results in Noctuid larvæ—*Aporophyla australis* (usually at seaside, Deal, Brighton, Portland, &c.), many Leucaniids, Agrotids, Noctuas (*sens. strict.*), &c.

Xylomiges conspicillaris imagines should now be sought at rest on trees, may also be beaten, and have been taken at sugar (Merrin).

During the last fortnight of April search the beds of striped grass in gardens for the larvæ of *Apamea ophiogramma*, which feed low in the stems, select the drooping or faded leaves ; gently pull the infested shoot from the bottom and avoid pressure as much as possible, and in the stem a larva will be found.

The larvæ of *Apamea unanimis* are to be found among grass in April when pupa-digging at the roots of willow, later the pupæ may be found spun up under the loose bark of willow. The moths emerge in late May and June (Machin). These larvæ hybernate and then spin their cocoons in late March and April ; the latter may be found not only under decayed bark on willows, but in stems of thistles, burdock, teasel, or, in fact, anything that affords sufficient concealment.

At the end of April (20th-30th) the nearly full-fed larvæ of *Agrotis* var. *ashworthii* may be found basking in the sun ; they also feed freely and crawl about the rock-cistus in the daytime, as well as at night. In captivity they will feed on hawthorn, primrose, dandelion (flowers) and sallow catkins. Penmaenmaur is one of the best known localities.

Throughout April is the time to sweep *Erica cinerea* for larvæ of *Agrotis agathina*. This should be done by night or in the early morning. The larvæ should be fed on *Erica cinerea* or *E. tetralix*.

On April 27th I swept 27 larvæ of *Agrotis agathina*, and a few days afterwards hundreds more (Norman).

To rear *Agrotis agathina* take a section of a 40-gallon paraffin cask, out of which the oil has been thoroughly burned, place it over a plant of *Erica* plunging it into the ground to a depth of six or eight inches, the topmost shoots of the plant should be about level with the top rim of the section. Then cover the whole with muslin. I use a light framework of cane to raise the muslin well above the tub and plants (Ash).

The larvæ of *Noctua sobrina* may be swept at night in April and May from *Vaccinium* on our moors.

Imagines of *Pachnobia leucographa* taken at sallow, placed in a large glass jar with a supply of plantain leaves, &c., will lay their eggs thereon ; feed with moistened sugar.

Pachnobia leucographa and *P. rubricosa* will lay freely on small bunches of thread-ends, if suspended or coiled up in the box or cover in which ♀ s are kept ; any bits of torn muslin or stray thread-ends will suffice, but the moths refuse a smooth surface.

Eggs of *Taeniocampa opima* may readily be found on the Cheshire sand-hills on the old dead stems of hound's-tongue and other plants, and being deposited in large clusters are conspicuous at a considerable distance (Porritt).

During April the small bushes in a recently cut wood should be searched by night for larvæ of *Aplecta tincta*, *Triphaena fimbria*, &c.

The larvæ of *Aplecta tincta* are best obtained by searching birch bushes in April by night with a lantern.

Larvæ of *Bryophila glandifera* and *B. perla* construct little cocoons in which they hide by day, made of silk with pieces of lichen, stone, mortar, &c., worked in, the whole having an appearance like that of the wall itself, but the slightly swollen surface betrays the presence of the cocoon ; the larvæ feed at night and early morning, and may be frequently found outside when the weather is wet and mild.

Eggs should also be obtained this month from *Xylina socia* (*petri-ficata*) and *X. semibrunnea*. Pieces of oak, birch, &c., should be given on which to lay.

The larvæ of *Cirrhoedia xerampelina* may be found at dusk with a lantern, crawling up ash trunks or feeding on the ash blossoms; they are concealed by day under moss on the trunk, in chinks of bark, or among grass at base of the tree; very retired and sluggish in habits; feed sparingly on ash-shoots and hawthorn (Merrin).

Towards the end of March beat the bare twigs of ash, either trees or hedgerow bushes, after dark, for larvæ of *Cirrhoedia xerampelina*. Feed up on the large unopened buds, until the trees break into leaf. By this means some 40 or 50 larvæ were obtained in the spring of 1897, in Suffolk, and a nice series of imagines was bred therefrom (James).

The larvæ of *Cirrhoedia xerampelina* are best obtained during April and May, when climbing up the trunk of ash-trees just at dusk. This is not only a less cumbersome method than beating, but a much more successful one, and three dozen is not at all an unusual number for a favourable evening's search; they appear to be most plentiful when the trees are in full bloom, and prefer the flowers to the buds for food. The larvæ that were collected gave 400 imagines and not a single ichneumon (Porter).

The end of March and first week in April—both at sugar and sallow—form the best period for hybernated *Dasycampa rubiginea*; they lay fairly freely, and the larvæ are not difficult to rear, the larvæ feeding on apple, plum, &c. (Mason).

Dasycampa rubiginea eggs are laid singly from the middle of April; the larvæ feed freely on apple, either in breeding-cage or sleeved; feed also on dandelion, but leave latter for apple. ♀ s taken after hybernation and kept for egg-laying should be fed on honey mixed with a few drops of sherry. Place all those captured in a bandbox with muslin cover, place twigs of apple scored with a knife, as the ♀ likes a niche in which to deposit her eggs (Robertson).

As soon as the sallows are in bloom every warm evening should be spent working those trees that are in sheltered spots. Living females of *Hoporina croceago* should be placed in large glass jars, fed with moistened sugar, and given a few oak twigs with old leaves attached, when eggs will be freely laid.

The eggs of *Tiliacea citrago* should be kept in a cool place all the winter, otherwise they will hatch before the leaf-buds of the lime are ready for them.

When sallows are in full bloom fill a bag with catkins from as many different localities as possible, place those from each district in a separate bandbox (or, preferably, a lard tub with a sheet of glass over it) and you will breed a great variety of really interesting insects— *Citria fulvago (cerago), C. flavago, Cleoceris viminalis*, &c.

The larvæ of *Citria flavago, C. fulvago*, &c., obtained from sallow catkins, should be supplied with sallow leaves as they grow larger, as they feed on these and a variety of low-growing plants.

At the end of April, and during May, pick the rolled-up leaves of sallow. *Epunda viminalis, Orthosia lota*, &c., will be bred.

ARCTIIDES.—Those who wish to pair *Spilosoma mendica* must, as in the case with many other species, put them in a cage through which a current of air blows.

PAPILIONIDES.—Watch a ♀ *Gonepteryx rhamni* at work egg-laying on the leaves and petioles of *Rhamnus frangula* ; collect the leaves and shoots afterwards, and you will readily find the spindle-shaped eggs, the larvæ from which are easily reared.

The young larvæ of *Apatura iris* may be beaten from sallows ; the lowest and most unpromising looking bushes are often productive of these larvæ (for details see *Ent. Rec.*, vi., pp. 146-147).

Larvæ of Argynnids—*paphia, adippe, aglaia, euphrosyne,* and *selene*— should now be sought on various species of *Viola* ; they feed in the day-time but are usually well hidden.

The larvæ of *Melitaea cinxia* are to be obtained in their restricted haunts in the Isle of Wight towards the end of April ; gregarious, and hence the capture of one usually means the capture of a brood.

Hedgesides, and the ridings of woods will give larvæ of *Pararge egeria, Pararge megaera, Enodia hyperanthus, Epinephele tithonus,* &c., in their respective localities ; these are best obtained by sweeping or searching with a lantern by night. Larvæ of *Arge galathea, Hipparchia semele,* &c., can similarly be obtained in their known haunts.

The larvæ of *Erebia aethiops* may be collected in abundance at night with a lantern in its local haunts.

MAY.

May, from which can hardly be dissociated early June, forms a very busy period of the year for the lepidopterist. The woods, fields, and hedges are clothed in their new verdure, and possibly one-third of the lepidopterous fauna must be collected during this period, if the species are to be obtained at all, the larval and imaginal stages being those in which the majority of species are now in existence. Larva-beating and larva-searching is at its height and large numbers of rare and local species are to be obtained ; at night, a good sweep-net for heather and rough herbage is invaluable, and it becomes necessary to consider the different habits of the various larvæ one is rearing so that the best results may be obtained in breeding them. Sleeving is undoubtedly one of the most successful methods for tree and shrub-feeding larvæ, but the creatures themselves are less distinctly under observation, and scientific results are less readily obtained. Above all things, fresh food, fresh air and plenty of space are absolute necessities. Various modes of day-work are in vogue, watching flowers for the " beehawks," beating for Geometrids, working with a long net around trees, or with a shorter around bushes, in the afternoon sunshine for Tortricids, trunk-hunting for Notodonts, Tineids and Geometrids, disturbing rough herbage for *Euthemonia russula,* &c., netting at sunset for *Nemeophila plantaginis,* &c., sembling for *Macrothylacia rubi,* &c., dusking for Geometrids and Hepialids, sugaring for Noctuids, and so on. The temptation to attempt more than one can deal with scientifically is very great, and one may here urge that a daily diary should be kept—as detailed as possible—that larvæ and pupæ should be dropped into spirit and labelled for future use, and that none of the little things that make collecting useful scientifically should be forgotten. Details as to killing, setting, &c., would be out of place here, but details as to sugaring are always worthy of remembrance. Never forget that moths have varying tastes and habits and exhibit them. The most attractive black treacle, mixed with the most alluring jargonelle, rum, or methylated spirit, is useless if the weather be wrong, and may be equally useless if the weather be right unless the worker attends to detail, and remembers that some affect the drips, others drink and walk round to the other side of the tree, and so on ; and that whilst some species come early, others come late and may be entirely missed by the worker in a hurry.   Flowers, too, are great during this season—Dianthœcias, Cucullias, Sphingids, and representatives of many other groups, coming in abundance on suitable evenings, as soon as dusk has really taken possession.   By the end of the month almost one-third of the British species of Geometrids and Noctuids are on the wing, but much must be allowed for the season, and late seasons may be quite two (or three) weeks behind a normal one, whilst the comparison of such years as 1888 (late) and 1893 (early), gives a difference of from six to seven weeks for the appearance of the same species—the middle of April produced abundantly, in 1893, the same species that occurred rarely in the middle of June 1888, and which should occur normally in the middle of May.   But Tortricids and Tineids are perhaps out in the greatest proportions, whilst, among the Pterophorids, only the double-brooded *Leioptilus microdactylus* and *Mimaeseoptilus bipunctidactylus* are as yet on the wing, almost all the remaining British species being obtainable still in the larval stage. Pupæ of the still increasing number of species that emerge in June are also available ; rolled-up leaves in profusion occur everywhere, containing pupæ or larvæ of Tortricids (and Tineina), and the larvæ of a small proportion of the Tineina alone would fully occupy all one's time and attention, and this, much more satisfactorily, than the promiscuous collection of imagines here there and everywhere, whilst the work among ova, larvæ, and pupæ still waiting to be done scientifically is enormous.   It would well repay an entomologist, in spite of the temptation, to lay down his net for two or three years and devote his time to hunting the lepidoptera in their early stages ; at the end of that time his observations and descriptions, properly digested and carefully prepared, would ensure his name being handed down to future generations of scientific workers, whilst his collection would be enriched with species quite unknown to the mass of collectors without souls.   At any rate, May and early June comprise one of the best periods of the whole year for work among the early stages, and the collection of imagines should be quite subsidiary to that of working for larvæ, pupæ, and the clearing up of difficulties relating to life-histories.   Canvas and linen bags of various sizes are much more satisfactory for carrying larvæ than larval tins, whilst closed tins should be eschewed altogether. The latter part of May (and early part of June) produce a large percentage of the species in some families in the imago stage—about one half of the genus *Eupithecia*, almost all the Notodonts, Acronyctids, a very large percentage of the Tortricids, Gelechiids, Gracilariids, Elachistids, Lithocolletids, and almost all the genus *Nepticula*—whilst at the same time almost all the Coleophorids, Pterophorids, Depressariids, &c., are in the larval stage.

MICROPTERYGIDES.—From mid-May into early June search the buttercup flowers in damp places for *Micropteryx calthella* ; the imagines are sometimes seen six or seven in a single flower feeding on the pollen.

Tineides.—The small cases of *Meessia richardsoni* (*vinculella*), resembling in shape unusually well-made cases of *Tinea pellionella*, may be found up to the end of May, on the undersides of stones, feeding on a microscopic lichen, but are hard to see, as they are so like their food in colour. This species occurs in Portland and Purbeck, and will not improbably be found elsewhere on limestone rocks on the coast, if carefully looked for. The imago is hardly ever seen at large.

Plutellides.—The larvæ of *Eidophasia messingiella* spin a few strands of silk across the young shoots of *Cardamine amara*, drawing them together. They should be collected in the middle of May.

The green spindle-shaped larva of *Plutella annulatella* lives amongst the flowers or buds of *Cochlearia*, spinning them slightly together. It is very inconspicuous but very lively when touched.

The semitransparent larva of *Plutella porrectella* feeds upon the flower shoots and leaves of *Hesperis matronalis* in May and June, and when full-fed spins a white silken cocoon under the leaves.

Coleophorides.—*Coleophora* cases should be collected now. Each species should be kept in different little glass jars, and the cases mounted with the imago for reference—*Coleophora troglodytella* makes blotches on fleabane and *Eupatorium*; *C. palliatella* on oak; *C. ibipennella* on birch; *C. genistaecolella* on *Genista anglica*; *C. vibicella* on *G. tinctoria*; *C. laricella* on larch; *C. saturatella* on broom; several species on elm, rose, sallow, &c. Those on heath must be swept.

The larva of *Coleophora chalcogrammella* was discovered in the larval state by Mr. T. Wilkinson, near Scarborough, feeding on the leaves of *Cerastium arvense* in May.

Cases of *Coleophora badiipennella* should be collected from elms and those of *Coleophora viminetella* from osier in late May and early June.

Old seed-heads of *Juncus maritimus* should be gathered in May for cases of *Coleophora obtusella*.

The cases of *Coleophora therinella* are to be found in May on growing *Cnicus arvensis*; the larvæ hybernate on the plant full-fed, pupate in May and the imagines emerge in June and July.

In May search or beat *Genista anglica* for larvæ of *Coleophora genistaecolella*.

In the middle of May cases of *Coleophora conspicuella* and *C. alcyonipennella* should be collected from *Centaurea nigra*.

Argyresthiides.—In early May the terminal shoots of juniper bushes should be searched for the feeding larvæ of *Argyresthia dilectella*.

Elachistides.—The full-fed larvæ of *Laverna raschkiella* are to be found in May (and again in July) mining the leaves of *Epilobium angustifolium*.

The larva of *Laverna ochraceella* mines longitudinally in stems of *Epilobium hirsutum*. About May it leaves the main stem, and mines up a leafstalk into a leaf, where it makes a long and rather tough cocoon in its mine, which may easily be found by examination.

In May the full-fed larva of *Batrachedra pinicolella* occupies a gallery on the surface of a twig of spruce (*Abies excelsa*). When full-fed it spins a slender, somewhat flattened cocoon, on the underside of the same or an adjoining twig.

During May watch carefully the flowers of stichwort (*Stellaria
holostea*). In the morning, when sunny, the flowers are frequented
in our southern woods by large numbers of *Asychna modestella*.

In May, the larvæ of *Chauliodus insecurellus* feed on the leaves of
Thesium humifusum (those of the second brood feed on the leaves,
flowers and unripe seeds in July).

In May and June, marshy places producing *Angelica sylvestris* and
Aegopodium podagraria should be visited for larvæ of *Chauliodus
illigerellus*, which live in crumpled leaves. (In August the larvæ of
the second brood eat round holes through the sheaths of the unexpanded
umbels, and feed on the immature flowers within.)

LYONETIIDES.—In early May the mines of *Bucculatrix maritima*
should be collected in leaves of sea-aster.

NEPTICULIDES.—The mines of *Nepticula weaveri* and *Lithocolletis
vacciniella* are worth working for in May on the moorlands ; the former
requires a lot of finding, although the larva mines the upper sides of
the leaves of *Vaccinium*, whilst that of the latter mines the underside.

TORTRICIDES.—By beating whitethorn hedges into an umbrella in
early May, the pupæ of *Spilonota suffusana*, *Sideria achatana*, *Sciaphila
nubilana* and other Tortricids may be obtained in abundance.

The larva of *Penthina corticana* feeds on birch and sallow, in May
and the early part of June.

The larvæ of *Penthina ochroleucana* feed on rose-trees, cultivated as
well as wild, in May.

Bright green larvæ collected in May, feeding in the shoots of *Salix
caprea*, and drawing the leaves together with a slight web, will most
likely produce *Penthina capreana*. The shoots should be kept in a
flower-pot in a cool place if you wish to rear imagines.

The larvæ of *Paedisca occultana* may be found feeding on
young larch shoots in May, and the imagines may be obtained flying
at dusk around the tops of small larch trees in early July. The larvæ
form silken galleries along the shoots ; they also feed on Scotch fir.

The conspicuously-twisted heads of bramble collected in May
produce *Aspis udmanniana*.

The larvæ of *Ditula semifasciana* feed in united shoots of sallow
in May and June, generally preferring the dwarf and stunted bushes,
and the imago appears in July.

Retinia posticana is reported to fly in May by day round the tops
of Scotch firs in Rannoch ; probably it has a much wider distribution.

In May and June the young shoots of stone pine (*Pinus pinea*)
should be collected for pupæ of *Retinia sylvestrana*.

At the commencement of May the side shoots of branches of
Scotch fir-trees should be searched for the light-brown pupæ of *Retinia
pinivorana*.

In May collect the young shoots of Scotch fir, which are tenanted
by larvæ and pupæ of *Retinia buoliana*.

At the end of May, working among the birches and heather on
heaths and commons gives an abundance of *Phloeodes tetraquetrana*
and *Phoxopteryx uncana*.

In late May and early June *Grapholitha lactana* flies among and over aspens, *Lobesia servillana* among sallows, *Capua ochraceana* among chestnut and hornbeam.

In May some curved structures, not unlike the cases of some Coleophorids, may be seen adhering to the twigs of various species of poplar and aspen. These consist of the material thrown out by the larvæ of *Spilonota aceriana*, which are feeding within the stem.

Towards the end of May and in early June the imagines of *Phoxopteryx upupana* fly high above birches and oak in the sunshine; it wants a long-handled net to capture them readily.

The larva of *Phoxopteryx siculana* is to be found on *Rhamnus frangula* in May; the imago flies at dusk, in late June and August, over the herbage where its food-plant grows.

In May the imagines of *Coccyx splendidulana* dance round the extremities of the branches of oak-trees in the morning and afternoon sun, often in large numbers.

Stigmonota nitidana flies high up around the projecting twigs of oak-trees in late May and June. Stand beneath the outer branches of an oak, and with a long net large numbers may be taken in the afternoon and until sundown.

The imagines of *Stigmonota weirana* fly in the sunshine around the tips of the branches of beech-trees in May and June.

The flowers of hawthorn are very attractive to *Pyrodes rhediana* at the end of May or early in June.

PYRALOIDES.—*Pancalia lleuwenhoeckella* resorts to daisy flowers in middle May, and may then often be captured in large numbers whilst resting motionless in the afternoon sun.

In May the larvæ of *Butalis senescens* make little web-galleries amongst moss at the roots of thyme.

The larvæ of *Butalis siccella* live in long galleries, composed of sand and silk interwoven, attached to half-buried stems of *Thymus serpyllum* and *Lotus corniculatus* in early May.

The larvæ of *Butalis variella* are to be found about the middle of May, making long tubes of silk and sand interwoven, and attached to half-buried twigs of *Calluna vulgaris* and *Erica cinerea*.

In May search or beat *Genista anglica* for larvæ of *Depressaria costosella*.

Towards the end of May the larva of *Depressaria nanatella* draws the two edges of a leaf of the carline thistle together and feeds in the roll thus made. The white shiny surface of the leaf being exposed makes it conspicuous.

Larvæ of *Depressaria conterminella* are to be collected from osier and sallow in late May and early June.

In the beginning and middle of May, larvæ of *Depressaria cnicella* are to be found on *Eryngium*, the dirty brown appearance on the crown of the plant often making their whereabouts known, but they are also found in rolled-up leaves. Stout gloves and a good knife wanted.

In May (at Southend) a fine plant of *Conium maculatum* gave larvæ of *Depressaria alstroemeriella*, which produced a fine series of imagines in July (Elisha).

The larva of *Lita suaedella* spins down the small fleshy leaves of *Suaeda fruticosa* to the stem, thereby concealing its presence. It is full-fed about the middle of May, and pupates in the sand or mud below the plant.

The larva of *Lita leucomelanella* feeds in the shoots of *Silene maritima*, sometimes spinning them down to the stones below, or burrowing down them for a short distance. It leaves its burrow before pupation and is full-fed in May.

By nipping off suspicious-looking *Lychnis* (*L. diurnea*) buds in May, the larva of *Lita viscariella* is to be obtained.

In the middle of May the roots of sea-plantain should be collected for larvæ of *Lita plantaginella*, it can, however, be obtained much more easily in rootstocks of *Plantago cornupus* than in those of *P. maritima*.

At the end of May heathery moorlands give *Gelechia ericetella* in abundance.

At the latter end of May, at Farmton, twisted shoots of *Lotus corniculatus* produced larvæ, from one of which, at the latter end of June, appeared a Gelechiid, which Mr. Stainton pronounced to be probably *Anacampsis cincticulella*, which feeds on the continent on *Genista* (Threlfall).

The twisted shoots of honeysuckle should be collected for larvæ of *Brachmia mouffetella*, which feed in a white silken gallery; the twisted shoots of *Genista tinctoria* for larvæ of *Gelechia lentiginosella*.

The larvæ of *Anarsia genistae* should be collected in June, when they are feeding on the shoots of *Genista tinctoria*. The larvæ of *A. spartiella* feed on shoots and flowers of furze at about the same time.

Drooping shoots of spindle contain larvæ of *Hyponomeuta plumbellus*, and the webs on the branches larvæ of *H. cagnagellus*.

PYRALIDES.—The larvæ of the first brood of *Botys asinalis* may be found on *Rubia peregrina* in May, but require a little searching for, though their eating is conspicuous. They will eat *Galium aparine* in confinement if their food-plant be not procurable.

The twisted tops of flea-bane (*Inula*) should be collected for larvæ of *Ebulea crocealis*.

CRAMBIDES.—The larvæ of *Crambus salinellus* should be searched for in May and June under stones resting on *Poa* grass. Turning over the stones exposes the tubular gallery attached to the lower whitish sheaths of the grass towards the roots, or to the stone itself.

At the end of May the blackish larvæ of *Phycis betulella* should be collected from the rolled-up leaves of birch.

The larva of *Rhodophaea marmorea* feeds on dwarf sloe in May and June, generally choosing low, stunted bushes, and spinning the leaves together in a web.

During the whole of the summer months (May-August) the larvæ of *Myelois pinguis* inhabit the living bark of ash, frequently pollard-trees, never affecting any dead or decayed portions of a tree nor penetrating into the wood. It does not eat far into the bark, however thick, and a few long black grains of frass block the entrance. This frass is characteristic, and should be looked for when searching a tree on any projecting bosses, as well as on the spreading foot, for stray grains of frass detected below afford a good clue to the situation of the mine above (Buckler).

Nephopteryx angustella appears in the beginning of May and continues on the wing till nearly the end of June ; in hot seasons a partial second brood appears in September and October. The egg is deposited on the fruit of the spindle-tree, generally on the underside, and frequently between two berries ; the newly-hatched larva at once bores into the berry, closing the entrance with silk, and is then difficult to find ; as it gets larger it passes to another berry, and its presence may then be detected by the frass protruding from the hole through which the larva has eaten into the fruit. By putting a supply of rotten wood into the breeding-cages for the larvæ to spin up in, almost every larva will produce a moth (Machin).

CYMATOPHORIDES.—When birch leaves are well out, in early May, search for rolled-up leaves. Many will contain larvæ of *Asphalia flavicornis.*

GEOMETRIDES.—Sleeving larvæ is perhaps one of the most economical, if not scientific, means of rearing lepidoptera successfully ; an excellent account of the *modus operandi* of dealing with the larvæ in sleeves is given by Mr. Alderson (*Ent. Rec.,* iii., pp. 64-65).

The larvæ of *Phorodesma smaragdaria* should be collected after hybernation during May. Essex marshes, Benfleet, Canvey, &c.

The larvæ of *Phorodesma smaragdaria* love the sun, and bask on all the sheltered plants ; in confinement they prefer southernwood to sea-wormwood, and reach quite normal size with plenty of food, sun, air and space.

The larvæ of *Geometra papilionaria*, almost full-fed, are to be found at the end of May firmly attached by the anal claspers to the twigs of birch, alder and hazel. Their resemblance to the catkins is remarkable.

In May search or beat *Genista anglica* for larvæ of *Pseudoterpna pruinata.*

The larvæ of *Pericallia syringaria* are to be found feeding on the low-growing honeysuckle in woods (also affect privet and lilac).

In May the yellow boat- or spindle-shaped cocoon of *Scoria dealbata,* very much like that of *Anthrocera,* but pointed at both ends, is to be found low down on grass stems.

The imagines of *Scodiona belgiaria* can usually be taken on the wing in May, when sweeping heaths for *Agrotis agathina* (Moberly).

Lobophora halterata emerges in May, and has no objection to sitting in the full rays of the sun on aspen trunks. It flies naturally after dusk, the males searching up and down the trunks of the aspens for the ♀ s, and only flying at a short distance from the trees. In the evening search the trunks for the species after pairing has taken place.

During the first week in May the young larvæ of *Scotosia vetulata* mine the young shoots of *Rhamnus catharticus,* their presence being indicated by the drooping condition of the young twigs. After a week or so they emerge from the mine, and live in the shelter made by spinning two or three young leaves together.

The larvæ of *Scotosia rhamnata* are to be found on buckthorn from the middle of May into early June, and should be beaten for. The young mining larvæ of *Scotosia vetulata* should be searched for in May on *Rhamnus,* at which time their presence is indicated by the flaccid and drooping condition of the succulent twigs, in a week or so they fasten two or three leaves together and reside therein ; [later it rolls a single leaf or uses two fastened face to face (Newman).]

Beat nettles into a newspaper at the end of May and beginning of June.   You will obtain larvæ of the Plusias, and a pea-green very ungeometer-like looking caterpillar with humped back.   This is the larva of *Eubolia limitata* (Arkle).

Junipers should be beaten in May and June for the larvæ of *Thera simulata*.   The imagines are to be taken on the junipers at night in July and August.

The larvæ of *Cidaria silaceata* from the May brood feed up well and rapidly on *Circaea lutetiana* and *Epilobium montanum* in June and July, many emerging in August, but some pupæ going over until the following May.

The imagines of *Phibalapteryx vitalbata* are on the wing in May, flying rapidly along hedges where clematis grows, at dusk.   A second brood flies similarly in August.   (Larvæ may be beaten in late June and July.)

At the end of May and June, in any locality where *Acidalia subsericeata* is known to occur, work well the herbage around the roots of dwarf bushes, low down near the ground ; the insect nearly always hides in such places during the day.

At the beginning of May the fullfed larva of *Eupithecia debiliata* spins two or more of the young spring leaves of *Vaccinium myrtillus* together.   It lives within the chamber thus formed.

Beat and search spruce in May for imagines of *Eupithecia pusillata*.

Towards the end of May the larvæ of *Eupithecia pumilata* may be beaten from furze blossoms.

Sallow catkins collected in May will sometimes give a plentiful supply of *Eupithecia tenuiata*.   (Select catkins from bushes growing in a variety of places.)

In the last week of May the flowers of maple (*Acer campestris*) should be first searched and then beaten for larvæ of *Eupithecia subciliata*.   The larvæ feed on and in maple bloom in May (ninety larvæ beaten May 22nd, in one year, and seventy on May 16th, another year, whilst the next year many were beaten May 30th).   They feed up rapidly, pupate in or near the suface of the soil or among the food-plants, and are very easy to rear.

DREPANULIDES.—Beating in May for "hooktips" is often profitable. *Drepana cultraria, lacertinaria, falcataria,* and *binaria* are abundant in most southern woods with the right growth.   The males of *D. cultraria* fly about freely in the sunshine, and sometimes those of *D. binaria* are to be seen flitting in the glades of the woods of our south-eastern counties.   The Fairmead district of Epping Forest is a good locality for *D. binaria*.

Imagines of *Drepana binaria* (*hamula*) may be passed over for *Orgyia antiqua* ; they fly in the hot sunshine up to midday around young oak-trees, and are sometimes common.

The eggs of *Drepana cultraria* (*unguicula*) may readily be obtained by enclosing a captured female in a muslin sleeve on a branch of a growing birch-tree.

PTEROPHORIDES.—The larva of *Cnaemidophorus rhododactylus* feeds in late May just beneath the leaf overlapping a rosebud, eating into the bud from the side, also in similar positions at the ends of the young rose-shoots.

A form of *Platyptilia pallidactyla* (*bertrami*), or a distinct species, feeds on the leaves of *Senecio* in May and June, and emerges in the latter month ; has been recorded as yet from Carlisle, Glasgow dist., Aberdeen dist., and various other northern localities.

During the first fortnight of May the larva of *Oxyptilus hetero-dactylus* (*teucrii*) is to be found full-fed on the leaves of *Teucrium scorodonia*. The larva eats a small round hole into the stem about two joints down, which causes the tip of the plant to droop, and then eats the growing leaves around.

The larvæ of *Œdematophorus lithodactylus* when young are to be found in the terminal shoots of *Inula dysenterica* ; later on they hide during the daytime, but may easily be got after dark, when they feed, exposed, on the surface of the leaves (South).

In early May search golden-rod in shady places for larvæ of *Leioptilus tephradactylus*. Their presence is easily detected, as they strip the leaves pretty successfully.

The early brood of *Leioptilus microdactylus* is well out among *Eupatorium cannabinum* in late May and early June.

The larvæ of *Aciptilia galactodactyla* are to be found on the underside of the central leaves of burdock in May ; the riddled state of the leaves indicates the plants affected.

SESIIDES.—Where the bark joins the wood of an oak stump (cut down two years previously) the thrown-out frass indicates where the larvæ of *Sesia cynipiformis* (*asiliformis*) has been at work. Dig out the cocoons in the middle of May, or saw off about four or five inches of the stump earlier in the year.

In early May the larvæ of *Sesia musciformis* (*philanthiformis*) are to be found feeding on the stems of sea-thrift (*Statice armeria*), which grows in profusion in many coast districts. The stunted plants scattered in the clefts of almost bare and waterworn rocks are those selected. A little red patch on the cushion of thrift betrays the work of the larva, and after a little practice affected plants may soon be recognised at a distance. The larva appears never to select plants growing luxuriantly in ordinary soil.

LACHNEIDES.—The larvæ of *Malacosoma castrensis* (small in May and still gregarious) are exceedingly abundant on a large variety of plants on the saltings of the Kent and Essex coasts, will feed freely on common garden chrysanthemums.

During May the larval nests of *Lachneis lanestris* should be searched for on hawthorn and blackthorn.

One of the most successful food-plants on which to rear *Poecilocampa populi* is alder. The larvæ thrive excellently on it (Bowles).

The larvæ of *Lasiocampa quercûs* should be searched for in May ; generally rests well up on branches of hawthorn, maple, &c., on hedges and outskirts of woods ; it wants looking for even when nearly or quite full-fed.

In May the larvæ of *Pachygastria trifolii* are to be found in abundance on Romney Marsh, on the tufts of a wiry grass growing on the shingle just above high-water mark. These larvæ give the rare yellow aberration only; the larvæ of the darker forms occur on the Wallasey sandhills, Lyndhurst (on heather), Devon and Cornish coasts, &c.

At the end of May the larvæ of *Eutricha quercifolia* should be sought on blackthorn, buckthorn, resting low down during the day ; it comes up and feeds by night, and is common on Wicken Fen.

"Cannock Chase.—*Gastropacha ilicifolia*, May 17th, in repose, clinging to a dead sprig of heather, apparently but lately emerged from the pupa. From its great resemblance to a withered leaf it would not probably have caught my eye, had I not luckily knelt down within a few inches of it to pin a small Tortrix.—W. S. Atkinson, *Zoologist*, p. 3396 (1852)."

SATURNIIDES.—The eggs of *Saturnia pavonia* are to be found on ling, sallow, &c., laid round and round a twig, and not at all unlike a small bunch of dried flowers of heather.

SPHINGIDES.—The last week in May, in early seasons, is the time for taking *Macroglossa bombyliformis*, hovering over flowers in the morning and midday sunshine ; *Ajuga reptans* is a specially favourite plant ; the ♀ s may sometimes be seen egg-laying on scabious.

The imagines of *Macroglossa fuciformis* fly in the morning and afternoon sunshine, in late May and early June ; specially fond of rhododendrons.

When rhododendrons are in bloom (end of May) they should be worked systematically at dusk for *Choerocampa porcellus*.

SIMÆTHIDES (CHOREUTIDES).—The larvæ of *Simaethis pariana* should be searched for in June on hawthorn and apple, on which plants it feeds, spinning a silken web on the surface of a leaf, under which it lives.

NOTODONTIDES.—During May it pays to search sallow bushes, not beat them, for larvæ of *Clostera reclusa*.

The birch trees on the Muckross Peninsula, at Killarney, are reported to be the exact haunts of *Notodonta bicolor* in Ireland. The imago appears in May, the larva in June.

Notodonta trepida is best taken from the middle of May to the middle of June upon the trunks of oak, about one to four feet from the ground (Mawson),

The trunks of trees in beechwoods should be well worked in late May and early June for *Stauropus fagi*.

NOCTUIDES.—An overhanging bank, a hole where a tree has been blown down, the edges of a quarry or chalkpit, a landslip, banks on sandhills, banks in lanes where the soil or gravel has fallen away, should be first searched carefully for lepidoptera, then gently scraped with a stick to disturb those overlooked (Gregson).

During late May and early June the reedbeds should be swept after dark for the larvæ of *Leucania straminea*. I have seen the larvæ near the tops of the reeds in large numbers, where they are rather con-spicuous, and used to pick off those within reach, and sweep for the more distant ones with a net. The larva pupates in or on the soil, and emerges without much trouble.

In May the full-fed larvæ of *Leucania littoralis* are to be found feeding by night on *Ammophila arundinacea*, and hiding by day under the sand at the roots of the same plants.

The larvæ of *Tapinostola elymi* feed on *Elymus arenarius* in coast districts and should be worked for during May. They can be shaken out of the plants by day.

Night-beating for larvæ, especially after 10 p.m., is the most productive means of capture possible, and many larvæ are obtained of species which do not always come to sugar, such as *Noctua ditrapezium*, *Aplecta tincta*, *Triphaena interjecta*, &c. I once beat out of the birches and sallows at Tilgate, over 500 larvæ in one night between 10.30 p.m. and 1 a.m., including sixteen *N. ditrapezium*, twenty *Triphaena fimbria*, eight *Aplecta tincta*, &c. (Fenn).

The larvæ of *Noctua ditrapezium* should be obtained by the middle of May (they can be collected earlier) ; they feed by night, late, and have been taken by searching with a lantern in almost all woods on the outskirts of London. The larva occurs on a variety of plants—birch, apple, whitethorn, sallow, bramble, bracken, stinging-nettle—on Hampstead Heath, Wimbledon Common, &c.

At the end of May, 1870, I beat three larvæ of *Agrotis subrosea* in the evening twilight from *Andromeda polifolia*, L. In captivity they also ate several species of willow (Berg).

Towards the end of May tracks made by crawling larvæ may be seen on the coast sandhills at the edges of the patches of dwarf willow (*Salix repens*). If one of these be traced, it will be found to end abruptly at a small upheaval in the sand ; under this is the larva of *Actebia praecox* (Almond). The larvæ are very frequently stung by ichneumons. Sea-birds (gulls) look for their nocturnal tracks on the sandhills and dig them up for food in the day-time ; full-fed about June 13th-20th (Kane)

From tufts of *Salix repens* on the Culbin sands, on May 24th, I dug out 156 larvæ of *Actebia praecox* and several of *Agrotis vestigialis* (Norman).

The heather can still be swept on the open moors, &c., for larvæ of *Agrotis agathina* and *Noctua neglecta*.

In sweeping heather for *Agrotis agathina* in May and June it is well to select the larger larvæ only as being less likely to be ichneumoned ; the parasites appear to leave the " stung " larvæ when the latter are about two-thirds grown (Collins)

In late May (and early June), *Hydrilla palustris* comes to light in Wicken Fen. It is generally supposed that that part of the Fen nearest to the village is the most productive for this species.

The imago of *Xylomiges conspicillaris* is to be found in May and early June on old stumps, fences and gate-posts, and looks just like a splinter of the wood on which it sits.

Imagines of *Calocampa vetusta* and *C. exoleta* come to sugar in April and May ; May 14th, for *Calocampa vetusta*, and *C. exoleta* on May 27th, are my latest records (Kane).

The larva of *Taeniocampa populeti* is to be found at the end of May between flatly united leaves of poplar. The larva of *Tethea subtusa* is to be fonnd at the same time and place, but in leaves folded upwards upon themselves.

C

In May the larva of *Taeniocampa populeti* spins two poplar (or aspen) leaves together, one upon the other, and lives between. The larva is easily seen against the light. It is very transparent, and almost colourless, of a yellowish-white tint with a black head. Often high up on tall trees.

During May the larva of *Taeniocampa gracilis* may often be found in marshy places, in tents formed of leaves of *Spiraea ulmaria* spun together.

During May it pays to search sallow bushes, not to beat them, for larvæ of *Tethea retusa*, which prefer sallows with leaves of thin texture; they are also found on willows.

In May the larvæ of *Dyschorista upsilon* can be obtained in abundance beneath the loose bark of willow trees, where they hide by day. They go up at night to feed. The larvæ of *Catocala nupta* may often be found in the same places.

At the end of May beat elms for the larvæ of *Calymnia pyralina*, *C. affinis*, and *C. diffinis*. The larvæ of *C. affinis* may be beaten from the lower branches, and though *C. diffinis* often occurs commonly in the same localities as its congener, the larvæ are generally much more difficult to obtain. Do they feed higher up the trees?

At the end of May, larvæ of *Mellinia gilvago* are usually abundant locally on elm.

The larva of *Cosmia paleacea* is to be found at the end of May or beginning of June on birch. Beat the lower branches hard. The larva is very sickly-looking, and, before I knew them, I used to throw them away thinking that they were ichneumoned (Corbett).

Full-fed larvæ of *Cleoceris viminalis* should be searched for in late May and early June when they live between sallow leaves or in spun-up terminal bunches.

The last week in May and the first week in June are the best time to work *Silene* flowers for *Dianthoecia caesia* and *D. capsophila*.

In the last week of May (and first in June) dusking over flowers of *Silene*, in the neighbourhood of Folkestone and Gosport, produces *Dianthoecia albimacula*.

The larvæ of *Polia nigrocincta* are to be found by day around the roots of *Statice armeria* or on the blossoms by night (Newman). We believe *Plantago maritima* is the more usual food-plant.

Heliothis dipsaceus flies abundantly at the blossom of *Lotus corniculatus* (with which the denes near Great Yarmouth are covered) in late May and early June (Harmer).

During May the almost full-fed larvæ of *Toxocampa pastinum* may be found feeding on *Vicia* by night, and by searching among the herbage at the roots by day.

The larvæ of *Catocala nupta* feed on willow and poplar, hiding under loose bark during the day, and crawl up to feed after dusk.

ARCTIIDES.—"To find the larvæ of *Nudaria mundana*, note some wall of loose stones (a 'dry stone dyke') where the imagines are abundant, and, in May, lift the upper stones and examine their undersides. The larvæ will be found feeding on a green confervoid growth that covers the stones" (F. B. White).

The larvæ of *Setina irrorella* feed on a ground lichen, which grows amongst the grass, often just outside the tide-mark. They afterwards feed fairly well on the grey lichens, which are not uncommon on apple and other trees (Crewe). Also occurs inland, at Box Hill, &c.

In May the larvæ of *Lithosia caniola* have been found commonly on Romney Marsh, on the low plants growing on the shingle just above high-water mark.

PAPILIONIDES.—In May the flowering stems of *Cardamine pratensis* and *Sisymbrium officinalis* should be collected for the orange-coloured eggs of *Euchloë cardamines*, which are usually laid (one on each flower-head) on the pedicel of a flower nearly over.

The long spindle-shaped eggs of *Leucophasia sinapis* are to be found in May on *Vicia cracca* and *Lathyrus tuberosus*.

Larvæ are exceedingly abundant this month. Among butterflies the Argynnids—*Argynnis adippe* (violet, &c.), *Brenthis euphrosyne* (violet, &c.), *B. selene* (violet, &c.)—*Polygonia c-album* (hop, nettle, currant), *Limenitis sibylla* (honeysuckle, often prefers small plants), *Apatura iris* (sallow), &c., are to be obtained.

The eggs of *Nemeobius lucina* are readily found on the underside of cowslip leaves, not more than four or five on a leaf, in woods where the butterfly occurs, at the end of May and beginning of June (Porritt).

JUNE.

June is, perhaps, *par excellence*, the month most beloved of the field-naturalist. Insect life is rapidly reaching its maximum in variety and number of species, and the lepidopterist is now exceedingly busy and alert. The first fortnight of June is always the most prolific in our southern woods, a couple of hundred species often falling to the collector in a single day's work in Kent—distributed over butterflies (*Melitaea athalia*, *Brenthis euphrosyne*, *B. selene*, *Nemeobius lucina*, &c.), Sphingids (about half the British species), Anthrocerids (*Adscita statices*, *Anthrocera trifolii*, *A. hippocrepidis*, St.), Arctiids (*Nemeophila plantaginis*, *Euthemonia russula*, *Lithosia aureola*, *Calligenia miniata*, &c.), Sesiids, Notodonts, Crambids, &c., although the bulk of the species will, if properly sought, be Geometrids (*Scoria dealbata*, Acidalias, &c.), Noctuids, Tortricids, and the various sections of the old Tineina. In the forest districts where the large butterflies still exist, the end of the month is very fruitful, and visitors to the New Forest perhaps prefer late June and early July to any other period of the year. On the chalk-hills almost all the British species of the Lycænids are on the wing, whilst, in the other superfamilies quite a new fauna is obtainable, *Adscita globulariae*, indeed, being confined to our southern downs. In early June the lepidopterist wants, indeed, to be everywhere—the early coast Noctuids—*Mamestra albicolon*, *Leucania littoralis*, *Agrotis ripae*, &c.—attract him ; the rare fen species—*Hydrilla palustris*, *Meliana flammea*, &c.—tempt him ; the marshes call him for *Leucania obsoleta*, *Senta ulvae*, &c., whilst the chalk-hills and the woodlands are teeming with life. The collector usually gets too much into a groove ; he is tempted, when he knows how, when, and where, to get a few good local species, to work for these year after year and to fill the blanks in his cabinet by exchange. It is much better to go afield, and the true naturalist should be equally at home on the southern downs

c 2

and northern moorlands, in the pine forest and beech-wood, on a bare, bleak coast-line, or on a savage mountain-side. Each locality wants learning, each requires a special knowledge, and he is the best naturalist who can wrest from nature her secrets, under the most varied conditions that she knows how to present to him so admirably. His own immediate home will always, of necessity, demand the most attention, but week-ends, and occasional holidays, should open up new ground much farther afield, and give the naturalist new views of nature under some one or other of the delightful phases that she everywhere presents to him. It is worth while to remember that sheltered spots, e.g., the sheltered sides of trees and palings, hollows on hillsides, &c., are the most productive situations, especially after heavy storms of wind or rain, and after bad weather, many rush, grass, and ground insects may be freely beaten from bushes, and from thatch, especially if the latter be on heaths or moorland where there is little shelter. In the middle of June the Breck district of Norfolk and Suffolk are particularly interesting, and the special fauna very attractive to the lepidopterist—*Acidalia rubricata, Lithostege griseata, Agrophila trabealis (sulphuralis)*, &c.—whilst the nearness of the fens—Wicken and Chippenham—is a further temptation to visit this interesting district. The woods of Northants are celebrated for their rarities— *Thecla pruni*, &c.—and the coasts of Cornwall and Devon for the much-persecuted *Lycaena arion*. Among the smaller fry in the imago state probably two-thirds of the whole of the British species in the superfamilies *Nepticulides, Lithocolletides, Elachistides, Coleophorides, Argyresthides,* and *Gelechiides* are on the wing at some period during the month. The collection of these, however, in the imaginal state, is, in Britain, of very little scientific interest, as most of the imagines are well known, and any increase in our knowledge must be largely in the direction of habits and detailed descriptions (on modern lines) of the insects in their earlier stages. Trunk-searching is an especially valuable mode of collecting some of these smaller species in the perfect state, but it also gives many of the larger species in abundance—*Stauropus fagi, Boarmia consortaria, Tephrosia extersaria, T. crepuscularia (biundularia), Macaria liturata*, &c., being among those that are best obtained in abundance by this mode of work. Flowers at dusk are also worth working ; red valerian is stated to be most attractive to the Plusias—*bractea, festucae, iota, pulchrina*—to the Dianthœcias—*capsophila*, &c.—whilst wild sage flowers attract swarms of Noctuids and Geometrids at Portland, among other good species being *Agrotis simulans (pyrophila)*. Sugar often fails in dry and hot weather, and flowers of grass, sedge, honey-dewed leaves, &c., should then be searched for Noctuids. Several scores of Noctuids may often be captured at sedge-blossoms on the Deal sandhills when the sugar proves blank. It is well, too, to carry a few small calico bags with you in which to place the seedheads of different plants collected for larvæ—*Silene* for *Dianthoeciae ;* toad flax (*Linaria vulgaris*) for larvæ of *Eupithecia linariata ;* flowers of foxglove (*Digitalis purpurea*) for *E. pulchellata* (especially if the flowers are partly closed up by being spun together) ; seed-heads of nettle-leaved bell-flower for *E. campanulata ;* seeding flower-heads of *Scabiosa arvensis* and *S. succisa* for larvæ of *Eupoecilia flaviciliana ;* cowslip seed-heads for *E. ciliella ;* and spun-together tops of sallow-shoots for *Peronea hastiana*, although most of these are certainly better obtained later in the year in normal seasons. The coast of the Isle of Man is most productive from the middle of June and on into July, and Birchall notes that, in 1876, at this season of the year, the flowers of *Silene maritima* growing in a little recess, about ten feet across, among the rocks on the shore near Douglas, of easy access, produced 105 *Dianthoecia caesia*, 32 *D. capsophila*, 4 *Plusia pulchrina*, 12 *Eupithecia venosata*, and many other insects during twenty evenings, the specimens being captured without moving from the spot, the captor, indeed, being seated on a stone most of the time. The hard-working collector, with resource, ought not, even in bad seasons, to complain of the results to be obtained during this month if his energy be put into entomology, and his heart be not set on obtaining further supplies of some rare or local species that may temporarily fail.

TINEINA (unclassified).—In the middle of June the tree-trunks should be searched for Micro-lepidoptera—often quite the best mode of collecting at this time of the year. In June the larvæ of *Gracilaria tringipennella* are in abundance in leaves of *Plantago lanceolata* at Castle Hill, Scarborough (and elsewhere), and the larvæ of *Coleophora alcyonipennella* on every plant of *Centaurea nigra* (Elisha).

TINEIDES.—In June collect hart's tongue for cases (with pupæ) of *Psychoides verhuella* ; keep dry and you will breed many ; I failed for many years by keeping them too damp (Hodgkinson).

ADELIDES.—In June the flowers of *Cardamine pratensis* should be worked for *Adela rufimitrella*.

PLUTELLIDES.—In early June the pale green larvæ of *Hypolepia sequella* may be beaten from maple. They are exceedingly active.

From the beginning to the middle of June the larvæ of *Plutella annulatella* are to be found on *Cochlearia anglica*.

The larva of *Harpipteryx scabrella* may be beaten into an umbrella from hawthorn early in June. It is not unlike that of *Cerostoma radiatella*, but has a white stripe down the back, and I do not think it is quite so lively. It is easy to feed up, and spins a beautiful cocoon of white silk, boat-shaped, with a triangular transverse section (Richardson).

ELACHISTIDES.—The flowering stems and leaves of *Epilobium* should be worked for the Lavernas—*Laverna conturbatella* (terminal leaves), *L. epilobiella*(in tops among leaves and flowers), *L. decorella* (in stems).

At the end of June the larvæ of *Laverna subbistrigella* may be obtained on *Epilobium montanum*. They live within the pods on the seeds. The infested pods are usually thickened and shortened, sometimes slightly distorted.

The larva of *Laverna miscella* mines the leaves of *Helianthemum vulgare* in June.

In June the larvæ of the first brood of *Chauliodus chaerophyllellus* are to be found on *Heracleum sphondylium* and *Pastinaca sativa*, mining at first, and making brown blotches in the leaves, and then feeding on the under surface of the large lower leaves beneath a slight web, and pupating among rubbish on the ground. The second brood feeds similarly in August and September.

In May and early June the larva of *Elachista scirpi* makes short broad mines in the upper half of the leaves of *Scirpus maritimus*.

In the middle of June the larva of *Elachista trapeziella* mines the leaves of *Luzula pilosa*.

The larvæ of *Elachista adscitella* are to be obtained in the stems of blue moor-grass (*Sesleria caerulea*) the second week in June (Hodgkinson).

In the middle of June the larva of *Nannodia eppelsheimi* makes conspicuous white blotches in the leaves of *Silene nutans*. This insect is still waiting to be added to the British fauna.

In June I find the larvæ of *Cosmopteryx druriella* in leaves of wild hop, at Sevenoaks (in a field adjoining the Bat and Ball Station) (Elisha).

GRACILARIIDES.—The larvæ of *Gracilaria auroguttella* feed in cones on *Hypericum perforatum* in June.

Towards the end of June the ash trees in the neighbourhood of Mickleham and Box Hill, give larvæ of *Zelleria hepariella*. Sometimes the close white cocoons are found on the leaves during July.

ARGYRESTHIIDES.—Examine well-grown plants of *Echium vulgare* on exposed slopes during the second week in June. *Douglasia ocneros-tomella* sometimes abounds, flitting about such plants, from 2 p.m. to 4 p.m.

The larva of *Argyresthia andereggiella* is to be found in June, beneath a web, spun near or at the end of the twigs of the crab-apple.

COLEOPHORIDES.—The black cases of *Coleophora vibicella* are readily found on the stems and leaves of *Genista tinctoria*, the blotches on the leaves are quite conspicuous, and the species, although very local, is sometimes exceedingly abundant.

The larvæ of *Coleophora genistaecolella* are to be found abundantly in June on *Genista anglica*.

At the base of the rocks at Whitbarrow, on the wild marjoram, during the first week in June the larvæ of *Coleophora albitarsella* may be obtained in plenty (Hodgkinson).

LITHOCOLLETIDES.—The larva of *Lithocolletis scopariella* makes an inflated mine along a broom twig. The pupæ should be obtained towards the end of June.

NEPTICULIDES.—The larva of *Nepticula poterii*, in early June, mines the leaves of *Poterium sanguisorba*.

TORTRICIDES.—The larvæ of the moorland Tortricids are now to be collected in great abundance—among others, *Tortrix viburniana* (*Myrica gale, Vaccinium*), *Peronea autumnana* (*Myrica*), *P. caledoniana* (*Myrica*), *P. maccana* (*Myrica*), *Grapholitha geminana* (*Vaccinium*), *Hypermoecia angustana* (sallows), &c.

The larvæ of *Tortrix piceana* are to be found in early June feeding on *Abies picea*.

Tortrix branderiana may be obtained freely at the end of June, flying from sunset to dark over the tops of aspens in woods.

The green larva of *Tortrix lafauryana* is to be found in June, making an upright tube of joined leaves on the top of a shoot of *Myrica gale*.

At the end of June visitors to Wicken Fen should collect the rolled-up leaves of dewberry (*Rubus caesius*), for larvæ of *Tortrix dumetana*.

The imagines of *Peronea shepherdana* are to be bred from rolled-up leaves of *Spiraea ulmaria*, collected in June.

The larva of *Peronea permutana* feeds in bound-together leaves of *Rosa spinosissima* generally on the lateral shoots, in June, July and August, spinning a white silken web among the leaves. It occurs at Wallasey and Penmaenbach (near Conway).

The larvæ of *Sciaphila conspersana* (*perterana*) feed in June, in the blossoms of *Chrysanthemum*, *Hieracium*, and other *Compositae*. Their presence may be recognised by the ray-florets being turned down so as to form a covering for the larva. The insect is more common in coast districts than inland.

The examples of *Sciaphila conspersana* (*perterana*), taken on the saltmarsh at Southend, "appear to be very different from those captured in the neighbourhood of Dover, on the chalk cliffs, not only in colour, but in texture. They may ultimately prove to be a distinct species." (Howard-Vaughan). [This has never been reported upon.]

The larva of *Ephippiphora grandaevana* feeds in June on *Tussilago*, and makes very long curious tubes in the sand (Zeller). The imago also occurs in July and August among *Petasites* in the Alps.

Ephippiphora grandaevana flies at the end of June and during July, at dusk, with a jerky flight of a few yards, very close to the ground. The species can be taken more freely by looking over the leaves of colts foot with a lantern, generally sitting on the top of the leaves of the smallest and most stunted plants (Gardner).

June (middle to end), *Ephippiphora gallicolana* is now out, flying swiftly over the tops of oak trees and bushes, from 2 till 7 p.m.

The imago of *Semasia woeberiana* flies in the sunshine among laurel, cherry, and other fruit-trees in June, and again in August.

The imagines of *Semasia spiniana* fly in June, in the afternoon sunshine around and above hawthorn hedges. May often be obtained abundantly in the morning by beating into an umbrella.

The larva of *Semasia janthiana* feeds on the berries of *Crataegus oxyacantha*, uniting them in twos or threes by means of a gummy substance, so that it can pass from one to the other without exposing itself. It eats the pulpy part of the fruit only without touching the skin. It pupates in August (Lafaury). Probably occurs later in Britain.

In June the larvæ of *Spilonota rosaeticolana* feed in shoots of rose, drawing together the leaves and eating out the young leaf-buds and flower-buds, and thus doing great damage in gardens.

Larvæ of *Spilonota aceriana* may be collected in the middle of June. They feed in the terminal buds of poplars and aspens. The presence of the larva is indicated by a tube of frass projecting from the end. Insert the cut twigs in damp sand.

In June the imagines of *Spilonota servillana* fly in the afternoon sunshine, and also at dusk among sallow bushes in the south of England. The larva makes a swelling in the twigs of sallows, turning to a pupa within the swelling thus formed.

The larvæ of *Grapholitha minutana* may be obtained in June feeding between flatly united leaves of black poplar.

The imago of *Phoxopteryx upupana* flies very swiftly over the tops of oaks, birches, and other trees in their immediate vicinity, from 2 p.m., or a little earlier, until an hour before sunset, in early June.

The blossoms of fir (*Pinus sylvestris*) should be collected in early June for larvæ and pupæ of *Sericoris bifasciana* and *Retinia sylvestrana*.

The last week of June and the first week of July are the time to find the larva of *Euchromia purpurana*. It lives underground in a silken tube, and gnaws the roots of *Sonchus arvensis* and *Taraxacum officinale* externally.

The imagines of *Penthina sellana* fly swiftly by day about grassy banks, where *Centaurea nigra* grows, during June and July. They may be easily overlooked for a *Dichrorampha*.

PYRALOIDES.—I think lepidopterists would often do better if they " smoked " for insects, especially the smaller ones ; it is possible on the coast sandhills to take thousands of specimens in a day if one only had the time to set them ; before I tried "smoking" I could only get 5 or 6 *Tachyptilia temerella* in a morning, with the smoke I can get 50 in an hour, and the same with other Gelechiids. I use old rags or brown paper, or in fact, any paper made into touch-paper, with saltpetre melted in water and dried again. A whiff or two of this sent among the roots of grass, etc., will make most things move (Baxter).

June is the great month for larvæ of Depressarias—flowers of furze, broom, and *Genista* give *Depressaria costosella, D. umbellella, D. assimilella,* &c. ; heads of *Centaurea scabiosa* produce *D. liturella, D. pallorella* ; heads of umbelliferous plants for *D. arenella, D. vaccinella, D. capreolella, D. angelicella, D. yeatesiella, D. applanella, D. granulosella, D. depressella, D. pimpinella, D. weirella, D. ultimella, D. nervosella,* &c. (*Anthriscus* and *Angelica* are the most frequently chosen.)

The larvæ of *Depressaria angelicella* are to be found twisting and crumpling the leaves of *Angelica sylvestris* in June.

The larvæ of *Depressaria atomella* are to be found in the shoots of *Genista tinctoria* about the beginning of June.

The larva of *Depressaria carduella* mines the leaves of *Cnicus lanceolatus,* and other thistles, in June, moving freely from plant to plant.

In the last fortnight of June the larvæ of *Depressaria albipunctella* and *D. chaerophyllella* should be obtained on *Chaerophyllum temulentum.*

The larvæ of *Enicostoma lobella* may be beaten from blackthorn in June. Loughton used to be a well-known locality.

In early June the unexpanded flower-buds of the common mallow are tenanted by a small white larva, which produces *Gelechia vilella.*

In the early part and middle of June the larva of *Anacampsis albipalpella* feeds on *Genista anglica* ; the larva draws several leaves together round the stem, and then eats them half through, thus discolouring them, and forming conspicuous clusters of yellowish-white leaves.

The larvæ of *Cladodes gerronella* may be found in the early part of June feeding on furze.

The larvæ of *Butalis grandipennella* are to be obtained on the furze-bushes, at Wanstead (and elsewhere), about the middle of June.

They make a web, placed along the stems of some years' growth, and a strong pair of cutting-pliers should be requisitioned for cutting off those portions of the stems containing the webs.

During June the shoots of *Lysimachia* should be collected for larvæ of *Doryphora morosa*. Wicken Fen is the best known locality.

In June the terminal shoots of furze which have turned brown should be collected for larvæ of *Anarsia spartiella*.

Spun shoots of *Lotus major*, collected in early June, will give larvæ of *Anacampsis vorticella*.

Near the chalk pit at Kemsing the larva of the very beautiful *Hypercallia citrinalis* (*christiernella*) lives in the shoots of *Polygala vulgaris*.

In June plants of *Sedum telephium* are often much covered with the webs of the larvæ of *Hyponomeuta vigintipunctatus*. The larvæ of the second brood are still more abundant in August.

CRAMBIDES.—Crambids are to be disturbed during the day but their usual time of flight is at dusk and after, June and July being the principal months, from 9 p.m.-10 p.m. *Crambus margaritellus* flies low from one grass blade to another in August. *Crambus furcatellus* and *C. ericellus* fly freely in the late afternoon and evening in late June. (Reid).

Crambus furcatellus was flying not uncommonly on June 21st, on the grassy slopes between Sprinkling and Styehead tarns, and was, in fact, found on almost every high hill ascended (Geldart).

In the middle of June, the barren denes near Yarmouth should be searched for the long cocoons (placed perpendicularly in the sand) of *Crambus fascelinellus*. They may be found just beneath the sand where *Triticum junceum* is growing. Cocoons are often exposed by the wind.

The imago of *Phycis carbonariella* prefers to rest on burnt places on heaths.

Examine stunted whitethorn and blackthorn bushes, growing on exposed slopes almost anywhere, during the first fortnight in June; you will find the silken galleries of *Rhodophaea suavella* spun close to the twigs.

In June collect the bunches of oak leaves that are spun together high up on oak-trees for larvæ of *Rhodophaea tumidella*. The bunches low down on the small bushes contain larvæ of *R. consociella*.

During the first week of June the larvæ of *Cryptoblabes bistriga* should be searched for on oak, or it may be beaten therefrom.

On June 27th, in the Fens around Norwich, in a part of the Fen ankle deep in water, creeping up from the tufts of a small rush, and fluttering among the reeds, was *Schoenobius mucronellus* in swarms—hundreds of them—nearly all males—their flight lasting from 6-8 p.m. (Barrett).

The males of *Chilo phragmitellus* fly freely at dusk by reed-beds, the females must be searched for on the reeds. The males also come freely to light.

PYRALIDES.—To find the larvæ of the *Scopariae* peel off the moss growing on the north side of shady rocks, large boulders and walls, in spring and early summer, and examine the underside of it. If larvæ be there the galleries of silk slightly spun upon the moss and the frass will indicate their presence. To rear them, place the tufts of moss in a jam-pot with ground top and cover with a piece of glass (F. B. White).

The rare *Botys repandalis* is reported as having been bred from larvæ found feeding in June, in the heads and young shoots of *Verbascum nigrum*, on the south coast of Devon.

The larvæ of *Pyralis glaucinalis* have been found during May and June in the nest-like bunches of twigs which may often be observed growing at the ends of branches on birch trees.

DREPANULIDES.—Stunted trees of beech should be searched for eggs of *Drepana unguicula*, laid on the leaves ; those of *D. lacertinaria* are laid on birch leaves or twigs.

CYMATOPHORIDES.—In June the larvæ of *Asphalia flavicornis* spin together two or three leaves of birch and live within. On the same bushes (and at the same time), the larvæ of *Phycis betulella* are to be found in twisted leaves. The larvæ of the first-named sometimes spin two leaves together, and sometimes fold one leaf. They sometimes pupate between two leaves and sometimes spin a leaf to the earth.

In June the larvæ of *Asphalia ridens* can best be obtained by looking up into oak trees, when the larvæ may be seen on the outside branches, lying half-curled between two leaves drawn slightly together. The larvæ must be searched for by day, or beaten by night.

BREPHIDES.—Beat (or search) aspen the first week in June for larvæ of *Brephos notha* ; the full-fed larva wants cork or rotten wood in which to make its cocoon.

The larvæ of *Brephos parthenias* are now to be obtained almost full-fed on birch (and those of *B. notha* on aspen) ; difficult to obtain during the day as they rest between united leaves.

GEOMETRIDES.—The long thin delicate cocoon of *Ourapteryx sambucata* is not difficult to find, spun up among ivy twigs ; much more difficult to detect on blackthorn, &c.

The imagines of *Epione advenaria* hide low down among bilberry and grass in wood ridings, but are easily disturbed.

The full-fed larvæ of *Boarmia abietaria* should be beaten from larch, spruce fir, &c., in the early part of the month ; they may be fed in confinement on many other plants—birch, &c.

At the end of June the pupæ of *Boarmia abietaria* should be carefully worked for just below the surface of the ground, at larch, spruce fir, &c.

Boarmia roboraria is best taken at rest in the early morning, but it comes to sugar late at night.

In June a freshly emerged ♀ of *Boarmia roboraria* assembles the ♂ s readily ; the latter put in an appearance about 12 p.m., and in two nights some forty were taken, about twice that number being attracted.

In June the larvæ of *Cleora glabraria* are to be found feeding on lichens on oak-trunks.

The cocoons of *Geometra papilionaria* are to be found in spun-together leaves of birch, and are most difficult to find ; occasionally the larva uses the edge of moss on tree trunks as a place to spin up.

In June the larva of *Pseudoterpna pruinata (cytisaria)* feeds upon furze (*Ulex europaeus*), *Genista anglica* and *Cytisus scoparius.*

Phorodesma bajularia flies high at sunset in oak woods during June and on to end of month, prefers broad rides and woodsides.

In early June the Scotch heaths want working for the various forms of *Fidonia atomaria*, which are exceedingly interesting to southern collectors.

The first brood of *Aspilates ochrearia (citraria)* is abundant in early June, in its favoured haunts, usually rough grassy tracts on sandhills near the sea. Eggs from these will give a second brood in August.

The larvæ of *Lobophora viretata* are most easily bred on ivy in captivity, eating every particle, whilst it keeps fresh, except the stem ; eggs readily obtained from captured ♀ s. The larvæ stick to a leaf like grim death, while there is a particle left, devouring ribs, stalk, and tissues alike, and do not wander—in fact they feed in uncovered glass jars (Tunaley).

The females of *Bupalus piniaria* are sometimes taken abundantly in June, clinging to grass stems under the trees on which the larvæ have fed.

Timandra amataria may be taken from June (middle) to July (middle) in abundance, flying at dusk in weedy lanes ; and after dark at rest on the grasses and hedgeplants.

The imagines of *Acidalia emutaria* begin to fly directly after sunset, settling on the herbage as soon as it is quite dark ; they are then easily found with a lantern. I have found them as early as the second week in June, and as late as the second week in August. They are abundant on the saltings of north Kent, along the ditch-sides of Deal and Sandwich, the tidal swamps of the Freshwater (Yar) district, and the bogs of the New Forest. In the Isle of Wight there are two distinct broods (June and August).

Acidalia rubricata flies most freely about half an hour before sunset, and one gets many in a locality where only individual examples can be walked up in the day or late afternoon (Wratislaw).

I sleeved some young larvæ of *Anticlea berberata* on an English barberry bush, and others on one of the foreign varieties ; leaving home for a short time I found on my return the former a good size, and the leaves a mere network, but the latter were all dead and not a leaf nibbled (Riding).

The larvæ of *Scotosia certata* not only feed on leaves of common barberry but they also occur in abundance on the holly-leaved *Berberis* of our gardens (Thornewill).

The second and third weeks in June are usually the most satis-factory in which to beat *Prunus spinosa* for larvæ of *Aleucis pictaria.*

Melanippe hastata flies at noon in June over the wettest part of Orton Moss, it is slow on the wing and easily captured (Armstrong).

In June *Eupithecia isogrammata* and *E. coronata* want a great deal of working for, beating the clematis is of no use at all; a stick must be thrust in and the bush regularly churned ; then wait for a few minutes and beat in the ordinary way (Kimber).

The larvæ of *Eupithecia pimpinellata* may be obtained by collecting the seed-heads of *Pimpinella saxifraga, Angelica sylvestris,* &c.

The larvæ of *Eupithecia pumilata* are to be obtained, sometimes in abundance, in flowers of gorse, broom, and many other plants. When the foxgloves are in flower, search the solitary plants growing in the open spaces in, or on, the outskirts of woods, for spun-up flowers containing larvæ of *Eupithecia pulchellata*.

The imagines of *Eupithecia plumbeolata* may sometimes be obtained freely among *Melampyrum pratensis*, in woods and meadows, flying at dusk in early June.

Eupithecia rectangulata swarms in some apple-orchards in June. As many as three dozen have been obtained in an hour by blowing on tree-trunks (Marsden); they are also sometimes to be taken in great abundance flying round the apple trees at dusk, in June ; also on the tree-trunks and adjacent fences by day.

The imagines of *Eupithecia helveticaria* are to be beaten in June, the larvæ beaten from juniper in early September.

The imagines of *Eupithecia subumbrata* may be freely beaten from yewtrees on all the Kent and Surrey downs (and possibly elsewhere) during June ; frequently accompanied by *E. satyrata.*

PTEROPHORIDES.—The pupæ of about one-half the British species of plumes can be obtained during this month, attached like butterfly pupæ, by their anal segments to their respective food-plants—*Platyptilia pallidactyla* on yarrow, *P. ochrodactyla* on tansy, *Mimaeseoptilus plagiodactylus* on scabious, *Leioptilus tephradactylus* and *L. osteodactylus* on golden-rod, *Aciptilia galactodactyla* on burdock, *A. spilodactyla* on *Marrubium*, &c.

During the first and second weeks in June the first brood of *Agdistis bennetii* is to be found on salt marshes where *Statice limonium* grows. The imagines cannot be taken much before eight in the evening.

Wherever *Marrubium vulgare* occurs in the Isle of Wight, there it appears in June and July is the larva of *Aciptilia spilodactyla*, whilst the imagines may be readily disturbed from the plants and surrounding herbage in late July and early August.

SESIIDES.—The imagines of *Sesia formiciformis* fly by day in the sunshine, around the willows in which the larvæ have fed up, and may sometimes be seen crawling over the leaves.

The first week in June is the time for *Sesia sphegiformis*, which assembles freely. It occurs in considerable abundance in Tilgate Forest, Basingstoke, and was so abundant in a Welsh locality—Dolau Cothy, Carmarthenshire—recently, that the larvæ destroyed almost all the alders growing there.

The pupæ of *Sesia musciformis* (*philanthiformis*) should be collected in early June ; the best situations are the detached rocks at the base of cliffs ; it is of no use searching among the luxuriant growth of *Armeria* on the mainland, but the stunted plants scattered in the clefts of water-worn rocks fringing the shore are those attacked ; a little red patch on the green cushion betrays the work of the larva (Birchall).

In late June search the stems of poplars near the roots in early morning for freshly emerged *Trochilium apiforme.*

ZEUZERIDES.—The imagines of *Macrogaster arundinis* are generally taken at light ; much more abundant in Chippenham than in Wicken

Fen; said to emerge from pupa at about 10 p.m., and hence the imago might be obtained when drying its wings, by searching the reeds where it occurs with a lantern at night.

From June 24th to the end of July is the time to search for *Zeuzera pyrina* in the London Parks and elsewhere.

COCHLIDIDES.—The male of *Heterogenea cruciata*, which is nearly black in colour, flies swiftly along the rides of Epping Forest, and is, in my experience, never beaten out (Battershell-Gill).

Kick the stems of young oaks in those parts of a wood free from undergrowth. Then look carefully at everything that falls—*Cochlidion limacodes* ♀ comes down half falling, half flying. The ♂ s fly swiftly over the trees in hot sunshine.

PSYCHIDES.—On June 22nd and July 4th (1827) I took a large number of larvæ and pupæ of *Psyche fusca* on the leaves of the hazel and young oaks growing in Hornsey Wood. I have also found them in Highgate Wood " (Ingpen). So many species still exist in these woods that one might almost hope that the cases of this insect would again be found.

ANTHROCERIDES.—In early June the imagines of *Adscita statices* are to be found resting on the flowers in the evening in those localities where they fly during the day.

In the first week in June the males of *Adscita geryon* fly among the long grass in the sunshine in hundreds at Dursley, the females hiding among the grass (Griffiths). Same habits elsewhere; also rests on flowers.

The almost barren terraces of limestone in south-western Galway and co. Clare produce *Anthrocera purpuralis* (*minos*) in immense numbers in the middle of June. The pupæ are to be found attached to stones.

At the end of June on the Clare coast, at Black Head, on the horizontal slabs of limestone at the very edge of the cliff, where nothing grows but a few stunted tussocks of grass and the rare *Adiantum capillus-veneris*, *Anthrocera purpuralis* (*nubigena*) occurs in such countless thousands, that when I passed my net along the edge of the cliff it came back full of the moths (Hon. Emily Lawless).

LACHNEIDES.—In June the larvæ of the fine northern form of *Trichiura crataegi* are to be found on the moorlands of Scotland.

During the first week in June the larvæ of *Trichiura crataegi* can be obtained by beating oak.

This is the great month for larvæ of *Malacosoma castrensis* on the Kent, Essex and Suffolk salterns. They are in amazing numbers in certain years on almost all the low marsh plants but are very uncertain in appearance.

In early June a sharp eye should be kept for the larval nests of *Lachneis lanestris* on hawthorn.

Macrothylacia rubi males are best assembled just before dusk; although on the wing all day they rarely assemble until evening.

In early June the larvæ of *Pachygastria trifolii* should be sought on the coast line between Lydd and Rye, and on the heaths at Lyndhurst.

Lasiocampa var. *callunae* is now on the wing on heaths and moorlands ; the males can be assembled in large numbers by a newly emerged female in the afternoon.

The parchment-like cocoons of *Cosmotriche potatoria* are to be easily found by anyone who searches carefully ; the sides of ditches are favourable haunts ; the position on a rush, grass or reed stem varies greatly.

The cocoon of *Eutricha quercifolia* (easily seen when hedges are leafless in winter) is spun up in the bottom of a hedge among the thick shoots in such a way as almost to defy detection when the foliage is on the bushes.

SPHINGIDES.—In early June *Choerocampa porcellus* and *C. elpenor* are abundant at and after dusk over flowers, whilst *Macroglossa fuciformis* and *M. bombyliformis* are abundant by day ; the three first-named are very partial to rhododendrons, the last-named to *Ajuga reptans*.

The eggs of *Macroglossa fuciformis* are not difficult to find on honeysuckle (underside of leaf) and that of *Macroglossa bombyliformis* on scabious. When the larvæ have hatched, those of the former species may be beaten or found by searching, those of the latter must be searched for carefully, as they manage to hide very successfully beneath the leaves.

DELTOIDES.—The imago of *Hypena rostralis* has been bred from a bright green, half-looping larva, found feeding on wild hop in Hackney marshes about the middle of June (Machin).

LYMANTRIIDES.—The larvæ of *Leucoma salicis* are exceedingly abundant in June in their special haunts, resting on the poplar stems by day, where they are easily detected ; they pupate in spun leaves on their food-plants in July ; also abundant in some localities on willow.

Throughout June the northern moors should be searched for cocoons of *Dasychira fascelina* spun up among the heather.

Imagines of *Dasychira pudibunda* are sometimes abundant at rest on the hop-poles in a hop-garden in the morning ; the larvæ feed on hop, and are known locally as " hop-dogs." The larvæ of *Orgyia antiqua* are known as " hop-cats."

Webs of *Porthesia chrysorrhoea* should be searched for in early June on hawthorn. The insect is very uncertain in appearance, and is now again abundant in Kent, Essex, &c., after the species has been practically absent from these districts for nearly 20 years.

NYCTEOLIDES (CHLOEPHORIDES).—Full-fed larvæ of *Hylophila bicolorana* are frequently found crawling over oak trunks in June ; also to be beaten from oak-trees.

NOTODONTIDES.—During the first week in June beat low elm-trees on the outskirts of woods, or on the borders of rides in woods. Large numbers of larvæ of *Asteroscopus sphinx* may thus be obtained. The larvæ require plenty of room in which to pupate ; they go deep, and unless only one or two are kept in the same pot are likely to disturb one another.

June is a great month for the eggs of Notodonts. Most of the species have pale-coloured eggs, and lay them scattered on the under-sides of the leaves of their respective food-plants.

The dark-coloured eggs of *Cerura furcula* are laid in twos and threes on the upper surface of leaves of sallow and willow ; those of *C. bifida* on the upper surface of leaves of poplar ; and those of *C. vinula* on those of willows, poplars or sallows ; they may be found with a little patient searching.

Spun-up leaves of poplar and sallow should now be overhauled for cocoons of *Clostera curtula*.

Trunks of beech-trees and those of other trees near beeches should be well searched for imagines of *Stauropus fagi* ; they do not always select large stems on which to rest.

Assembling in June with a virgin ♀ of *Stauropus fagi* sometimes gives good sport, the time of flight being from a little before 11 p.m. till a little after midnight (for details see *Ent. Rec.*, i., p. 67).

NOCTUIDES.—Moths may not only be found upon or below the sugar, but a foot or two above, or on the other side of the tree, or hanging on the twigs nearest to the sugar.

In June the flowers of *Cotoneaster microphylla* are very attractive to Noctuid moths.

The young larva of *Apatela aceris* sits curled beneath a leaf of horse-chestnut, sycamore, or maple, and eats the lower parenchyma between the veins. In its second stadium it coils itself in a circle and still leaves the veins and upper cuticle, and it is not till it is in the third stadium that it eats the whole thickness of the leaf.

The eggs of *Acronicta leporina*, though laid separately, are usually placed to the number of a score or thereabouts in close proximity ; consequently, when young, the larvæ are usually to be found, several near together, on adjacent leaves or branches, so that one being found, others should be searched for. The young larvæ invariably select the underside of a leaf to rest on, and, in feeding, leave the veins and upper cuticle, but when older they eat through the whole thickness of the leaf.

Moma orion comes to sugar at late dusk from 9 p.m.-9.20 p.m. during the first fortnight in June, and settles, with wings closed, at the top of a patch, looking remarkably like the green lichen, which, in the New Forest, covers the trees.

In June the eggs of *Moma orion* are laid beneath an oak-leaf in batches of fifty or more, regularly disposed in close order like those of many Noctuids.

The young larvæ of *Moma orion*, in the first and second stadia, eat the lower side of the leaves of oak, leaving the upper cuticle and veins untouched, so that a leaf often presents a very perfect skeleton, but with the upper cuticle (Chapman).

In pairing *Jocheaera alni* in confinement the ♀ should be kept for two nights regularly fed before the ♂ is put into the cage ; the result is then more favourable, and, if the object be to obtain a brood, the sacrifice of a few specimens must be made cheerfully ; warm nights and patience are two conditions necessary for success (Smith).

In June, shaking marram grass which hangs over a bank on coast sand-hills often gives a supply of *Leucania littoralis* and *Mamestra albicolon*. In July and August, *Agrotis tritici*, *A. cursoria*, and *Actebia praecox* may be obtained in the same way.

Imagines of *Agrotis ripae* have been found in abundance under pieces of wood lying on the sand by the seashore, just above the reach of high tide, in June.

From June 7th-20th is about the best time to work for *Hydrilla palustris*. Most of the examples that have been taken in Britain have been captured at light at Wicken and in the adjacent fens, but York and Carlisle have also produced specimens, so that there is no doubt the insect is widely distributed and only wants working for.

The females of *Chortodes arcuosa* are seldom to be seen on the wing, but may easily be found at rest on *Aira caespitosa* after dark, in June (Porritt).

Chortodes morrisii (*bondii*) has been bred from larvæ found in June feeding in the grass tussocks of *Arrhenatherum avenaceum*. The sickly-looking stems break off close to the roots when gently pulled, and usually below them a larva or pupa is to be found.

In early June the imago of *Dianthoecia caesia* is to be obtained flying over the newly-opened flowers of *Silene maritima*, whilst the larva is to be found a fortnight later feeding on the flowers of the same plant (and those of *S. inflata*) in the coast districts bordering the Irish Sea.

During the first week in June at Witherslack the little *Phothedes captiuncula* was sunning itself upon the flowers of the oxeye daisy (Shuttleworth).

The imagines of *Phothedes captiuncula* fly swiftly in the sunshine (2 p.m.-5 p.m.) in June at Whitbarrow, near Bowness ; in the middle of July, on the Morsden Rocks, near South Shields.

I have to record the capture, by myself and two friends, of over 300 larvæ of *Xylophasia scolopacina* in woods at Hampstead and Highgate, between the 1st and 3rd of June (Lockyer).

Where *Apamea unanimis* occurs search fences, trunks, palings, and low plants about 8 a.m. for the newly emerged imagines drying their wings.

Apamea ophiogramma is a genuine dusk flyer, about half an hour being the time in which one has to get them ; they fly quietly, look very light on the wing, settle on different flowers—*Scrophularia* and nettles for preference—and are very quiet in the net and easy to box (Farren).

The scarcity of thistles, or any suitable blossoms, to sugar on the Freshwater downs necessitates the use of bundles of cut blossom, of which that of *Heracleum sphondylium* is the most attractive ; a well-sugared bunch on June 14th, 1892, was a magnificent sight, being almost covered with *Agrotis lunigera* in the grandest condition.

In June and early July *Agrotis simulans* can generally be found in certain parts of Scotland—Forres (and probably elsewhere)—within doors. Norman visited the various rooms and outbuildings just after dark and found the moths fluttering on the inside of the glass windows.

Common white horehound (*Marrubium vulgare*) and black horehound (*Ballota nigra*) sugared on the Freshwater coast are very attractive to Noctuids, especially *Agrotis lucernea* (Hodges).

About June 20th search for larvæ of *Actebia praecox*. They feed on sallow at night, and hide themselves beneath the surface of the sand during daytime, whence they have to be brought out by raking about the roots of the food-plant.

Agrotis agathina larvæ full-fed on ling, June 23rd ; imagines found *in copula* September 23rd. Other females laid ova on September 13th (Kane).

The larvæ of *Pachnobia hyperborea* are to be found under moss during early June in fir-woods where *Vaccinium myrtillus* grows (Staudinger), but in Scotland the larvæ feed on crowberry and bilberry, preferring the former (Meek).' The imagines of all the Dianthœcias may be obtained freely at and after dusk at flowers of *Silene, Lychnis,* &c., and the eggs on the flowerheads of their respective food-plants.

Silene flowers on the west coast of Britain should be worked at dusk in June for *Dianthoecia luteago* var. *barrettii*. [It should be sought at Howth during the first fortnight in June (Birchall).]

About the middle of June (June 15th), in the evening, wet or dry, *Dianthoecia conspersa* will take wing, quick in motion, but not flying far, and dropping soon on the flowers of *Lychnis flos-cuculi* (Armstrong).

Place at the bottom of an old chip hat-box two inches of sand. Collect the seed-capsules of *Silene inflata, S. campestris,* &c., and place on the sand. Cover the top of the box with gauze by means of an elastic band. Put in a few fresh capsules occasionally, and you will breed *Dianthoecia conspersa, D. carpophaga* and *Eupithecia venosata* (Hall).

During the first fortnight of June the capsules of *Silene maritima* may be collected for larvæ of *Dianthoecia conspersa*. Later in the month those of *Lychnis vespertina* should be collected for *D. carpophaga,* and those of *Silene cucubalus* (also pinks, sweetwilliams, &c., in gardens) for *D. capsincola*. It is better to search by night than by day*.

The larva of *Dyschorista upsilon* is sometimes to be found in the greatest profusion under the loose bark of willows in June.

Imagines of *Hyppa rectilinea* and *Plusia interrogationis* are to be captured when resting on stumps and stones in dull weather, in June and early July.

Early in June search the sallows and willows for spun-up chambers containing the green larvæ of *Cleoceris viminalis*. As soon as these disappear, examine the folded leaves towards the end of the twigs for larvæ of *Tethea retusa*.

When searching for larvæ of *Clostera reclusa,* do not overlook larvæ of *Tethea retusa ;* they fold a leaf or two of sallow round them much after the manner of *C. reclusa,* but the larvæ themselves much resemble those of *Epunda viminalis*.

Tethea subtusa uses a single poplar leaf to hide in, turning over one side and fastening it firmly round the edge with silk. *Taeniocampa populeti* invariably uses two leaves, one fastened above the other. The best way to find the latter is to get below the branches of a tree, when the fastened leaves are readily seen, and, in bright weather, the larvæ can often be distinctly seen through. The larva of *Taeniocampa populeti* also feeds between two united aspen (as well as poplar) leaves.

Larvæ of *Tethea subtusa* are to be found in June "in pockets at the edges of the leaves of poplar, made in an almost identically similar manner to those formed by the young larvæ of *Tiliacea citrago.*"

---

* Mr. Prout remarks of this note :—" *Lychnis* for *D. capsincola, Silene* for *D. carpophaga* and *D. cucubali.*"

D

Larvæ of the red form (ab. *rufa*) of *Taeniocampa gracilis* are to be obtained from bog-myrtle in the New Forest in June.

The pupa of *Dicycla oo* is to be obtained at the roots of oak : but the species is most uncertain in its appearance in all the British localities.

The rather delicate cocoons of *Cirrhoedia xerampelina* should be worked for at the roots of ash-trees at the end of the month ; the larva, like that of *Hoporina croceago* (at base of oak), remains in cocoon unchanged for some weeks.

Early in June the wych-elm should be beaten for larvæ of *Mellinia* (*Xanthia*) *gilvago :* usually locally abundant.

During the first week in June beat the low elm trees on the outskirts of woods, or on the borders of rides in woods. Large numbers of larvæ of *Calymnia affinis* may thus be obtained.

The larvæ of *Polia nigrocincta* may be found in June with a lantern at night feeding on *Plantago maritima* and other plants on the coast rocks of the Isle of Man (Porritt).

Larvæ of *Cucullia chamomillae* are nearly full-fed on *Matricaria* at the end of June. Search for them when the morning sun is on the plants.

Mullein plants should be searched in late June for larvæ of *Cucullia verbasci*. The presence of the larvæ is readily shown by the turning back of the rough surface of the undersides of the leaves ; the larvæ themselves are conspicuous enough if the leaves be turned over.

In early seasons during the last fortnight of June (and in July) search *Verbascum* well for young larvæ of *Cucullia lychnitis*.

Ballota nigra is most attractive to *Plusia iota*, *P. chrysitis* and *Habrostola urticae*.

The cocoons of *Plusia festucae* want considerable finding when spun up in the broad leaves of grasses, iris, reeds, &c., by ditch-sides, although conspicuous enough when once detected.

Throughout June the northern moors should be searched for cocoons of *Plusia interrogationis* spun up among the heather.

The imagines of *Agrophila trabealis* (*sulphuralis*) occur in June and July, taking short swift flights from one place to another when disturbed. The species is somewhat difficult to discover when at rest, and is not unlike a *Coccinella* in some positions ; about 5 p.m. is the most natural time of flight.

In hunting for *Agrophila trabealis* " a switch, for the purpose of brushing the herbage is of great advantage, and in capturing the moth, the net should be quickly placed over it as soon as one can get within reach " (F. Bond).

About June 10th is the best average date for *Banksia argentula :* at this time it abounds in Chippenham Fen.

The larvæ of *Toxocampa pastinum*, full-fed in early June, are to be obtained by day at the roots of *Vicia cracca*, &c., and by night more abundantly feeding on the same plant ; the larvæ sometimes spin up high among the purple vetch, some spin the cocoon to the ground, others make a cocoon in the ground.

In June search in crevices or under the loose bark of willows and poplars for larvæ of *Catocala nupta*.

ARCTIIDES.—During the second week in June larvæ of *Lithosia caniola* and *L. complana* are to be found at Howth (Birchall).

The larvæ of *Lithosia caniola* should be sought during June on the southern and western shores of England and Ireland, feeding on the flowers of *Lotus corniculatus*, &c.

To obtain *Lithosia sororcula* (*aureola*), in early June, beat the outskirts of a wood with a long pole (9 ft.-10 ft.), on a really hot day, from 5 p.m.-7 p.m. They flutter out gently from oak, maple, blackthorn, ash, &c. (Raynor).

The larvæ of *Lithosia deplana* are in some years very abundant, and may then be beaten commonly from both oak and beech ; most of the larvæ obtained in the New Forest are from trees on which there is no lichen (Sellon).

Lithosia mesomella flies freely just at dark ; it is possible on a still evening in late June or early July, to net large numbers in a favourable spot, mostly ♂ s, hence probably the favourable spot indicates the nearness of a ♀.

In years when *Œnistis quadra* is common the cocoons are to be found in the crevices of the bark of tree-trunks, on palings near trees, &c., at the end of June ; sometimes most abundant in the New Forest.

The cocoons of *Setina irrorella* form a slight web, are placed under stones, oyster-, cockle-, or mussel-shells, &c., covered with ground lichen, on the coast just above the tide mark (Merrin). Also in quite inland localities, where the larvæ feed on ground lichens and spin up under stones.

In June *Euthemonia russula* must be walked up among rough herbage. The ♀ s are best disturbed in the morning between 9 a.m.-10 a.m.

PAPILIONIDES.—During the last week of June search the plants of *Cardamine pratensis* and *Sisymbrium officinalis* for larvæ of *Euchloë cardamines*.

The gregarious larvæ of *Eugonia polychloros* should be sought on elm, willow, sallow, aspen, &c., the eggs are laid in the spring by hybernated females and the presence of hybernators in April and early May in a locality should lead to a search for larvæ in June.

Thecla pruni affects the privet blossom in Barnwell Wold. The blossoming of the privet is usually a good sign as to the date to obtain this species.

During the first week in June beat low elm-trees on the outskirts of woods, or on the borders of rides in woods. Large numbers of larvæ of *Thecla w-album* may thus be obtained.

During June bushes of blackthorn should be beaten for larvæ of *Zephyrus betulae :* stunted ones are often most prolific.

The larvæ of *Zephyrus quercûs* can sometimes be beaten in large numbers from oak in early June, as also can those of *Thecla w-album* from elm.

At the commencement of June search the underside of primrose and cowslip leaves for the eggs of *Nemeobius lucina*.

JULY.

This period is the heyday of summer, nature at its fullest, and the hands of the lepidopterist more than full ; the freshness of the early

spring departed, the solid work of the year to be accomplished—
now or never. Late June and early July produce the harvest of the
lepidopterist; three-fourths of the more local species of British
butterflies are on the wing; the chalkdowns by day, and still more
strikingly at early dusk, are alive with insects; sugar frequently pays
the diligent collector, and almost all well-known methods of work are
productive. As a rule, the lepidopterist is too busy for evening work
at flowers, but the rarer Plusias, and, if there have been many immigrants arriving, many of the rarer Sphingids, may be thus taken.
Larva-beating, too, is more often neglected now than at any other
season of the year, owing to the temptation to hunt the imagines that
abound everywhere. The Broads are at their best in the last week of
July or in early August—*Lithosia muscerda*, *Leucania brevilinea*, and
many other rare species being on the wing, and larvæ of *Nonagria
cannae* ready for collecting. In the fens, too (Chippenham, Wicken,
&c.), *Macrogaster arundinis* frequently comes to light in amazing
numbers, as also *Nascia cilialis* and other rare species—large and
small. In riverside marshes, *Leucania obsoleta* of mid-June is replaced
by *L. straminea* in early and mid-July, and other good species are
fully out; while on the coast sand-hills, choicest and most reliable of
all sugaring grounds, a new fauna delights the inland worker. This,
too, is the great season in Scotland, where everything, except a few
rare spring and autumn species, is huddled into two months' collecting, especially on the hills and higher mountains. Coast work wants
experience; many a species is restricted to a small area, here and
there, and many exceedingly abundant species in one locality are
absent from an apparently similar one, for no obvious reason. There
are the local Lithosiids—*L. lutarella* var. *pygmaeola*, at Deal and Sandwich, and *L. caniola*, at Rye, Torquay, Howth, &c.—*Acidalia ochrata*
and *Nyctegretes achatinella*, at Deal and St. Osyth, *Agrotis lunigera*, on
the coast of the Isle of Wight, Howth, and a few other chosen haunts;
while *Agrotis simulans*, at Portland and in various parts of Scotland,
offers a peculiar range of distribution. *Agrotis obscura* (*ravida*) is
probably more abundant on the outskirts of Wicken Fen and along
the reaches of the Humber than elsewhere. We have tried most kinds
of localities, but doubt whether anything interests us more at this
time of the year than the exhilarating day-work of the chalk-hills, or
the excitement of light in the fens. Light in July, when it is paying,
is a marvellous tonic to a lepidopterist—a close, warm, cloudy night,
when one can feel the excitement of the air, due perhaps to
the rolling thunder in the distance. One has to wait for
these nights, but when they come you feel that it has been
worth while to wait. On such nights *Macrogaster arundinis*,
Eutricha quercifolia, *Cosmotriche potatoria*, *Malacosoma neustria*, and
Spilosoma fuliginosa throw themselves at the light, or with a heavy
vibration of the wings buzz around you. You must be wary, or you
will miss *Schoenobius mucronellus*, *Catoptria expallidana*, *Sericoris fuligana* and other rare micros that run up and down the front of the
lamp, or the *Pharetra alborenosa* that will go to the back of the sheet,
while before the light is put on you net the rare Gelechiids and Tortricids, for which the fens are celebrated. Hunting in the forests and
large woods for *Limenitis sibylla*, *Apatura iris*, and the larger fritil-
laries provides excellent sport for the collector pure and simple or the
aberration-hunter; and to hunt the marram-grass blossom on a

favourable night on the Deal sand-hills, with a lantern, for Noctuids, and the *Hippophaes* stems for *Nola centonalis*, makes a pleasant change to the town lepidopterist. Of the value of rush- and sedge-flowers as an attraction on certain nights, one cannot speak too highly. Skepper states that during the last fortnight in July, on the low marshy ground near Lowestoft, he went out with a lantern every night, and found the moths swarming from 9-10 p.m.; so much so as to make the rushes (*Juncus effusus*) look full of variously coloured flowers. Hundreds could have been taken every evening, for they sat perfectly still, extracting something from the heads of the rushes, then past flowering, and all one had to do was to make a selection and box what one wanted. The turn-cap lily is reported to be a special favourite, among garden flowers, of the July moths, attracting many species in great numbers. Quite distinct is the collecting in our great woods, and a visit to the New Forest is not only productive of rare imagines, but of larvæ innumerable, and the lepidopterist who will not beat in July is likely enough to lose *Notodonta trepida*, *Stauropus fagi*, and other equally good insects. July is also a great month for the larvæ of Rhopalocera, Geometrids, and Noctuids ; the first named generally to be obtained by searching, the others by beating. Larvæ of *Papilio machaon* are usually to be obtained in abundance on the fens (Wicken, &c.), those of *Euchloë cardamines*, on the flower- and seed-stalks of various species of *Cruciferae*—*Cardamine*, *Erysium*, &c.; *Gonepteryx rhamni*, quite full-fed, in the early part of the month on buckthorn ; *Colias edusa* and *C. hyale*, if there has been a late spring or early summer immigration, on leguminous plants ; *Euvanessa antiopa* and *Pyrameis cardui*, also immigrants, the larvæ of the former on willow, the latter on thistles, &c.; *Vanessa io*, gregarious and very conspicuous on nettle ; *Pararge egeria* and *P. megaera*, readily swept at night when feeding on grass ; *Zephyrus betulae*, full-fed, early in the month on sloe ; *Callophrys rubi*, on the flowering shoots and blossoms of broom, bramble, *Helianthemum vulgare*, &c., *Nemeobius lucina*, on the undersides of primrose and cowslip leaves, and several Lycænids—*Polyommatus astrarche*, *P. icarus*, *P. bellargus*, *Lycaena arion*, *Cupido minima*, &c. Sweeping and beating for Geometrid larvæ always pay—beating, by day and night, and sweeping, as a rule, only by night—for now *Pericallia syringaria*, all the Ennomids—*Ennomos autumnaria*, *alniaria* (*tiliaria*), *fuscantaria*, &c.—are feeding, whilst *Amphidasys strataria*, *Tephrosia consonaria*, the Zonosomas, &c., are full-fed; in fact, all the autumnal Geometrids with a single generation, as well as those that have an autumnal (as well as vernal) brood, and, in addition, many that feed up now and go over the winter in the pupal stage are at this time to be obtained. The larvæ of such species as *Acidalia ornata*, *A. marginepunctata*, *Aspilates ochrearia* (*citraria*), &c., must be swept, whilst collecting the seed-capsules of *Lychnis dioica* gives larvæ of *Emmelesia decolorata* and *Eupithecia venosata* (the latter also on various species of *Silene*), foxglove flowers give *E. pulchellata*, flowers of *Valeriana officinalis*, *E. valerianata*, those of *Melampyrum pratense* give *E. plumbeolata*, of clematis, *E. isogrammata* and *E. coronata*, whilst larch must be beaten for larvæ of *E. lariciata*, Scotch fir for *E. indigata* and *Thera variata*, spruce for *E. pusillata*, oak for *E. dodoneata*, *E. abbreviata* and *Cidaria psittacata*, and *Sisymbrium* must be searched at Tuddenham for larvæ of *Lithostege griseata* ; beating, searching, and sweeping for Noctuid larvæ must also be prosecuted, whilst almost all the Notodonts are now to be obtained. We would insist on larva-beating in July; so many lepidopterists look on August and September as the regular months for this work, and are so busy during July with imagines that they systematically overlook many species undoubtedly abundant in their districts. If one were to cast aside the net entirely in July and simply collect larvæ one would possibly find a new fauna in the old haunts, and great rarities would, in many instances, be found common. July, too, is the great month in which to collect the mines of Lithocolletids and Nepticulids in order to obtain the autumnal broods of these interesting superfamilies. In breeding and rearing larvæ it may be well to note that cork is frequently of great service in order to obtain the satisfactory pupation of many species, but when larvæ do thus pupate in cork, be careful to isolate each one as it prepares to bore, otherwise two or more are almost sure to enter the same burrow, and only the last will have any chance of escape. As for imagines, July produces probably more species in the imaginal state than any other month in the year, and not only are Geometrids, Noctuids, and Tortricids abundant, but Pyralids, Crambids, Pterophorids, Gelechiids, and Coleophorids are also especially numerous.

ADELIDES.—*Nemotois schiffermillerellus* occurs freely, resting in the flowers, on the made roads in Chippenham Fen, in the afternoon sun during July and August.

COLEOPHORIDES.—The cases of *Coleophora flavaginella* are to be found during July and August on the leaves, flowers, and fruits of *Suaeda maritima*.

NEPTICULIDES.—The larvæ of *Nepticula betulicolella* feed in early July, in small contorted galleries, in birch, filled with brown excrement.

ELACHISTIDES.—The larvæ of *Laverna raschkiella* mine the leaves of *Epilobium angustifolium* in July, whilst those of *L. conturbatella* feed in the shoots, drawing them together with a slight web.
During July and August the twisted heads of *Epilobium* should be collected for larvæ of *Laverna fulvescens* (*epilobiella*).
In July the bushes of *Cornus sanguinea* should be searched for mines of the larvæ of *Antispila pfeifferella*.

GRACILARIIDES.—In July the cones of *Gracilaria populetorum* are to be found on birch. The cone occupies an entire leaf, in which the green, rather transparent, larva feeds.
At the end of July the full-fed larvæ of *Gracilaria omissella* blister the leaves of *Artemisia*. The moths usually emerge about three weeks later.
Rolled-up leaves of maple collected in July give larvæ of *Gracilaria semifasciella*.

GLYPHIPTERYGIDES.—At the end of July the larva of *Aechmia dentella* is to be found spinning together, with a slight web, the seed-heads of *Chaerophyllum temulentum*. It feeds on the seeds.

At the end of July the larva of *Tinagma resplendellum* mines in the leaves of alder, making blotches.

TORTRICIDES.—The imagines of *Tortrix transitana* (*diversana*) fly at dusk in July, over the tops of birch and elm trees, on which the larvæ feed in spring.

The larva of *Peronea cristana* feeds between united hawthorn leaves in July.

In early July the green larvæ of *Peronea caledoniana* feed upon *Myrica gale* (Pears).

The imago of *Ditula semifasciana* flies at dusk in July and August, over the tops of sallow growing in damp and marshy places, it will also come to sugar. The larva feeds in the shoots (and catkins) of sallow in April and May.

In the last week of July, on the banks of the river Bure, I found *Sericoris doubledayana*, flying gently among *Lastrea*, *Myrica*, and reeds, in the late afternoon sunshine. It is necessary to look for the moths, so little do they rise above the undergrowth (Meek).

The tops of the shoots of *Euphorbia amygdaloides*, if the central leaves be spun together, should be collected in July; the larva of *Sericoris euphorbiana* lives within, and bores down the stem some distance; the pieces picked off should therefore be of good length.

Sweep the flowering heads of *Daucus carota* in July for the imagines of *Semasia rufillana*.

Stigmonota orobana flies when the sun is setting, and seems quite merry for a few minutes in the low sun's rays (Farren).

During July and early August, *Coccyx nanana* can be obtained by beating spruce fir in the afternoon, preferably in sunshine.

The last few days of July and the first week of August, should be spent in looking over the seed-heads of *Jasione montana*. The reddish larvæ of *Eupoecilia pallidana* feed within.

During the last week in July, the berries of *Rhamnus frangula* are sometimes found fastened together with silk. The larvæ of *Eupoecilia ambiguana* clear out the berries after thus spinning them together.

In the beginning of July the larvæ of *Eupoecilia ciliana* may be obtained feeding on the seeds of cowslip, they require cork or rotten wood in which to pupate.

The larvæ of *Eupoecilia atricapitana* feed in July (second brood) within the growing stems of *Senecio jacobaea*, eating the pith and stopping the growth of the central shoot so that it becomes thickened and covered with a bunch of leaves, while the side shoots grow up past it (Barrett); usually in terminal shoots.

At the end of July collect the seed-heads of the common blue-bell (*Scilla nutans*) for larvæ of *Eupoecilia maculosana*. It is necessary to open the seed-vessels to find them, as the larvæ give no outward indication of their presence.

In July the imagines of *Argyrolepia hartmanniana* may be obtained by brushing plants of *Scabiosa succisa*.

The heads of *Centaurea scabiosa* should be collected in July and August for larvæ of *Catoptria fulvana* and pupæ of *Conchylis alternana* (*gigantana*), very abundant on cliffs near Dover, where pupa-skins of latter are often seen sticking out of the capitula. Keep in very tightly-fitting box, and you will breed plenty of imagines.

The larvæ of *Conchylis stramineana* feed in July, and again in September at the base of the flower-heads of *Centaurea nigra*, eating the young seeds, and lying curved in the cavity formed, or (if disturbed) retreating into the stem (Barrett).

Imagines of the beautiful *Chrosis rutilana* used to be taken freely, flying by day in July and August, among juniper on the Surrey downs.

The imagines of *Dichrorampha simpliciana* fly by day in July and August, among mugwort (*Artemisia*). The larvæ feed in the roots during the autumn and winter months.

PYRALOIDES.—Collect heads of *Umbelliferae* for larvæ of *Depressariae*.

During the first and second weeks of July, the deep green larvæ of *Depressaria capreolella* are to be found feeding on leaves on the higher shoots of *Pimpinella saxifraga*. Through the plant being usually buried among taller herbage, the larvæ are somewhat difficult to find.

Gelechia boreella and *Hysilophus juniperellus* are to be beaten from junipers in July in Moray (Horne).

The imagines of the rare *Doryphora palustrella* are to be taken when it is just getting quite dark flying over *Sparganium*, &c., in marshy and boggy places ; also comes to light later ; occurs in late July and early August.

CRAMBIDES.—The males of *Schoenobius mucronellus* are to be taken at light directly after dark in late July and early August. They may also be obtained sparingly by dusking ; so taken at Sandwich and Wicken.

Dusking along the reed-beds in marshes is always very productive during July, *Chilo phragmitellus* may thus be captured, frequently in abundance, as the ♂ s assemble to newly-emerged ♀ s ; they may also be taken freely, *in cop.*, on the reeds after dark, with a lantern.

Crambus furcatellus occurs on the scanty turf which covers the highest parts of Helvellyn and the adjacent mountains, early in July. Most of our specimens of late years, however, have come from Scotland and Wales.

Ragonot says that the larva of *Crambus alpinellus* is still unknown, the moth flying on dry sandy pasture-lands, where heath, broom, and *Artemisia campestris* grow in July and August. It is not an alpine insect, nor does it specially frequent fir woods. Occurs on Deal sand-hills, &c.

During the last week in July search should be made for small aborted cones of common spruce fir and *Abies douglasii*. The aborted cone usually forms one of a bunch of three or four, the others being well developed. These will be found to contain larvæ of *Euzophera* (*Cateremna*) *terebrella*.

During July, August and September, some white web over the mid-rib of elm leaves indicates the presence of the plain green larva of *Rhodophaea formosa*, which feeds openly by day on the upper surface of the leaves, and probably uses the web for shelter only at night.

The chance capture of a specimen of *Homoeosoma sinuella* in early July would always result in the capture of many, if the species were properly worked for ; the imagines are most easily disturbed in the late afternoon, when they often fly freely, and continue to do so until dusk.

PYRALIDES.—The imagines of *Aglossa cuprealis* are common in July and August, in the barns and stables attached to all the farms about Wicken (probably elsewhere). They sit on the woodwork among swarms of *A. pinguinalis* and *Pyralis farinalis*.

The larvæ of *Agrotera nemoralis* feed on *Carpinus betulus* during July, living in a loosely spun web on the underside of the leaves.

In July and September the white patches formed by the larvæ of *Botys asinalis* on the dark green leaves of *Rubia peregrina* are very conspicuous.

The full-grown larvæ of *Botys terrealis* live in July on *Solidago virgaurea*, they eat the flowers, often strip the entire spike, spinning a slight web among the flowers, out of which they wriggle when disturbed (Newman).

Ravula sericealis is to be disturbed in large numbers in late July from the undergrowth, bushes, &c., in Chippenham Fen.

Searching lichen-covered trunks and fences for *Scopariae* is advisable; *Scoparia ambigualis*, *cembrae*, *mercurella*, *crataegella*, &c., are almost sure to be thus taken.

CYMATOPHORIDES.—The larvæ of *Cymatophora or* and *C. ocularis* are now to be obtained feeding between the leaves of poplar, those of *Asphalia flavicornis* in rolled-up leaves of birch, and of *A. ridens* between united leaves of oak. Although these can be found, sometimes commonly, by beating and searching by day, they are much more readily obtained at night when they have left their day domiciles.

At the end of July on the Culbin sands larvæ of *Asphalia flavicornis* are not uncommon on small birch trees (Horne).

GEOMETRIDES.—The imagines of *Epione apiciaria* always used to occur in abundance late at night (11 p.m.-12 p.m.) round sallow bushes in Wicken, in late July and early August. We used often to make up a bag with this species on the way home.

The imagines of *Geometra papilionaria* occur in early July among birch, appearing regularly about 11 p.m., and remaining on the wing till midnight.

At dusk in July, work round the edges of big oak-trees for *Phorodesma bajularia*. The species must be bred to get really fine specimens.

Beat furze during the daytime, in the first week of July, for *Nemoria viridata*.

Ellopia prosapiaria may be found commonly in early July just emerged from the pupa, either on the ground at the roots of the pine-trees or on the bark of these about six inches above the ground (Carlier).

After dusk in July the males of *Geometra vernaria* "assemble" freely to the females that are hidden in the clematis bushes. They may occasionally be disturbed during the day when working for *Phibalapteryx tersata*.

The imagines of *Psodos coracina* fly by day in July in the sun, on the grassy slopes of our Scotch mountains at a considerable elevation.

On rough heath-covered ground in Scotland at a moderate elevation, where there is an abundance of bare rocks, *Dasydia obfuscata* is found in July, resting on the lee side of the rocks, and creeping into the shelter of an overhanging part (for full details see *Ent. Record*, i., p. 70).

In July the imagines of *Fidonia brunneata* (*pinetaria*) are easily disturbed during the day, and fly sometimes quite freely in the afternoon sun.

In the middle of July (also in May) the trunks of mountain-ash should be searched for *Venusia cambricaria*.

In its Bristol habitat, *Acidalia dilutaria* (*holosericata*) is abundant in the first week in July ; it flies early in the evening, and is very feeble on the wing.

The imagines of *Acidalia rusticata* are sometimes to be taken in abundance in early July, sitting on leaves of ivy, pellitory, &c., growing at the foot of the rocks near the sea-shore, or at the foot of old hedges.

Imagines of *Acidalia contiguaria* may be found, sometimes not uncommonly, at rest in the daytime in July, on rocks on the mountains in North Wales, very often a pair a few inches apart, in which case the eggs from the female may be relied on as being fertile (Porritt).

In July *Larentia caesiata* flies off rocks and trunks in clouds as one approaches them in the moorland districts that it most frequents.

The imagines of *Lobophora sexalata* fly swiftly over the tops of sallow bushes at dusk in July. Sometimes very common at Wicken, Ranworth, Horning, New Forest ; it appears also to be very generally distributed in Kent.

The larvæ of *Emmelesia decolorata* are to be obtained by collecting the capsules of *Lychnis dioica* in July.

The imagines of *Anticlea cucullata* (*sinuata*) are to be disturbed from among *Galium* (at Tuddenham and elsewhere) during the daytime in early July ; the species is better obtained by searching for larvæ in August and September.

To obtain *Phibalapteryx tersata* by day thrust a stick into the clematis bushes, and churn them gently so that every part of the leafage is disturbed. It is most abundant during the first week in July.

Towards the end of July, at Tuddenham, examine the heads of *Sisymbrium sophia* which are just seeding. Many of the apparent seed-pods will be found to be full-grown larvæ of *Lithostege griseata*.

Scotosia rhamnata and *S. vetulata* fly freely at dusk around clumps of buckthorn in woods, hedgerows, fens, &c., during July.

In July the imagines of *Collix sparsata* (in Askham Bog, Wicken, New Forest, &c.) swarm at early dusk flying about the buckthorn bushes at Wicken with *Scotosia rhamnata*.

Cidaria picata is to be obtained in July by searching tree-trunks in our southern woods, preferably fairly early in the morning.

The larvæ of *Chesias rufata* (*obliquaria*) should be searched for after dark, with a lantern, on broom.

Collect seed-capsules of *Silene* in July for larvæ of *Eupithecia venosata*. Seed-heads (unripe capsules) of *Linaria vulgaris* in July and August contain larvæ of *E. linariata*. Flowers of *Digitalis purpurea* (foxglove) in July contain larvæ of *E. pulchellata*. Flowers of *Valeriana officinalis* succour larvæ of *E. valerianata* in July. Seed-pods and flowers of *Melampyrum pratense* in July and August produce *E. plumbeolata*.

The larvæ of *Eupithecia venosata* feed during July inside the capsules of *Silene*, much in the same manner as the Dianthœcias ; keep in well-covered flower-pots, with a good supply of sand at bottom.

From solitary flower-heads of foxglove in open spaces in woods, the larvæ of *Eupithecia pulchellata* may be obtained in the spun-up flowers in July.

In July and August, collect seed-heads of *Linaria vulgaris* for larvæ of *Eupithecia linariata*.

Larvæ of *Eupithecia fraxinata* may readily be found on small ash-trees and especially on the young shoots which often spring from near the bases of large trees, at the end of July and beginning of August (Porritt).

The flowers of *Melampyrum pratense* should be collected in July and August for larvæ of *Eupithecia plumbeolata*.

Towards the end of July and early in August, the earliest larvæ of *Eupithecia campanulata* are to be obtained by knocking the plants of the nettle-leaved bell-flower (*Campanula trachelium*) against the sides of an open umbrella. The larvæ feed in the seed-capsules. In confinement they will feed upon garden species of *Campanula*. In some localities almost every dry corolla-tube contains one or more larvæ.

The larvæ of *Eupithecia subumbrata* (*scabiosata*) feed on flowers of *Daucus carota* in July. The larvæ of *Spilodes palealis* feed later in heads of the same plant.

The larvæ of *Eupithecia valerianata* (*viminata*) feed on flowers and seeds of *Valeriana officinalis*, usually growing in damp woods and osier beds. They become full-fed from the middle of July until the end of August.

The full-fed larvæ of *Eupithecia lariciata* may be beaten from larch and spruce-fir during July ; they pupate in an earthen cocoon.

PTEROPHORIDES.—*Aciptilia tetradactyla* and *A. baliodactyla* are abundant among wild thyme, can be disturbed by day but fly much more abundantly at dusk. *Marasmarcha phaeodactyla* can also be disturbed freely by day, but is much more abundant at dusk. *Cnaemidophorus rhododactylus* is a night flyer, and occurs most freely after dark, when it can be taken with the aid of a lantern flying about rose-bushes.

The pupæ of *Marasmarcha phaeodactyla* can readily be found attached by the anal segment to stems and leaves of *Ononis*.

The pupæ of *Aciptilia spilodactyla* are usually found attached to the upper surface of leaves of *Marrubium vulgare*.

Œdematophorus lithodactylus rarely flies till after dark, it then sometimes abounds in places where a specimen is quite unobtainable during the day or early evening.

During July the flowers of common centaury (*Erythrea centaurea*) should be gathered for larvæ of *Mimaeseoptilus loewii* (*zophodactylus*).

The second brood of *Platyptilia gonodactyla* feeds in July and August in a loose web on the underside of the leaves of coltsfoot (*Tussilago farfara*).

HEPIALIDES.—During the first week in July, *Hepialus hectus* sometimes swarms for about half an hour just before dark, flying swiftly in wood-ridings, &c., after dark the pairs may be found suspended from grass culms.

SESIIDES.—During the first fortnight of July, visitors to the south-west coast (Lizard, &c.) should look out for *Sesia musciformis (philanthi-formis)* flying along the flowery earth-walls, and settling on the thyme and seathrift ; *S. ichneumoniformis* is also there at the same time. Both fly in the bright sunshine, and prefer the morning sun.

In July sweep *Lotus corniculatus* in the early morning or late afternoon (after 5.30 p.m.), for *Sesia ichneumoniformis* ; sometimes very abundant in Folkestone Warren, &c.

Search trunks (near roots) of poplar trees in July for newly-emerged *Trochilium apiforme*. The moths may often be found in numbers just emerged from about 7.0 a.m.-8.30 a.m., the empty pupa-cases sticking out of the trunks just beside them.

Similarly at the roots of sallows and willows, also often higher up, but on comparatively small stems and shoots, the imagines of *Trochilium crabroniforme (bembeciforme)* may be taken.

The flat, dark-brown ova of *Trochilium crabroniforme (bembeciforme)* are to be found on leaves of sallow.

COCHLIDIDES.—The males of *Cochlidion limacodes* fly high up over the oaks at mid-day, in the hottest sun in July, and require a long-handled net for their capture. The females (and males in dull weather) are best obtained by jarring the small oak-trees growing in wood clearings, when they tumble to the ground.

LACHNEIDES.—The larvæ of *Malacosoma castrensis* are now to be found full-fed, crawling in the sunshine on the saltings of the Thames and eastern coast; they spin up from about the middle of the month and produce imagines (exceedingly variable) in August. They are abundant on Wakering Marsh, &c., have been observed to feed on *Statice limonium*, *Atriplex portulacoides*, *A. littoralis*, *Artemisia*, and coarse grasses ; in confinement they show a marked preference for rose and birch.

The larvæ of *Malacosoma castrensis* can be reared readily on chrysanthemums ; cover a plant with a muslin net, leave the larvæ until they spin up, then clip off the cocoons and put in a breeding-cage (Button).

The nests of *Lachneis lanestris* are now conspicuous objects on whitethorn and blackthorn, and should be collected for experimental purposes, especially with regard to the elucidation of the problem of the lying-over of the pupæ for many successive years.

The cocoon of *Trichiura crataegi* is to be found on the surface of the ground at the base of any of its various food-plants ; sometimes also under pieces of bark at base of trunk, &c.

The eggs of *Cosmotriche potatoria* are often laid in irregular batches on grass or reed-culms ; sometimes they are laid more regularly round the culm, the white colour and distinct greenish oval rings surrounding them make them unmistakeable.

DIMORPHIDES.—Larvæ of *Dimorpha (Endromis) versicolora* now large, and resembling big birch catkins sticking from a twig, are to be found in July ; when younger they are gregarious, dark-coloured, and likely to be passed over as sawfly larvæ.

At the end of July, on the Culbin sands, larvæ of *Dimorpha (Endromis) versicolora* are not uncommon on small birch trees (Horne).

SPHINGIDES.—Towards the end of July examine bed-straw (*Galium*) from 10 p.m. to 12 midnight with a lantern ; by this means the larvæ of *Choerocampa porcellus* can often be found in considerable numbers.

In mid-July and on through August, the larvæ of *Macroglossa stellatarum* should be searched for low down on *Galium*.

Larvæ of *Macroglossa fuciformis* may be readily obtained on honey-suckle, nearly or quite full-fed, those of *M. bombyliformis* on scabious, particularly fond of hiding on the underside of leaves.

SIMÆTHIDES (CHOREUTIDES).—The larvæ of *Simaethis scintillulana* are easily to be obtained by searching *Scutellaria galericulata* in July and early August.

DELTOIDES.—The pale yellow, rather large, globular (slightly oval and indented on upper side) eggs of *Hypena proboscidalis* are laid on nettle (leaves and stems) and do not hatch until March (Merrin).

NOLIDES.—From the middle to the end of July, at night, not earlier than 10.30 p.m. to early dawn, search with a light, in grassy places on the south-eastern coast, especially if dwarf bushes of sea-buckthorn occur, and you will probably find *Nola centonalis* sitting quietly on the grass or leaves of the buckthorn. They do not fly freely, and require a close search, or they may be readily overlooked, and have been found on the plants of *Hippophaes rhamnoides* on the Deal sandhills, also at Folkestone, Hastings, Freshwater, &c.

As soon as it is really dusk, the rare little *Nola albulalis*, like a soft, white, animated snowflake, flies gently about its restricted habitat in north Kent. [Please spare the females.]

LYMANTRIIDES.—During July search the grass, heather, &c., in the proper localities for imagines of *Dasychira fascelina*, which are frequently taken *in cop.* by this method.

The batches of fluff-covered eggs of *Porthesia chrysorrhoea* are readily to be found on the twigs of sloe, whitethorn, *Hippophaes*, &c., in its selected habitats.

The eggs of *Leucoma salicis* are to be found on poplar leaves (usually on underside) in batches covered with a salivary-looking material.

NYCTEOLIDES (CHLOEPHORIDES).—The larvæ of *Sarrothripa undulanus* (*revayana*) may be beaten from oak in July.

NOTODONTIDES.—The circular, smooth, white, opalescent eggs of *Lophopteryx cucullina* are to be found on maple on the underside of leaves in shady places in woods, from about the 10th to the end of the month ; they are generally laid singly, but sometimes in twos or threes (Merrin).

The young larva of *Lophopteryx cucullina* must be searched for from the middle of July, on chalky hillsides sloping to the north, in the densest shade of beech woods, on stunted maple bushes, and often within a few inches of the ground. A likely bush for the larva may be known by the blotches on the leaves, caused by the larva having eaten, when young, the under surface of the leaf (Bernard Smith).

The eggs of *Leiocampa dictaea* are distributed on poplar, sallow and willow ; those of *L. dictaeoides* on birch.

Notodont larvæ should now be beaten (most lepidopterists wait for August and lose many)—*Pterostoma palpina* on willow, *Lophopteryx camelina* on oak, beech, &c., *L. cucullina* on maple and sycamore, *L. carmelita* on birch, *Leiocampa dictaeoides* on birch, *L. dictaea* on willow, &c., *N. dromedarius* on birch, hazel, &c., *Drymonia chaonia* and *D. trimacula* (*dodonea*) on oak and birch.

The larvæ of *Notodonta ziczac* may be beaten from sallow, aspen, and poplar, from July to October.

The larvæ of *Cerura vinula* are usually very abundant on poplar, sallow, and willow the first week in July, and are readily obtained by searching.

The larvæ of *Clostera reclusa* must be searched for in rolled-up leaves of sallow, willow, &c.

NOCTUIDES.—When sugar fails, as it often does in hot and dry weather, instead of going home empty-handed and grumbling, search by means of a light, flower-heads, grass-stems, rushes, or honey-dewed leaves, when frequently you will realise a rich harvest. My best captures have been so obtained (Tugwell).

The yellow form of the larva of *Acronicta leporina*, on birch, resembles a dead yellow leaf or two, with some spinning attaching them to the living leaf, and the black tufts that often persist in this form resemble bits of grass and other dark chips that are entangled in such vacated lodgings (Chapman).

The green white-haired form of the larva of *Acronicta leporina*, when full-fed and sitting curled-up near the middle of the underside of an unilluminated alder leaf, is invisible from above, but looking up from below it looks like a gleam of light falling through on a portion of a leaf (Chapman).

Give the full-fed larva of *Acronicta leporina* some rotten wood, elder-pith, or cork to burrow in, in order to make their puparia. The pupal stage sometimes lasts over three winters.

During July the larva of *Pharetra albovenosa* is not at all difficult to find by searching the reeds growing in Wicken Fen.

The young larva of *Craniophora ligustri* always rests under a leaf, and when full-grown along the central petiole of an ash-leaf, which is its usual food ; the tapering of the larva at either extremity makes it readily overlooked.

The larvæ of *Craniophora* (*Acronycta*) *ligustri* are, however, sometimes to be beaten from privet hedges, in July and August.

The larva of *Arctomyscis* var. *myricae* is found on the moors at Rannoch ; in Aberdeenshire it frequents the coast, roadsides, edges of fields, &c., wherever there is an abundance of sorrel and plantain which are its chief food-plants (also feeds on ragwort, bramble, Scotch thistle, &c.).

The larva of *Jocheaera alni* is to be found throughout July and August ; it is conspicuous, sits on upper side of leaf (in wet weather lies under the leaf), is easily seen and easily dislodged ; it likes sunshine, and mounts to the top of a bush or hedge to enjoy it ; the sunny side of a hawthorn bush is a likely place for it ; full-grown larvæ should be provided with elder stems, or raspberry cane, ready bored and dried, to pupate in (for details as to breeding see *Ent. Rec.*, i., pp. 136-137).

The larva of *Jocheaera alni* appears to be of a thirsty habit, and in dry weather cannibalism is apt to show itself; the use of a syringe in the evening before the larvæ begin to feed for the night is obvious; the same habit has been remarked of the larva of *Notodonta trepida* (Smith).

To pupate, the larva of *Moma orion* likes to get under a dead leaf, or some similar object, and make a cocoon of loose material on the surface of the ground; in confinement the larvæ will spin up in sawdust, always near the surface, and often aggregate their cocoons together (Chapman).

The larvæ of *Cuspidia menyanthidis* are to be obtained freely in moorland districts on *Menyanthes*, heath, sallow, *Myrica*, &c.

Dusking along the reed-beds in marshes is always very productive in July; *Calamia phragmitidis*, *Leucania straminea*, &c., are thus to be captured.

During the first week or so in July, after dusking for *Calamia phragmitidis*, which commences to fly systematically between 8.15 p.m. and 8.30 p.m., and *Leucania straminea*, which waits till dark and can always be recognised by its white appearance, I used always to search with a lantern for the paired imagines, which remain from about 10 p.m.-12 p.m. seated on the reed-culms about half-way up, and frequently took large numbers of both species. They often drop into the water and swim well to the nearest reed.

Although *Leucania straminea* is best taken on the wing from 9 p.m. to 10 p.m. from about July 1st-August 1st, a drop of sugar in the centre of a flower of the dwarf thistle, common on the side of the dykes in the marshes, will frequently attract it.

In early July on the north denes at Great Yarmouth, known as the "Marrams," *Leucania littoralis*, *Tapinostola elymi*, and many other species abound at the flowers of *Arundo arenaria* (Harmer).

The imago of *Tapinostola elymi* appears to be confined to the sandhills of the eastern coast of Britain; it abounds locally among *Elymus arenarius* and marram, from which it may be taken freely by night with a lantern, or shaken therefrom by day in early July.

The larvæ of *Calamia lutosa* feed in roots of *Arundo phragmites*, and should be looked for early in July; pupæ, end of July to middle of August. The larva feeds very deep in the roots, but when about to change, leaves the plant and pupates in the soil, two or three inches from the surface. The presence of the larva may be readily detected from the bleached appearance of the plants which have been attacked; the pupa, however, will be much more easily found than the larva, for which I have dug as deeply as a foot, and then not reached it (Gardner).

The larva of *Nonagria sparganii*, which feeds very similarly to that of *N. arundinis* (*typhae*), should be sought in the marshes around Hythe, Deal, &c., in July and August.

Messrs. Webb and Jeffrey were in search of lepidoptera, in Kent, on a wild gusty day in July, 1878, when they turned their attention to the bored stems of the yellow flag (*Iris pseudo-acorus*). The result was the addition of *Nonagria sparganii* to the British list. The familiar green larva was found in the summer of 1879, in bur-reed

(*Sparganium ramosum*). The insect probably occurs in all the marshes from Deal and Sandwich to Hythe.

About the beginning of July, before the healthy reeds overtop the affected ones, look over reed-beds for reeds having the top shoot of the reed yellow and withered. About the end of July a circular scar will be observed well down where the larva of *Nonagria geminipuncta* has gnawed through the reed-stem, except the outer skin, in readiness for the emergence of the imago. Cut low down and keep in water, or empty the pupæ out upon damp flannel.

Coenobia rufa is to be found in late July and early August, but frequents low-lying marshy ground, flies rapidly through the low herbage for an hour or so at dusk, and may afterwards be found at rest on the stems of rushes and grass.

Chortodes arcuosa flies abundantly just after dusk among grass, *Aira caespitosa*, in early July. The ♀ s rest on the grass, and must be sought with a lantern.

In sugaring for *Agrotis lunigera* at Freshwater during July and early August, select plants growing near the edge of the cliffs or among the loose chalk rubble in the crevices on the seaward slopes. The best plants for this purpose are *Onopordum acanthium*, *Carduus acaulis*, long upright stems of *Beta vulgaris*, *Heracleum sphondylium*, &c. Occurs also on the Beer Cliff near Seaton, Portland, Torquay, &c.

During the first three weeks in July *Agrotis simulans* is sometimes abundant at Portland on wild sage flowers ; the latter are very attractive to Noctuids and Geometrids so long as the blossom lasts.

In July and August the females of *Stilbia anomala* are to be found early in the evening, sitting on the flowers of ragwort.

Acosmetia caliginosa flies in the grassy rides of Stubby Copse, Brockenhurst, through July. It is best obtained by gently sweeping a net over the herbage as you walk along by day, and it comes freely to light at night. *Hyria auroraria* occurs at the same place, but flies in the sunshine.

The larva of *Celaena haworthii* is grey, spotted with black, and feeds in the stems of cotton grass, *Eriophorum* ; a small hole is made in the stem, from which the excrement is ejected in profusion ; the larva is full-fed at the end of July (Edleston).

The larva of *Hydroecia petasitis* feeds on the subterranean stem of *Petasites vulgaris*, in which it makes long excavations during July. The large plants in a dry situation are the most likely to be affected. The pupæ should be dug up about the third week in August.

During the latter half of July and August, look out for any tall thistles (especially *Carduus palustris*), which show, by the drooping and sickly flower-heads, the effects of the internal ravages of a larva. Cut about the end of August and the pupa of *Gortyna ochracea* (*flavago*) will be obtained.

In late July, the imagines of *Triphaena interjecta* may be found flying over bramble blossom in the late afternoon (4 p.m.-6 p.m.).

Old walls, by or near the seaside, should be searched during July and August for the imagines of *Bryophila muralis*, the early morning is perhaps the best time.

The larvæ of the *Dianthoeciae* can still be collected in the seed-heads of *Lychnis*, *Silene*, sweet-william, and other Caryophyllaceous plants.

Collect during July and August, the seed-heads of the white campion for larvæ of *Dianthoecia capsincola ;* the larvæ often have their bodies hanging half-way out, and round holes in the capsules denote the affected ones.

In July and early August the larvæ of *Dianthoecia cucubali* are to be taken from the flowers of ragged robin (*Lychnis flos-cuculi*).

The larvæ of *Dianthoecia irregularis* may be swept in July from *Silene otites.* Later in August the full-fed larvæ are best obtained by scratching away, with the fingers, the earth near the roots of the food-plant, as here the larvæ lie hidden during the day.

In July collect the seed-capsules of *Silene inflata* for larvæ of *Dianthoecia conspersa.* Southern collectors would be glad if the Scotch collectors could supply them with more of their forms of this species.

On the south-eastern coast, or wherever the Nottingham catch-fly (*Silene nutans*) grows, the larvæ of *Dianthoecia albimacula* may be found in the evening by searching the plants and then gently beating them into a net. Place larvæ so obtained in a calico bag or bandbox, with a supply of seed-heads, and they will feed on the unripe seed. It is better to collect the pupæ from these receptacles, as often you have other lovers of pupæ confined with them, who will gladly make a meal off them (Tugwell).

Full-fed larvæ of *Dianthoecia caesia* may readily be found in the Isle of Man in the daytime, by turning over the large clumps of *Silene maritima,* which grow over and lie upon the coast rocks during July (Porritt).

Towards the end of July and throughout August the imagines of *Eremobia ochroleuca* are to be found sitting in the centre of a scabious bloom, or that of *Centaurea scabiosa,* in the afternoon sunshine.

Larvæ of *Panolis piniperda* feed on fir, closely resemble the needles, and are best obtained by beating.

Pupæ of *Hoporina croceago* at the roots of oak, *Taeniocampa populeti* at roots of poplar, *Tethea retusa* and Citrias—*flavago* and *fulvago*—at roots of sallow, *Cirrhoedia xerampelina* at roots of ash, *Tiliacea aurago* at roots of beech, *T. citrago* at roots of lime (often between leaves), *Calymnia diffinis* and *C. affinis* among *débris* at roots of elm or in crannies of bark, holes in trunk, &c., should now be collected.

The imagines of *Dyschorista fissipuncta* sometimes fly in swarms about the higher branches of willows at dusk in July (also come to sugar).

The larvæ of *Xylina semibrunnea* are to be found on willow and ash in July (Croydon and Hackney Marshes) (Machin).

The larva of *Cucullia gnaphalii* is to be found feeding on golden-rod, from the end of July to the end of August. The woods above Sevenoaks and Seal are well-known habitats for this rare species.

During July mullein should be searched for the conspicuous destruction caused to the leaves by the larvæ of *Cucullia verbasci ;* the undersides of the leaves must be searched for the larvæ.

The full-fed larvæ of *Anarta myrtilli* may be swept from *Calluna vulgaris* in July and again in September (Newman).

During July search *Ononis arvensis* for the larvæ of *Heliothis peltigera.* They prefer the flowers and green seed-pods.

Towards the end of July the larvæ of *Banksia argentula* can be obtained by shaking the high grass stems, *Poa annua, Poa aquatica,*

D

Poa pratensis, &c., over a sheet of white paper.   (Chippenham Fen is the best-known locality.)

Plusia interrogationis has the habit of resting on birch-stumps in places where the heather has been burned down the previous year, and the females deposit their ova on the young heather growing in such spots.   The larvæ may (earlier in the season) be swept off the older heather (McArthur).

The flight of *Plusia interrogationis,* in early July over heather, is very like that of *P. gamma.*

Catocala nupta larvæ, now full-fed, are to be found hiding under the loose pieces of bark of willow and poplar, as well as in the cracks of the bark.

Toxocampa pastinum can readily be disturbed during July wherever its food-plant (*Vicia cracca*) grows in Chippenham Fen and elsewhere; the plants and surrounding herbage want a complete churning with a stick.

ARCTIIDES.—The eggs of *Arctia villica* are often to be obtained from a variety of plants, always laid in true Arctiid fashion in regular batches, on the underside of the leaves.

Nudaria senex literally swarms in early July in a damp meadow near Clevedon ; an evening following a day that has been hot and fine, and on which the dew begins to be deposited on the grass at about 7 p.m., is perfect for them ; they commence to fly at this time in scores, the grass and rushes become full of them, fluttering up the rush-stems, and then taking flight for a short time ; in half an hour or so all are again at rest on the herbage and not a single one on the wing (Mason). We have seen them in hundreds at Wicken and Sandwich crawling up the grass and rushes, then fluttering off, threading their way among the low herbage for about half an hour.   After that their flight is over, and they rest on the plants, where they may be found by searching with a lantern.   They also come freely to light.

In late July and early August the imagines of *Nudaria mundana* can be obtained freely by using a hand-lamp in the " droves " and meadows just outside the fens (Wicken), and netting them as they come within the range of the light.

Lithosia var. *molybdeola* is to be taken in July and August on grass stems by searching after dark with a lantern (Collins).

Lithosia lutarella var. *pygmaeola* is best obtained when sitting at rest on the marram grass after dark.   On some evenings it comes quite freely to light.   Can also be found in very early morning resting on grass, or gently flying, or the males sometimes " assembling " to a newly-emerged female in numbers.

Eulepia cribrum may be obtained all through July on the lichen-covered heather, two or three miles out of Ringwood, on the Bournemouth Road.   Gently sweep or brush your net over the heather as you walk along, and *E. cribrum* will start up and fly rapidly a short distance.   Mark down and stalk it.   They may be disturbed all through the day, but early evening is the best time.

PAPILIONIDES.—The newly-emerged imagines, progeny of the spring-immigrating broods of *Colias edusa* and *C. hyale,* commence to appear the last week in the month; ♀ s enclosed over growing plants of clover and lucerne and placed in the sun, will give an abundance of ova, producing another brood in October.

Pupæ of *Gonepteryx rhamni* in the middle of the month may be found attached to stems and the petioles of leaves of buckthorn.

The second and third weeks in July usually prove the best time for *Dryas paphia*, *Limenitis sibylla*, *Argynnis adippe*, and *A. aglaia* in the New Forest, but in early seasons they may be quite a fortnight earlier.

The eggs of many species of Rhopalocera may be obtained this month—*Limenitis sibylla* (a most beautiful microscopic object) on honeysuckle ; *Apatura iris* on sallow (I have seen a dozen collected in a single morning by watching a female, where larvæ could never be beaten owing to the density of vegetation) ; *Dryas paphia* on violet, *Plebeius aegon* on *Ornithopus perpusillus*, &c.

Find a privet-bush in full bloom where *Melitaea athalia* and *Thecla w-album* are known to occur ; on fine, hot, sunny days, quite at the beginning of the month, you will find the species congregated there.

Cossus-infected trees are a sure trap for Vanessids from July to September—*Eugonia polychloros*, *Euvanessa antiopa*, &c.

Apatura iris is usually fairly on the wing towards the end of the second week of the month ; flies lower early in the morning, 9.30 a.m.-10 a.m., and after 3 p.m. also comes down ; frequents puddles to drink.

Select an oak (on the outskirts of a wood preferably) where the ends of the broad spreading lower branches come down to within six or seven feet of the ground ; with a long-handled net *Zephyrus quercûs* may often be captured in numbers. Larvæ should be beaten from such trees in June.

AUGUST.

August is the beginning of the end, but a beginning scarcely noticeable until the heavy dews of the last week are fairly upon us. The first fortnight of the month is, perhaps, the busiest of the year in many districts, whilst the last fortnight may, in exceptionally inclement seasons, represent the close of active collecting in the more exposed mountainous districts of Scotland; but, after all, there is still much sport to be obtained in the first fortnight of the month in the Highlands, and, of the various forms of sport, none, perhaps, is more exhilarating than rock-hunting. Massive boulders on the moors are the resting-places of certain species of Geometrids, which collect in amazing numbers, a dozen *Dasydia obfuscata*, perhaps, on a single boulder, and the upright cliffs on a mountain side over which a waterfall bounds from the higher levels, afford, just out of reach of the rushing water and spray, a marvellous haunt for *Cidaria immanata*, *Larentia salicata* and other interesting species, whilst the trunks in a large pine wood will sometimes be alive with the slowly undulating wings of the former species, which fly off at one's approach. The Norfolk Broads are at their best during the last week of July and the first week of August, but light must be used for the rarities—*Leucania brevilinea*, *Lithosia muscerda*, &c.—or disappointment is sure to follow. In other districts, notably the woods of Kent, the Fens, &c., it is an off season and the Kentish collector wends his way to the chalk-hills and rough waste pastures, whilst the Cambridge collector chooses the flowery downs of Tuddenham and Brandon rather than fenland for his treasures, but towards the middle of August a great change takes place; in the fens the second broods of many species appear—*Papilio machaon*, *Pharetra alborenosa*, &c., whilst *Tapinostola hellmanni*, *Helotropha leucostigma*, *Noctua umbrosa*, &c., abound, and one is busy enough. But the coast districts are still most attractive at the commencement of the month. On the Lancashire and Kent (and almost all other suitable) coast lines, thousands of Agrotids swarm at sugar— *Agrotis cursoria, tritici, nigricans* and *vestigialis*, in boundless profusion and variation, afford material for careful study—whilst, now, if ever, the autumnal Coliads and Vanessids will provide the sportsman with excitement. We know a little, not much though, about the migratory *Colias edusa, C. hyale, Pieris daplidice, Argynnis lathonia* and *Euvanessa antiopa*. Possibly at this season of the year nothing is more attractive than waving heads of *Eupatorium cannabinum* in full flower, espe- cially in the evening, when a little thin sugar, carefully sprinkled on the flowers, will attract Lithosiids, Geometrids and Noctuids, sometimes in abundance, whilst *Tritoma uvaria* is by many awarded quite the first place as a plant to be grown for attracting the August (and September) species—many Noctuids and Geometrids being quite stupefied and easily boxed. Speaking of beds of *Eupatorium* reminds one, too, that this is the favourite resting-place of *Callimorpha hera*, the imagines of which appear during the first half of this month, the species being generally distributed now over a considerable part of Devonshire. Both in Scotland and at Howth, and possibly in many other districts, the flowers of ragwort are the most enticing means of capturing Noctuids during this month, many Geometrids also being attracted. Flowers of *Calluna vulgaris* and over-ripe blackberries, rather late in the month, have also a good reputation. Many partially double-brooded Geometrids appear in August, the more or less complete character of the brood being determined by favourable or unfavourable meteorological conditions, whilst regularly double-brooded species, *i.e.*, those that produce a second brood, it would appear almost independently of weather conditions, are also now on the wing. Among others, the Drepanulids—*falcataria, binaria, cultraria (unguicula)*—are partially double-brooded, as also the Notodonts—*camelina, dictaea, drome- darius, ziczac*, &c.—in favourable seasons. But August always appears to us to be the important Noctuid season, due in great part, undoubtedly, to the fact that we have worked more regularly and systematically through this month than any other over a long series of years. At the beginning of the month the Agrotids, Triphænids, Noctuids and Plusiids are among the most abundant, whilst towards the end of the month the Catocalids, Xanthiids (and their allies), *Aporophyla australis, Epunda lutulenta, nigra* and *lichenea, Xylina conformis*, &c., are accounted among the good things to work for, and *Leucania albipuncta* and *L. vitellina* usually put in an appearance before the end of the month. A good night with *Tapinostola hellmanni* and *Holotropha leucostigma* on Wicken Fen is not to be despised in mid-August, nor a night on the Broads at the commencement of the month with *Leucania brevilinea*. Harding considers that for collecting at sugar, the wind should be south or south-west, dark, light rain, slight breeze, and the result will be good, but if one point to east or north no good; north-west, if any wind, no good, if calm a little may be done; south or west, still and warm are breeding nights, sugar little or no good ; north or east, with a light wind, little good, with a strong wind, no good. No time is good just before rain, but sultry weather just before a thunderstorm is good, but not after ; in general, entomologise after rain but not before. August, too, is a moderately good month for the Crambids, whilst many Tortricids are almost restricted to a few weeks at this time of the year, the genus *Peronea* providing a large percentage of specimens—*sponsana, schalleriana, comparana, caledoniana, cristana, hastiana, maccana, ferrugana, shepherdana,* &c. The Depressarias and Gelechias, too, are quite among the strongest groups so far as the number of imagines is concerned, and Lithocolletids and Nepticulids are almost as numerous as in June, second broods of almost all the species being obtainable, whilst Pterophorids are still fairly numerous. Flower-beds, planted out in spring and early summer, will now yield their reward, and plants of honeysuckle still in bloom should be carefully worked for Sphingids and Plusias. Possibly petunias and *Nicotiana affinis* share first place for *Sphinx convolvuli, Deilephila livornica* and *Choerocampa celerio,* all our autumnal specimens of which are British-born, but from alien immigrants of the early part of the year. Gas-lamps are a fruitful source of specimens in some seasons, but the work wants a great deal of skill to produce really successful results, so many details being necessary not to miss the best-hidden species. Larva-beating is, in August, usually most productive, and the species to be obtained are so numerous, and the likely localities so diverse, that no one can fail to be successful. Keeping to a special kind of tree is much more useful than promiscuous beating, and if one will pay sole attention to elm for a time, then to oak, then to sallow, &c., and keep the victims from each separate, the result cannot fail to give satisfaction to the collector. The small willows that fringe the sides of ditches, streams, railway banks, &c., should be carefully searched for larvæ of *Smerinthus ocellatus, S. populi, Cerura vinula, Pterostoma palpina, Notodonta ziczac, Earias chlorana,* &c., before being beaten, the larger larvæ being easily injured, and larvæ of *Acherontia atropos* are to be enquired after before the end of the month. Pupa-digging may also be carried on with a certainty of success, and Norgate describes a novel and ingenious mode by which larvæ can be trapped and pupæ secured where suitable trees exist in one's grounds. Fasten a band round the trunk below the boughs, the band to be made of rough cork or oak bark, also a second band below the cork, the band to be made of sacking, its lower edge fastened lightly to the trunk, its upper edge separated from the trunk by a galvanised wire which has been first twisted (round a ruler) into a spiral form, and fastened round the trunk by hooking its ends together. The sacking-band may then be filled with earth or cocoa-nut fibre for descending larvæ to pupate in. Poplar is a specially good tree. The chief difficulty is to find all the pupæ, as the cork is so easy to penetrate that the borings of the larvæ are scarcely visible. In August, too, one can commence to collect the seed-heads of gentian, *Pimpinella saxifraga,* wild carrot, &c., which should be placed in a flower-pot, and exposed all winter ; in July next year you will probably breed *Semasia rufillana, Œcophora flavimaculella* and *Asychna profugella* therefrom. Tugwell advises the lepidopterist to search low plants by the sides of marsh ditches during August for larvæ of *Spilosoma urticae,* fond of sunning themselves ; willow-herb and water-bedstraw for larvæ of *Choerocampa elpenor*, and white or yellow bedstraw on dry banks, sandhills, &c., for larvæ of *C. porcellus* and *Macroglossa stellatarum*. whilst if in woods you notice the leaves of honeysuckle eaten on the low trailing branches, turn them over for larvæ of *M. fuciformis*.

TINEIDES.—During August and September the seed-heads of *Artemisia absinthium* should be collected and tied up in a linen bag. In this way the larva of *Tinea ferruginella* was discovered clearing out the seeds from the dried flower-heads (Bignell).

ADELIDES.—The larva of *Nemotois scabiosellus* lives singly in the heads of *Scabiosa arvensis* and *S. columbaria;* later, in September, when the heads are falling, it makes a broad case in which it winters, remaining on the ground or on the lower leaves.

The Larva of *Nemotois minimellus* lives in the heads of *Scabiosa succisa*, and, in late autumn, makes a case (from the old florets) in which it winters, remaining either on the ground or on the radical leaves. The case is contracted medially at the edges, and is dark brown in colour.

NEPTICULIDES.—In August and September the larva of *Nepticula woolhopiella* mines in birch leaves. This species was successfully bred by Dr. Wood, who supplied the larvæ with earth and kept them out of doors all the winter.

The larvæ of *Nepticula minusculella* may be bred from pear leaves collected in August.

ELACHISTIDES.—The larva of *Chauliodus chaerophyllellus* makes large brown blotches on the leaves of *Heracleum sphondylium* during the summer and autumn months.

In August, larvæ of *Asychna terminella* mine in the leaves of *Circaea lutetiana*, in dark places in woods (Threlfall).

The second week in August is the best time to collect the galls made on the stems of *Polygonum aviculare* by larvæ of *Asychna aeratella*. These should be exposed to direct rays of sun.

About the middle of August search the upright stem of *Epilobium* for a swelling near or at one of the joints. Within it the pupa of *Laverna decorella* is to be found.

ARGYRESTHIIDES.—The larvæ of *Argyresthia aerariella* feed in the berries of mountain-ash, in August.

LITHOCOLLETIDES.—The long mines of *Lithocolletis frolichiella* are to be found on the underside of alder leaves, sometimes as many as four mines in one leaf.

LYONETIIDES.—*Lyonetia clerckella* feeds on apple, pear, hawthorn, mountain-ash, birch, sallow (*Salix caprea*) and *Cotoneaster affinis* (Fletcher).

Lyonetia padifoliella occurred at Worthing in some plenty in 1893, feeding on apple, *Cotoneaster affinis* and *Prunus japonica* (*siniensis*) in August-September. On apple, the larvæ patronised the topmost leaves of the shoots of the year. The lepidopterist, therefore, should not summer-prune his apple-bushes (Fletcher).

The larvæ of *Phyllocnistis suffusella* are to be found mining leaves of poplar in August. The mine gives the leaf the singular appearance of a snail having crawled over it.

The larvæ of *Cemiostoma lathyrifoliella* are to be found mining leaves of *Lathyrus sylvestris* in August.

GRACILARIIDES.—In August the cones of *Gracilaria phasianipennella* should be collected on *Polygonum hydropiper*; they produce imagines in September.

PLUTELLIDES.—The beautiful *Cerostoma sequella* may be found in August, at rest on trunks of ash and beaten from the foliage; it is very local. Also comes to sugar.

COLEOPHORIDES.—In August and September, the cases of *Coleophora wilkinsonella, C. limosipennella* and *C. paripennella* are to be found on birch. The larvæ hybernate full-fed, but pupate next year without further feeding.

The cases of *Coleophora adjunctella* are to be found during August and September, in salt-marshes, on the seed-heads of *Juncus gerardi*.

The larvæ of *Coleophora inflatella* feed on the seeds of *Silene inflata*, in August, and appear to prefer the dry seed-heads to the unripe pods. Common in the lanes about Croydon.

The larvæ of *Goniodoma limoniella* are to be swept from flowers of *Statice limonium* in September, when they use an empty flower for a case. When full-fed they bore into the stem, dropping the flower, and close the hole with silk. The larvæ hybernate in the boring, and in May the old stems of *S. limonium* should be collected for them; they pupate in June and the imagines emerge in July (Fletcher).

TORTRICIDES.—In the middle of August collect the twisted sallow and willow tops for larvæ of *Peronea hastiana*; the terminal leaves are preferred, but not the only ones selected. Enclose in a bandbox and the imagines will appear without further trouble.

Twisted leaves of *Viburnum lantana* and *V. opulus* should be collected at the end of August for larvæ of *Peronea tristana*.

In early August *Paedisca occultana* haunts Scotch fir, swarming sometimes, and flying off in great numbers from the twigs on which they rest and which they much resemble in colour. This date is rather late for the south of England.

The imagines of *Paedisca semifuscana* fly at dusk among sallows in August and September. The larvæ are to be obtained in May feeding on terminal shoots of sallow.

The imagines of *Paedisca sordidana* are to be obtained in August and September, in damp places among alder and *Myrica gale*. The larva is reputed to feed on both these plants in the early part of the summer.

In the beginning of August the full-fed larvæ of *Sericoris euphorbiana* are to be found feeding in the shoots of sea-spurge, which are drawn together, the larvæ then eating out the hearts of the shoots.

In August the stems of *Stachys palustris* (occurring in marshes and fens) should be collected for larvæ of *Antithesia carbonana*.

Rolled-up sallow and willow leaves will give larvæ of *Phoxopteryx biarcuana, P. diminutana, Grapholitha campoliliana, Phloeodes crenana*, &c.

The larva of *Grapholitha microgrammana* is to be found in August and September, in the still green seed-pods of *Ononis spinosa*.

The pods of *Astragalus glycyphyllos* should be collected in August for larvæ of *Stigmonota pallifrontana*.

The flower-heads of *Knautia arvensis* should be collected in August for larvæ of *Eupoecilia flaviciliana*.

In August and September the larvæ of *Eupoecilia degreyana* feed in the seed-pods on the unripe seeds of *Linaria vulgaris*.

Seed-heads of cowslips should be collected in August for larvæ of *Eupoecilia ciliana*. The larvæ want cork in which to pupate.

During August (sometimes as late as September) the seed-heads of *Picris hieracioides* should be collected for larvæ of *Eupoecilia hybridellana* (Barrett).

PYRALOIDES.—The larvæ of *Teleia scriptella* fold the leaves of maple in August.

In early August, *Doryphora palustrella* flies directly after dusk, *i.e.*, when it is actually getting dark, and may be then taken flying over the reeds, *Sparganium*, *Typha*, *Iris*, and other plants that go to make up a veritable ditch flora. Later on it comes to light. We have taken it from 10.30 p.m. to 12.30 p.m. by this means.

The larvæ of *Teleia triparella* feed between united oak leaves at the end of August. They are not uncommon on scrubby oak bushes, in lanes near Wanstead.

Orthotelia sparganella may be found in early August, where *Sparganium* abounds, but is rather a local species ; it flies just at dusk and soon settles on its food-plant.

The turned-down leaves of *Sorbus aucuparia* give larvæ of *Epigraphia steinkellneriana*.

The larva of *Anesychia funerella* is to be found feeding on comfrey at the beginning of August.

CRAMBIDES.—In August and September the larvæ of *Homoeosoma binaevella* may be found in the flower- and seed-heads of *Carduus lanceolatus*, forming a large cavity at the base of the flower-head, and feeding on the young seeds.

In August and September the berries of the spindle-tree should be examined for larvæ of *Alispa (Nephopteryx) angustella*. Frass protrudes from the hole through which the larva enters the berry. A supply of rotten wood for the larva to pupate in is very necessary.

In August and September the larvæ of *Nephopteryx splendidella* feed on the cones of spruce fir (*Abies excelsa*). The larvæ pupate in rotten wood in September, and emerge the following year.

The larvæ of *Gymnancycla canella* should be collected at the end of August when feeding on the unripe seeds of *Salsola kali*.

PYRALIDES.—Flower- and seed-heads of *Eupatorium cannabinum* should be collected for larvæ of *Perinephele (Botys) lancealis*, of *Melampyrum* for larvæ of *Botys fuscalis*, of golden-rod for those of *B. terrealis*, of madder for those of *B. asinalis*, of wild mustard for those of *Orobena (Pionea) margaritalis*.

The imagines of *Acentropus niveus* sit on the slimy masses that collect on the surface of stagnant pools in August, and fly over the surface late at night.

In August search the flowers of *Solidago virgaurea* for a Pyralid larva, living in a slight web. You will probably thus obtain the caterpillar of *Botys terrealis*.

The larva of *Perinephele lanccalis* feeds in a web amongst the leaves and flower-heads of *Eupatorium cannabinum* in August and September. When full-fed it spins a cocoon, but does not change to a pupa until the following May.

By poking a haystack (near Martin Mill Station) with a stout stick the first week in August, I disturbed considerably more than a hundred *Pyralis glaucinalis*, which came out two or three at a time and were easily caught.

During August the larvæ of *Orobena* (*Pionea*) *margaritalis* are to be found (often two or three together) under a web among the seed-heads of *Sinapis alba* and *S. arvensis*. They are best found, however, by night, when they are feeding on the seed-pods. Those plants found near growing corn are most frequented.

In August and September the plants of *Teucrium scorodonia* should be well shaken (or beaten) for the larvæ of *Ebulea verbascalis*. The latter should be given sand in which to pupate.

The larvæ of *Spilodes palealis* are to be found at the end of August, feeding on the umbels and flowers of *Daucus carota*.

CYMATOPHORIDES.—The larva of *Cymatophora fluctuosa* may be beaten from birch in August and September. It still occurs in Darenth Wood.

United leaves of birch should be collected for larvæ of *Cymatophora duplaris*, of poplars for those of *C. or* and *C. ocularis*.

GEOMETRIDES.—Pupæ of the Ennomids should now be collected— *Ennomos tiliaria* among *débris* at roots of oak, birch, &c., *E. fuscantaria* in leaves of ash, &c., *Selenia lunaria* at roots of oak ; also *Amphidasys strataria* just below surface at roots of elm and oak. Pupæ of Hybernias—*Hybernia rupicapraria* (oak, blackthorn, &c.), *H. aurantiaria* (birch, &c.), *H. marginaria* (hawthorn, birch), *H. defoliaria* (oak, hazel, &c.), *H. leucophaearia* (oak), also obtained at the same time, in cocoons on, or just below, the surface of the ground.

From the latter end of August onwards, the collector should be on the lookout for the autumn " Thorns," which will often be found at rest on, or flitting round, gas-lamps, and may also be beaten from trees and hedges in the daytime, or found resting on fences.

The pale larvæ of *Hemerophila abruptaria* should now be collected on privet and lilac, usually abundant in suburban gardens where these plants are grown.

The larva of *Phorodesma smaragdaria* should be searched for on *Artemisia maritima* on the marshes at the mouth of the Thames, in August and September (and after hybernation in May). It covers itself with small pieces of its food-plant and is difficult to detect without experience. The larva can be hybernated on common southernwood.

The second brood (often only very partial) of *Acidalia emutaria* is now on the wing at dusk; in many localities a week or so later *Phibalapteryx lignata* follows it.

Search carefully any suitable rocks, walls, &c. (especially near the sea) for the imagines of *Acidalia marginepunctata* (*promutata*). This species is on the wing from June to September (most abundant in August and September).

The imagines of *Gnophos obscurata* fly freely at dusk. After dark they settle again, and may then sometimes be taken in large numbers with a lantern.

The seed-heads of *Silene* and *Lychnis* should be collected for larvæ of *Emmelesia affinitata* and *E. alchemillata;* those of yellow-rattle for *E. albulata.*

The larva of *Emmelesia decolorata* may be found in the flowers of campion (*Lychnis dioica*) in August. Its presence is first made known by the half-eaten petals, and it is usually found within the calyx of the unopened flower bud. Later on it re-enters the seed capsule.

The larva of *Asthena blomeraria* feeds on wych elm, in or near hilly woods, remains on the underside of the leaf (Merrin).

Curled-up leaves of alder should give larvæ of *Hypsipetes impluviata.*

The spun-up leaves of *Myrica gale* should be collected for larvæ of *Melanippe hastata.*

The plants of *Galium verum* should be well searched, in August, for larvæ of *Anticlea cucullata* (*sinuata*).

At the end of August or beginning of September the larva of *Eucosmia undulata* is to be found feeding in a silken web on the upper surface of leaves of sallow. As it gets older it spins two or three leaves together.

Oporabia filigrammaria is best collected with a lantern after dark, when the moths can, on favourable nights, from the end of August to the middle of September, be boxed in numbers as they sit paired on the heather-twigs on moors (Porritt).

At the end of August examine the trunks of *Pinus sylvestris* in the afternoon, for the freshly emerged imagines of *Thera firmata.*

The larvæ of *Cidaria sagittata* feed on the seed-heads of meadow-rue in August, in Burwell Wicken Fen (and possibly elsewhere).

Towards the end of August the second brood of *Phibalapteryx vitalbata* may be found amongst its food-plant, *Clematis vitalba*; it is an insect easy to rear, often laying its ova in scores on the setting-board, the perfect insect also frequents flowers of bramble, if any are in the vicinity of its food-plant (Mason).

A ♀ of *Camptogramma fluviata* taken in August or September should be kept for eggs. The larvæ feed up well on knot-grass and dock, and the imagines will emerge in November.

In August and September the larva of *Pelurga comitata* is usually abundant on *Chenopodium.* We used to get immense numbers of this, and the larvæ of *Eupithecia subnotata*, on Greenwich marshes, by shaking the plants at night.

The larvæ of *Collix sparsata* are to be found in August on *Lysimachia vulgaris.*

The flowers of *Solidago virgaurea, Eupatorium cannabinum, Angelica sylvestris,* and *Scabiosa succisa,* should be beaten into an umbrella, in August, for *Eupitheciae* larvæ.

Flowers should be worked for *Eupitheciae*—toadflax (*Eupithecia linariata*), ragwort, golden-rod, *Angelica,* &c. (*E. centaureata* and *E. castigata*), *Achillea, Artemisia* (*E. succenturiata*), campanulas, gentians, &c. (*E. subumbrata*), *Melampyrum pratense* (*E. plumbeolata*), *Campanula trachelium*, &c. (*E. campanulata*), clematis (*E. coronata* and *E. isogrammata*), scabious and *Calluna* (*E. minutata*), *Calluna* (*E. nanata*), golden-rod (*E. virgaureata*). Searching is perhaps the best method to ensure success, but where practicable beating the flower-heads on the sides of an old umbrella is a much quicker method of obtaining a number of larvæ.

Towards the end of August near Queenstown, I beat the flowers of *Senecio jacobaea*, *Angelica sylvestris*, *Solidago virgaurea* and *Eupatorium cannabinum*, growing at the edge of a wood, into an umbrella. On *Senecio* the larvæ of *Eupithecia virgaureata* were much more common than those of *E. absynthiata* which also occurred on the *Eupatorium* together with those of *E. pumilata*, *E. coronata* and *E. castigata*; on the *Angelica* were plenty of small larvæ of *E. albipunctata* (Crewe).

The larvæ of *Eupithecia extensaria* may be beaten from *Artemisia maritima*, at the end of August ; they feed freely on this food if planted in pots, and females deposit eggs freely thereon in captivity (Porritt).

In August and September the larvæ of *Eupithecia subnotata* can be obtained, sometimes in great abundance, on *Chenopodium*.

About the middle of August the larvæ of *Eupithecia campanulata* are sometimes to be found freely, feeding in and upon the seed-capsules of *Campanula trachelium*.

In August imagines of *Eupithecia subciliata* are to be taken among maple in woods (Torquay, Bury St. Edmunds, &c.), best captured between 10 a.m. and 1 p.m. by beating the trunks and branches of the old trees, when they are easily dislodged, flying off rather quickly (Fox).

At the end of August the larvæ of *Eupithecia fraxinata* may be beaten out of ash.

The larva of *Eupithecia helveticata* feeds on juniper in August and September.

In August look over the clematis flower-buds ; pick off those with a little round hole in them, and pack away in a large flower-pot with earth at bottom ; the larvæ will pupate on surface ; and hundreds of imagines of *Eupithecia isogrammata* can be bred with little or no trouble.

The larvæ of *Eupithecia satyrata* var. *callunaria* may be obtained in August and September, by sweeping the flowers of *Calluna vulgaris*.

In August and September the seeds of *Pimpinella saxifraga* should be collected for larvæ of *Eupithecia pimpinellata*.

At the end of August sweep the flowers of *Calluna vulgaris* for the larvæ of *Eupithecia nanata*.

The larvæ of *Eupithecia coronata* feed on the flowers of *Clematis vitalba* and ragwort in August.

LACHNEIDES.—Sembling for males of *Pachygastria trifolii* pays well at dusk with a newly-emerged ♀—Rye coast, Wallasey coast, Lyndhurst, &c.

DIMORPHIDES.—The cocoons of *Dimorpha* (*Endromis*) *versicolora* are now to be found spun-up among the *débris* about the roots of the alder and birch trees and bushes on which the larvæ have fed.

SATURNIIDES.—The full-fed larvæ of *Saturnia pavonia* should be collected in early August; on the heaths they prefer *Calluna* and sallow, but in the fens usually select *Spiraea ulmaria*; larvæ abound in Wicken Fen during the first fortnight of the month.

SPHINGIDES.—The larvæ of *Macroglossa fuciformis* on honeysuckle, and those of *M. bombyliformis* on scabious, are full-fed early in August.

The larvæ of *Choerocampa porcellus* may be found on *Galium*; larvæ of this species and those of *Macroglossa stellatarum* are in some years exceedingly abundant on the stunted plants growing on the shingle or near the sea-shore on the Kent coast, the latter readily found well down in the day, the latter only found at all easily after dark with a lantern.

The sides of ditches are the favourite haunts of *Choerocampa elpenor*: we have found them stretched at length on *Galium palustre*, &c., in the day, basking in the sun, and also, with a lantern, feeding voraciously at night.

On the sea-shore (or near the sea) the larvæ of *Deilephila galii* feed from the beginning of August till the end of October on *Galium verum* and *G. elatum*: they seem to prefer scrubby plants, are nearly always found in pairs, may be traced from the frass on the sand-hills, and vary much in appearance (Merrin).

After the second week in August look for *Sphinx convolvuli* at flowers of *Nicotiana affinis*, and bedding geraniums; it is on the wing just before dusk, and continues flying for two or three hours.

Sugared bouquets of flowers, consisting chiefly of honeysuckle, geraniums (pale-coloured) and petunias are a most attractive bait to *Sphinx convolvuli*; white and pale-coloured flowers are far more attractive than those of deeper hues (McRae).

The larvæ of the three Smerinthids should now be collected; they are not difficult to find—*Smerinthus ocellatus* on willow and sallow, *S. populi* on poplar, &c., and *S. tiliae* on elm, &c.

PTEROPHORIDES.—In early August search the *Stachys* in woods and by hedges for larvæ of *Amblyptilia cosmodactyla* and *A. acanthodactyla*.

The second brood of *Leioptilus microdactylus* occurs freely among *Eupatorium cannabinum* in early August. They are sometimes on the flowers during the afternoon, but if not can easily be disturbed by walking among and shaking the plants, and fly freely towards dusk.

NYCTEOLIDES.—During August collect the twisted tops of osiers for the larvæ of *Earias chlorana*. There can be no mistaking the affected heads so conspicuous are the silk-fastened bunches. Enclose securely in a bandbox and they will spin their boat-shaped cocoons without further trouble.

LYMANTRIIDES.—During August the larvæ of *Demas coryli* may in some years be beaten in large numbers from oak and beech.

The white spittle-like patches on the underside of willow and poplar leaves cover the eggs of *Leucoma salicis*; the young larvæ hybernate when quite small, and with very little preliminary feeding.

NOTODONTIDES.—The Cerurid larvæ are now practically full-fed, and should be beaten or searched for (if practicable), *Cerura bicuspis* on alder, *C. furcula* on sallow and willow, *C. bifida* on poplar; *C. vinula* is a fine object on the upperside of a poplar or willow-leaf.

The larva of *Pterostoma palpina* feeds on poplar and sallow in August and September, and when full-fed forms a rather large silken cocoon usually at the foot of the tree.

Larvæ of *Ptilophora plumigera* buried in a dry sandy soil; after three weeks the pupæ were removed and laid on a perfectly dry surface, with a little moss thrown over them ; they were subject to a high temperature during the latter part of summer and, autumn, without any moisture whatever, yet all emerged well (Gascoyne).

The larvæ of *Clostera curtula* are to be searched for by day (beaten by night) in spun-up leaves of poplar and sallow.

Noctuides.—I find searching the ragwort shortly after dusk, say from 9 p.m.-10.30 p.m., more productive than any other method for capturing Noctuids at Howth in August (Hart).

The larvæ of *Pharetra albovenosa* are to be found, some years abundantly, on the rough low herbage on Wicken Fen ; conspicuous and easily seen.

Nonagria arundinis (*typhae*) is in the pupal state by the end of the month ; it should be sought for in stems of *Typha latifolia* ; the infested plants may be known by the central leaves being yellow and withered ; cut the plant low down, strip off some of the green outside leaves and the outlet from which the moth will emerge will then be seen ; shorten the stems, leaving five or six inches above and below the point of exit, insert in wet sand in a deep flower-pot, covering the whole with a piece of muslin ; examine every morning or late in the evening, when the imagines will most likely be found clinging to the stems or sides of the pot.

Tapinostola fulva occurs towards middle or end of August in low-lying marshy ground, flying for a short time before dusk and afterwards resting on the stems of rushes and grass.

On August 9th-12th, 1882, forty examples of *Nonagria neurica* were taken at Salhouse (a place a little above Horning), the majority rather worn, so that a few days earlier would evidently have been the proper time (Harmer).

Tapinostola hellmanni, usually supposed to be a purely Fen species, occurs abundantly in Monk's Wood (Hunts) in August. It is much more abundant on the outskirts of Wicken Fen than in the Fen itself.

In August, after dark, the palings and fences around pasture lands and meadows should be searched for the imagines of *Luperina testacea*.

In August cut down the tall thistles that show drooping flower-heads. You will obtain the pupæ of *Gortyna ochracea* (*flavago*). They are also to be found in burdock stems.

Larvæ of *Mamestra albicolon* are now best obtained on sandhills, feed chiefly by night, and appear to prefer *Atriplex*, &c.

Heliophobus hispidus is found at Portland from the beginning of August till early in October, being most abundant during the first half of September ; it is generally distributed, but occurs most frequently on some half dozen steep slopes, covered by long grass, the imagines being found at night sitting on the stems thereof ; they are quiet and lethargic allowing themselves to be boxed off the grass and remaining quietly on the sides of the box.

The globular eggs of *Noctua depuncta* are laid on sorrel and other low plants, pale, straw-coloured, with a brown micropylar spot and encircled at some distance with brown (Merrin).

Triphaena interjecta is on the wing in August, and will be seen dashing about the sides of hedges as early as four or five o'clock in the afternoon ; comes to lavender flowers at night, is also found at rest on foliage covered with honeydew ; is rather difficult to take on the wing. (All the Triphænas want care in setting, as the wings are apt to split with the setting-needle).

Agrotis cursoria occurs in August on the ragwort flowers on the extreme edge of the sandhills nearest the sea ; the other sandhill Agrotids (*Agrotis restigialis, A. tritici, A. nigricans, Actebia praecox*, &c.) will be found further inland, where scarcely a specimen of *A. cursoria* will be taken (Porritt).

Agrotis obelisca rarely appears before the end of August, it then comes freely to sugar and is most abundant in early September, abounds in some years on the Freshwater coast in the Isle of Wight.

In late August and early September the females of *Stilbia anomala* sit on the flowers of ragwort in the early evening (Jäger).

The second week in August is the best time to capture *Agrotis agathina*. " The best plan to capture it is to light the lantern and watch the places among the heather which are partly sheltered with trees. The insect appears to fly for about a quarter of an hour briskly over the heather, after which it settles for half an hour or so, during which time it may be found on the heather bloom ; the slightest shake, however, causes it to fall like a stone, when it is usually lost. After this half-hour's rest it flies again, and must be taken with the net and lantern. This period yields by far the greatest number of moths " (G. Norman).

On the sea-coast, beneath the plants of *Atriplex littoralis* and *Salsola kali*, the larvæ of *Agrotis ripae* may be obtained in large numbers, resting, when young, on the stems and leaves of the food-plant, but afterwards tunnelling under the sand, where they hide during the day. They are best obtained from the middle to the end of August, when they are nearly full-grown, and can generally be found simply by passing the fingers through the sand.

The larvæ of *Dianthoeciae*, if not collected the preceding month, should be taken now. A bag full of seed-heads of *Lychnis silene* (from as many localities as possible) should be obtained ; at the same time, it must not be forgotten that, whilst some of the species, e.g., *Dianthoecia capsincola*, bury themselves in a large capsule when full-fed, others, e.g., *D. irregularis*, feed by night, and hide by day in a little cavity hollowed just below the surface of the ground.

On the coasts of Devon and Cornwall the capsules of *Silene inflata* and *S. maritima* should be collected in August for larvæ of the dark aberration of *Dianthoecia conspersa*.

In early and middle August, the seed-capsules of *Silene maritima*, in the Isle of Man, Ireland and the west coast of Britain, should be shaken and collected for larvæ of *Dianthoecia capsophila* and *D. caesia*.

The dimorphic (green and red forms of the) larvæ of *Hadena pisi* are frequently abundant in August and September on broom, bracken, sallow, &c., and although they are to be found somewhat freely in the daytime, they may be collected much more commonly after dark.

Hadena pisi feeds on ling, *Myrica gale* and bracken ; *Calocampa vetusta* feeds sometimes on *Myrica gale* (Kane).

During the first fortnight of August, search carefully the seeding plants of lettuce in gardens. The variable larvæ of *Hecatera dysodea* will be found stretched at full length, during the day, on the blossoms and seed-heads.

The pupæ of Tæniocampids are best obtained now—*Panolis piniperda* at roots of fir-trees, *Taeniocampa gothica*, *T. incerta*, *T. stabilis*, *T. munda*, *T. miniosa*, *T. pulverulenta* at roots of oak, &c., *T. populeti* at roots of poplars, *T. gracilis* at roots of willow, *T. opima* at roots of dwarf sallow.

Pupæ of *Taeniocampa opima* are to be found at roots of dwarf sallow and *Rosa spinosissima* ; this species prefers coast sand-hills in some districts, and the pupæ are then found just below the sand.

Pupæ of *Panolis piniperda* are to be found in Scotch and other fir woods, preferring the margin of, or open places in, these woods ; sometimes in crevices of bark, or under moss and fallen needles, rarely within 2ft. of the tree and from one to two inches deep (Merrin).

The imagines of *Tiliacea citrago*, besides coming to light and ivy at the end of August, frequent the leaves and twigs of lime-trees at night (Merrin).

The somewhat fragile cocoon of *Cirrhoedia xerampelina* must be searched for at the roots of ash in August and early September.

Search trunks of ash-trees in late August and throughout September, from 10 a.m. to 3 p.m. (sometimes even till dusk), for the freshly-emerged imagines of *Cirrhoedia xerampelina*. The grass around the trees should also be very carefully worked, also dead leaves around. Detached trees (*i.e.*, those not growing in hedgerows) are the best.

Verbascum plants should be well worked for larvæ of *Cucullia lychnitis*, *Tripolium* and golden-rod for *C. asteris* (exceedingly abundant some years on the Thames salterns), golden-rod for *C. gnaphalii*, and *Artemisia absinthium* for *C. absynthii*.

The larva of *Cucullia umbratica* hides by day near the ground under leaves of *Sonchus*, lettuce, &c.; feeds on the upper leaves and flowers at night (Merrin).

During August, the larvæ of *Cucullia absinthii* may be obtained by beating *Artemisia absinthium*.

Towards the end of August the larvæ of *Cucullia asteris* are frequently very abundant on sea starwort.

In August, *Cloantha solidaginis* may be collected from bilberry and heather with the aid of a lantern after dark on northern moors, or in the day-time at rest on walls and pine-trunks in moorland districts (Porritt).

In August the larvæ of *Chariclea umbra* (*Heliothis marginata*) are sometimes abundant on *Ononis*.

The larva of *Anarta myrtilli* should be swept for on heaths (covered with *Calluna*) with a circular net, which is better than beating into an umbrella for this species. August and September are the best months. Does this species still occur on Wimbledon Common (the side near the park) ?

The cocoons of *Catocala nupta* are spun-up among the leaves, in crevices of bark, or under loose bark, of willows and poplars.

The larvæ of *Erastria fasciana* (*fuscula*) may be found feeding at night about half-way up the culms of *Molinia caerulea* (Merrin).

Larvæ of *Euclidia mi* occur on sandhills feeding on the marram-grass ; drop off when disturbed and twist into grotesque attitudes and feign death ; the white mottled colour and shape exactly imitate the exuviæ of snails (*Helix nemoralis*) (Kane).

ARCTIIDES.—The low plants growing along the sides of marshy ditches should be carefully searched during August and September for larvæ of *Spilosoma urticae*.

The imagines of *Lithosia caniola* fly at dusk ; come to sugar at dawn, and the males may be attracted by a newly-emerged female.

Imagines of *Lithosia griseola* and var. *stramineola* frequently come freely to sugar in July and August in the fens.

SEPTEMBER.

September, dependent upon meteorological conditions, may be, and sometimes is, one of the most interesting months to the field lepidop-terist. The downs and clover-fields may be alive with *Colias edusa* and *C. hyale*, the Vanessids are in their freshest beauty, the pupæ of *Polygonia c-album* are sometimes to be obtained in great numbers from the western hop-gardens, the second brood of *Polyommatus bellargus* is usually in abundance, and *Argynnis lathonia* and *Lampides boetica* may be obtained in our southern counties. The late Sphingids are on the wing, it is *par excellence* the month for *Sphinx convolvuli*, *Deilephila livornica* and *Choerocampa celerio* at beds of *Nicotiana affinis*, petunias, and late flowers of honeysuckle. If *Deiopeia pulchella* immigrants have appeared in May and June their progeny will be now on the wing, and light will repay the dil'gent worker—*Trichiura crataegi*, *Ennomos autumnaria*, *E. alniaria*, *E. fuscantaria*, *E. erosaria*, *E. quercinaria*, *Himera pennaria*, *Oporabia dilutata*, *Camptogramma fluviata*, *Cidaria psittacata*, *Gortyna ochracea*, *Luperina cespitis*, &c. In the marshes *Nonagria arundinis* (*typhae*) and *Calamia lutosa* often abound, and on waste ground *Hydroecia micacea* (among dock) and *H. petasitis* (among *Petasites*) ; at sugar our coast districts give us *Aporophyla australis*, *Peridroma saucia*, *P. ypsilon*, *Agrotis obelisca*, and *Actebia praecox*, whilst *Epunda lutulenta*, *E. lichenea*, *Heliophobus hispidus* and *Leucania albipuncta* also appear to be confined to our coasts, and *Euperia paleacea* (*fulvago*), *Mellinia gilvago*, *Trigonophora flammea* (*empyrea*) and *Aplecta occulta* prefer woods, whilst *Calocampa solidaginis*, *Noctua glareosa*, *Agrotis agathina*, *Celaena haworthii*, &c., prefer the heaths and moorlands. But ivy is the attractive sport, although ivy usually carries us well into October, and the fortunate collector who bottles *Dasycampa rubiginea*, *Orrhodia erythrocephala* or *Xylina furcifera*, will perhaps not care for *Orthosia macilenta*, *Anchocelis rufina*, *A. litura*, *Orrhodia vaccinii*, *O. ligula* (*spadicea*), *Calocampa vetusta*, and *C. exoleta*, although he will not despise *Xylina socia* or *X. semibrunnea*. Often when ivy fails altogether these species will abound at sugar, as also will *Hoporina croceago*, *Tiliacea aurago*, *Polia flavocincta*, whilst *Catocala nupta* often gives variety and excitement to the sugar patches, and *C. fraxini* may always appear. Ripe yew-berries sometimes attract the autumnal moths in large numbers—*Dasycampa rubiginea*, *Hoporina croceago*, *Orthosia macilenta*, *Anchocelis rufina*, *A. pistacina*, &c., in fact, almost all the ordinary frequenters of ivy-bloom are to be thus obtained. Among the Tortricids, the Peroneas are the chief game, although *Paedisca* still provides *ophthalmicana* (aspen), *occultana* (fir), *solandriana* (birch), and *sordidana* (alder), in their special haunts. Of the remaining groups the Depressarias are probably the strongest (most of these hybernate as imagines), whilst Gracilarias are also well represented. There is, also, during this month, a number of species that produce a partial second-brood in favourable seasons, and the larvæ from these, although frequently reared successfully in confinement, must, in an early and cold autumn, undergo annihilation, in nature, if it become necessary for them to reach the pupal stage before hybernation. Where there are ripe plums in autumn one should find some available means of access for night work, for all the moths in the district will feed by night, where flies and wasps feast by day—*Hadena protea*, *Anchocelis pistacina*, *A. litura*, *Tiliacea citrago*, *Citria flavago*, *Luperina testacea*, *Gortyna ochracea*, *Polia flavocincta*, *Mellinia gilvago*, *M. circellaris*, *Miselia oxyacanthae*, &c., are among the moths to be captured. Pupa-digging will always be one of the favourite modes of collecting during autumn, perhaps more so in October than September, when the beating-stick is so busy. So much has been written about the most likely positions, the most likely trees, and the most favourable conditions under which pupæ are to be obtained, that one would suppose that everything that can be learned is already known ; all we can say is, that, in practice, the most cherished theories fall to the ground. We have known the solitary tree produce no pupæ, and a tree in a crowded wood give many ; we have worked trees in sandy localities without success, and obtained pupæ from trees growing on a clay soil, thus setting all the accepted theories at defiance. One rule alone holds universally, and that is, that the lepidopterist who works hardest, keeps at it longest, tries the most varied methods, and visits the most widely differing localities will score most heavily at the end of the year in the number and variety of species discovered (see *Ent. Record*, xi., p. 51). With regard to keeping pupæ through the winter, we would refer our readers to the notes in *Ent. Record*, i., pp. 263, 264 ; iii., p. 242 ; ix., pp. 324-326, and suggest that extreme drought, extreme moisture, mould, and living enemies (chiefly invertebrate) are the greatest opponents of success. Larva-beating is the most universally recognised mode of collecting in September ; every active lepidopterist beats for larvæ this month, however much he may neglect it throughout the remainder of the year. Most of the butterfly larvæ now feeding (chiefly Satyrids, Lycænids and Hesperids), as also the Arctiids and the Lachneids are better obtained after hybernation ; the Sphingid larvæ (which do not hybernate in this stage) are best obtained now—*Sphinx ligustri*, *S. convolvuli*, *Acherontia atropos*, *Deilephila galii*, *Macroglossa stellatarum*, &c.—as also the Lymantriids—*Dasychira pudibunda*, *Notolophus gonostigma*, &c.—the Geometrids—*Eurymene dolobraria*, *Selenia tetralunaria*, *Boarmia cinctaria*, Zonosomas, Acidalias, and above all the Eupithecias, the latter alone forming quite an autumn's work in the collection of the various species. Notodontid larvæ, too, are sometimes frequent (especially if double-broods have been usual), as also are those of Drepanulids, Cymatophorids, Acronyctids, and a large number of Noctuids, many of the latter, however, being

F

usually obtainable after hybernation. Nor must it be forgotten that larva-beating, however well it may pay by day, should be regularly prosecuted by night; birch alone gives a host of good things, and many of the birch-feeding larvæ cling well, *e.g.*, those of *Cerura bicuspis*, *Leiocampa dictaeoides* and *Notodonta dromedarius;* others pass the day in spun-up leaves coming out to feed at night, *e.g.*, *Cymatophora fluctuosa*, *C. duplaris*, *Drepana falcataria*, &c.; these are difficult to obtain during the day but they tumble readily enough into the beating-tray at night when they are feeding. Tortricid larvæ are less frequent during this month unless they be those species obtainable throughout the winter, and, of the smaller lepidoptera, the Nepticulids, the Lithocolletids, the Gracilariids (including *Ornix*) and the later Coleophorids should now be collected. If the collection of mined leaves for Nepticulids and blotched leaves for Lithocolletids be put off till the leaves have dropped, comparatively few will be obtained, especially of the Nepticulids which do not always spin their cocoons on the leaves which they have mined. In late September and early October a visit to the New Forest should produce among others the following larvæ—*Gnophria rubricollis*, *Heterogenea cruciata*, *Apatura iris*, *Demas coryli*, *Lobophora sexalata*, *Eurymene dolobraria*, *Tephrosia extersaria*, *Drepana falcataria*, *Notodonta dromedarius*, *Chloephora prasinana*, *Zonosoma linearia*, *Z. punctaria*, *Acronicta leporina*, *Cleora glabraria*, *C. lichenaria*, *Lithosia helveola*, &c. The flower-heads and seed-heads of many plants should be collected now and tied up in linen bags, *e.g.*, *Solidago*, *Artemisia*, *Senecio*, *Angelica*,&c., whilst flower-heads of *Saponaria officinalis* are reputed to give a good result, as larvæ of many species feed thereon during this month. It is time, too, to commence to beat thatch (of hay, hopbine, reeds, pea-haulm, &c.), for Depressariids and other hybernating species, a mode of work we have already suggested as being advantageous during the whole of the winter.

LYONETIIDES.—The larvæ of *Bucculatrix cidarella* are to be found in September on leaves of alder. They spin thin ribbed cocoons on the stem.

GLYPHIPTERYGIDES.—The full-fed larvæ of *Acrolepia autumnitella* are to be obtained in September, in mines that make conspicuous greenish-white blotches in the leaves of *Solanum dulcamara*.

'ELACHISTIDES.—At the end of September, from a hedge composed principally of dogwood, a large number of nearly full-fed larvæ of *Antispila treitschkiella* was collected ; the mined leaves were put in a flower-pot, covered with a glass cylinder, and, as the larvæ cut out their cases, the leaves were removed. The pot was kept exposed to the full influence of the weather till the emergence of the moths in July, when a large number was reared (Machin).

The first fortnight in September is the time for the larva of *Cosmopteryx schmidiella*, which whitens the leaves of *Vicia sepium*.

In September and October the larvæ of *Elachista taeniatella* may be found in the leaves of *Brachypodium sylvaticum*.

LITHOCOLLETIDES.—Although the work can be done later, mined leaves should be collected this month for Lithocolletids. Mined oak-leaves for *Lithocolletis hortella*, *L. irradiella*, *L. lautella*, *L. querci-*

foliella, L. hegeeriella, and *L. cramerella ;* elm-leaves give *L. tristrigella, L. schreberella;* nut-leaves give *L. nicelliella ;* alder-leaves give *L. frolichiella, L. stettinella, L. kleemannella, L. alnifoliella;* poplar-leaves, *L. comparella;* honeysuckle, *L. trifasciella, L. emberizaepennella;* horn-beam-leaves, *L. tenella, L. carpinicolella;* osier- and sallow-leaves, *L. viminiella, L. spinolella, L. viminetella, L. salicolella, L. quinqueguttella;* furze, *L. ulicolella,* &c.

COLEOPHORIDES.—During the first week in September the cases of *Coleophora melilotella,* which are very like the dark seeds of *Melilotus,* are to be found on the latter plant.

The best time for collecting the larva of *Coleophora fuscocuprella,* is in September and the beginning of October. It feeds on nut, and is said to show a preference for the undersides of the little leaves at the termination of a bough.

Thistles in sheltered (and frequently open) situations give, in early September, the cases of *Coleophora therinella.*

TORTRICIDES.—The larvæ of *Tortrix viburniana, Grapholitha geminana, Phoxopteryx myrtillana, Coccyx vacciniana,* &c., may be collected freely from *Vaccinium.*

In early September sallow- and willow-leaves fastened to the stem, or the spun-together terminal leaves, will still often contain larvæ of *Peronea hastiana.*

In September and October, *Peronea lipsiana* and *P. maccana* love to sit for an hour or two in the afternoon sun before flight, on the upper sides of bracken leaves and bilberry ; *P. rufana* on the leaves of sweet-gale and sallow ; *P. mixtana* on the heather, &c.

Collect the seed-heads of wild carrot, in September, for larvæ of *Semasia rufillana.*

In September the dirty-whitish-green larvæ of *Penthina postremana* are to be found within the stems of *Impatiens noli-me-tangere.*

The larvæ of *Sericoris euphorbiana* are to be found in the closed heads of *Euphorbia amygdaloides* during the first fortnight of September.

Beech nuts should be gathered in September and October for larvæ of *Carpocapsa nimbana.* This larva pupates among moss on the trunks of beech-trees.

The larva of *Carpocapsa juliana* feeds on acorns in September, and is full-fed about the time that the acorns fall.

The larva of *Catoptria tripoliana* feeds in the seed-heads of *Aster tripolium,* and is full-fed towards the end of September.

The larvæ of *Catoptria aemulana* feed in the seed-heads of golden-rod in September and October.

The larvæ of *Phoxopteryx derasana* are easily detected on buckthorn in September by the leaves being folded over and fastened together for the whole extent of the leaf. They remain in the larval state till spring, and then, after wandering about for a day or two, pupate in rough cork.

The larvæ of *Stigmonota weirana* should be searched for in September and October, when they feed between united beech leaves.

The larvæ of *Stigmonota orobana* are to be found in pods of *Vicia cracca, V. sylvatica, Genista tinctoria,* and *Orobus tuberosus* (Merrin).

The seed-heads of burdock (*Arctium lappa*) should be collected in

September, for the larvæ of *Argyrolepia badiana*, which pupate among rubbish at the roots of the plant.

The larvæ of *Eupoecilia curvistrigana* should be searched for in September, in flowers of *Solidago virgaurea*, eating out the young seeds and, passing from one flower to another, they spin up finally among rubbish on surface of ground ; larvæ of *Eupoecilia subroseana, E. implicitana* and *Catoptria aemulana* may also be obtained at the same time.

Gather seed-heads of *Picris hieracioides* for larvæ of *Eupoecilia hybridellana*, also heads of *Daucus carota* for larvæ of *Semasia rufillana;* both these species should have old stems or rubbish in which to pupate, and must be kept all the winter in rain and sun.

COSSIDES.—Larvæ of *Cossus ligniperda*, large and fullfed, are often found wandering on paths, roadsides, &c. Put a large piece of rough cork in a flower-pot, cover the latter with glass ; the larva will bore into the cork, make a winter cocoon, remain therein without feeding all the winter, pupate the following April or early May, and emerge some three weeks afterwards.

PYRALOIDES.—In early September the conspicuous white mines made by larvæ of *Nannodia naeviferella* are to be seen in leaves of *Chenopodium*. At the same time, and on the same plant, the less conspicuous mines of *N. hermannella* are also to be found.

In early September the blotched appearance of the leaves of sallow betokens the presence of the larvæ of *Teleia notatella*. On the same plant, and at the same time, the cones of *Gracilaria stigmatella* (or their white, silvery-looking cocoons, on the underside of the leaves) are to be found.

The larva of *Teleia scriptella* is to be found feeding between united maple leaves in the early part of September. The larva changes to a pupa, in a slight silken web in the fold of a leaf, towards the end of the month.

The seed-heads of burdock should be collected in September, for the larvæ of *Parasia lappella*, which pupate in the seed-heads.

Seed-heads of *Daucus carota*, collected in September, sometimes give imagines of *Depressaria depressella* the following month.

The larvæ of *Enicostoma lobella* live in turned-down leaves of blackthorn in September. Imagines may be beaten in June.

In September the larvæ of *Ptocheuusa inopella* occur in seed-heads of *Inula dysenterica*.

CRAMBIDES.—Nests of humble-bees and wasps collected in September and October, will often give large numbers of the larvæ of *Aphomia sociella* (see *Ent. Record*, pp. 288-289).

Examine heads of ragwort, especially near the coast, for a gregarious Tortricid-like larva that feeds therein ; if you can bring such through the winter you will probably breed *Homoeosoma nimbella* or *senecionis*.

PYRALIDES.—Larvæ of *Pyrausta punicealis* are to be found during September, feeding on the flower-heads of *Nepeta catarina*, under a confused covering of silken threads. They spin tough cocoons about the middle of October, and pupate irregularly in the spring.

In September the larvæ of *Spilodes palealis* are to be found feeding in the umbels of wild carrot which they draw together with a web. They pupate on the ground and the cocoons are best kept out of doors during the winter (Simmons).

In the middle of September the larva of *Ebulea stachydalis* feeds on the leaves of *Stachys sylvatica*, living in a sort of tube, formed either by turning down the top of a leaf and folding it closely to the under-surface with a quantity of silk, or else by drawing together a fold of the under-surface, and covering it over with a thick silken web, in either case leaving an opening at the end. It comes out at night to feed. The larvæ are usually found low down on the plants, eating large holes through the substance of the leaves, but leaving the margin and veins untouched (Buckler).

During September work among coltsfoot for *Scopula lutealis*, when it is pretty easily disturbed.

DREPANULIDES.—The larva of *Drepana harpagula* (*sicula*) is to be obtained in Leigh Woods, near Bristol, by beating, in the middle of September.

The larva of *Drepana falcataria* is common in September and October on birch and alder. The work of the larva is conspicuous on *Alnus glutinosa* owing to its bending the sides of the leaves upward with a few silk strands. If not within this tent, it may usually be found on the upperside of some neighbouring leaf.

The larvæ of *Drepana lacertinaria* are now to be found between united birch-leaves, those of *D. binaria* (*hamula*) on oak and those of *D. cultraria* (*unguicula*) on beech ; beating by night or searching by day will give these in their known haunts.

GEOMETRIDES.—The larva of *Pericallia syringaria* hybernates small, and may be found in September and October on honeysuckle, privet and lilac (also in May after hybernation).

In September *Ennomos alniaria* (*tiliaria*) is often very abundant on gas lamps. It flies at dusk, and normally rests on the trees, after dark, in woods, &c.

The ova of *Ennomos fuscantaria* are readily obtained in confinement from a gravid female ; in nature the flat eggs are laid along the twigs of privet and ash in short rows. Those of *E. alniaria* (*tiliaria*) are laid on oak, birch, &c.

The pupæ of *Odontopera bidentata* are frequently found under moss on trees of various kinds.

The cocoons of *Hemerophila abruptaria* are to be found by close searching on the twigs or between the forks of privet and lilac, or on fences near where these trees are growing; the species is very common in the London suburban districts.

Larvæ of *Boarmia roboraria* may be sleeved successfully during the winter on birch and will eat the bark of the twigs before the leaves appear. As they sometimes kill the young shoots, early removal in spring to another branch is necessary.

The young larva of *Geometra papilionaria* prepares a silken web, to which it attaches itself by its anal claspers, and remains thus for hybernation during the winter ; it looks much like a small twig of the alder or birch to which it is attached, and is brown in colour, the bright green coat being assumed in spring.

The larvæ of *Phorodesma smaragdaria* are to be found on the Essex salt-marshes, in September, by searching *Artemisia maritima*. They are strange little atoms, with scraps of the food-plant gummed over their bodies.

The larvæ of *Abraxas sylvata* (*ulmata*) are to be beaten, during this month, from wych elm. The larva of *Asthena blomeraria* is also to be beaten from the same food-plant and is said to prefer the undersides of the leaves, if searching be preferred to beating.

The pupa of *Zonosoma linearia* is fastened to a leaf of beech, the margin of the leaf often curling over and concealing the pupa; the pupa of *Z. porata* is similarly fastened to an oak leaf; that of *Z. punctaria* to a leaf of oak or birch, and that of *Z. pendularia* to birch (Merrin).

The flowers of *Lysimachia vulgaris* should be well searched and beaten for larvæ of *Collix sparsata*, sometimes occur also on the leaves.

Cut off, during the last week of September and the first of October, a bag full of the unripe seed-capsules of *Odontites rubra* (*Bartsia odontites*), turn these into bandboxes, with muslin over the top, remove from the muslin as they come up and place in cage or flower-pots with fresh food and sand and they will soon pupate. Do not let the seed-capsules get too dry whilst the larvæ are feeding. *Bartsia* grows as a weed by roadside hedges, on waste land, rough pastures, &c.

Search the umbels of *Angelica sylvestris* and *Pastinaca sativa* for larvæ of *Eupithecia albipunctata* and *E. trisignata*; and those of *Pimpinella saxifraga* for larvæ of *E. pimpinellata* and *E. oblongata*, from the middle to end of September.

In September, search out a locality where ragwort abounds. Beat the ragwort heads against the side of an old umbrella. Several species of *Eupithecia* larvæ are thus to be obtained; among others *E. satyrata*, *E. centaureata* and *E. absynthiata* are usually the most abundant.

On the second Saturday in September we could, by seeking after dark with a lantern, see the larvæ of *Eupithecia succenturiata* in twenties on the upper parts of the mugwort plants, whilst during the daytime they are only to be found on the lower portion of the plants, on, or amongst, the twisted dead leaves (Gregson).

Collect the flowers and leaves of yarrow (*Achillea millefolium*) in September and October, and you will breed *Eupithecia subfulvata*.

The larvæ of *Eupithecia arceuthata* feed on wild juniper from the end of September to the middle of November. They are seldom full-fed until the end of October.

The larva of *Eupithecia helveticata* feeds on wild juniper, and is full-fed from the beginning to the middle of September.

The larvæ of *Eupithecia pimpinellata* may be found in scores on the seeds of *Pimpinella saxifraga* in September. Some of the larvæ were green when the seeds were green, but later in the season, when the seeds were brown, the larvæ were chiefly brown (Hodgkinson).

Beat *Solidago virgaurea* from the middle to end of September for larvæ of *Eupithecia virgaureata*, and also from now until middle of October for those of *E. expallidata*.

Spruce-fir cones should be collected during September for the larvæ of *Eupithecia togata*. They feed between the scales of the cone, upon the ripe seeds at the base. The protruding frass makes an affected cone quickly recognisable.

The larvæ of *Eupithecia satyrata* may be found feeding upon the flowers of *Scabiosa succisa* in early September (Crewe).

Small and full-fed larvæ of *Eupithecia pygmaeata* are to be obtained during the whole of September, and the imagines mostly emerge the next year from June till the middle of August, some going over until a second year.

Larvæ of *Eupithecia albipunctata* are abundant in September feeding on *Pastinaca sativa* and *Angelica sylvestris* (common at Box Hill, Caterham, Askham Bog, &c.).

In September the larvæ of *Eupithecia trisignata* are often very abundant, feeding on the flowers of *Pastinaca sativa* (particularly common at Box Hill, in the York district, &c.).

In September the larvæ of *Eupithecia assimilata* may be searched for on, or beaten from, hop.

The larvæ of *Eupithecia minutata* and *E. nanata* can be swept from *Calluna* in September and October.

The large beds of *Chenopodium*, which are to be found flourishing on the waste places near river-banks in most districts of England, should be beaten in September and October for the larvæ of *Eupithecia subnotata*. Many larvæ fall to the ground and want searching for there.

The flowers of *Senecio jacobaea*, in September and October, give a plentiful supply of larvæ of *Eupithecia absynthiata* and *E. satyrata*.

The pupæ of *Eupithecia fraxinata* can be found enclosed in a cocoon, all through the winter, under moss and loose bark of ash-trees, from September till May.

The imagines of *Thera firmata* are to be taken in September and October on the boles of fir and larch trees, just out of pupa.

The imagines of *Cidaria psittacata* are to be obtained at ivy, or hiding in outhouses in September and October; they hybernate later.

CYMATOPHORIDES.—The larvæ of *Cymatophora duplaris* and *C. fluctuosa* are still to be found between united birch leaves, and those of *C. ocularis* between those of poplar, better to beat by night.

COCHLIDIDES.—The almost full-grown larva of *Cochlidion limacodes* (*testudo*) is to be beaten from oaks, from the first week of September onwards in our southern woods.

The larva of *Heterogenea cruciata* (*asella*) must be obtained by searching the beeches from the middle of September until the end of October. It is a very uncertain species, apparently absent in its best locality (Lyndhurst) in some seasons, quite abundant in others.

SPHINGIDES.—The morning twilight in September is said to be better than the evening twilight to capture *Sphinx convolvuli*, at the blossoms of petunias. Most captures, however, are made in the evening when dusk has fully fallen.

The larvæ of *Acherontia atropos* should be persistently sought during this month. The men at work in the potato-fields will, in some seasons, get a very large number and want but little coaching in the art of handling them carefully and carrying them safely. Rough handling otherwise is responsible for the death of many when the pupation stage is reached. Pupæ, too, are often obtained, and without previous training injury is still more likely to be done to the captives.

During this month the larvæ of *Deilephila galii* will be found freely on *Galium*, whenever there has been an immigration of imagines in June; not difficult to find during the day, either low down among the plant or basking; and often traceable by the large and conspicuous pellets of frass. The small plants on the shingle at Deal, Walmer and Kingsdown, and the larger beds distributed over the sandhills at Deal, Wallasey and the Essex and Suffolk coasts usually produce a large number in its special seasons. Larvæ of *Macroglossa stellatarum* are usually taken abundantly at the same time.

About the second or third week in September specimens of *Choero-campa celerio* may sometimes be seen hovering over the flowers of scarlet geraniums; on the wing as early as 7 p.m.; in 1885 captures were made September 20th and 24th, two being observed during the latter evening (Mason).

The isolated patches of *Galium* that grow on the shingle in our coast districts, often harbour large numbers of the larvæ of *Macroglossa stellatarum* in September.

LYMANTRIIDES.—The larvæ of *Demas coryli* are to be beaten from beech, in the middle of September, but larvæ can generally be obtained in most districts where the imagines have been observed in June and July.

In the third week of September the larvæ of *Notolophus gonostigma* may be beaten from oak and birch in Sherwood Forest (Porritt).

NOTODONTIDES.—The larvæ of the Notodonts should now be worked for—*Lophopteryx camelina* (oak, beech, &c.), *L. cucullina* (maple), *Leiocampa dictaea* (poplars, willow), *L. dictaeoides* (birch), *Notodonta dromedarius* (birch, nut, &c.), *N. ziczac* (poplar, willow, &c.).

The larva of *Stauropus fagi* is now to be beaten; it prefers beech, but is also found on other plants—oak, birch, hazel, &c.

In early September alder is one of the best paying trees to beat for larvæ. Visions of the full-fed larva of *Cerura bicuspis* are always well to the fore.

The cocoon of *Cerura vinula* is to be found spun-up on trunks of willows, poplars and sallow, about 3ft. to 4ft. from the ground (Merrin). Most entomologists search for the Cerurid cocoons in October.

NOCTUIDES.—The beginning of September will be found a good time to commence searching for pupæ of *Agriopis aprilina*, *Hadena protea*, &c., at the foot of oak trees; all pupæ taken at oak should be specially cared for, as a host of good insects feed on this tree and pupate round the base; the trunks should also be well examined during September and October in hopes of finding *Poecilocampa populi*, which spins up in the crevices of the bark; it must be borne in mind that trees standing singly in parks and fields are the most prolific (Mason).

The larvæ of *Apatela aceris* are not uncommon in the London district; usually seen crawling on trunks, fences, &c., when they have left the trees and are seeking a suitable place for pupation.

The larvæ of *Cuspidia megacephala* are still to be found coiled up by day, on the trunks of poplars, where they rest when full-fed and just before pupation; they are usually conspicuous enough if carefully looked for.

SEPTEMBER. 89

A warm overcast afternoon in middle September will give plenty of *Celaena haworthii* flying over the mosses ; a clear sunny afternoon prevents their flight, but given dull calm and moist weather the males fly freely (Finlay).

Celaena haworthii must be worked for on mosses and bogs in September, between 6.15 p.m.-7 p.m. Before 6.15 p.m. there may not be a specimen visible, but a few minutes after there may be hundreds (Stott).

The larvæ of *Agrotis ripae* are to be found by day just below the sand where *Salsola kali* grows in September and October.

In the beginning of September *Agrotis agathina* is to be found feeding at the blossoms of *Calluna vulgaris* by night.

Ragwort bloom in September sometimes gives an abundance of imagines of *Noctua glareosa.*

The carpet-like layer of needles and moss at the foot of a pine tree wants rolling back, when the pupæ of *Panolis piniperda* are readily found at about a foot from the tree (Norman).

Towards the middle and end of September, when the ordinary autumnal sugaring is generally considered finished, *Epunda lutulenta, Aporophyla australis* and other local species, often occur freely in coast districts.

Epunda lichenea occurs in September at Portland, but does not begin to fly until about midnight.

Larvæ of Cucullias should be sought for early in the month—most are better obtained, however, in August—*Cucullia asteris* is still obtainable on golden-rod and *Tripolium*, *C. absynthii* on wormwood, *C. gnaphalii* on golden-rod, &c.

Just as we come up to the wood from Kemsing, are high banks on each side of the road. The golden-rod (*Solidago virgaurea*) on these should be carefully examined for larvæ of *Cucullia asteris* and *C. gnaphalii* (Carrington).

Some of the Dianthœcias are more or less partially double-brooded, and their emergence is continued over a long period ; late larvæ are, therefore, often obtained during this month, *e.g.*, those of *Dianthoecia capsophila* on seed-heads of *Silene maritima*, *D. capsincola* on *Lychnis* and *Silene*, *D. carpophaga* on *Lychnis*, &c.

In September the imagines of *Citria fulvago* and *C. flavago* often frequent the dry feathery-looking grass heads by the sides of the rides of woods ; they are also attracted to honey-dew on sallow leaves much more frequently than to sugar.

The eggs of *Cirrhoedia xerampelina* are laid on ash, probably in the chinks of the twigs, buds, &c. (Merrin).

In early September, *Ononis* should be swept after dark for larvæ of *Chariclea umbra (marginata)*.

Larvæ of *Chariclea umbra (marginata)* and *Heliothis dipsacea* feed well on the green pods of scarlet runner beans, *Hadena pisi* and the Plusias eating them just as freely. Possibly they would do well for *Dianthoecia irregularis*, &c. (Norgate).

During September sweep ling (*Calluna vulgaris*) for larvæ of *Anarta myrtilli*. The larva of this species is very difficult to obtain in any other way, its similarity to its food-plant being most striking.

During September (and August) the larva of *Erastria fasciana* (*fuscula*) is to be swept by night from *Molinia caerulea* (and perhaps other grasses). The larva pupates about the end of the month.

ARCTIIDES.—The larvæ of *Spilosoma urticae* should be searched for in marshes and fens where *Iris*, *Pedicularis*, *Trifolium* and *Mentha aquatica* grow abundantly in the first week in September. They feed on all these plants, and are to be found, extremely low down, on plants overhanging water-holes and ditches.

PAPILIONIDES.—Sleeve larvæ of *Apatura iris* out on sallow, so that they can get on a thick branch: they must be removed from the sleeves every day till they settle down on a twig, as none ever hybernate successfully if left on the sleeve (Hewett).

The eggs of *Cyaniris argiolus* (second brood) are laid on the flower-stalks of holly, the young larvæ burrow in the unexpanded buds.

OCTOBER.

October is generally a most uncertain month entomologically. So much depends on the meteorological conditions, on whether sugar will pay, and whether the ivy-blossom is in just that condition on suitable evenings to attract a multitude of insect visitors. If the weather be really satisfactory, much out-of-door work can be done even with imagines, but it is rather to pupæ that the hopes of the lepidopterist turn, and many are the details that a successful pupa-hunter must grasp before he can become at all proficient at the work. Among other things, the tops of moss-covered stumps of trees that have been cut down should be carefully peeled for pupæ, and the moss growing on, and around the roots of, trees must be thoroughly worked, but pupa-digging, undoubtedly, enables a collector to employ his leisure in winter and spring in the most profitable manner; a trowel and a three-pronged fork with prongs bent backwards for pulling up turf are useful tools for the purpose. In October and November the various enemies of pupæ have not had time to find a very large proportion of them. Search under moss on (or at roots of) trees for pupæ of *Demas coryli*, *Eurymene dolobraria*, *Odontopera bidentata*, *Eupithecia consignata*, *E. fraxinata*, &c., dig round the roots for pupæ of Notodonts, also for those of *Cymatophora ocularis*, *Selenia bilunaria*, *Amphidasys strataria*, *Tephrosia bistortata*, *T. crepuscularia*, *Asthena blomeri*, &c., especially working with the fingers where the surface of the earth meets the trunk; search under loose pieces of bark for cocoons of *Poecilocampa populi*, *Triaena tridens*, *Acronicta leporina*, &c., and search the trunk itself for those of *Cerura bicuspis*, *C. furcula* and *C. bifida*. Dead leaves about the roots of trees give cocoons of *Heterogenea cruciata* (*asella*), *Cochlidion limacodes* (*testudo*), *Dimorpha* (*Endromis*) *versicolora*, *Lachneis lanestris*, *Stauropus fagi*, *Clostera curtula*, *C. pigra* (*reclusa*), *Cymatophora or*, *C. fluctuosa*, *Lobophora sexalata*, &c., and pupæ of *Zonosomae* (fastened by tail to leaves). For the smaller species the collection of the seed-heads of various plants will give a good return for the time spent. The seed-heads of gentian, wild carrot, *Pimpinella saxifraga*, &c., collected in the autumn, placed in covered flower-pots, and stood out of doors, will give imagines of *Semasia rufillana, Œcophora flavimaculella, Asychna profugella, Eupithecia pimpinellata*, &c., the following spring. It is well to bear in mind in rearing larvæ of Noctuids, Agrotids, &c., that feed through the winter, that sliced potato or carrots, buried in a tub of sand, make very acceptable winter fare. Some thought of the following year also is necessary, and Mason states that he has found many moths particularly fond of the flowers of lavender, and recommends any entomologist who has a garden at command to plant a short row of it ; cuttings put in during autumn, and kept in a cold frame, will root during the winter, and be ready to plant out the following May. Another point that must not be forgotten by hasty collectors is that it is not wise to proceed hurriedly in throwing out one's entomological breeding-cage rubbish without very close inspection, as so many species go over two (or more) winters in the pupal stage—the Cucullias are specially noted for this peculiarity—but many other species are equally persistent in going over—*Cidaria reticulata, Emmelesia unifasciata*, &c.—and pupæ of some species will go over four or five winters.

NEPTICULIDES.—The mines of Nepticulids are best collected in October ; later the larvæ leave their mines in order to find a suitable place in which to spin their cocoons, which in many cases is not on the leaf in which they have fed up. Many species will be missed unless the mines are collected early. Oak-leaves give *Nepticula atricapitella, ruficapitella, basiguttella, quinquella, subbimaculella ;* birch gives *N. betulicola, argentipedella, woolhopiella, confusella, continuella, luteella, lapponica, distinguenda ;* elm, *N. viscerella, ulmivora, marginicolella ;* on *Rosa canina* one finds *N. anomalella, angulifasciella :* on *Rosa arvensis, N. fletcheri ;* on *Rosa rubiginosa, N. centifoliella ;* on hawthorn, *N. pygmaeella, N. ignobilella, N. atricollis, N. gratiosella, N. regiella, N. oxyacanthella ;* on *Pyrus malus, N. pomella, N. desperatella, N. malella, N. pulverosella ;* on *Pyrus communis, N. minusculella, N. pyri ;* on *Pyrus aucuparia, N. aucupariae, N. nylandriella, N. sorbi :* on *Rubus fruticosus, N. aurella, N. auromarginella ;* on *R. caesius, N. rubivora, N. splendidissimella ;* on *Agrimonia eupatoria, N. agrimoniae, N. aeneofasciella, N. fragariella ;* on *Fagus sylvatica, N. turicella, N. basalella :* on *Populus tremula, N. argyropeza*, and *N. assimilella ;* on *P. nigra, N. trimaculella ;* on *Lotus corniculatus, N. eurema, N. cryptella*, &c. [For a complete history of the British and Continental species, the foodplants, mode of feeding, &c., see the author's *British Lepidoptera*, vol. i., pp. 162 *et seq.*]

The larvæ of *Nepticula lapponica* feed in broad serpentine mines in birch-leaves, and should be collected at the same time as those of *N. betulicola, viz.,* from October 1st-20th (Threlfall).

During October the yellow larvæ of *Nepticula aeneofasciella* may be obtained in the blotches in leaves of *Agrimonia eupatoria.*

Mines of *Nepticula catharticella* often occur in great quantities in leaves of *Rhamnus catharticus*, but are difficult to see until one's eye becomes accustomed to them.

Larvæ of *Nepticula myrtillella* are sometimes common in moorland districts in leaves of *Vaccinium myrtillus.*

The larva of *Nepticula desperatella* mines in the leaves of wild apple-trees in October. It prefers the leaves on the small incon-spicuous shoots that grow close to the ground.

In October the larva of *Nepticula headleyella* makes a long and very narrow gallery in the leaf of *Prunella vulgaris*. The affected leaves are of a dull purple colour.

The orange-coloured larvæ of *Nepticula pomella* are to be found in October and November in the leaves hanging from the lower branches of apple-trees. Sometimes there are as many as half-a-dozen larvæ in a single leaf.

Mines of *Nepticula alnetella* and *N. glutinosae* both occur in alder-leaves.

LITHOCOLLETIDES.—Mines of *Lithocolletis stettinella* occur in alder-leaves ; the larvæ are partial to the small leaves at the end of the twigs.

In October and November collect the leaves of *Vicia sepium* and *Orobus tuberosus* for mines of *Lithocolletis bremiella*.

The mines of *Lithocolletis anderidae* made in birch-leaves should be collected in October.

In some localities the mines of *Lithocolletis comparella* are common in leaves of many species of *Populus*, but prefer the garden one with white undersides to its leaves. The moths emerge almost at once.

Larvæ of *Lithocolletis viminetorum* should be sought for in mines at the edges of leaves of *Salix viminalis*.

COLEOPHORIDES.—The following Coleophorid larvæ should be searched for :—*Coleophora limosipennella* and *C. wilkinsoni*, on birch ; *C. paripennella* on bramble, rose, sloe, &c. ; *C. fuscocuprella* on hazel ; *C. argentula* on *Achillea millefolium* ; *C. potentillae* on bramble, *Potentilla tormentilla*, *Spiraea filipendula*, *Poterium sanguisorba*, &c., *C. badiipennella* on elm.

Larvæ of *Coleophora murinipennella* should be collected through the autumn on seeds of *Luzula pilosa*. The species occurs freely at West Heath, Hampstead (Elisha).

The cases of *Coleophora argentula* are sometimes to be obtained in great abundance in October, on the seed-heads of yarrow.

In October the larvæ of *Coleophora salinella* should be collected from *Suaeda maritima* growing on the coast salterns.

The full-fed larvæ of *Coleophora wilkinsoni* should be collected from birch in September-October. They hybernate full-fed, crawl about in early spring without feeding, then pupate, and the imagines emerge about the end of June.

GRACILARIIDES.—Bladder-like leaves of *Artemisia vulgaris* should be collected in October for larvæ of *Gracilaria omissella*.

ELACHISTIDES.—The brown mines of the larvæ of *Antispila treitschkiella* should be obtained on dogwood the first week in October.

During October the larva of *Elachista dispunctella* mines in *Festuca ovina*. It hybernates in old grass-stems and pupates later in spring (Threlfall).

TORTRICIDES.—During the winter the **dry stems of** *Umbelliferae* should be opened for larvæ. Many species retire therein to pupate, that feed on other plants.

Peronea tristana and *P. ferrugana* are to be beaten from their respective food-plants, *Viburnum lantana* and *Betula*. Both are very variable, so should be carefully examined.

Collect fallen and diseased acorns; place in a shallow box containing leaf mould and dead leaves. Stand out of doors during the winter until June, *Carpocapsa splendana* will be bred in plenty.

Dictyopteryx lorquiniana may be bred in abundance from the flower-heads of *Lythrum salicaria* collected in October, and from shoots of the same plant found in June (Farren).

During the autumn collect the common beech mast for the larvæ of *Carpocapsa grossana*. The full-fed larvæ avail themselves of cork and rotten wood in which to bore. They remain during the winter as larvæ in the cocoons, and often go over two winters in this condition. They should be placed out of doors until the following spring, when they can be brought into the house.

Collect diseased hips from wild rose. The larvæ of *Stigmonota roseticolana* emerge from them from the middle of September to the middle of October; place in the jar with the hips a few moderate-sized pieces of rotten wood, and cover well so as to prevent the larvæ from escaping. When all have buried themselves in the rotten wood, place out-of-doors during the winter, bringing them in again in May. The imagines should be reared in plenty in June.

The cocoons of *Stigmonota regiana* are to be obtained during autumn and winter under the bark of old sycamore trees ; the larva appears to remain in its cocoon unchanged until spring.

From October until April the larva of *Ephippiphora nigricostana* may be found in the stem and root of *Stachys sylvatica*, climbing in April into the dried flower-stems, where it pupates just below a joint in the stem.

From October to April, split the stems of *Impatiens noli-me-tangere*, for the active whitish-green larva of *Penthina postremana*. When the affected stems can be detected without splitting them, do not do this, as it interferes with the hybernation of the larva.

The larvæ of *Catoptria candidulana* feed on the seeds of *Artemisia maritima* during the autumn months.

In October dig up whole plants of ragwort where they are known to contain larvæ ; plant in shallow boxes, a dozen or two in each box, and place in garden through the winter ; if brought indoors in June *Ephippiphora trigeminana*, *Eupoecilia atricapitana*, &c., will be bred.

Cut off shoots of mugwort six or eight inches from the ground and plant them in seedling boxes, two or three dozen in each ; place in garden through the winter and bring indoors in June when *Ephippiphora fœnella*, *Dichrorampha simpliciana*, &c., will be bred in July.

Roots of wild carrot, yarrow, thistles, &c., yield very good results in the same way, and they are all plants that defy extermination.

Collect heads of teasel in October (in the better cultivated parts of the country they are destroyed before spring) ; tie in bundles and suspend out-of-doors during the winter ; put in a band-box in June, *Eupoecilia roseana* and *Penthina gentiana* will be bred.

Collect upper two-thirds of stems of wild parsnip, and treat in same way (as last) for *Conchylis dilucidana*. Take care that the stems are placed out of the reach of earwigs.

Golden-rod, aster, tripolium, *Anthemis*, &c., collected and tied up in similar bags, and treated similarly, give good results.

The first week in October is the time to collect the full-grown larvæ of *Argyrolepia maritimana*; at that time they are to be found mining in the roots of *Eryngium maritimum*, sometimes to a depth of six or eight inches below the surface of the sand; the long pipe-like roots, for about three inches of sound root beyond the mine, must be taken out very carefully, and the mined roots should not be opened but buried upright in sand.

Larvæ of *Conchylis smeathmanniana* are now about full-fed, and are to be found in seed-heads of *Achillea millefolium*, connecting the seed-heads with a silken gallery. [Butterfield suggests keeping them through the winter in bags made of the material in which Australian mutton is exported.]

Roots of *Picris hieracioides* should be dug up for larvæ of *Chrosis tesserana*; plant in tubs or large flower-pots and leave out all the winter.

The seed-heads of golden-rod should be collected in October, and kept out-of-doors during the winter. In spring they should be put under cover, and in this way it is possible to breed *Eupoecilia subroseana* freely in May and June.

PYRALOIDES.—Beating thatch is a most profitable employment in the autumn; large numbers of local *Depressariae* and other species are more readily obtained in this than in any other manner.

The larvæ of *Lita acuminatella* are to be found in October mining the leaves of thistles; the imagines emerge freely the following spring; common in Hackney Marshes.

Collect heads of burdock for larvæ of *Parasia lappella*; of *Centaurea nigra* for larvæ of *P. metzneriella*; of *Carlina vulgaris* for *P. carlinella*; of *Cirsium acaule* for *P. neuropterella*.

At the end of October the larva of *Nothris verbascella* is to be found eating out the undeveloped leaves at the heart of plants of *Verbascum pulverulentum*. It feeds slowly through the winter, and full-fed specimens are to be found throughout the spring until the middle of June. In July the pupa is to be found in a slight web on the underside of the large lower leaves, generally in the angle of two ribs, or in the turned-down edge of a leaf.

Beat hedges containing spindle for *Pteroxia caudella* before it hybernates.

CRAMBIDES.—During the winter months the dried stems of thistles should be collected for the larvæ of *Myelois cribrum*.

During the winter the hybernating larvæ of *Anerastia farrella* are to be obtained in sandy localities, where *Anthyllis vulneraria* grows, in their sand-balls, by passing the sand through a sieve (Schleich).

During the winter months (October to March) the full-fed larvæ of *Homoeosoma sinuella* may be found feeding in the root-stocks of plantain. They spin cocoons in March, but the larvæ do not pupate until May and early June.

Larvæ of *Euzophera cinerosella* (*artemisiella*) feed during the winter and spring in the stems and root-stocks of *Artemisia absinthium* (Barrett).

ORNEODIDES.—Beat thatch of old summer-houses in gardens and *Orneodes hexadactyla* may be obtained in plenty.

PTEROPHORIDES.—The pupæ of *Amblyptilia acanthodactyla* are to be found in October and November attached by the anal segment to the flower spike of *Stachys sylvatica*, between two of the whorls of seed-vessels, where they look like dried-up flowers.

CYMATOPHORIDES.—In October, beating for larvæ by night often pays better than beating by day ; *Cymatophora fluctuosa* and many other birch-feeders hide in spun leaves by day, but fall readily into the beating-tray at night.

The roots of old poplars always provide a prolific hunting-ground for the collector of pupæ. One of the best finds is *Cymatophora ocularis*, which may be found spun-up under bark, moss, among rubbish, or spun up among ivy leaves three or four feet from the ground, attached to old stumps near, in fact, in any corner, but they want carefully searching for, as they are easily crushed (Robertson).

At the roots of oaks one finds, among large numbers of common Tæniocampid and other pupæ, those of *Cymatophora ridens*.

DREPANULIDES.—During October the spun-together leaves of birch should be collected for larvæ and pupæ of *Drepana falcataria*, similarly fastened-together leaves of beech will give larvæ and pupæ of *D. cultraria* (*unguicula*). The larvæ of these species are, however, much more readily obtained by beating at night, as they hide in their little leaf domiciles by day but leave them to feed at night.

GEOMETRIDES.—The imago of *Himera pennaria* flies at dusk in October and November (comes also very freely to light). Many may be sometimes found at rest on the bare twigs of bushes afterwards, and at rest they very closely resemble dead leaves.

Larva-beating is often most successful during this month—*Selenia tetralunaria*, *Amphidasys betularia*, *Zonosoma linearia*, *Z. annulata*, *Asthena luteata*, &c., are thus to be obtained.

The moss at the roots of fir-trees should be carefully rolled back for the pupæ of *Macaria liturata* and *Fidonia piniaria* ; they are usually obtained at only a short distance below the surface of the ground.

Larvæ of *Melanthia ocellata*, *Larentia salicata* and several other moths are full-fed in October, spin a silken web, but do not change to pupæ until the spring. Care should be taken not to interfere with such once the puparia are made, as disturbance in this stage is usually fatal.

The larva of *Uropteryx sambucata* hybernates well either in the open air or under cover, in a shed or outhouse ; if in the open air, a strong muslin bag tied to a branch of *Prunus spinosa*, in a sheltered position, forms a good winter cage ; if under cover the twigs should be placed in a bottle of water and the larvæ secured by a muslin bag tied round the neck of the bottle ; the larvæ should be supplied with fresh food so long as a vestige of green leaves is obtainable (Grapes).

In October and November the regular worker at ivy-bloom will probably get *Camptogramma fluviata* and *Cidaria miata* among the usual common visitors. Always keep any ♀ *C. fluviata* that falls into your hands for eggs. Autumnal eggs soon hatch, and, in confinement, the larvæ usually feed up quickly and produce another brood in a very short time.

After dark *Thera juniperata* is to be found in great numbers resting on juniper bushes.

You can still collect the seed-heads of *Bartsia odontites* for larvæ of *Emmelesia unifasciata*. Tie up in linen bags or keep in band-box, and give fresh heads until pupation has taken place.

The pupæ of *Emmelesia albulata* will be found within the spun-up sepals of yellow rattle.

Beat juniper during September for larvæ of *Thera simulata* and *Eupithecia arceuthata*.

Collect flowers or seed-heads of *Achillea millefolium* for *Eupithecia subfulvata*, golden-rod for those of *E. expallidata*, ragwort for *E. absynthiata*, *Knautia arvensis* for *E.* var. *knautiata*, &c.

The larva of *Eupithecia arceuthata* feeds on wild juniper from the middle of September to the middle of November ; it is seldom full-fed until the end of October (Crewe).

The pupa of *Eupithecia fraxinata* is to be found enclosed in a cocoon under moss, on the trunks of ash-trees, from September until May ; as many as fifteen were taken in 1898, in one small hollow in the trunk of a pollard ash (Robertson).

Eupithecia campanulata larvæ feed on the seeds of *Campanula trachelium*, and may also be found in gardens on cultivated varieties of Canterbury bells. These may be collected, according to the season, from late July until October.

You may still collect the flowers of golden-rod (*Solidago virgaurea*) in October ; put into a large band-box or breeding-cage, and you will breed *Eupithecia virgaureata* and *E. expallidata* the following year.

The eggs of *Cidaria immanata* must be kept out-of-doors all the winter if one wishes to ensure the hatching of them (Porritt).

The imagines of *Chesias spartiata* are best obtained by searching the broom bushes after dusk with a lantern ; the larvæ may be beaten in May and June.

LACHNEIDES.—The pupæ of *Poecilocampa populi* are to be found in October either under loose bark or spun-up among grass and rubbish at roots of oak, elm, willows, and poplar (Robertson) ; under bark of willow-trees seems to be a specially favourite situation.

SPHINGIDES.—The larvæ of *Acherontia atropos* are still to be collected, and the help of men who are potato-digging in fields or market-gardens should be requisitioned both for larvæ and pupæ ; when potato-leaves are not available for late larvæ—jasmine, tea-tree, *Lycium*, and other food-plants may be obtained. Pupæ rarely pass successfully through the winter, and most success is obtained by forcing them at a fairly high temperature ; once the forcing process has been begun, there should be no slackening, and the temperature should not be allowed to fall very appreciably even at night.

The best way to treat the larvæ of *Acherontia atropos* in confinement, is to procure an old tea-chest, or some other similarly sized box, which will give plenty of room to hold the potato haulm, without letting it touch the sides of the box ; then get some phials with large mouths, so as to hold the stalks of the haulm, keeping the stalks firm in the bottles by filling up the necks with moss, which also prevents any chance of the caterpillar tumbling into or becoming saturated with the water. Then fill the box with fresh earth up to the level of the tops of the necks of the bottles, so that the caterpillars can crawl from one batch of haulm to another, which they might not be able to do unless the earth be level with the top of the phials ; cover the whole chest with muslin or strong net, which will let the air through and not exclude the light, and place it somewhere in the open air where the sun is not too hot upon it, but where it gets free light. Thus you will have secured for them, as far as possible, the same surround- ings as though they were in a state of nature. The droppings should be cleaned from the top of the earth every day, and fresh potato-haulm given every night and morning whether they have consumed it all or not, as fresh food is one of their first requirements, and, if possible, the haulm should be always gathered from the same plot, or rather from the same sort of potato. There, then, the caterpillars remain until they reach their full-growth, but directly they begin to change colour and show a restless disposition, crawling round and round the bottom of the chest, they should at once be removed and each cater- pillar placed by itself in a rather large flower-pot filled with fresh damp earth, leaving only about an inch of room to spare between the surface of the soil and a heavy book placed upon the top of the pot. This answers well as they one and all bury directly. On each pot chalk the date of the disappearance of the caterpillar beneath the soil and let them remain untouched for a fortnight, after which time care- fully remove the earth and put the chrysalis into an incubator ; in no case should the pot be shaken or moved wherein the larva has buried. Moving them backwards and forwards causes the walls of the slender cell which the caterpillar forms under the earth to collapse, and the upper part of the pupa, which for several days remains very tender, becomes flattened out of shape (Morres).

The pupæ of *Smerinthus ocellatus, S. populi, S. tiliae, Sphinx ligustri, Choerocampa elpenor*, in fact all our strictly native species, are best kept throughout the winter and allowed to emerge naturally at the usual time of appearance in the imaginal state. Those pupæ resulting from immigrant parents—*Macroglossa stellatarum, Deilephila galii, Acherontia atropos, Sphinx convolvuli*, &c.—should be forced, as they rarely emerge satisfactorily if kept through the winter.

My incubator, or rather incubators, for I had so many pupæ that one would not hold them all, for forcing pupæ of *Acherontia atropos*, consisted in each case of an earthenware crock, some two and a half inches deep, and about twelve or thirteen inches wide. In this I first laid a thin layer of gravel to act as drainage, and on that, some two inches of moss, well pressed down, so as to afford a soft but firm foundation for the pupæ to rest on ; for if the moss be too loose, there is a danger of the moth on emergence getting entangled, and working downwards instead of upwards ; in which case the wings would never fully develop, although the moth in the end, might be able to extricate itself. On this firm substratum of moss, lay your pupæ, handling

G

them as tenderly as you can, and then cover them over with an inch or more of loose moss, so that they can easily make their way through it on emergence. Then, on the top of the moss, balance a tripod of fairly large sticks for the moths to climb up upon directly they change. This is most important, as the wings of the moth must hang down perpendicularly from the body until they are fully developed (which takes about an hour and a half to effect). Next cover the whole apparatus with a bell-glass some twelve or fourteen inches high and just wide enough to fit inside the rim of the crock, and your incubator is complete. Before, however, the moss is put into the crock, you should soak it in boiling water and wring it out as dry as possible, then when it has grown cool enough to lay the back of your hand comfortably upon, deposit the pupæ on the top of it; the pupæ will wriggle themselves into a suitable position. Then cover them over with more hot moss, place the glass over all, and there is nothing more to be done. The soaking of the moss, however, in hot water should be renewed about every third day, so as to keep the temperature within the bell-glass of a consistently moist, as well as warm, heat. In order to keep the incubators of an uniform warmth, place them one in each corner of the dining-room grate, inside the fender, and make it a rule never to sit down to any meal without turning the incubator round; you thus ensure that each side of the crock shall receive an equal share of warmth, and, as a precaution against over-heating, cover the whole apparatus with a thick piece of brown paper on the side nearest the fire. Having duly carried out these instructions, patience alone has to be exerted and the reward will come (Morres).

The full-fed larvæ of *Sphinx convolvuli, Daphnis nerii*, &c., pupate well between two or three layers of flannel arranged in the bottom of the box in which they are placed, instead of earth. They spin a slight cocoon, are in no risk from the falling in of the material normally forming the roof of their slight earth-cocoon, and so injuring the changing larva or newly-formed pupa (Morres).

SIMÆTHIDES.—*Xylopoda pariana*, in sunny weather, sometimes occurs freely on flowers of various *Compositae;* it also frequents thatch.

NOTODONTIDES.—In October and November work round the roots of willows, shaking the turfs and débris there collected, for the cocoons of *Leiocampa dictaea* and *Pterostoma palpina*, which will be found spun up among the roots or rubbish; often common on the small willows by the railway banks (Deal, Sandwich, &c.); also stated to prefer the side nearest the water when the bushes are near ditches, &c.

At the roots of oak one finds among hosts of common pupæ, those of several rare species such as *Drymonia trimacula (dodonea)*, *D. chaonia*, &c.

The spun-up leaves of poplar should be searched for the full-fed and pupating larvæ of *Clostera curtula*.

Larva-beating will give many good species, among others, it is not yet too late at the commencement of October, for belated examples of *Cerura bicuspis*.

Cocoons of *Cerura bicuspis* may be found until April, on birch and alder trunks; in shape they resemble a Brazil nut; rarely above 2ft. from the ground (Merrin).

The cocoons of *Cerura bifida* are placed irregularly, sometimes on the bark at the level of the soil, sometimes well up, usually well within the line of vision; sometimes the larvæ go from the food-plants (poplar and aspen) to an adjacent tree or fence to pupate (but always near the food-plant).

NOCTUIDES.—In October and November, the regular worker at ivy bloom, will possibly get among hosts of commoner species—*Epunda lutulenta*, *Xylina semibrunnea* and *Orrhodia erythrocephala*.

Ivy bloom during late October and November should be searched for *Dasycampa rubiginea*. (November 15th, 1891, four specimens boxed between 6 p.m. and 7.30 p.m. by Mr. Bankes, others November 16th, 1891, and November 3rd, 1883.)

The imagines of *Dasycampa rubiginea* are to be taken at sugar as well as at ivy blossom in October and November; they hybernate and occur again at sallows and at sugar in the spring.

During October and November the most profitable field-work in suitable weather is undoubtedly pupa-digging. Those who prosecute this mode of work regularly usually get very gratifying results. At the same time work well under moss on trunks for cocoons spun-up there. The ash-feeding larvæ are particularly fond of moss in which to spin up.

A small trowel is, on the whole, the best implement to use for pupa-digging, a pupa-digger having the disadvantage of maiming most of the pupæ it happens to touch, though it is most useful in pulling away the grass round trees. A bark ripper is objectionable on the ground of the havoc it makes with trees, yet it can be used to advantage for picking off small pieces of bark (Hunter).

Search the trunks of ash trees under moss, or under the stones on walls near, for the hard cocoons of *Craniophora ligustri*, also under the coping-stones of walls near privet hedges, where they are often placed.

In October, the imagines of *Calamia lutosa* are to be found by searching the reeds after dusk with a lantern. The species has often been found to be abundant, when searched for in this manner, in localities where its presence was not before suspected.

The middle of October is the time to sugar for the very rare *Leucania unipuncta*.

The larvæ of *Caradrina morpheus* are to be found commonly in October, on various low plants—wild hop, *Chenopodium*, &c., and in gardens on horseradish, &c.

The larvæ of *Cerigo matura* should be wintered on *Poa*, or coarse grasses, growing in a garden seed-pan filled with mould and protected by a muslin cover. It is advisable to examine the receptacle for predatory insects occasionally (Grapes).

Eggs of *Peridroma saucia* laid in autumn hatch in a short time, and the larvæ will feed on dock all the winter and produce imagines in May; the larvæ are very easily reared in large flower-pots filled with clean, loose sand, into which they can burrow and bury themselves by day.

Larvæ of *Noctua dahlii* and *N. umbrosa* will similarly feed throughout the winter in confinement, but are not usually so rapid in completing their metamorphoses.
G 2

The roots of old poplars are always a prolific hunting-ground for the collector of pupæ; *Smerinthus popnli*, *S. ocellatus*, *Taeniocampa populeti*, &c., may be found.

The larvæ of *Aplecta nebulosa* should be hybernated on *Rumex*, planted in a circular wooden vessel a foot or two in diameter, and covered with muslin tied tentwise to a central support (Grapes).

The larvæ of *Plusia chrysitis* may be wintered on *Lamium album* planted in a large size flower-pot, secured by muslin tied round the rim and to a central support. On the approach of winter the larvæ cease to feed, secrete themselves in the folds of the fallen leaves, where they remain throughout the winter; they recommence feeding very early in the spring (Grapes).

ARCTIIDES.—The larvæ of *Lithosia rubricollis* are in some years abundant, full-grown at the end of October, feeding on lichens and mosses growing on elm-trees and on walls near those trees.

NOVEMBER AND DECEMBER.

Field work is practically finished for the year and the work of differentiating, arranging and labelling the captures of the year, has to be done. It may, perhaps, be well to note in cleaning boxes, that naphthaline melts at a temperature of 80° C. and the fusion may readily be done in a test-tube over a spirit lamp. If a small brush be inserted in the test-tube and the liquid applied hot to the sides, corners or crevices of pocket-, postal-, relaxing-boxes, &c., it hardens immediately where applied, and is most efficient in keeping boxes so treated, free from mites. Pupa-digging and pupa-collecting are still by far the most profitable out-of-door work—the pupæ of *Smerinthus ocellatus, Pterostoma palpina, Notodonta ziczac* and *Tephrosia bistortata* may still be obtained at the foot of willows ; *Smerinthus populi, Leiocampa dictaea, Cymatophora ocularis, Cuspidia megacephala* and *Taeniocampa populeti* around poplars ; *Smerinthus tiliae* at elm, *Selenia bilunaria, S. lunaria, Phigalia pedaria, Amphidasys strataria, Eupithecia exiguata, Lophopteryx camelina, Notodonta trepida, Drymonia chaonia, D. trimacula (dodonea), Taeniocampa miniosa* and *T. munda* at roots of oak, *Moma orion* in crannies of oak bark, *Tephrosia consonaria, T. bistortata, T. punctulata, Lophopteryx carmelita, Leiocampa dictaeoides, Notodonta dromedarius* at foot of birch trees. Spun-together leaves under birch trees will give cocoons of a variety of species, *e.g., Endromis versicolora, Drepana falcataria, Cymatophora duplaris, C. fluctuosa, Asphalia flavicornis, Acronicta leporina,* &c.; trunks of trees must be searched for cocoons of Cerurids, Acronyctids, &c., and almost all *débris* under and around the base of trees is likely to prove productive. On warm evenings many Noctuid (and other) larvæ come out to feed ; they are especially numerous by sheltered woodsides, in wood-ridings and wood-clearings, by the edges of sandhills, at foot of hedges, &c. Other work has already been suggested, *viz.*, to collect roots of golden-rod, tansy, &c., many species being obtained abundantly by this means, the last-named contains larvæ of many species—*Dichrorampha tanacetana, D. alpinana, D. sequana, D. petiverana,* &c. Elisha says that during the winter months the salt-marshes at Southend and Canvey (and undoubtedly almost all our salt-marshes on other parts of the coast) give a variety of species, *e.g.*, larvæ of *Coleophora salinella* on *Atriplex portulacoides* and *Suaeda maritima*, those of *Coleophora tengstroemella* on *Chenopodium*, of *Coleophora artemisiella* on *Artemisia maritima*, of *Semasia rufillana* in seed-heads of *Daucus carota*, of *Conchylis francillonana* and *Argyrolepia zephyrana* in the stems of *D. carota*; whilst in the heads of teasel, larvæ of *Eupoecilia roseana* and *Penthina gentiana* are abundant; larvæ of *Conchylis dilucidana* occur in the stems of wild parsnip, larvæ of *Gymnancycla canella* on *Salsola kali*, cases of *Coleophora argentulella* are plentiful on seed-heads of yarrow, whilst the larvæ of *Dichrorampha petiverana* are in the roots. All these may be collected in the dead season and will in due course produce imagines. Flower-heads of yarrow, golden-rod, aster, umbellifers, &c., should be collected and tied up in linen bags, and roots of yarrow, mugwort, &c., should be planted in boxes, where the insect tenants will remain perfectly satisfied till spring arrives. Beating stacks and thatch will give many hybernating species— Depressarias, Gracilariids, &c.—in much better condition now than when they leave their winter-quarters; whilst the stems of *Umbelliferae* should be split open for larvæ, as those of many species retire therein to pupate after feeding up on other plants. It now becomes a matter for serious consideration as to how pupæ are best kept through the winter; the summarised opinions of a great number of our most successful breeders of lepidoptera will be found in *Ent. Record*, ix., pp. 324-326. As supplementary to these we quote the remarks of the Rev. Bernard Smith, one of the most successful breeders of lepidoptera, who writes (*Ent. Rec.*, i., p. 264) : "Larvæ frequently seek a sheltered place for pupation, and such species should be given the shelter from their various foes, which they require; a room with a fire is certainly the wrong place, but a place offering the degree of dampness possessed by a cellar or larder seems most suitable. As to the dampness of the earth in or on which pupæ are laid, a distinction should be made according to the season. In winter, when hard frost is to be looked for, the earth should be moderately dry, but when spring arrives, to put one's cages out in a warm drizzle occasionally is most desirable, and more especially with insects like *Stauropus fagi* and *Notodonta dromedarius*, is moisture required as the period of emergence approaches. Then, as to the quality of the earth in which the pupæ should be kept. Mix light garden mould with silver sand and cocoa-nut fibre, in equal parts, and press the compost down rather firmly. Moss is objectionable as a covering to pupæ as encouraging insect pests. When it is necessary to bake the compost in order to destroy these, boiling water should be first added, and then the whole sub-mitted to a heat rather above that of boiling water, in a kitchen oven, for twenty minutes. This does not make the earth caustic but kills all insect life. The cages in which to keep pupæ should be made of zinc (called "wash-ups" at the ironmongers), and they can be stored one upon another to save room. When pupæ have been unearthed, do not as a rule bury them in earth again, but a slight covering of cocoa-nut fibre should be used to prevent them drying up. Some pupæ require more moisture than others, among which may be mentioned *Petasia nubeculosa*, *Asteroscopus sphinx*, *Notodonta dromedarius*, *Craniophora ligustri*, *Arctomyscis* var. *myricae*, *Stauropus fagi*," &c. It may be well, perhaps, here to remind collectors that micro pupæ must be placed out-of-doors if one is to be successful in keeping them healthy through the winter. With the Nepticulids this is even more important than with the Lithocolletids, although better success is attained with the latter when they are kept outside than when they are kept indoors. Coleophorids, too, do much better out-of-doors, and, as a general rule, micro-lepidopterists will do well to keep their pupæ exposed to the weather during the winter rather than nurse them with the idea of taking care of them. Some of the macros do not seem to get on so well when thus exposed, and Richardson thinks that exposure is more important when larvæ spin their cocoons in autumn, but do not become pupæ until spring. Many lepidopterists, however, consider it advisable to force certain pupæ; success in this is not always due to a mere matter of temperature, and some pupæ cannot be induced to give up their imagines by an increased temperature; a large flower-pot half filled with loose friable mould or cocoa-nut fibre, well-drained, answers well; lay a thin layer of moss on the mould, on this place the pupæ, then another thin layer of moss, the pot being covered with a piece of calico, damp regularly every morning, and place on a kitchen mantel-shelf, and many pupæ of Sphingids, Noctuids, &c., will emerge throughout the winter. Holland advises lepidopterists not to fail to put on sugar every possible night during November for *Dasycampa rubiginea* and suggests that it would probably be heard of in many more districts, if collectors did not give up the sugaring too early. A good place to try is in or near an oak wood. If ivy be not near or convenient to search, you are recommended to cut off branches with bloom and place in favourable-looking spots, and it will double your chance; and even if you do not get *D. rubiginea* or *Orrhodia erythrocephala*, other species as *Xylina semibrunnea*, *X. socia* (*petrificata*), *X. rhizolitha*, *Calocampa vetusta*, *C. exoleta*, *Hoporina croceago* and *Cidaria psittacata* may be attracted, sometimes till the end of November, in fact, in many seasons, ivy bloom is not really at its best until November has arrived.

ADELIDES.—The larva of *Adela fibulella* lives in a flat case on leaves of *Veronica chamaedrys* (Merrin).

COLEOPHORIDES.—The larvæ of *Coleophora virgauraeella* should be collected in November, from the seed-heads of golden-rod. They are easy to rear if kept exposed to the influence of the weather.

In breeding *Coleophora therinella*, plant a thistle in a large flowerpot, low down in the pot, and leave the larvæ in this all the winter out-of-doors, with a piece of muslin tied over the top of the pot; the larvæ spin their cases on the sides of the pot and come out well, requiring no food in the spring.

TORTRICIDES.—The larva of *Coccyx strobilella* feeds in the cones of spruce-fir, which should be collected during the winter.

The first specimens of *Carpocapsa nimbana* obtained in this country were " bred by Lord Walsingham, from larvæ found hybernating in cocoons under moss on beech trunks, in Buckinghamshire " (Barrett).

The larvæ of *Semasia obscurana* (*gallicolana*) live through the winter in the oak apples or galls of *Cynips terminalis* fixed on the twigs of young oaks, preferring those of the preceding year's growth; the oak-apples should be collected in winter and early spring.

Stems of wild parsnip with conspicuous little nodules of white frass, showing just above the lower nodes of the stem, will contain larvæ of *Conchylis dilucidana* ; cut off the stems containing larvæ into convenient lengths and place in damp sand.

Machin used to breed *Conchylis dipoltana* freely from seed-heads of yarrow, collected at Southend in the autumn.

The pine-twigs bearing the large resinous cones in which are the larvæ of *Retinia resinana* should be placed out in the open, stuck into the ground, with no protection whatever during the winter. Bring them in, in April or May, place in very large flower-pots (better than a wooden box), tie a piece of calico over the top ; almost every larva will pupate successfully, and the imagines emerge in due course.

PYRALOIDES.—Throughout the winter, stacks of thatch, heath, turf, reed, and bracken, should be beaten for hybernating Depressarias and other moths.

By collecting the seed-heads of wild marjoram in November, and keeping them exposed to the weather during the winter, I bred a fine series of *Ptocheuusa subocellea* the following year (Elisha).

CYMATOPHORIDES.—Spun-together birch leaves should still be most carefully searched, for cocoons of *Cymatophora duplaris*, *C. fluctuosa*, *Asphalia flavicornis*, &c.

Almost all the *Cymatophora ocularis* in collections are bred from pupæ which are found spun-up at the roots of various kinds of poplars.

GEOMETRIDES.—Pupa-digging gives many useful species among the Geometrids—*Tephrosia bistortata* at the foot of willows, sallows, &c., *Tephrosia crepuscularia* at the foot of birch, oak, &c., *Amphidasys strataria*, *A. betularia*, *Nyssia hispidaria*, and many others should alos be obtained.

The hybernating larvæ of *Geometra vernaria* rest on the food-plant, *Clematis vitalba*, throughout the winter. They are then brown in colour, exactly like the dead stems, and drop at a touch, quite rigid. They should be supplied with fresh twigs, in confinement, as soon as the buds commence to swell in spring. After feeding a short time they moult and become green, and are then not easily to be found by searching, but must be beaten.

The pupæ of *Eurymene dolobraria* are to be found under moss on the bark of beech and oak trees, often directly under the edge of the moss, so that care should be taken in loosening the latter.

The male *Hybernia aurantiaria* sits on the leafless twigs of oak and birch, or hangs from the twigs and branches, or on the grass below, from the beginning until after the middle of November, after dark ; pairing takes place about 9.30 p.m. At the same time *Cheimatobia boreata* also may be taken, sitting on the birches, but this species pairs earlier in the evening, directly after it is dark. Both species are common in the larval stages in May.

The pupæ of *Hybernia leucophaearia* are usually found at tree-roots among tufts of grass.

Around the base of *Silene* and *Lychnis* plants, pupæ of *Eupithecia venosata*, &c., may be obtained.

The pupa of *Eupithecia dodoneata* may be found through the winter months under loose bark on hawthorn trees.

The slight earthen cocoons of *Eupithecia abbreviata* are sometimes spun at the root of a hawthorn tree, where they may be found during the winter.

LACHNEIDES.—The black cocoon of *Poecilocampa populi* is to be found in a variety of situations, more usually firmly fixed to the inside of a piece of loose bark, or spun up closely among decayed leaves and rubbish at or near the base of a tree. Should be obtained in early November.

Two large lamps about ten feet from the ground and about the same distance apart; about fifteen feet behind them hang a large dark sheet; *Poecilocampa populi* when attracted will rest quietly on the sheet (Holland); from middle of November into December, whenever the weather is mild and otherwise favourable.

About the middle of November look on the dark supports or framework of lamps for *Poecilocampa populi*.

The eggs of *Trichiura crataegi* may be found laid along hawthorn and sloe twigs, and those of *Poecilocampa populi* on hawthorn, oak, &c.

To hybernate *Macrothylacia rubi*, plant a root of heather out-of-doors; knock the bottom out of a cheese crate, put it round the heather and cover with perforated zinc. The larvæ come up in March and spin up without eating (Hewett).

Young larvæ of *Lasiocampa quercüs* may be fed on bramble to the so-called hybernating stage (middle of November), but they will, if possible, go on feeding throughout the winter in mild weather, and then bramble leaves picked from sheltered places should be given. In cold weather they remain on the stalks of the food-plant motionless, but on warmer days will often move sluggishly and nibble their food, until in April they commence feeding again in real earnest. Keep in roomy cage and they are easily reared.

SATURNIIDES.—The cocoons of *Saturnia pavonia* may be readily seen as soon as the heather has become thin on the moors, and in the hedges as soon as the latter have become leafless.

SPHINGIDES.—The pupæ of *Smerinthus tiliae* should be searched for at the roots of elm and lime. Hollows in an old trunk are a favourite place for the pupation of this species.

LYMANTRIIDES.—In November and on through the winter, the cocoons of *Dasychira pudibunda* are to be found among the loose rubbish collected about the roots of various trees or under loose bark. When fresh the cocoon is somewhat conspicuous, but afterwards is not at all easy to find.

The cocoons of *Demas coryli* may be found, sometimes plentifully, under moss, at the roots of beech trees, very rarely on the trunk. all through the winter.

NOTODONTIDES.—Cocoons of *Cerura furcula* should be sought on trunks of willow or sallow; those of *C. bifida* on trunks of poplar and aspen; those of *C. vinula* on sallow, willow and poplar.

Trunks of birch and alder must be carefully searched for cocoons of *Cerura bicuspis*. (The lepidopterist who works for this species should read Dr. Chapman's instructive article *Ent. Record*, vii., p. 73).

The cocoons of *Cerura furcula* may be found by searching the dead wood of sallows, either on the tree or littered round the roots. They are generally at the base of a divergent branch, but sometimes on the straight, and very often at the hollow where a branch has been pulled off (Hewett).

Pupa-diggers will find most profitable occupation this month. Among the Notodonts—*Pterostoma palpina, Leiocampa dictaea, Notodonta ziczac* at foot of willows, sallows, poplars ; *Drymonia trimacula (dodonea)* at oak, *Notodonta dromedarius, Leiocampa dictaeoides, Lophopteryx carmelita* at birch, &c.

November is the best month in which to find the delicate and weak cocoon of *Lophopteryx camelina* under moss, on trunks of oak, beech, elm, &c.

Pull off the bark from willow-trees and rough fences near, and you will frequently find the cocoons of *Pterostoma palpina*.

The cocoon of *Notodonta trepida* is to be found in November, at the roots of oak. The larva prefers a sandy or at least a friable soil in which to pupate.

The dry turfs collected in the corners at the roots of oak trees should be well searched, also the angles themselves, for cocoons of *Drymonia chaonia* and *D. trimacula (dodonea)*. Feel very carefully along the trunk, especially where the earth meets it, for adherent cocoons.

Spun-up poplar leaves should be collected for pupæ of *Clostera curtula*, and those of willows and sallows for *C. reclusa* ; the pupæ of *Stauropus fagi* are to be found in spun leaves of beech, birch, oak, &c.

About the middle of November, look on the dark support or framework of lamps for *Asteroscopus sphinx*. [See also note for attracting *Poecilocampa populi (anteà* p. 104). Imagines of *Asteroscopus sphinx* generally attracted at the same time, also rest quietly on the sheet.]

NOCTUIDES.—On warm evenings many larvæ come out to feed; they are especially numerous by sheltered woodsides, in wood-ridings, at the edge of sandhills, at foot of hedges, &c.

The pupæ of all the Tæniocampids may still be dug for in November. A detached tree usually (but by no means always) gives much better results than those growing close together.

The pupæ of *Taeniocampa populeti* are to be found deep down at the roots of various kinds of poplars.

To keep eggs through the winter in a natural state of dampness, " I get a clean smooth piece of that velvety moss which grows on old walls and cottage roofs, and, having carefully sprinkled the eggs over it, place it in a flower-pot, together with the food-plant ; as the eggs sink into the moss, they cannot get shifted about ; and, moreover, the moss will not entangle the legs of the newly-hatched larvæ. Of course the moss should be growing " (J. Hellins).

The eggs of *Cirrhoedia xerampelina* hatch from December to February (sometimes in November). The young larvæ feed only by night, on ash, though they will eat hawthorn and guelder rose. Sometimes they attain a fair size before the spring, whilst others remain quite small ; in the case of early-hatching larvæ the buds of hawthorn and guelder rose will be found useful as a substitute for ash (Merrin).

Around the base of *Silene* and *Lychnis* plants, pupæ of *Dianthoecia carpophaga, D. capsincola, D. conspersa,* &c., may be obtained.

Turn over stone heaps, &c., in November, for *Dasypolia templi,* a widely distributed insect in the north, but retiring in its habits. Where stone-walls are made from flat stones, the latter are often allowed to lie about in heaps, and by turning them over, one or two *D. templi* may be found in every heap (Robson).

Pull off bark from willow trees and rough fences near, for larvæ of *Apamea unanimis,* which hide away early; you will sometimes find pupæ of *Pterostoma palpina* when doing this.

The larvæ of *Noctua rhomboidea* make no objection to sliced potato, as food during winter. Many Agrotids will feed during the winter on carrots, buried in sand in a tub.

Sugar throughout November if the weather be suitable. You may get *Dasycampa rubiginea,* and, if not, such species as *Xylina semibrunnea, X. socia, X. rhizolitha, Calocampa vetusta, C. exoleta,* &c., will be attracted until the end of November.

At ivy *Dasycampa rubiginea* is usually found on blossoms partially hidden, and wants well searching for on bushes that cannot be beaten. It also appears to have a partiality for small detached bushes with but few heads of bloom (Mason).

At the latter end of November and beginning of December it was unusually mild, and the ivy bloom over early, so I sugared regularly for a few days, and obtained four *Dasycampa rubiginea,* two on November 28th, and two on December 1st (Mason).

PAPILIONIDES.—The eggs of *Thecla w-album* may be found on the twigs of elm, those of *T. pruni* on sloe, and those of *Zephyrus quercûs* on oak in their respective localities.

Leave the pupæ of *Papilio machaon* on the sticks on which they spin up; loose pupæ always emerge rather badly; the best success is obtained when the sticks bearing the pupæ are pinned up vertically. The same is true of the pupæ of *Euchloë cardamines.*

ADDENDA ET CORRIGENDA.

Page 39, line 43, for "*rosaeticolana*" read "*rosaecolana.*"
Page 40, line 24, read "flowers *and leaves* of "; line 26, read "heads *and leaves* of "; line 27, read "heads *and leaves* of."
Page 40, line 41, for "larvæ" read "imagines."
Page 41, line 39, for "larvæ" read "imago."

Modern Works on British Entomology.

Our entomological magazines are largely filled with repetitions of already well-known facts made by lepidopterists who still appear to consider that books published almost, or quite, half-a-century ago contain all the up-to-date information obtainable on the subjects they discuss. For advanced work of the most recent kind, the more scientific and educated lepidopterist should use the following works as his text-books.

THE NATURAL HISTORY OF THE
BRITISH LEPIDOPTERA
By J. W. TUTT.
Vols. I and II.

Demy 8vo., strongly bound in Cloth.

Vol. I consisting of 560 pp.   Vol. II of 584 pp.

Price £1 each Volume (net).

This work is the most important that has ever been offered to working and scientific lepidopterists. It contains series of exhaustive monographs which no working ento-mologist should be without, and if he be dealing with the same superfamilies he will find in these volumes a mass of detailed information such as never has been got together before on all the branches relating to the species treated. At the same time, the discussion of general biological problems and the great amount of detail relating to such problems give the books a value for a class of workers quite apart from the British and Continental lepidopterists for whom they were primarily intended.

The British Noctuæ and their Varieties.

By J. W. TUTT, F.E.S.

(Complete in 4 volumes.   Price 7/- per vol.).

The four volumes comprise the most complete text-book ever issued on the NOCTUIDES. It contains critical notes on the synonymy, the original type descriptions (or descriptions of the original figures) of every British species, the type descriptions of all known varieties of each British species, tabulated diagnoses and short descriptions of the various phases of variation of the more polymorphic species; all the data known concerning the rare and reputed British species. Complete notes on the lines of development of the general variation observed in the various families and genera. The geographical range of the various species and their varieties, as well as special notes by lepidopterists who have paid particular attention to certain species.

The first subscription list comprised some 200 of our leading British lepidopterists, and up to the present time some 500 complete sets of the work have been sold. The treatise is invaluable to all working collectors who want the latest information on this group, and contains large quantities of material collected from foreign magazines and the works of old British authors, arranged in connection with each species, and not to be found in any other published work.

Monograph of the Pterophorina.

(Demy 8vo., 161 pp., bound in Cloth. Price 5/-).

This book contains an introductory chapter on "Collecting," "Killing," and "Setting" the Pterophorina, a table giving details of each species—Times of appearance of larva, of pupa and of imago, food-plants, mode of pupation, and a complete account (so far as is known) of every British species, under the headings of "Synonymy," "Imago," "Variation," "Ovum." "Larva," "Food-plants," "Pupa," "Habitat," and "Distribution." It is much the most complete and trustworthy account of this interesting group of Lepidoptera that has ever been published.

The Entomologist's Record and Journal of Variation.

EDITED BY

J. W. TUTT, F.E.S.,

ASSISTED BY

Professor T. HUDSON BEARE, B.Sc., F.E.S., F.R.S.E. ; M. BURR, F.Z.S., F.L.S., F.E.S. ; Dr. T. A. CHAPMAN, F.Z.S., F.E.S. ; H. St. J. K. DONISTHORPE, F.Z.S., F.E.S. ; L. B. PROUT, F.E.S.

An illustrated magazine of general Entomology. Published monthly. Subscriptions, 7s. per annum, including all Double Numbers and Special Index (with every reference carefully indexed). This magazine has taken quite a leading position among Entomologists, and has a wide circulation among foreign specialists. The leading articles are written by first-class entomologists, and, besides these, each number contains many shorter notes. These are classified as "Scientific Notes and Observations" "Variation," "Notes on Collecting," "Notes on Life-histories, Larvæ, &c.," "Practical Hints," "Current Notes," &c. The articles and notes relating to Coleoptera are under the special charge of Mr. H. St. J. K. Donisthorpe and Professor Beare, and those relating to Orthoptera under the care of Mr. M. Burr.

A complete set of the Back Volumes of this important Magazine (vols. i.-xiii.) forms one of the very best reference libraries for the use of the working entomologist. The leading articles deal with the most recent knowledge of the subjects treated, whilst the shorter notes contain a fund of information which cannot possibly be obtained elsewhere. These volumes can be obtained from Mr. H. E. PAGE, "Bertrose," Gellatly Road, Hatcham, S.E. Price 7s. 6d. per volume.

Important Notice to Librarians and others.

Melanism and Melanochroism in British Lepidoptera.

By J. W. TUTT, F.E.S.

(Demy 8vo., bound in Cloth).

The last few Copies of this important and interesting brochure are now on sale at 5/- per copy. The work deals exhaustively with all the views brought forward by scientists to account for the forms of Melanism and Melanochroism; contains full data respecting the distribution of melanic forms in Britain, and theories to account for their origin; the special value of "natural selection." "environment," "heredity," "disease," "temperature," etc., in particular cases. Lord Walsingham, in his Presidential address to the Fellows of the Entomological Society of London, says, "An especially interesting line of enquiry as connected with the use and value of colour in insects, is that which has been followed up in Mr. TUTT's series of papers on "Melanism and Melanochroism."

Rambles in Alpine Valleys.

Price 3/6.

Bound in Cloth, with Map and Photographs of District.

This book contains a series of essays dealing with the colours of insects, and suggestions as to the relation in past time between the Alpine and British fauna. Many new facts are brought forward, &c. entomological problems discussed from various standpoints.

Woodside, Burnside, Hillside and Marsh.

Price 2/6. 242 pp. and 103 Woodcuts and full-page Illustrations.

A series of Entomological Essays based on the insects to be found in various well-known entomological localities. Deals with a great many of the more philosophical subjects now before the entomological public.

Notes on the Zygænidæ.

Price 1/-.

A few copies. These papers contain a full and scientific account of the synonymy, variation, distribution, and habits of several species common to Britain and the Alps. There is also a description of a new species hitherto confounded as a variety of *Zygaena trifolii*.

To be obtained from H. E. PAGE, "Bertrose," Gellatly Road, Hatcham, S.E.

PRACTICAL HINTS

FOR THE

FIELD LEPIDOPTERIST

BY

J. W. TUTT, F.E.S.

Editor of *The Entomologist's Record and Journal of Variation ;*
Author of *A Natural History of the British Lepidoptera, British Noctuae
and their Varieties, British Butterflies, British Moths,* &c.

II

Price 6s. net (interleaved).

October, 1902.

LONDON:
ELLIOT STOCK, 62, PATERNOSTER ROW, E.C.
BERLIN:
R. FRIEDLÄNDER & SOHN, 11, CARLSTRASSE, N.W.

PREFACE.

The success of Part I of *Practical Hints for the Field Lepidopterist*, published some eighteen months since, is the only excuse for publishing a second Part. As the preliminary remarks for each month were in Part I especially directed to the possibilities of field work, so in Part II additional information on the subject has been given, and, in addition, extended notes on the various methods of—Rearing larvæ in confinement, Sleeving, Breeding-cages of several kinds, Treatment of young larvæ, Hybernating larvæ, Food, Feeding larvæ, Special Treatment of internal feeding larvæ; Obtaining eggs in confinement from butterflies and moths, Keeping eggs through winter, Management of eggs before hatching; Special treatment of pupæ, Keeping pupæ through the winter, Best modes of keeping underground pupæ, Forcing pupæ, Treatment of Sphingid pupæ ; Sugaring (in all its various forms), Sallowing, Beating, Assembling (as practised with species of different habits), Light, Thatchworking, Fen and Marsh work, and a host of similar important topics to the entomologist. The result, as may be seen, has been to make a book considerably larger than Part I.

The collection of such a mass of accurate and detailed facts is necessarily difficult. No one man, however assiduous a collector, could possibly amass the exact information here given, on such a large number of species. Not only is there the quarter of a century's experience of the author in the field brought into requisition, but the published and unpublished observations of a large number of our foremost field workers have been included and freely drawn upon. Besides the obligations we owe to so many lepidopterists for hints here included, we have especially to thank Messrs. Bower, Dollman and Montgomery, for a very direct share in adding to or overhauling the contents for publication.

We would again reiterate what we insisted on in the publication of Part I, *viz.*, that these hints have been put together in the hope that the young scientist will be enabled rapidly to make a sufficiently large number of observations to aid him to make the necessary generalisations for his work to have a scientific value. For the lepidopterist who uses these notes only to enlarge his collection and to help to exterminate our fauna, and whose work has no scientific basis whatever, we can only express a sorrow bordering on contempt.

The following ladies and gentlemen share with the author the responsibility of publication :—

Rt. Hon. Earl Waldegrave (2), Sir J. T. D. Llewelyn, Bart., M.P. (2) ; Messrs. F. J. Hanbury (2), T. A. Lofthouse (2), A. Sich (2) E. F. Studd (2). Miss E. M. Alderson, Miss N. M. Jermain, Miss M. Prideaux ; Hon. N. Charles Rothschild ; Revs. C. Chichester, F. B. Cowl, T. B. Eddrup, E. Grose Hodge, A. M. Moss, G. H. Raynor, J. E. Tarbat ; Major F. D. Bland; Capt. E. W. Brown ; Messrs. H. C. Arbuthnott, F. R. Atkinson, E. A. Atmore, E. R. Bankes, C. Bartlett, H. Beadnell, E E. Bentall, E. R. Bush, E. D. Bostock, T. Carlyon, N. Chamberlain, A. H. Clarke, C. W. Colthrup, J. W. Corder, J. Cotton, G. O. Day, W. Daws, J. C. Dollman, J. Drysdale, H. M. Edelsten, F. Emsley, T. B. Fletcher, J. E. Gardner, P. T. Gardner, G. R. Garland, A. T. Goodson, F. W. Goodson, E. D. Green, T. Green, G. C. Griffiths, G. D. Hancock, A. Harrison, O. Harrison, W. Hewett, C. M. Holt, M. F. Hopson, J. R. Johnson, B. Jones, W. D. Kearfott, H. Main, J. A. Malcolm, G. W. Mason, F. Merrifield, H. Murray, John F. Musham, L. Naniot, L. W. Newman, F. Norgate, J. Ovenden, W. T. Page, R. C. Paton, D. H. Pearson, J. Peed, E. Petrie, J. R. Pickin, John Porter, R. W. Proud, E. Ransom, W. Raeper, C. Rea, W. Rich, P. W. Ridley, E. A. Rogers, G. B. Routledge, A. Russell, V. E. Shaw, R. S. Smallman, A. M. Smallpiece, W. Smart, G. B. Smith, W. Hawker-Smith, A. M. Swain, R. Tait, jun., E. H. Thornhill, T. Tunstall, H. J. Turner, F. Wallace, C. J. Watkins, A. J. Willsdon.

PRACTICAL HINTS

FOR THE FIELD LEPIDOPTERIST.

JANUARY, FEBRUARY AND MARCH.

Following on our remarks in Part i., pp. 1—2, as to the work to be done in the winter months, there is not, perhaps, much to add, but some details may be worthy of notice. *Poecilocampa populi* is still to be found on gas-lamps during the earliest days of the new year. Captures of this species have been recorded when the roads have been sheets of ice, the atmosphere foggy, and the thermometer at freezing-point, but with the wind, in spite of the temperature, westerly. As soon as a few mild days set in, however early in the year it may be, the Hyberniids will be on the move. That park-fences pay to work at such times is proverbial, and, through January and February, continuous searching on satisfactory days usually results in a fair measure of success. At Calcot, in 1890, fresh specimens of *Hybernia leucophaearia* were found abundantly from January 16th to March 6th in such a situation. The males usually emerge about 11 a.m., and many may be found drying their wings about this time. Unsatisfactory weather conditions frequently break the continuity of emergence, but with mild weather success is certain. The females are rarely to be obtained on fences, but must be searched for on the tree-trunks in the neighbourhood, old oaks and elms, perhaps, producing most specimens. On the same fences the males of *Phigalia pedaria* are usually abundant, and females of this species must also be obtained on the oak-trees about. Occasionally in January and early February, when working along hedgesides or woodsides for the Hybernias, with a lamp, on mild evenings, one comes across *P. pedaria* males flying in large numbers in a limited space, probably attracted by a virgin ♀. but the ♀ is rarely found. During the winter, sallow and alder are cut down in the Hereford district in large numbers for hop-poles (in Kent ash is generally chosen), and these are frequently found to be tunnelled and inhabited by the larvæ of *Trochilium crabroniforme (bembeciforme)*, sometimes two or even three larvæ in one pole. As the season goes on, larvæ become plentiful on mild evenings, coming out soon after dark to feed, and those who live in the neighbourhood of the localities for rare and local species may work out their life-histories and obtain examples for their friends. One of our most local coast species is *Heliophobus hispidus*, the larvæ of which may be found plentifully at Portland, Torquay, and a few other localities, from the middle of February to the end of

April. Searching with a lantern should commence about the middle of February, on the grass, in the localities the species affects. The larvæ are very lethargic, not falling off their foodplant when touched, and requiring considerable persuasion to make them quit their hold of the blade of grass on which they may be feeding, and it is best to break (or cut) off the grass-blade on which they may be. They are generally found at the bottom of the slopes, on the tufts of grass overhanging places where the soil has crumbled away owing to the action of the weather. In confinement they bury by day, coming up to feed after dark on grass, which should be obtained fresh every day, or a growing sod supplied, and they are generally fullfed by the middle of March. Where a sod is given them, freshly cut grass should be placed on top when they have cleared the former, as breaking up the sod to find the larvæ is dangerous work, and usually results in the injury of some of them. A similarly interesting species, taken in the same localities as *H. hispidus*, but with a much wider distribution all round the coast, is *Epunda lichenea*, the larvæ of which occur on a variety of plants from mid-February to the end of April. They are usually taken, however, on the top of grass-stems, generally in little colonies of five or six, in sheltered spots, at the foot of banks, beside large boulders, or in crevices of a rock. They may be fed up on chickweed, groundsel, &c., and mature rapidly in April, all going down during the latter half of this month and beginning of May. The larvæ of this species will fall off the grass (or any other plant on which they may be resting) at the slightest touch, and the lid of a larva-tin should be held underneath to catch them. The lepidopterist will see from these instances that a careful study of locality, time and place, as well as weather, is absolutely necessary to ensure success with any local species, and as there is scarcely any locality in which some really good local insects do not occur, it should be a part of the collector's work to find out all about these particular species, and to publish a full account of their life-histories and habits. As spring goes on vegetation progresses, and with it the number of insects obtainable, and we have already shown (pt. i., pp. 9—10) that, in March, the condition of the vegetation influences largely the number of species that may be obtained and the work that may be attempted. So great is this difference that, in some seasons, sallowing may be well on by the middle of March and over before the end of the month, whilst, in other seasons, the sallows are not out until April is well in, and the moths are usually correspondingly late. Sallowing is the first real hard work of the lepidopterist in spring. The condition of the bloom necessarily determines the right time to commence work, and one knows that, with the first fully opened catkins, the earliest Tæniocampids will appear and the hybernating imagines will commence to feed again. Glass-topped or glass-bottomed boxes, a good lantern, and a large, strong umbrella are the only impedimenta wanted. A good knowledge of the locality to be worked is essential. In some places the sallow bushes are fairly easy to get at, in others they are most difficult. With small and fairly large bushes the handle of the umbrella is thrust carefully into a bush, the most important branches are hooked, and a sudden and dexterous twist of the handle brings down the moths (and over-ripe catkins). On a good evening the captures will include *Pachnobia leucographa*, *P. rubricosa*, *Taeniocampa populeti*, *T. incerta (instabilis)*, *T. munda*, *T. gothica*, *T. stabilis*, *T. miniosa*, *T. gracilis*, *Panolis piniperda*, *Hoporina croceago*, *Orrhodia vaccinii*, *O. ligula*, *Xylina socia*, *X. semibrunnea*, *X. ornithopus*, *Gonoptera libatrix*, *Hypena rostralis*, *Eupithecia abbreviata*, *Lobophora lobulata*, *Tephrosia bistortata* and sundry hybernating Depressarias. The visits of *Dasycampa rubiginea* are infrequent. The species attracted to sallow-bloom occur in varying proportions in different years, and one sometimes gets a " *leucographa* " year, a " *miniosa*" year, a "*populeti*" year, and so on. The green-grey form of *P. piniperda*, the black aberration of *T. populeti*, and the dark red forms of *T. gracilis* should always be carefully looked for. *Hoporina croceago* must be saved for eggs; both sexes should be placed together in a large glass jar and supplied with a few twigs, bearing dead leaves of oak, and fed with moistened sugar, when a good supply of eggs will be obtained. *Xylina semibrunnea* and *Dasycampa rubiginea* should also be retained for eggs, whilst a few female *Pachnobia leucographa*, imprisoned with a supply of leaves and old flower-stalks of plantain, and treated similarly to the mode suggested for *H. croceago*, will usually give an abundance of ova. In some localities where the trees are large, climbing has to be resorted to for shaking the branches, and a large sheet has to be spread beneath the tree, whilst in Bishop Wood, near Selby, the lepidopterists cut the branches of sallow-bloom from the more or less inaccessible bushes, and hang them on the branches of other trees in the rides of the wood; *T. leucographa* and *T. populeti* are especially abundant in this wood, and *T. opima* is sometimes taken, although it is at the catkins of the dwarf sallows growing on the Lancashire, Cheshire, and other coast sandhills that the latter species is usually most abundant. Here, searching with back bent double is necessary, and large numbers of specimens of this species are frequently captured. It sometimes happens that, in selecting a district for sallowing, one is chosen in which there is a great abundance of bushes; often, however, such localities are less remunerative than those in which the sallows are more scattered and the moths more concentrated on the available bushes. One finds the female catkins just as attractive to the moths as the male catkins; the former, however, are less conspicuous in the dark, and are liable to be passed over. Whilst waiting for dusk when sallowing, *Lobophora lobulata* may sometimes be found on tree-trunks, or *Brephos notha* (often paired) knocked from the tall aspen trees by a smart kick, a net being held beneath to catch the falling moths. *Tephrosia bistortata*, too, is worth searching for, and sometimes *Eupithecia abbreviata* is on the wing. As bearing on the methods of taking and rearing *Dasycampa rubiginea*, Robertson details *(Ent. Record*, i., p. 107) how, in the spring of 1888, he took 7 examples at sallow-bloom and sugar, and 4 others in 1889, obtaining ova both years, by keeping the imagines in a bandbox covered with muslin, and feeding them on honey mixed with a few drops of sherry. Placed inside the box were twigs of apple, which had been scored with a knife, the females appearing to like a niche in which to oviposit, the eggs being laid singly between the middle of April and the middle of May. The larvæ were fed on apple, and were reared first in a glass filter, and afterwards sleeved on an apple-tree. Hewett succeeded equally well by feeding the larvæ exclusively on dandelion, which they will neglect, however, for apple. The localities for this species are often exceedingly localised, confined, maybe, to a few hundred square yards, and not to be found in all the district around. Before March is out the birches must be searched for *Asphalia flavicornis*, which, however, vary considerably in the time of their appearance from year to year. With the advent of sallow-bloom, too, other flowers are becoming attractive, and the flowers of *Berberis* will produce *Tephrosia bistortata*, *Hybernia marginaria (progemmaria)*, *Anticlea derivata* and *Panolis piniperda*. Sugaring should be attempted side by side with sallowing. It some-times happens that when sallows are unattractive moths may be taken at sugar, and *vice versa*. We are indebted to Alderson for some excellent notes on "sugaring" *(Ent. Rec.*, i., p. 140). He states that he always uses coarse brown sugar, when procurable, but has found old black treacle quite as effective. Blackberries, gathered when ripe, and boiled down with sugar, form an excellent substitute, and is especially attractive to the Xanthiids. When laying the sugar on, it is advisable to add a little rum (methylated spirit, jargonelle, and other substitutes are also used by various collectors), every dozen trees or so, rather than to mix the whole previous to starting. It is also desirable to commence in sufficient time to allow finishing the last tree on the round before dusk, the first hour being as a rule the most productive. A long thin line (the width of the brush) almost to the foot of the tree is better than a small patch, one advantage of the former method being that the insects are not clustered so closely, and fewer escape, especially if one commences by throwing the light first at the bottom of the tree. On a windy night the majority of the moths are frequently on the lowest part of the sugar, the upper portion being almost deserted, whilst those Noctuids that fly nearest the ground are much more likely to be attracted. It is also always worth while to sugar a quantity of small-limbed trees, as these frequently pay well when the bare trunks of large trees are little patronised. It is advisable, too, to shake the brush over low-growing shrubs, and also to carefully let a drop of sugar fall on the centre of thistle and other composite flowers, or to sprinkle a little over the flower-heads of *Eupatorium*. Be careful also to keep at work on the same ground, the moths appear to congregate more on a round that is continually worked, and the trees kept constantly sugared. Changing ground does not usually prove particularly productive, especially for the first few nights, besides, the fact that more sugar is required on a new round is important, an old round wanting very little to freshen it up. Boxes should always be ready when renewing old sugar, as many species, early on the wing, will be found already at the bait, and, the spirit having evaporated from the latter, these early-comers are usually exceedingly wary. Pupæ are sure to occupy a fair portion of the lepidopterist's time during the slack season of the year. The various modes of pupa-collecting must be worked for all they are worth in the autumn, but their care during the winter is a matter of continuous and anxious thought, and failure is so easy. It may be here noted that Gordon suggests searching the tops of moss-covered stumps of trees that have been cut down, as favourite resorts for certain species of larvæ when about to pupate, and notes that they generally make their puparia between the moss and the flat surface of the wood. The moss-covering is readily removed by the hand, and when dense and thick its under-surface should be carefully examined, as well as the part of the stump from which it was raised, as it is apt then to carry pupæ with it. Larvæ are also found hiding in these localities, the place having been possibly selected as being higher and drier than the surface of the ground. In order to get as much work as possible done in the slack season of the year, many lepidopterists, after exposing their pupæ to the temperature of an outhouse until January, bring them indoors and place them in a warm position in order to force them, *i.e.*, induce them to emerge before their normal time. One of the best rough-and-ready methods of forcing pupæ is to get a large flower-pot, half-filled˙ with well-baked mould, sand, or cocoanut fibre, drain well at bottom, make surface fairly hard, strew a thin layer of moss on the mould; on this moss place the pupæ, cover them with a thin layer of moss, tie a piece of calico over the top, and place on a kitchen mantel-shelf. Every morning sprinkle the surface of the top layer of moss well with water, and a very fair percentage of species will be persuaded to move. Above all things, pupæ should not be placed on loose mould, so that little pieces of earth can get into the movable incisions, for if this happens, death usually ensues Heat without moisture appears detrimental to most pupæ, yet sometimes dry heat would appear not to be injurious, and the following notes on forcing pupæ of *Notodonta trepida* is very puzzling. Nineteen pupæ of this species (that had all changed to pupæ on the same day in 1888) were placed in a small forcing-house in one of the hottest parts of a hothouse in January, 1889. Eight of these produced imagines in May, but 11 remained in the pupal stage. During the summer of 1889 they were in the forcing-house in the driest possible soil. Many days in the summer the heat was intense, and no more *N. trepida* were expected to appear; however, on March 17th, 1890, two fine males emerged, and before the end of the month five others—all fine specimens. Merriman notes that to force Sphingid pupæ, he makes use of a tortoise-stove in a greenhouse. On the top of the stove four bricks are placed, two on each side (one above the other), on these a square foot seed-pan, the finish being a good-sized fish-bait tin; an aperture left between the bricks allows the hot air to circulate freely all round; in the tin moss is placed, moistened with warm water sprinkled over the pupæ every other day, temperature kept at about 90°F.—100°F. On November 17th, 1885, 6 pupæ of *Manduca atropos*, 6 of *Eumorpha elpenor*, 8 *Theretra porcellus*, 12 *Celerio (Deilephila) euphorbiae*, 4 *Smerinthus ocellata* and 3 *Sphinx ligustri* were placed in one. By January 20th, 6 *E. elpenor*, 9 *C. euphorbiae*, 2 *T. porcellus* and 1 *M. atropos* had emerged—the other pupæ of *M. atropos* having died owing to some mismanagement, the remaining pupæ, however, being healthy; pupæ of *Hemaris fuciformis* introduced on December 21st, 1885, had emerged by the same date. A method practised by Elisha for forcing Micro-lepidoptera is worth noticing. The apparatus consists simply of a box, 10 inches square and 6 inches deep, open at the top and lined with thin zinc; a zinc tray is made to fit the top, 1½ inch deep, to contain damp sand; underneath on the tray is soldered a much smaller tray, 1 inch deep, which forms the boiler; a short piece of pipe is soldered in the upper tray through which to fill the boiler; the tray is then put on the box, the edges being made larger prevent it falling through, and underneath is placed a spirit lamp or gas-jet, the flame being barely a quarter of an inch long, which is quite sufficient to give a great and regular heat; a square hole is cut in the side of the box in front to put the hand through and to regulate the light, and in the opposite side, just underneath the tray, a few holes are drilled in the box for ventilation, or the light will go out. Above the tray and resting on the damp sand is a square zinc glazed case 8 inches high; the top square of glass is loose to lift out, for placing the bottles or glass jars that contain the pupæ in, and also to regulate the heat. When all is ready, fill the boiler nearly to the top with water, then fill the tray with damp sand to give a moist atmosphere, and put on the glazed case, after which put in the glass jars containing the pupæ, and, in the centre, suspend a small thermometer, and light the gas or lamp, which can be regulated with ease, to keep the heat between 60° and 70°F.; it is then no further trouble, and will repay any one for getting it in order. During February and March one may thus without difficulty breed long series of the following Lithocolletids, often more than usually large specimens—*Lithocelletis spinicolella, L. faginella, L. corylella, L. salicicolella, L. carpinicolella, L. tenella, L. ulmifoliella, L. tristrigella, L. emberizaepennella, L. nicelliella, L. schreberella, L. lantanella,* &c. *L. lantanella* being a hybernating larva, might raise doubts, but after being forced for five or six days they will be found to have pupated, and in the following week the imagines will appear. Of the Cosmopterygids, one can force *Cosmopteryx lienigiella* and *C. druriella.* Many of the Nepticulids that may be thus reared appear three months ahead of their usual time. Tortricids also are easily reared, indeed all the species that have been experimented upon, and that appear normally in May and June, have emerged without trouble, and thus the work of these latter months may be eased considerably, and the setting done at a slack season of the year, leaving one quite free for collecting at the busiest period of the year. One has to be especially careful with regard to pupæ that have gone over their usual allotted time, especially if they have come from a northern locality. It would appear a very common habit for many species that are quite regular in their appearance in our southern counties to go over two or more winters in more exposed and colder habitats, and the progeny of moths with this habit maintain it largely, even if brought south and reared under what appear to us to be quite improved conditions; thus, in 1888, a number of Scotch pupæ of *Eupithecia togata* produced many imagines in April, 1889 several, however, did not come out until March and April, 1890, although the pupæ had been kept indoors for over 18 months. Scotch pupæ of *Saturnia pavonia,* and Shetland pupæ of *Eupithecia venosata, Emmelesia albulata* and *Heydenia auromaculata* frequently go over two or three years, and one suspects that many pupæ in Scotland must normally pass two years in the pupal stage. At any rate, lepidopterists should be particularly careful not to throw away pupæ which do not emerge the first year at the normal time.

UNCLASSIFIED.—To force *Macrothylacia rubi* and *Spilosoma fuliginosa*, place the larvæ (after having been exposed to the weather for part of winter) in box with damp moss, and the box in a warm greenhouse or kitchen; keep temperature up to about 80° F. ; the larvæ will spin up almost at once without further feeding, and the imagines will emerge in from two to four weeks (Moss).

From the catkins of a fine female plant of *Salix caprea* I took a large number of larvæ of *Eupithecia tenuiata* at Llanferras, at end of March, and with them plenty of young larvæ of *Grapholitha nisana*, and, of course, no end of eggs and young larvæ of *Citria flavago* and *C. fulvago* (Gregson).

A large American cheese box with the top knocked out, leaving a ring by means of which a piece of leno can be stretched over the top of the box makes a very good breeding-cage (Bate).

TINEIDES.—In January and February work beech-woods for the large fungus *E. boletus* which is generally found growing on beech-trees. This fungus is affected by the larvæ of *Scardia boleti*. Do not, however, take the fungus now, but note it to be collected during the first week of May, as then you will be pretty sure to breed the species (Elisha).

ELACHISTIDES.—Coarse grasses growing by hedgesides sometimes show very evident signs of mining ; these mines usually contain the dark green larvæ of *Elachista megerlella* (Elisha).

The long broad mines that are to be seen running nearly the whole length of the long fresh-looking grass blades on hedge-banks in March will, if held against the light, be found to contain the pale yellowish larva of *Elachista rufocinerea* which feeds on the parenchyma between the upper and lower cuticle (Elisha).

TISCHERIIDES.—The larvæ of *Tischeria emyella* are to be obtained in March in the leaves of brambles, sheltered under hedges.

LITHOCOLLETIDES.—Under oak-trees many mined leaves containing pupæ of the rarer Lithocolletids are to be found in January and February. Many of the rarer ones feed in the higher parts of the tree, and can only be obtained when the leaves have fallen (Elisha).

COLEOPHORIDES.—The cases of *Coleophora virgauraeella* were frequently seen at Llanferras, at end of March, attached to sticks and dead plant-stems, whilst I was searching among the wild marjoram and golden-rod, gröwing together, for the hybernating larvæ of *Leioptilus osteodactylus* (Gregson).

In March, boggy pieces of ground, where rushes abound, should be worked for the whitish cases of *Coleophora caespititiella* which are to be found mostly between the seeds at this time.

The blotches made by the larvæ of *Coleophora discordella* on the leaves of bird's-foot-trefoil *(Lotus corniculatus)* are conspicuous in March and tell very plainly the whereabouts of the larvæ (Elisha).

Towards the end of February examine carefully the leaves of

ground ivy, *Glechoma hederacea*, for leaves showing whitish blotches; in this way the larvæ of *Coleophora albitarsella* may be found (Elisha). The earliest leaves of the sallows are blotched in March by the larvæ of *Coleophora viminetella* which have already begun to feed and whose cases are not difficult to find.

Where *Genista anglica* grows freely traces of the work of *Coleophora genistaecolella* are generally evident in March but they should be left till later for they are difficult to get through if taken too small.

The leaves of the lower branches of larch trees are often bleached in March by the larvæ of *Coleophora laricella*, whose small whitish cases are readily found on the underside of the leaves.

On the underside of the leaves of *Stellaria holostea*, the slender cases of *Coleophora solitariella* will be found, the larvæ having recommenced feeding after their long hybernation (Elisha).

Old plants of *Statice limonium* brought from the marshes near Southend yielded between June 30th and July 18th more than 60 examples of *Goniodoma limoniella* (Whittle).

GRACILARIIDES.—Young larvæ of *Gracilaria tringipennella* were indicating their presence on *Plantago lanceolata* leaves, in warm corners, at Llanferras, at end of March (Gregson).

TORTRICIDES.—*Leptogramma scotana* is to be found on birch trunks in spring; by sleeving the ♀ on birch trees long series may be bred. It is a mistake to let the larvæ spin up in the muslin, as they are long in turning and ichneumons sting them through the meshes.

In early March among oak scrub in almost all woods, *Tortricodes hyemana* is to be seen flying literally in hundreds in the sunshine; in cloudy weather they scarcely move, settling at once on the twigs or herbage and are then rarely to be disturbed.

Fill a bag with catkins and terminal shoots of *Alnus glutinosa*, in and on which larvæ of *Grapholitha penkleriana* are feeding.

In January and February collect as many oak-galls and oak-apples as possible and store them in a large roomy box. From them you will breed magnificent specimens of *Ephippiphora obscurana, Heusimene fimbriana, Coccyx splendidulana, Œcophora lunarella,* and many other species (Elisha).

The aborted buds that are conspicuous in March on the topmost shoots and main branches of young fir-trees indicate the position of the larvæ of *Retinia turionana* inside the stem.

Living females of *Retinia duplana* from Scotland, 1892, sleeved on a small Scotch fir at Worthing, laid eggs, the larvæ fed in the young shoots, the imagines emerging indoors during March, 1893 (Fletcher).

To winter the resinous cocoons of *Retinia resinana*, put some sand in a square earthenware pan, with holes in bottom (as used by gardeners); stick the twigs in the sand and place them out of doors for the winter, covered with wire for protection; the larvæ pupate in the resinous mass and the pupa-skin is left sticking out when the imagines emerge (Elisha).

In January and February search the ground under spruce fir-trees for misshapen cones, for it is these only that contain the larvæ of *Coccyx strobilana* (Elisha).

During the last week of March and first three weeks of April, the imagines of *Steganoptycha pygmaeana* are to be obtained by working spruce hedges. Here the ♂ flies freely for about half an hour at mid-day, and after this time can be beaten from the sunny side of the hedges from 12.30 p.m.—4.30 p.m. The females, however, when disturbed drop to the ground like a stone, where they are not difficult to find. April 15th is probably a good average date for the species to be well out (Durrant).

The larva of *Euchromia purpurana* spins together the terminal leaves of *Vaccinium vitis-idaea* in March and April, eating the inside surface of the leaves, making much black frass, spinning a white web among this *débris* at the end of April and turning to a pupa therein.

The withered stems of *Stachys sylvatica* should be collected in January and the following months for the larvæ of *Ephippiphora nigricostana* (Elisha).

The larvæ of *Grapholitha nigricana* are to be found in the early spring, feeding in the buds of *Abies picea*; they appear to prefer the terminal buds of the side shoots, whilst their whereabouts are indicated by the covering of silk spun over the spot at which they have entered.

The larvæ of *Grapholitha paykulliana* are common feeding in the catkins of birch in the early spring.

Roots of *Achillea ptarmica* growing in a damp spot in Epping Forest in early March, contained larvæ of a bone-white colour, with light brown heads, the imagines from these appeared in June, and proved to be *Dichrorhampha alpestrana* (Thurnall).

Fill a large bag with teasel-heads in January and February for they contain the larvæ of *Eupoecilia roseana* which eat through the seeds, and *Penthina gentiana* which feed on the pith. Empty in large bandbox and they will emerge in due course (Elisha).

A good supply of catkins of the white poplar should be collected in March for they contain the larvæ of *Eupoecilia nana* and *Grapholitha nisana* (Gregson). [Does *E. nana* ever feed on poplar?]

Long series of *Argyrolepia zephyrana* may be bred from larvæ feeding in roots of *Daucus carota* collected in January on the coast between Herne Bay and Whitstable (Elisha).

The pupæ of *Conchylis francillonana* can be obtained throughout January, February and March in the stems of wild carrot *(Daucus carota)*, little holes in the stems readily denoting their whereabouts— very abundant in Folkestone warren, &c. In the same way and at the same time the pupæ of *C. dilucidana* are to be found in the stems of wild parsnip (Elisha).

PTEROPHORIDES.—In February examine the dead stems of *Eupatorium cannabinum*. If small holes be detected in the sides of the stems they should be carefully cut and examined for the larvæ of *Leioptilus microdactylus*. By cutting a small bundle of these stems and keeping them out in all weathers until near the time of emergence fine series may be bred (Elisha).

The larvæ of *Oxyptilus heterodactyla (teucrii)* hybernate on the underside of the leaves of *Teucrium scorodonia*, and have at this time all the appearance of little oval tufts of whitish hair.

In sheltered positions, the flower-heads of coltsfoot can be collected in late March for larvæ of *Platyptilia gonodactyla*. Keep in a large roomy bandbox.

Young larvæ of *Mimaeseoptilus bipunctidactylus (scabiodactylus)* were plentiful at Llanferras, at end of March in the " cases " of *Scabiosa columbaria*, whilst, in the terminal shoots of *Teucrium scorodonia*, growing in sheltered corners, the larvæ of *Oxyptilus heterodactyla (teucrii)* were just beginning to feed (Gregson).

The young larva of *Platyptilia ochrodactyla (dichrodactyla)* mines down the stem into the roots of tansy in the autumn, and there remains until the fresh shoots are thrown up the following spring, up which the larva works as the plant grows, throwing out frass from the joints, causing the whole plant to droop, and producing an effect very like that produced by the larva of *Exeraetia allisella* in the stems of *Artemisia vulgaris;* the larva is fullfed about the end of June. In confinement the larvæ will readily leave the old stems when they commence to wither and enter new ones provided for them ; the mouth of the mine is placed between the axil of a leaf and the stem, and here an accumulation of frass becomes very conspicuous and betrays the presence of the larva.

PYRALOIDES.—Examine dead and decaying wood, and search under rotten bark for the larvæ of *Dasycera sulphurella* in January and February. Many species of *Œcophora* also feed in this manner and their larvæ should be obtained at the same time (Elisha).

Under the loose bark of elm-posts and the rails by the sides of fields, etc., the dirty blackish larvæ of *Œcophora fuscoaurella* may be found in February living under a slight web, mixed with frass and gnawed wood. The larvæ may also be found through March and April (Elisha).

On heaths, examine the shoots of furze for a mass of white web binding the shoots together ; these webs contain larvæ of *Butalis grandipennis*.

Larvæ of *Parasia lappella* can still be obtained throughout January and February in the heads of burdock *(Arctium lappa)* (Elisha).

In the seedheads of the carline thistle *(Carlina vulgaris)* the larvæ of *Parasia carlinella* may still be obtained through January and February (Elisha).

In late March the fast-growing *Stellaria holostea* will already be found to have some of its topmost shoots drawn together and distorted by the larvæ of *Lita (Gelechia) tricolorella* (Elisha).

In March the bushes of common broom, frequently found by the rides in woods, &c., will have some of their twigs joined together by a slight web ; on pulling them apart the brown larvæ of *Depressaria assimilella* will be found. It is best to cut cff good-sized pieces of the twigs with the larvæ and keep them in a cage with plenty of ventilation and out in the open air, otherwise the broom will very soon go mouldy.

Exeraetia allisella is to be bred from roots of *Artemisia vulgaris*, which may be collected in the lanes about Banstead Downs in February ; these larvæ in the spring work up the new shoots, causing them to droop and die (Elisha).

EPIGRAPHIIDES. — The imagines of *Epigraphia avellanella* are common in March and April at rest on birch trees. Large numbers are obtained in Rannoch by the collectors who take the species when they are searching for *Petasia nubeculosa*.

PYRALIDES.—Under the tufts of Tortulæ and Hypnums, which grow so freely round Llanferras, I obtained young larvæ of *Scoparia muralis*, *S crataegalis* and *S. mercurialis;* also a few Noctuid larvæ that were hiding there (Gregson).

CYMATOPHORIDES.—Imagines of *Asphalia flavicornis* are to be taken on the birches of Paul's Cray Common, near Chislehurst, during the latter end of March (Carr).

The egg of *Asphalia flavicornis* is laid singly or in pairs in the angle formed by the growing bud with the twig. At first white, it soon changes to a bright and lovely pink, darkening in hue as the time for hatching arrives—usually in about a month (Merrifield).

GEOMETRIDES.—The larvæ of *Geometra papilionaria*, sleeved out on birch, nibble the bark and buds in March, as do those of *Boarmia roboraria*. They want removing to a new branch early, as their nibbling often tends to kill the twigs.

The larvæ of *Geometra papilionaria* rest during the winter in a nearly straight position at an angle with the twig, or bent like a bow with their heads by the side of the twig and always towards its tip. Search birch, hazel, and alder.

When the young larvæ of *Geometra papilionaria* are on the move after hybernation, they may be found very near the tops of the twigs on which they are resting. They rarely remain on the same twig two days in succession, appearing to change their position between 9 p.m. and 11 p.m. At that time, turning their heads towards the trunk of the tree, they travel till they arrive at a fork in the branch when they will diverge towards a fresh position. They are always attached to a twig by their claspers, never to any portion of a leaf.

Boarmia repandata is a very interesting species to breed, the larvæ may be found on mild evenings in early spring feeding on all sorts of low-growing plants—ivy, honeysuckle and bramble appearing to be those most frequently selected (Mason).

The larvæ of *Boarmia roboraria* may be beaten from oaks when beating for larvæ of *Cleora lichenaria* in February before the oak-buds show the slightest sign of leaf. They must loose their hold of their silken pad surprisingly early; the larvæ also feed on sallow, whitethorn and birch, feeding up well in confinement on sallow (Moberly).

Larch plantations, especially if near or among beech and oak woods, are the favourite haunts for *Tephrosia bistortata*.

The best time to search for imagines of *Amphidasys strataria* is directly after 4 p.m., and for *A. betularia* after 5 p.m. (Bate).

A freshly emerged ♀ of *Amphidasys strataria*, towards the end of March, suspended in a suitable " cage " from an oak, will on a good night attract a large number of males, usually before 9 p.m.

Tie a piece of cotton round the abdomen (close up to thorax) of a ♀ *Amphidasys strataria* and tether her to a tree in a satisfactory position; males will readily assemble to such a female in March and April.

Imagines of *Nyssia hispiaaria* are to be taken in late February and March, newly emerged and drying their wings between 4 p.m. and 5 p.m. on the lower part of the trunks of oaks.

In March (late February in early seasons) *Nyssia hispidaria* is sometimes exceedingly abundant; between 9 a.m. and 2 p.m. large numbers may sometimes be found drying their wings on tree-trunks in Epping and Chingford ; they usually rest from 4ft.-5ft. up the trunks, some much higher. They press themselves very closely into crevices in bark, are not easy to find ; the majority rest on oak, a few on hornbeam, beech, &c., if a ♀ be on the bark one or more males are sure to be present, a period of seven weeks has been noticed as the time over which emergences are spread in some seasons.

The ♀ s of *Phigalia pedaria* and *Nyssia hispidaria*, which are comparatively seldom seen in the day-time, may sometimes be taken freely by examining the trunks of trees in woods with a light after dark in March (Porritt).

Nyssia hispidaria is easy to pair in captivity. A ♂ that has been out for a day or two, placed in a fairly large box with a freshly emerged ♀, is sufficient to ensure fertile eggs. Copulation usually takes place in the evening, and does not last more than about 15 minutes.

The males of *Nyssia hispidaria* are readily obtained by assembling. Six or seven freshly emerged ♀ s in a small gauze cage, about 5ft. from the ground, on a warm and windy evening, brought the first ♂ at 6.45 p.m., others following in twos or threes till 7.30 p.m., when they ceased; but the males became active again from 10.30 p.m.—11 p.m.

The larvæ of *Nyssia hispidaria* feed readily on birch, hawthorn, and hornbeam in confinement; knowledge of this fact is often useful as these trees are usually earlier in forward seasons than oak in putting out their leaves.

The eggs of *Anisopteryx aescularia* are laid necklace-like round a twig of birch, &c., each batch consists of some 200 or more eggs, and the whole batch is covered with down from the anal tuft.

In February and March, *Hybernia leucophaearia, Phigalia pedaria, Nyssia hispidaria, Amphidasys strataria*, are found commonly by trunk-hunting on the outskirts of the London district—Richmond Park, Chingford, etc.

Imagines of *Hybernia leucophaearia* are common on trunks of oak-trees in February and March, some drying their wings during the forenoon in the sunshine, the lantern revealing them, however, still freshly emerging after dusk. *Phigalia pedaria* may be found at the same time, and are often seen hanging with limp wings on the oak trunks at about 8 p.m.

The spring Hybernias are all common in Epping in February and March, very fine aberrations of male *Hybernia leucophaearia*, and also ♀ s, are to be obtained on fences and tree-trunks in the day-time ; *H. rupicapraria* on every hawthorn hedge, the ♀ s, about an hour after dark, are found at the bottom of the hedges, later on they get higher, and are seen on the outside as well as towards the middle of the bushes ; the ♀ s of *H. marginaria* are to be found freely on tree-trunks, after dark, with a lantern, some of the ♂ s are very fairly dark.

The larva of *Acidalia imitaria* feeds on *Stellaria media;* it is difficult to keep through the winter even on plants growing in pots. The larvæ recommence feeding pretty early and are fullfed by the middle of May.

BREPHIDES.—The last week in March will give an abundance of *Brephos parthenias* at Pauls' Cray Common, near Chislehurst, in sunny weather; from 2 p.m. until 3.30 p.m. they fly in greatnumbers. In dull weather, imagines are to be found resting on the small brown birches, in the forks of two branches, with the wings pressed tightly against them (Carr).

At the end of March, in a long grass drive in Sherwood, on a bright sunny morning, countless numbers of *Brephos parthenias* may be seen flying around the birch trees; but to make a bag, look on the ground, for *B. parthenias* loves the sunshine, and may be seen every few yards basking in it on the ground, sitting with fully expanded wings on the bare sandy patches of the grass-grown drives, the orange of the hindwings contrasting brilliantly with the sober colour of the surroundings. In this way it can easily be taken; but if it should be missed, it is quite useless to give chase into the woods, as it is quite impossible to keep it in sight for more than a few yards (Alderson).

Brephos notha occur in March on the outskirts of a wood near Broxbourne, in which are a few aspens; they begin to fly about 10 a.m., at which time they are, like *B. parthenias*, very sluggish and easy to capture; at midday they retreat into the wood but about 2.15 p.m. they appear again in numbers on the sheltered side of the wood (Battley).

DIMORPHIDES.—At the end of February place your cocoons of *Dimorpha versicolora* in the sun during the day, and in a warm kitchen at night and during dull weather; few pupæ will then die that are ready to emerge. The imagines will only pair in the sun or in a fairly high temperature, but given these they will do so and lay well.

By the middle of February, the pupæ of *Dimorpha versicolora* kept in a warm room, will commence to emerge from their cocoons, often coming right out and lying exposed for a few days before the emergence of the imagines. Many pupæ go over to a second or third year. All those that will emerge during the year are fully formed some time before emergence; the rest are still fluid.

The females of *Dimorpha versicolora* should be allowed to pair more than once, otherwise a fair number of eggs will be sometimes found to be infertile; the ova should be kept in the shade or at least not exposed to the full rays of the sun until they begin to hatch.

In confinement place fertilised ♀ s of *Dimorpha versicolora* on birch twigs; they bungle badly sometimes in attempting to lay in a cardboard box, but go ahead steadily, laying their eggs in little batches, on a large fresh birch-twig.

At the end of March and through April *Dimorpha versicolora* is on the wing. The males fly till about 3.20 p.m., and will do so without sun if the temperature be suitable.

The ♀ s of *Dimorpha versicolora* commence to call about three hours after their exclusion, and if a ♂, preferably a day or two old be available, pairing takes place at once ; they prefer to pair between 11 a.m. and 3 p.m., and usually separate about dusk, the ♀ commencing to lay almost immediately.

The eggs of *Dimorpha versicolora* found in the woods, are laid in little batches of 6 to 8 or so, in double rows on small outside twigs of birch, 2ft. to 4ft from the ground ; the eggs are pale yellow at first, but, after a day or two, darken to a purple-brown colour, just the tint of the birch twigs.

ATTACIDES.—The cocoons of *Saturnia pavonia* are easily found on the moors when the heather is wintered ; they are either on the heather or on the ground loose, the latter situation being almost as common as the former ; the important factor for success is to search for them when the vegetation is dead (Arbuthnott).

Cocoons of *Saturnia pavonia* may also be found in whitethorn hedges in winter, spun up on the lower branches about a foot from the top of the bank on which the hedge grows. They are most difficult to see when the foliage is on the trees, and, although exposed when the leaves fall, the colour is so like that of the surrounding branches that it is easy to miss them (Ransom).

LACHNEIDES.—Larvæ of *Macrothylacia rubi* can easily be found in early spring (middle of March and April) crawling on heather when the sun shines. Dull days, even if warm, are no good, but a sunny day, even if preceded by very cold weather, will tempt the larvæ out (Arbuthnott).

I can get the imagines of *Lachneis lanestris* to emerge freely. I put the cage containing the cocoons in the sun in early March, and they will come swarming out like flies (Thornewill).

Forcing in February will bring out those pupæ of *Lachneis lanestris* that are ready to come out, but will not affect those that have not yet matured and which intend to go over until another year.

The eggs of *Poecilocampa populi* are laid in November, December or January in little rows of five or six, placed closely side by side, with their long axes parallel on the twigs of their foodplant. Keep in cool place as they sometimes hatch as early as February indoors, although the end of March to middle April is the normal time out of doors.

Early hatching larvæ of *Poecilocampa populi* may be fed up on lettuce ; eggs often hatch indoors before the out-of-doors foodplants are available.

The eggs of *Trichiura crataegi* are usually laid the day after the female has paired, on the bark or on a twig of *Crataegus oxyacantha*, *Prunus spinosa*, or other of its foodplants ; this takes place in August and September. They are glued firmly to the twig by the long side, in contact with each other so as to form a ribbon 8 to 10 in number, with the long axes parallel ; and are covered with coarse dark-grey silky hairs from the abdomen of the ♀. They may be found all the winter, and hatch towards the end of March or in early April, Doubleday and others noticing that the eggs of a batch do not hatch all at once, a few of the larvæ appearing daily for two or three weeks.

The larvæ of *Trichiura crataegi* ab. *ariae*, the moorland form of the species, are to be found on the Scotch moors in the early spring; the eggs of this race hatch in the autumn, the young larvæ hybernate small the first winter, feed up and pupate the following summer, some emergences taking place in autumn, whilst other pupæ go over the winter and do not produce imagines till the following summer.

The larvæ of *Lasiocampa quercûs* are to be found on warm days in winter and spring on hedges; they nibble throughout the winter on sunny days, but remain small, and do not increase much in size till the spring is well advanced.

NOTODONTIDES.—Great care must be taken in March to keep the ova of *Ptilophora plumigera* in a cool spot, otherwise they will insist on hatching before their regular foodplants are ready for them, and they are not satisfied with any other plant, the leaves of which can be obtained in late February and early March. They will feed readily on sycamore as well as maple.

NOCTUIDES.—At the end of March, we adopted the plan of cutting branches of the best sallow bloom from the inaccessible parts of Bishop's Wood, and hung them up on the branches of trees in rides, on each of these the moths swarmed and we had an umbrella literally covered with Tæniocampids after each shake (Walker).

The sap exuding from freshly-cut dogwood stems proved of such superior attraction to the usual sallow-loving insects, that the sallows were practically blank, and each stem of cut dogwood was covered with moths (Thornhill).

During the first week of February I searched the stone walls and dykes that separate the fields in Aberdeen, &c., for cocoons of *Arctomyscis* var. *myricae*, and in a few hours picked up a dozen around Pitcaple; the snow, however, was very deep and the cold intense, and these factors much interfered with a more successful hunt (Reid).

In January cut open the old dead stems of *Arctium lappa;* inside, the eggs of *Gortyna ochracea* are laid, in batches of about 50, in the old burrows made by the larvæ. They are laid about half-an-inch down from the opening in that side of the stem from which a moth has previously made her exit. The young larva presumably feeds firstly on the old pith, and later descends to the new growth of the plant (Shaw).

During the winter, the quite round, shiny, smooth eggs of *Nonagria cannae* are to be found on *Typha latifolia*. They are laid singly, remain as eggs all the winter, the young larvæ appearing about the second week in April; the young larvæ feed well in the sheathing-leaf of *Iris kaempferi* when young (Edelsten).

The haunts of *Senta maritima* are usually coincident with those of *Nonagria geminipuncta*, the larvæ of the former wintering in the old burrows of the latter. The larvæ are most easily obtained in a hard winter, when one can more readily examine the dead reeds that fringe the sides of lakes and moors. They have also been found in the old galleries of *Nonagria sparganii* and *N. cannae* in *Typha latifolia* (Dadd).

To rear *Dasypolia templi*, prepare a few boxes of earth, in which plant roots of *Heracleum sphondylium;* as soon as the larvæ hatch, transfer three or four into the axils of the growing leaves on the head of each plant and leave them.   About the middle of July, cut down the plants to within a few inches of the root, and cover with canvas, to prevent the escape of the larvæ when they leave the plant to pupate, as they frequently roam at this period.   The garden parsnip goes very soft, and is, therefore, not so good to rear this species on as *H. sphondylium* (Tugwell).

Working at night at Llanferras, at end of March produced larvæ of *Leucania conigera* and *L. lithargyria* feeding freely at dusk; and near midnight larvæ of *Agrotis lucernea* were found stretched at full length on rock-faces, one or two feet from the ledges on which their food grows.

The larvæ of *Leucania albipuncta* feed only at night, but may be found by day concealed on the ground, under leaves of low plants —mullein, hemlock, teasel, &c. (Brahm).

The larvæ of *Tryphaena fimbria* are to be taken in spring at night on the newly-expanded leaves of osiers in company with those of *T. ianthina*, *Noctua triangulum* and *N. augur*.

In March and April, when one larva of *Agrotis ashworthii* is found, keep to the near neighbourhood, and work day after day (or rather night after night) for others; they are best seen soon after changing their skins.

The larvæ of *Agrotis ashworthii* recommence feeding after hybernation on the first warm days of February, and continue through March, and are then to be found singly under stones, and may be obtained by turning over the stones of carboniferous limestone, found strewn about in their locality.   The food, most preferred in spring, consists of the male catkins of *Salix caprea* (Gregson).

On March 25th, 1873, I went to Llanferras, Denbigh, and on Pen-y-garra Win and Pant Moen, took larvæ of *Agrotis ashworthii* and *A. cinerea* amongst mixed herbage.   The latter seems to prefer *Festuca ovina*, feeding downwards from the extreme tips of the grass and stumping the tufts down that it has fed upon, afterwards hiding away in the tufts (Gregson).

Epunda lichenea larvæ, of all colours, from light green to dark, chequered brown-olive, and of all sizes from three-eighths of an inch long to full-fed, were also found feeding on *Sedum acre* and *S. reflexum*, or stretched out on various plants or on the rocks (Gregson).

The larvæ of *Epunda lichenea* are to be taken regularly on the sandhills of the Lincolnshire coast, north of Mablethorpe, where they are invariably found on *Cynoglossum officinale* which grows in some abundance along the banks on the landward side of the hills. These feed up well in confinement on dock.   Larvæ have also been found in Cornwall feeding on stinging-nettle, and on the coast of the Isle of Man feeding on dock and *Plantago maritima*, and occasionally on thrift (Ash).

The larvæ of *Epunda lichenea* may be taken, when small, about nine lines in length, and being then of a vivid green they are easily seen at night on the grass stems; some being stretched out flat and others resting with the head and fore part of body tucked inwards, and looking somewhat like a note of interrogation (?).   They are best searched for when small and green, as later on they entirely change their habits and no longer rest on the grass stems, but remain concealed at the roots of low plants, and are then hard to find. I have taken over 200 in the green stage, but have not found more than half a dozen after they had assumed their mottled olive suit as described by "Newman." They fall off the grass at the slightest touch, making it necessary to hold the lid of the larva-tin underneath to catch them (Brown).

The larvæ of *Epunda lichenea* are to be found in March and April at Wallasey feeding on stonecrop. This is an excellent winter and spring foodplant for those larvæ that will eat it (Woodforde).

The imagines of *Taeniocampa populeti* come to sallow bloom when the latter is near aspens, but they can also be obtained freely by searching the aspen twigs after dark.

The eggs of *Taeniocampa populeti* are laid in batches on the twigs of poplar and aspen in March and early April; they soon hatch, and the young larvæ are to be obtained from the young leaves (Battley).

To rear *Taeniocampa miniosa* successfully, fill an 8- or 10-inch flower-pot with cocoanut fibre; in this plunge a jar containing the foodplant—in my case bramble—and carefully pad the mouth with cotton wool; insert a light bamboo-cane to support a gauze sleeve, fastened round the rim of the flower-pot with elastic, place the pot in the open air where it will receive plenty of morning sun; these arrangements nearly approach the natural conditions of the larvæ, which are dry feeders, and love the sunlight.

In collecting young larvæ from sallow in early spring it will be found, upon examination, that the young leaf-shoots at the tops of the twigs almost invariably contain a larva. In this way I have bred *Tethea retusa*, *Eurymene dolabraria*, and other species which will possibly not be gathered with the catkins alone (Dollman).

The young larvæ of *Tiliacea citrago* live in the young lime buds, but in early May are to be found between two flat leaves of lime, fastened by silk; by standing under the trees, so as to get the leaves between the eye and the sky, the larvæ may be readily detected; they come out to feed at night and can then be beaten; the larva forms its cocoon in early June, but does not pupate for nearly two months after doing so.

Eggs of *Tiliacea aurago* laid in autumn hatch at the end of March or beginning of April, the young larvæ can be sleeved on beech and will be found to be fullfed at the end of May and early June.

The larvæ of *Tiliacea aurago* hatch during the last week of March, and take to the buds of the beech from which the outer coverings have been removed; they creep into the buds, are practically invisible until they are at least three weeks old, and only show signs of their existence by frass.

As a substitute for the usual food-plant of larvæ of *Tiliacea aurago*, which sometimes hatch very early, it may be noted that sycamore buds are very acceptable, the larvæ feeding on them without hesitation (Butler). The larvæ grow more rapidly and vigorously on sycamore than on beech and the former is earlier in leaf; the larvæ are fullfed in May but remain unchanged to pupæ in their cocoons until July (Prideaux).

The imagines of *Hoporina croceago* pair after hybernation, come freely to sallows, lay their eggs on dried oak leaves and twigs (freely

in confinement if fed with a little thin sugar); the young larvæ rest
on the veins of the newly-developed oak-leaves and feed up readily if
sleeved in gardens ; full-fed at end of May.

To obtain eggs of *Orrhodia vaccinii* and *O. ligula (spadicea)* place
♀ s captured in spring at sallow in large chip boxes, which have been
previously scored well with a penknife, or cracked so that they can
find some place to hide their eggs in ; they will then readily oviposit
(Robertson).

The larvæ of *Cirrhoedia xerampelina* are to be found in the Perth
district in March and April crawling up from the burn sides, at the
roots of old stunted ash-trees, mounting the trunks of those that have
very prominent flower-buds, in the first week of April (Bush).

Polia nigrocincta is reputed to be difficult to breed in confinement
in the south, but has been fed up in the open on potted plants of
narrow-leaved plantain (Whittle), and on sallow (Bower), the moths
emerging in September.

The eggs of *Catocala promissa* hatch in March and April and can
be fed on split oak-buds until the early leaves are procurable.

From February until April, search after dusk in the localities
where *Agrotis simulans* occurs for the larvæ of this species which hide
low down about the roots of grass.

In Scotland in the early spring when the sap begins to rise, it is a
common practice to bore holes with a good sized gimlet in the trunks
of birch-trees. The sap spurts out freely and after dark the three
insects, *Asphalia flavicornis*, *Calocampa exoleta* and *C. vetusta*, as well
as many Tæniocampids come to feed on this sap, even more freely
than, later on, insects come to sugar. I have never heard of this being
tried in the south, but doubtless in districts where birch is common,
nearly all the moths which hybernate could be thus taken. Experiment can only prove when the sap is running, but I feel sure one
could forestall the sallow blossom, say from March 7th onwards
(P. C. Reid).

The hybernating imagines of *Calocampa exoleta* are to be taken
at sallow-catkins in March and April; the sexes pair about the
middle of April, and the ♀ s lay some one to two thousand ova apiece.

January and February, if mild are the months to sweep
for the larvæ of *Triphaena subsequa*. It feeds at night, but is
out on the blades and stems of grass in the afternoon, stretched at
full length; it frequents dry, sandy banks, especially where dense beds
of *Dactylis glomerata* appear. It appears to be entirely a grass-feeder
in its natural state, though it will eat other herbage in confinement.
At large, I have never found it feeding on anything else but *Dactylis
glomerata* and *Triticum repens* (Williams).

The larvæ of *Noctua triangulum* are to be found in March and
April on bramble, although they will eat almost anything in confine-
ment. They are full-fed by the end of April, and pupate in early
May.

The larva of *Noctua brunnea* is to be found in March on bramble,
hornbeam, &c., though in confinement it will eat dock and other low
plants, is fullfed about the middle of April, buries under the earth and
pupates, the imagines emerging in middle June.

Larvæ of *Anchocelis rufina* may be fed up well on the partially-
opened buds of *Crataegus oxyacantha*, attaining their full growth very

rapidly in February and March on this pabulum (Bryant).

PAPILIONIDES.—The larvæ of *Chrysophanus phlaeas* are sometimes very common on *Rumex pulcher* in February and March; they are difficult to see as their bodies are about the same size as, and the crimson dorsal line and broader spiracular stripe render them very like, the young curled-up leaves in the centre of the plant (Bate).

Hybernated *Eugonia polychloros* are to be taken in March. They are never worth setting and should be retained for eggs. Both sexes should be sleeved together for this purpose as the earliest caught ♀ s after hybernation are rarely fertilised.

Females of *Polygonia c-album*, captured in late March and early April, will lay freely on hop, currant and nettles, if carefully sleeved on plants that obtain a fair amount of sunshine.

APRIL.

During the spring, night-beating for larvæ, especially after 10 p.m., is particularly productive, and many larvæ are obtained in numbers, the imagines of which do not always come at all freely to sugar, *e.g.*, *Noctua ditrapezium, Aplecta tincta, Triphaena interjecta*, &c. Fenn has recorded that, on one occasion, at Tilgate, between 10.30 p.m. and 1 a.m., he beat from birch and sallow more than 500 larvæ in one night, including, among others, 16 *Noctua ditrapezium*, 20 *Triphaena fimbria*, 8 *Aplecta tincta*, whilst the larvæ of *Boarmia repandata, Noctua baia, N. brunnea, N. triangulum, Aplecta nebulosa, Triphaena ianthina*, &c., were exceptionally abundant. Where *Boarmia repandata* ab. *conversaria* occurs, the larvæ should be searched for at night about the end of April along wood-ridings or on hedges, the pale larva being then easily found or beaten into an umbrella, and is more usually discovered on hawthorn, blackthorn, nut and birch. Now, too, is the time to see that the flower-beds are in order, and that shrubs and other plants with flowers attractive to moths are well placed and thriving. As border plants, *Solidago canadensis* and *Tritoma uvaria* are remarkably useful, the rose-coloured varieties of bedding geraniums and *Salvia pratensis* are also specially attractive. *Eumorpha elpenor* will come to the various garden forms of *Iris;* laurustinus flowers attract the Tæniocampids and other sallow-loving moths after the sallow catkins are over, so do the flowers of common laurel, those of the Portugal laurel being far less attractive. Lavender is considered by many quite the most attractive of all plants to moths, which literally swarm at the blossoms in favourable seasons, whilst *Althea frutea*, rhododendron, weigelia and petunia are well-known attractions later in the year for *Agrius convolvuli*. [For a list of the species taken at a row of lavender plants some 40 yards long, close to the outskirts of a large plantation, see *Ent. Record*, ii., p. 65.] *Spur valerian* is said to be especially attractive in Lancashire, doubtless, also elsewhere, and long lists of species captured at the flowers of this plant have been published, some being attracted in great abundance. We have already given a few general hints on sugaring *(anteà*, p. 4), but to any one with a fair-sized garden, or who can gain access to a private plantation or wood, so that the traps may not be interfered with, the following method has been suggested (*Ent. Record*, i., p 141)—Make several stout cloth bags, about a foot long, two inches or so in diameter, weighted at the bottom to prevent them swaying with the wind, a wire ring fixed round the top to keep the bags open, and a hook with which to hang them on the branch when filled with liquid.   Such bags do not require attention more than once a fortnight even in hot weather ; it saves no end of time as they are always ready for use, and only want a little rum administered each evening, a difficulty easily met, by squirting some on from a scent spray, whilst making the first round. The objection to these bags is the havoc made by wasps on them in some summers, a difficulty that has been met by making the sus-pended bags out of old coffee canisters, perforated to allow the sugar to ooze through into flannel which tightly covers the tin. The wasps are unable to get in these, and hence can only utilise so much sugar as they can obtain from the outside, in what may be considered a legitimate way.   " Assembling " should begin this month.   The species usually first attracted by this means in the spring is *Amphidasys strataria*, the males of which are readily "assembled" by a virgin female in suitable weather.   The female should be placed in a small white muslin bag, about 3 inches in diameter, hung in a sheltered position by the side, or in a riding, of a wood, when, if the males are to be attracted, they will be observed coming up to, and crawling upon, the bag soon after dark, and will continue at least until midnight.   The bag should be watched from below, so as to get the bag between the eye and the sky, and the moths are better seen on white, than on darker-coloured, muslin.   They usually come from above and should be allowed to settle on the bag and crawl round it a few times before an attempt is made to capture them when they may be taken either with a bottle or by means of a chip-box.   The females are rarely of use a second night, appearing attractive only when quite freshly emerged, and one finds it best to allow a male to pair with the captive at the end of the first night's sport, as after this she seems too sluggish and the males will often take no notice of her.   This method of assembling is applicable to many species ; the habits and peculiarities of the different species are, however, exceedingly variable and must be carefully observed if success is to be obtained.   In April, one of the rarest species, likely to be obtained by trunk- and fence-hunting is *Lophopteryx carmelita*.   Such situations near birch are usually those favoured, and the birch-trunks themselves must always be carefully overlooked.   In the localities where *Lobophora viretata* is to be found, one may obtain, during the last week in April, the eggs of this species, on the terminal shoots of holly, parti-cularly those at the top of the trees bearing flowers.   The larvæ hatch in from seven to twelve days, feeding first on the flowers, if there are any, next on the green berries, and lastly on the young leaves.   In confinement they take readily to common privet, and devour the bark as well as the leaves.   The larvæ are difficult to beat, a sheet spread on the ground and a ladder being necessary.   In emerging, the perfect insect is somewhat erratic in confinement, some appearing as a second brood in August and September, others from the same batch remaining through the winter in the pupal state. Here we may, perhaps, safely indulge in a few words about the rearing of lepidoptera, which is to be encouraged on every ground that appeals to the lepidopterist. To the collector it provides finer specimens than he can possibly catch; to the naturalist it gives material for the study of the habits of the species in their earlier, and, in nature, less observed, stages; to the biologist it furnishes the material for close and intimate study of structural peculiarities and comparison with the allied kinds that may, in the field, be confused with it; breeding has settled absolutely the specific or varietal rank of many species whose exact values as species or varieties were uncertain. No doubt, for one or other of these reasons, breeding will always be considered one of the most important occupations of the lepidopterist. Breeding from ova saves a host of disappointments to which one may be subjected when one obtains larvæ. The latter may be injured by the beating-stick, or parasitised by their numerous enemies. One must always, however, in breeding from the egg know within a little when the eggs should hatch; some eggs remain as eggs for eight or nine months, others for only a few days. A glass-topped box or a small wide-mouthed bottle is convenient to place eggs in that are near hatching. One should always count the young larvæ before putting them on the foodplant—one knows then how many larvæ to look for in the first change of foodplant, a most desirable detail that saves considerable time. It must, before all things, be borne in mind that newly-hatched larvæ must be kept in close quarters; they readily wander, and do this most frequently if the food commences to wither; they are more difficult to find when changing the foodplant if scattered over a considerable amount of the food and in a large receptacle; besides, they may get off the food and be unable to find it again, and so die. Glass tubes and chip-boxes are, perhaps, most frequently used by beginners, and here two evils are sure to present themselves to the young lepidopterist: (1) He will find that perfectly air-tight tubes tend to keep the food fresh, but that the transpiration of the leaves causes moisture to collect on the inside of the tube, and the larvæ sometimes get drowned. (2) The usage of chip or wooden boxes allows a most rapid withering of the leaves, and the young larvæ find themselves starved for want of sufficiently succulent food. To get rid of the difficulties caused by the accumulation of moisture on the inside of the glass or tube in which the larvæ are being reared, the following practical idea has been suggested: Take two wide-mouthed bottles, the larger of the two with a mouth sufficiently wide to allow the smaller to pass quite inside; place the young larvæ inside the smaller bottle with some recently-gathered food, but carefully avoid all moisture on the surface of the leaves (wet leaves are often quite fatal to young larvæ), and tie a piece of washed-out calico over the mouth of the bottle. The smaller bottle can now be turned upside down, and placed mouth downwards inside the larger, a piece of calico tied over the mouth of the larger bottle. On reversing this, the smaller bottle stands mouth uppermost, and by this means the condensation of moisture takes place on the side of the larger, i.e., outer bottle, leaving the smaller inner bottle free from it. This double casing of glass is said to answer

admirably, the condensation, if it occurs at all, always taking place
on the outer bottle, and ceases therefore, to be a source of danger to
the young larvæ. Complicated as this may appear from the description, in practice it is comparatively easy. In changing the food
many dodges have to be adopted. One of the best means is to have
a duplicate tube ready. Turn the larvæ and food out of the one first
used upon a piece of white paper, shake the used food free from
frass, then drop it with any adherent larvæ into the fresh tube, pick
up other larvæ with a piece of sharp-edged notepaper and put in tube,
then put in it the fresh food, tie up and place in outer bottle as before ;
the larvæ will soon crawl on the fresh food. After one or two changes
of this kind the old food can be dispensed with, and, under all conditions, take care that no leaves that have a trace of mildew are allowed
into the tube with the larvæ. The following method for attending
to newly-hatched larvæ has also been suggested : A drawing-board
covered with cartridge-paper, an enamelled iron basin, a tin of dry sand, a
lump of cotton wool, a box of rubber bands, bottle-cages with spare
covers, glass-topped metal boxes, together with scissors, forceps, brushes
and a teaspoon. Transfer the newly-hatched larvæ at once to bottle-
cages (described in detail *posted*), making a careful note of numbers
hatching each day. Do not put too much food into the cage, because the
water, condensing on the inside of the glass, will drown the larvæ. Do
not put too many larvæ in a cage. Food used will keep some time without
changing if the ends of the stems be clipped off, a fresh surface exposed,
and the water in the gallipots changed. The large metal glass-topped
boxes also make first-class breeding-cages for newly-emerged Arctiid,
Noctuid, Lachneid and Geometrid larvæ. A few fresh leaves put
on the old ones every day prevent handling, as the larvæ soon
crawl on the fresh food, after which the old ones can be removed
and the larvæ and fresh leaf placed, by means of the forceps, in a
fresh clean box ; rub out the used box and it is ready for the change
next day. Stray larvæ are best moved with a camel-hair brush.
The boxes containing the larvæ must on no account be placed in the
sun, as the concentrated heat will kill all the larvæ in a few minutes.
The larvæ can be manipulated thus until large enough to be moved to
a breeding-cage, of which that designed by Young, of Rotherham,
is by far the most satisfactory. So much for very young larvæ.
In selecting food, care must be exercised ; the natural sap of the
plant is required, leaves containing such are much more likely to
serve as a satisfactory foodplant than leaves with an excess of water,
the result of keeping the foodplants standing in water. Food should
be collected preferably in the early morning and again preferably
should be carried home in tins ; by this means the natural sap is
retained within the plant the longest time possible ; food gathered in
the middle of the day will have had much of the sap already
exhausted by the sun, and it will then wither very quickly. Well-
developed wood, with sound foliage is, as a rule, better than young
succulent shoots, the matured parts of a plant keeping their freshness
longer and providing the best food. Very succulent food, especially
if afterwards kept in water, is likely to bring on diarrhœa. For
Micro-lepidoptera that do not require earth, especially for leaf-rolling,
leaf-mining and seed-eating species, jam-pots are the most widely
used as breeding-cages by lepidopterists. The jam-pots are ground a

little at the top and covered with a flat piece of glass so as to exclude the air. They form a very clean and simple sort of breeding-cage. Some lepidopterists maintain that it is not advisable to grind the edges and cover with glass, stating that it is better to simply tie over it a piece of muslin and then cover it with a piece of ordinary window-glass, the little inequalities of the surface allowing a certain amount of air to penetrate but not enough to dry up the food. This mode of feeding is often spoken of as the "close" or "dry" method, so called because the foodplant has not the stems immersed in water, but is simply put in the jar when freshly gathered. Not being exposed to currents of air it usually keeps fresh for nearly a week. Wide-mouthed sweet bottles are exceedingly useful, and the method is known to be exceedingly satisfactory for small Geometrid as well as micro larvæ. For Noctuid larvæ I prefer large flowerpots with plenty of drainage at bottom, well sifted and baked sand for two or three inches above this ; the food has simply to be put in the pot and the top covered with a piece of muslin. In this way large broods of *Peridroma saucia* on dock, *Pachnobia leucographa* or *Plantago lanceolata*, *Hoporina croceago* on oak, &c., have been reared with scarcely a death. Be sure and clear off all surface frass carefully.

Unclassified.—When working for larvæ, striking the limbs of a tree that are firm in growth is the best means of dislodging larvæ which have a light hold, but among plants of slender growth, or low-growing habit, jerking and stirring the herbage are the only hope of success (Dollman).

Beating upper branches of trees—There is still a reputation to be made by the man who devises a good method of working for larvæ with this habit.

In the third week of April 1892, my catch at Kinloch-Rannoch included *Lobophora lobulata* and its aberrations, *Petasia nubeculosa*, *Asphalia flavicornis*, *Brephos parthenias*, *Semioscopus avellana*, *Depressaria ciniflonella*, etc., (Reid).

Some of the flowers that have been found very attractive to moths are *Nicotiana affinis*, red valerian, rhododendrons, honeysuckle, cherry-laurel, raspberry blossoms, lavender, barberry, laburnum, etc. (Riding).

Newly-emerged moths must not be killed too quickly ; before killing such an insect one must make quite sure that the wings are firm and the body walls hardened. One can make sure by seeing that the wings are not limp and the body not swollen.

Female moths that have emerged with wings crumpled or are otherwise unfit for cabinet use should be placed on a tree in a garden (or elsewhere) ; they will usually attract a male and the collector can often thus obtain a good male of the species and make sure of a fertile batch of eggs.

To secure a falling larva, pass your hand under the place the larva was in when first seen ; by doing this the hand comes in contact with the silk thread which most larvæ spin when falling, and by following up the thread the larvæ can almost always be found.

As birch and alder do not keep well in water I have reared the larvæ of *Dimorpha versicolora*, *Notodonta dromedarius*, *Brephos parthenias*, &c., on *Corylus avellana*, and have been quite successful even when rearing them from the egg (Borkhausen).

ERIOCRANIIDES.—Birch-bushes should be systematically worked throughout April—especially on still sunny days—for *Eriocrania purpurella, E. semipurpurella, E. unimaculella, E. sangiella*, &c., whilst *E. subpurpurella* occurs in May among oak.

TINEIDES.—The larvæ of *Meesia richardsoni* live in small cases of the colour of the very fine powdery microscopic lichen on which they feed ; the cases are to be found by turning over stones covered with this lichen in Portland, and are usually discovered on the sides or under the stones ; the favourite haunt is among the loose piles which are so abundant in Portland. They may be collected in April and May, the imago in July.

ADELIDES.—The last week in April is the time to search for the larvæ of *Nemotois schiffermilleriella*. They feed on *Ballota nigra*, draw themselves into their cases as soon as touched and drop to the ground, and must be looked for upon the ground where their flat, oblong, oval cases, drawn in medially like a figure 8, are not difficult to see.

ELACHISTIDES. — Leaves of *Helianthemum vulgare* showing a bleached appearance should be collected in April for larvæ of *Laverna miscella*. Dry chalk slopes, or chalky hedge-banks are the usual habitats.

The larvæ of the April brood of *Elachista cerussella* feed on the broad leaves of the reed-grass or reed canary-grass *(Phalaris arundinacea)* ; at this time the leaves of the common reed *(Arundo phragmites)*, on which the August brood of the larvæ feed, have not yet grown up, so that the two broods appear to have different food-plants (Barrett).

LYONETIIDES.—The larvæ of *Bucculatrix aurimaculella* are to be found in April mining the leaves of the ox-eye daisy *(Chrysanthemum leucanthemum)* in long galleries which they will soon quit to feed on the leaves externally. Dig up the plants and pot them ; the larvæ will soon spin their beautiful reticulated cocoons (Elisha).

LITHOCOLLETIDES.—The larvæ of *Lithocolletis scabiosella* are to be found in April feeding on the radical leaves of *Scabiosa columbaria* in sheltered places (Elisha).

GRACILARIIDES.—In April, large bladdery-like lines on the leaves of plantain contain larvæ of *Gracilaria tringipennella* (Elisha).

In April, cones on the leaves of *Hypericum* should be collected for larvæ of *Gracilaria auroguttella* (Elisha).

COLEOPHORIDES.—In April, on sea-wormwood, cases of *Coleophora maritimella* are abundant ; they are studded with small particles of grit like those of *C. laripennella*.

The brownish keeled cases of *Coleophora olivaceella* are to be found locally on *Stellaria holostea* in April when the larvæ are feeding.

The leaves of sallow are being blotched by the larvæ of

Coleophora viminetelli in April; the cases are easily discovered; they are best searched for at the end of April and beginning of May, being much less easily found when the sallow leaves become larger and more numerous.

Bleached leaves ot wild-rose in April should be carefully examined for the flat serrated cases of *Coleophora gryphipennella.*

The tips of the radical leaves of *Echium vulgare*, growing on dry chalky slopes, that appear to be withered, should be carefully searched for the cases of *Coleophora onosmella.*

The leaves of *Lotus corniculatus* will be blotched in April by the larvæ of *Coleophora discordella;* the cornucopia-shaped cases are not difficult to find (Elisha).

The nearly straight, irregularly cylindrical-oval, brown case of *Coleophora glitzella* is to be found on *Vaccinium vitis-idaea* in April at Rannoch. The young larva is said to mine a leaf of its foodplant in June and hybernate in the mine until the next spring, when it feeds up rapidly, mining a broad gallery in the leaf, following the margin; on leaving the leaf it makes a new case in which it travels to a fresh leaf—spins up for pupation at end of April, and emerges about the middle of June (in Scotland).

When sweeping for larvæ of *Coleophora juncicolella*, choose the damper parts of the heath you are sweeping for this species. Patches of ling growing near rushes form the best spots.

Ground-ivy, growing under the shelter of a bramble-bush or at the foot of a fence among grass, may be examined in April and early in May for larvæ of *Coleophora albitarsella.*

Early in April mark the white spots in the leaves of *Ballota nigra* and turn up the leaf for larvæ of *Coleophora lineolea.*

ARGYRESTHIIDES.—In April the trunks of birch trees are often covered with silk threads, and on such the larvæ of *Argyresthia goedartella* are to be found in great numbers burrowing into the crannies of the birches, many huddled together into very small depressions. The silk suggests that the larvæ feed at the top of the tree, and only come down for pupation.

Edleston notes *(Ent. Ann.,* 1858) that the larvæ of *Argyresthia glaucinella* are to be found, in April, feeding solitarily on the sound bark of oak and horse-chestnut trees, revealing their retreats by protruding a little reddish frass from the hole. Do they thus feed? or are they simply there for pupation?

TORTRICIDES.—Throughout April and May the pines should be well searched for the larvæ of *Tortrix piceana;* the species is moderately common in the Esher district.

Imagines of *Peronea mixtana* are to be driven out of the heath-bushes in Delamere Forest by puffing tobacco-smoke under them, in April.

The larva of *Ditula woodiana* feeds on mistletoe, mining as a rule into the thickness of the leaf and eating the cellular tissue. It is able, however, if its mine does not last until the larva is fullfed, to spin two leaves together, and it then clears out the cellular tissue of the second leaf where the leaves are in contact, but this does not appear to be a very usual habit.

The larvæ of *Sericoris littorana* are to be taken at the end of April on sea-thrift on the Kent and Essex saltmarshes.

The roots of scabious should be worked in April for the larvæ of *Xanthosetia zoegana*.

The drawn-together shoots of wild-rose should be collected in April for the dull brown larva of *Spilonota roborana*.

Coccyx cosmophorana lays its eggs in any fissures in the bark of *Pinus sylvestris* and *P. picea;* the young larva then bores through the tender bark into the young shoots, when resin oozes out and is built up into resin-galls; these increase in size with the growth of the larva, but are always smaller than those of *R. resinana* (Kaltenbach).

The larvæ of *Coccyx cosmophorana* feed in various ways, and the imagines may be bred from the resinous excrescences of *Retinia resinana*, from those of *Phycis abietella*, from galls, from small resinous excrescences on firs, and from resinous exudations on fir-trunks.

Pupæ of *Coccyx cosmophorana* are to be obtained in one-year-old resinous nodules that have been tenanted by *Retinia resinana* the previous year (Horne).

The larvæ of *Coccyx strobilana* occur in fallen spruce fir cones in April. By gathering a quantity of the fallen cones at this season, a large number may sometimes be bred.

Heathy ground, chiefly in northern localities, gives imagines of *Amphisa prodromana*, which fly freely in the sunshine in April.

The pupæ of *Eupoecilia affinitana* are to be found, in April (also in February and March), in the upper part of the previous year's flower-stalks of *Aster tripolium* on the Kent and Essex saltings.

The distorted dry stems of *Senecio jacobaea* may be collected in spring for larvæ of *Eupoecilia atricapitana*, which live thereon throughout the autumn and winter. Some spin up and pupate within the burrow; others, however, leave the stem to spin up elsewhere. The moths emerge in May and June.

The larvæ of *Eupoecilia rupicolana* are to be found in April in the old (often prostrate) stems of *Eupatorium cannabinum*.

The imagines of *Tortricodes hyemana* are sometimes to be obtained in numbers in April (and March) by searching the trunks of oak-trees after dark (Thurnall).

ORNEODIDES.—The eggs of *Orneodes hexadactyla* are laid in April on the flower-buds of honeysuckle, when the corolla is about a quarter of an inch in length; the eggs are deposited in almost any part of the corolla or on the margin of the calyx, and are at first white, then yellow and lastly orange in colour; they are readily seen and easy to find on the growing heads of flower-buds of the honeysuckle, often more than one to be found on a head.

PYRALOIDES.—The larvæ of *Dasycera olivierella* are to be found in April and May in oak stumps when searching for those of *Sesia cynipiformis;* the imagines emerge about the middle of June.

In mid-April collect thistle-heads for larvæ of *Parasia carlinella;* the imagines appear in the beginning of August (Miller).

The carline thistle growing on chalky slopes have the leaves drawn together by the larvæ of *Depressaria nanatella* which feed on the upper surface (Elisha).

On damp hedgebanks, the hemlock *(Conium maculatum)* must be closely examined for leaves with the tips folded over. These must be collected for larvæ of *Depressaria alstroemeriana* and *D. weirella.*

The larvæ of *Depressaria cnicella* may be found feeding in shoots of sea-holly from April onwards at Shoeburyness, Deal, &c.

The larvæ of *Lita fraternella* is to be found in April, drawing together the shoots of *Cerastium vulgatum* ; affects a chalky habitat.

In April the imagines of *Butalis incongruella* are to be found flying in the sun over heather on Carrington Moss ; none were observed unless birch trees were near (Edleston).

In the early part of April the larvæ of *Lita instabilella* mine the leaves of *Atriplex*, completely eating out the fleshy inside in patches, making the leaf appear whitish-green, also spinning up the leaves against the stalk to a slight extent.

PLUTELLIDES.—The leaves and shoots of garden-rocket, *Hesperia matronalis*, are often seriously attacked in gardens by the larvæ of *Plutella porrectella* in April.

PYRALIDES.—The larvæ of *Scopula olivalis* are to be found in a web on the underside of the leaves of *Glechoma hederacea.*

CYMATOPHORIDES.—The eggs of *Cymatophora ridens* are laid on oak-twigs in late April and early May ; each egg is jammed carefully between a terminal twig and an unexpanded oak-bud ; they are whitish when laid, turning to scarlet, and hatch in about a fortnight.

BREPHIDES.—*Brephos notha* sometimes has an extended pupal stage ; Chapman notes how four completed a second winter in the pupal state, and were very nearly thrown out as useless, because they had changed colour, from brown to black, a change often implying decay, but which was taken, in these, to mean partial development of the imago, from some cause arrested, probably by death. These, however, produced four imagines, after being in pupa three winters.

During the third week of April work for the imagines of *Brephos notha*. Before 11 a.m. a large number are to be found on the ground, and these are easy to catch ; after that hour, however, they get up in the trees, beginning to fly fairly freely at 2 p.m. (when the sun is hot), but very rapidly, and they are hard to catch. On the only cloudy day, they flew very freely when the sun was hidden, ceasing immediately during the short hot intervals of sunshine (Woodforde).

GEOMETRIDES.—The eggs of *Ennomos angularia* commence to hatch in early April, not all at once, but in small numbers over a long period. The larvæ feed well on oak and birch, preferring the latter, and when fullfed spin a slight cocoon between the leaves of their food-plant (Lockyer).

In early April, a newly-emerged ♀ *Selenia tetralunaria* will attract ♂ s freely on a suitable night favoured with a slight breeze. Hang the ♀ in a small gauze cage ; the first arrivals come about 7.30 p.m., and continue till 8.15 p.m. They come singly, and can be boxed easily if a lantern be not used, but if a lantern be used the ♂ s are startled and fly off. Pairing takes place in confinement about midnight (Bacot).

The eggs of *Epione apiciaria* hatch in April and May; sometimes the emergence of a batch extends over as long as a month.

During the last week of April fullfed larvæ of *Ellopia prosapiaria (fasciaria)* are to be beaten from Scotch fir; they spin up during May, and the imagines emerge throughout June.

The spring-feeding larvæ of *Phorodesma smaragdaria* are great lovers of sun, basking on all the sheltered plants. They will feed up readily on southernwood and seawormwood, and can be bred to full size, provided the larvæ have an abundance of food, sun, air and space.

In April and May the larvæ of *Geometra vernaria* (green in colour after the first spring moult) can be freely beaten in many localities from *Clematis vitalba.*

Larvæ of *Geometra vernaria* may be obtained in April, but they are still small, dark in colour, rigid and quiet in the tray, so like the mass of rubbish that falls with them as to be readily overlooked, and so readily injured that they are much better worked for a few weeks later (Dollman).

In the first week of April larva-beating will give *Cleora lichenaria* (oak, &c.), *Metrocampa margaritaria* (elm, birch, oak, &c.), *Thera variata* (pine, Scotch fir), *Ellopia prosapiaria* (Scotch fir), *Geometra papilionaria* (birch and alder), still in their brown coats, &c.

The last week in April is the time for *Boarmia cinctaria;* Holmsley is its best known habitat; it refuses to rest on trees which grow closely together and prefers the medium-sized stunted Scotch firs in the most boggy parts of the heath (Carr).

The females of *Boarmia cinctaria* deposit readily in chip boxes, beneath the rough wood of the box or between the rim of the lid and the outside of the box.

A female *Biston hirtaria*, kept in a box, laid about 180 eggs the first night after pairing. These were placed as a thin cake, two eggs thick, in a crevice left by the starting of the edges of a box in which the female was confined. The eggs are small, bright bluish-green in colour, very smoothly laid, and remind one of a piece of German bead-work. The same ♀ laid above 500 eggs, all in a similar manner in any available crevice (Merrifield).

At the end of April, 1892, on Wallasey sandhills, although a drizzling rain was falling, we found the imagines of *Nyssia zonaria* exceedingly abundant; the females outnumbered the males and literally swarmed; the males in some cases rested flat on the short herbage, others were found down on the stems of coarse grass (Tait).

The larvæ of *Nyssia hispidaria* when fullfed bore deep into the soil in order to pupate, in shallow pans they are restless and often come to the top again, great care being needed to make sure that they do not escape.

The middle of April is the best time to work for imagines of *Aleucis pictaria*, which may often be secured, although the blackthorn bushes may not yet be in blossom; they are best obtained by searching the low hedges about an hour after dusk with a lantern, although they may be captured in fewer numbers by dusking.

About the middle of April imagines of *Larentia multistrigaria* are common at dusk flying over the heather in the Monkswood section of Epping Forest.

Sloe blossom is attractive in the Chingford district to imagines of *Eupithecia pumilata*, *Cidaria suffumata*, *Auticlea badiata* and *A. nigro-fasciaria*. The same species are to be found flying around the blooms on the sloe bushes at dusk at Chattenden so probably these are general visitors in the south.

Early imagines of *Eupithecia virgaureata* sleeved in April and May on hawthorn, lay eggs freely, the larvæ feeding up, pupating, and producing a second brood of imagines in August (Vivian).

The larva of *Eupithecia pygmaeata* feeds on the petals and anthers of *Stellaria holostea*, and in confinement will also eat greedily the petals and stamens of *Cerastium tomentosum*. The larvæ are fullfed in June and July (Crewe).

PTEROPHORIDES.—The hybernated Sphinx-like larva of *Agdistis bennettii* may be found in late April and early May, and again in July, at rest upon *Statice limonium* from which they fall on the slightest disturbance. It is only by searching the plants that show signs of being eaten by the lavæ that one can find them.

The larvæ of the early brood of *Platyptilia isodactyla* should be sought in April and May burrowing into the young shoots of *Senecio aquaticus*.

The larva of *Platyptilia pallidactyla* feeds in and on the young shoots of *Achillea millefolium* and *A. ptarmica*, and can be found from April to early June. It attacks the top of a young shoot, eats out the heart, mines downward for a distance into the young stem, and then leaves it to attack another. It seems to prefer a central shoot where obtainable. The pupa is to be found spun up to a leaf or on the stem, usually of its foodplant, being only attached by its anal segment.

The larvæ of *Cnaemidophorus rhododactylus* are to be found from April to June, feeding on wild rose, beneath the leaf overlapping the rosebud, eating into the unexpanded bud from the side; others, however, are to be found in similar positions at the tips of the young shoots. When fullgrown, those that have been feeding on the buds affix themselves to the side of the leaf close by the bud and draw the leaf and the bud together by means of a few silken threads; the others draw together, in a similar way, several leaves at the end of the young shoots.

The light green larvæ of the first brood of *Mimaeseoptilus bipuncti-dactylus* feed on *Scabiosa columbaria* in April and May, eating down into the heart of the plant before the flowering stem is thrown up, and thus utterly destroying it. The affected parts soon become covered by the strong healthy shoots so that the larvæ are not at all easy to find.

The young larvæ of *Oxyptilus heterodactyla* are to be found in late April and early May on *Teucrium scorodonia;* they eat the young leaves at first, but after the first moult (about first week in May) they go down the stem until they get within about $1\frac{1}{2}$ inches from the bottom, when they eat the stem just half-way through, causing the parts above to bend down and soon to become half dead and very soft ; on this part the larvæ feed ; the plants around soon over-top it and cover it, and hence one has to search carefully for the affected plants beneath the surrounding foliage. When fullfed the larva descends to below where it has bitten, and, on the sound stem there, attaches itself for pupation, head downwards.

PSYCHIDES.—The long, slender larva-cases of *Taleporia tubulosa* are to be found on tree-trunks, fences, &c., in April, but the larvæ never come up from their winter-quarters on the ground until they are quite fullfed, and the cases are usually much more abundant the middle of May, when they may be collected for breeding purposes. The ♀ s come outside the case for copulation.

The long, slender cases of *Taleporia tubulosa* are to be found in late April and early May, on old palings, &c.; the cases should be carefully scraped off the paling, pinned to the side of the breeding-cage by the end that was attached to the wood of the fence, for the imago emerges from the top end of the case.

The short, round, mealy-looking cases of *Narycia monilifera* (*melanella*) are to be found on lichen-covered trunks of trees in open parts of woods (Elisha).

ZEUZERIDES.—*Zeuzera pyrina* larvæ prefer branches and young trees of abut 8 inches in circumference; in trees of larger growth, elms for instance, they affect the upper branches of about that circumference; often to be taken from branches not more than 2½ inches in circumference—willow, elm, sycamore, pear, lilac, plane, &c. (Quail).

ÆGERIIDES.—Search for the larvæ of *Sesia chrysidiformis* early in April. The roots of dock and sorrel affected sometimes contain from four to six larvæ in each. The slopes on the sea-face of the cliffs are most productive, and those plants which have a sickly appearance or stunted growth are those that contain larvæ.

In digging the roots of dock and sorrel for larvæ of *Sesia chrysidiformis* gently move the surrounding mould so as not to damage the larvæ, whose presence is easily discovered by the mines and frass. If the roots do not contain larvæ replant them, as the next year these replanted roots are usually the most productive.

Transfer the dug roots of sorrel and dock containing larvæ of *Sesia chrysidiformis* to a breeding-cage. A very useful form for this purpose resembles a fern-case, with glass sides and ends, the top covered with a sheet of perforated zinc; fill the bottom loosely with a mixture of silver-sand, and the calcareous soil from the locality where the roots are dug. In this the roots should be planted, watered from time to time, and exposed freely to the rays of the sun. Early in May each larva sends up a conical case from the roots, composed of small particles of the fibres, and varying from an inch to an inch and a half in length (Russell).

LACHNEIDES.—In April the young, newly-emerged larvæ of *Trichiura crataegi* spin a slight web over a part of the foodplant, and are gregarious for a short time; in their second stadia they rest in small groups, and appear to prefer a twig with no smaller twigs branching from it, covering such part of it as they rest upon with silk, and though they leave it to obtain food they return to rest on the same twig; they love to bask in the morning sun.

DIMORPHIDES.—Keep pupæ of *Dimorpha versicolora* out of doors, and the chief emergence will then take place the 1st week in April. This will synchronise with the emergence of the wild imagines. A fresh ♀ taken to the breeding-grounds will then attract the ♂ and pairing will follow. Be on the ground not later than 11.30 a.m. Pairing does take place up to 3.30 p.m., but then with a much lessened chance of success (J. Clarke).

In mid-April two bred virgin ♀ s of *Dimorpha versicolora* were placed, without covering, on a birch-bush at 3.15 p.m. ; one commenced calling at once when a ♂ came up against the wind and paired with her ; the second commenced to call at 4.50 p.m., when another ♂ came up against the wind and paired directly. The branches the paired moths were on were then cut off and carried home in muslin bags. The insects remained paired till about 8.30 a.m. Next day the ♂ s were both mated with fresh ♀ s (Butler).

The eggs of *Dimorpha versicolora* are laid in small batches or from ten to twenty on the small twigs of birch, and are not difficult to find in April, when the haunts of the species are known.

The females of *Dimorpha versicolora* should be allowed to pair more than once, as frequently one pairing is insufficient to fertilise all the ova. The species can be inbred for a few years, but fresh blood should be introduced, otherwise the moths gradually dwindle in size and become less fertile.

Careful searching in Wyre Forest generally gives a few small batches of ova of *Dimorpha versicolora*, ten or a dozen being placed in a little cluster at the end of a thin twig.

To obtain the best results in breeding *Dimorpha versicolora*, cut off the twigs on which the ova are laid and tie such twigs to the living birch ; some 18 to 24 ova will be sufficient to place in a yard muslin sleeve (Clarke).

ATTACIDES.—In April and May the eggs of *Saturnia pavonia* are laid round and round a twig of some dead or living plant—ling, broom, gorse, &c.—close together and placed uprightly. On a heather stem they bear a most remarkable resemblance to a small bunch of dried *Calluna* flowers, a resemblance still more striking when seen among natural surroundings, in spite of the dissimilarity of individual eggs to individual blossoms. From 150—170 eggs are often placed in a single batch, although Holland observes that about a third of her eggs are placed by the ♀ in each of three different places. They hatch in about 4 weeks.

LIPARIDES.—By the end of April the almost fullfed larvæ of *Dasychira fascelina* are very abundant on the sallows on the sandhills of the Lancashire and Cheshire coasts ; they may be found at pretty well any time during the day and night, and two hundred forms a fair day's take.

The larvæ of *Dasychira fascelina* can be sleeved on sallow in April and May, and will spin up, requiring no attention except that of moving the sleeve when a branch has been cleared. They commence pupating in the middle of May.

The larvæ of *Dasychira fascelina* occur freely on heather and birch in the Carlisle district, but feed up well in confinement on hawthorn.

NOTODONTIDES.—If the season be forward, start searching the beech trunks for *Stauropus fagi* the last week in April. If a fertile ♀ be then found, a second brood should be obtained in late July and early August. By forcing these, there should be a fair chance of getting through a third brood by November (J. Clarke).

Petasia nubeculosa is attracted to light in both sexes during early April (Slevogt).

Pairings of *Petasia nubeculosa* are not difficult to obtain in confinement if the moths be placed in a large box, and out in the open air or near an open window. The moths do not copulate till the fourth or fifth night after emergence; the eggs are scattered over the gauze covering of the box in which the moths are kept (Maddison).

NOCTUIDES.— Larvæ of *Leucania littoralis* are the worst larvæ I know for escaping from anything in which they may be confined. They will not thrive unless given a lot of sand and exposed to the sun, I find the best plan is to keep them in a fern-case and feed them on marram-grass growing in a pot, if this be done, nearly all will go through all right excepting those ichneumoned.

Hybernated females of *Dasypolia templi* lay their eggs in early March. A batch consists of some three hundred eggs, and these hatch in five weeks. Placed on their foodplant, the young larvæ enter the root-crowns, eating their way in with ease; in a month they remove from old to new plants, eating the crown-leaves before they burst, and thence down into the roots (Gregson).

The larvæ of *Apamea ophiogramma* migrated from ribbon-grass to carnation-pinks in April, 1892 (Battley).

The last week in April is the time for larvæ of *Apamea ophiogramma*. Here and there the striped grass shows a drooping or faded leaf. Find the bottom of the infected shoot and pull, avoiding pressure as much as possible. In the stem, generally head upwards, lies the larva. It feeds only as far as the solid or semi-solid part of the stem goes. When that is gone, it leaves for another stem which it enters by making a large irregular hole near the ground (Burrows).

On the evening of April 20th, 1896, at Castle Moreton, I captured a beautiful specimen of *Xylomiges conspicillaris* whilst feeding on plum blossom (Fox).

The larva of *Celaena haworthii* is grey spotted with black, feeds in the stems of *Eriophorum*; a small hole is made in the stem, from which the excrement is ejected in profusion; fullfed in July. Pupal stage three weeks (Edleston).

The larvæ of *Noctua castanea*, common on heath, are to be obtained most freely by sweeping at dusk, yet many are to be obtained in the morning and late afternoon. During the middle of the day they appear to fall to the roots of the heather. They feed voraciously on hawthorn in captivity.

The larvæ of *Noctua sobrina* should be swept for from the end of April until the first week of June, when they are quite fullfed. The imagines commence to emerge the last week in July and continue till the second week of August. The larvæ feed on bilberry only, at least that is the foodplant in some places, although heath and various grasses are growing amongst it.

The larvæ of *Noctua festiva* feed well in confinement on violet, are fullfed in April and May, when they bury and pupate, the imagines emerging in June.

Search for larvæ of *Agrotis agathina*. You get fewer than by sweeping, but many captured thus are more or less injured by the sweeping and better results are obtained with larvæ taken by searching (Tugwell).

To rear *Agrotis agathina* larvæ successfully, take a large earthen propagating pan that has been planted beforehand with a good supply of *Calluna vulgaris* all round the edge of the pan, leaving some 4ins. or 5ins. in the centre clear of plants for renewing food, which should be done once a week. A few healthy plants of young and well-leaved *Erica tetralix* should be dug up weekly and plunged into a small earthen pan to occupy the space left free in the larger pan. By this means the food is changed with very little disturbance of the larvæ. The whole should be covered with canvas on a wire frame (Tugwell).

Imagines of *Pachnobia rubricosa* are sometimes plentiful at blackthorn bloom in Portland (Partridge).

The sandhills of Wallasey should be worked about the second week of April, on a calm warm night, for imagines of *Taeniocampa opima;* most are to be obtained from the sallow blossom, but many are to be found at rest on the marram and dead stems of hound's-tongue, ragwort, &c. The greater part of these latter will be found to be females engaged in the work of oviposition.

Eggs of *Taeniocampa opima* placed in a fine calico sleeve on sallow hatch well, and the larvæ feed therein quite satisfactorily till about one-third grown ; then divide the batch into smaller companies and move to larger sleeves, shift about every fourth day to provide plenty of fresh food and air ; remove when fullgrown to breeding-cage or large flower-pots for pupation.

To rear *Taeniocampa opima*, feed on sallow, keep in shade, with a little sunshine before 7 a.m., but none afterwards. Treated thus they pupate well in loose shallow trays which fit the bottom of my breeding-cages, into which the larvæ are removed from the bell-glasses about a fortnight before going down (Mason).

To rear *Taeniocampa opima* keep the larvæ in a large roomy box in a cool place, supply with plenty of fresh sallow (with hawthorn and plum to choose if they wish), and remember that after the last moult they are most ravenous feeders. They are fullfed from June 20th to end of month (Bisshopp).

Larvæ of *Taeniocampa pulverulenta* will feed up freely in confinement on whitethorn. Fullfed about the end of May.

When newly-hatched the larvæ of *Tiliacea aurago* bury themselves in the buds of hornbeam ; later they spin the young leaves together and live two or three together in the little tents thus formed. They are fully fed about the middle of May, spin their cocoons among leaves and rubbish, but do not pupate for about six weeks.

Tightly cork the young larvæ of *Tiliacea aurago* (which, if the ova have been kept out of doors, may be expected to emerge the first week in April) in tubes, and feed on maple or sycamore buds. In changing the food, search most carefully through each bud leaf by leaf, otherwise some will be lost. Their habit is to bury themselves right up to the centre of the bud, and thus many escape notice.

The larva of *Mellinia gilvago* leaves the egg in April, feeds at

E

first on the seeds of wych elm, then on the leaves, from which it can be beaten in May and June, and when fullfed its similarity to the larva of the much commoner *Mellinia circellaris* makes it difficult to distinguish from the latter.

The larva of *Mellinia circellaris* hatches in April and May and also feeds on the seeds of wych elm, from which it is to be beaten in company with the larvæ of the much more local and highly-prized *M. gilvago* in May and June.

A female *Dasycampa rubiginea* captured April 4th by beating sallows, placed in a large cardboard box and fed with thin syrup, laid, between April 18th and May 8th, 123 eggs, rarely more than from four to sixteen a night. The first larva emerged on April 27th, the last on May 19th, the egg-stage lasting about ten days ; the young larvæ fed well on apple (refused dandelion when young, but ate it after second moult) ; they were fullfed in July, and remained some time in the cocoon without pupating, 71 imagines emerging between October 21st and November 9th, from noon to 4 p.m.

The larvæ of *Polia* var. *nigrocincta* feed by night, and thrive in confinement on sallow and groundsel, and like a warm, dry situation ; growing plants in pots, covered with glass-cylinders, are best for rearing them in. They are fullfed in June, and pupate under small pieces of stone or fine sand ; the pupæ should not be disturbed (Murray).

The eggs of *Polia* var. *nigrocincta* hatch at the end of May or beginning of April, feed upon plantain, sea-plantain, violet, dandelion, &c., taking later on to *Statice armeria*, preferring the flowers. Towards the end of June the fullfed larvæ may be found by night on the latter plant on the coast of the Isle of Man. The larva appears to be strictly nocturnal in its habits.

During the second week of April the larvæ of *Epunda lichenea* are to be found freely at night on *Sedum acre* on the coast sandhills of Lancashire and Cheshire.

Searching tree-trunks in April produces a fair number of *Xylocampa areola (lithorhiza)* in the Epping Forest district.

In the New Forest in April, *Xylocampa areola (lithorhiza)* usually rests on birch and fir-trunks, generally very low down, and not in places where the trees grow thickly together.

Hybernated females of *Calocampa exoleta*, taken at sugar or sallows in early April, should be fed and kept for eggs ; they often live on into May before ovipositing.

Females of *Xylina socia* captured in April deposit ova that hatch in early May. Young lime leaves and buds appear to be preferred to half-opened birch leaves. The larvæ feed up well on lime, keeping to the underside of the leaves. The species is frequently taken on sallow in the spring. I think that the moths always lay their eggs at this season, and not in autumn (Kane).

In April and May, when young, the larvæ of *Plusia moneta* feed two or three together in a web on *Delphinium* and *Aconitum*, but later feed independently, and may easily be shaken out of their foodplant (Hall).

ARCTIIDES.—The larvæ of *Arctia villica* are frequently to be obtained in April, stretched at length on furze bushes, basking in the sun. They seem to nibble the young shoots of the furze, but frequently they appear to return to the low plants about the roots of the furze bushes, when the necessity for real feeding again occurs.

PAPILIONIDES.—At the end of April and in early May the eggs of *Cyaniris argiolus* are laid singly on the underside of the calyx of holly buds, so that when the flowers open the petals fold over the egg, hiding it altogether from sight ; the young larvæ hatch in about ten days, commence at once to feed in the buds and flowers, after attacking the young tender leaves and shoots upon which they thrive. The larvæ also feed well on young ivy leaves.

The larval colonies of *Melitaea aurinia* are found in April on scabious and honeysuckle, but the eggs always appear to be laid on the former. The pupæ are to be found in May spun up under dead oak-leaves, or suspended from grass culms, reeds, bilberry stems and other low-growing plants.

During April and the early part of May the larvæ of *Melitaea aurinia* are usually very abundant in their now-restricted haunts. The imagines are usually out by the end of May or in early June. The larvæ feed up well in confinement on honeysuckle.

In confinement, the larvæ of *Melitaea aurinia* appear to be very susceptible to warmth, collecting in the hottest part of the cage, and becoming lively when the sun is on them. They are much better fed up, however, on a growing plant than in a breeding-cage.

The hybernated females of *Eugonia polychloros* should be sleeved on sallow and fed with syrup-soaked pieces of bark ; they will then lay freely. Larvæ fed upon sallow are said to produce larger imagines than those fed on elm.

Females of *Polygonia c-album*, captured in early April, and sleeved on nettle and currant, will give eggs that hatch in due course, the pupæ producing imagines usually in July.

In the spring, the larvæ of *Apatura iris* feed up quickly ; they eat at night and rest in the day-time on the mid-rib of the upperside of the leaf, the head towards the base of the leaf. One ought to be able to find them by the leaf hanging down with the weight of the larva, otherwise they are almost invisible (Hewett).

In the last week of April and early May search for the larva of *Limenitis sibylla* ; it is then in its brown stage, and rests on the brown stem of the honeysuckle, just below the green shoot, generally low down on the bush, in a sheltered position. Sometimes the larvæ may be found on the green leaves where they are much more conspicuous than on the stem.

MAY.

The large amount of field work that has to be done in May has already been pointed out (pt. i, pp. 23-24) and there is little to add. For the sportsman, pure and simple, a few hours at the end of May and June should be devoted to the rhododendrons in some good locality. By day the Hemarids—*Hemaris fuciformis* and *H. tityus*—and by night the Eumorphids—*Eumorpha elpenor* and *Theretra porcellus*—give excitement enough in their favoured haunts to the most exacting, whilst if it be a " *celerio*" or " *livornica* " year, the immigrants will join the Eumorphids at the feast. There is one line of work, however, which has attracted much attention of late and will possibly attract still more notice in the future, and that deserves the attention of the more scientific breeders of lepidoptera, *viz.*, the rearing of hybrids between allied species. Among the easiest of these to procure is *Smerinthus* hybr. *hybridus*, *i.e.*, *Smerinthus ocellata* ♂ × *Amorpha populi* ♀, the reverse cross being, however, very difficult to obtain. To get crossings between two allied species, place a male of one of the species and a female of the other in the same cage, and, in close proximity, but in another cage, a female of the same species as the male in the first cage. The attractive influence of this second female will excite the male, and he will then pair readily with the female with which he finds himself shut up. One can obtain a normal pairing of any of our three British Amorphids—*Smerinthus ocellata*, *Amorpha populi*, *Mimas tiliae*—pretty readily by placing a newly-emerged female quite free on a tree in a garden, and leaving her all night, when, usually, a male will be found paired with her in the early morning. Often to make quite sure that the female shall not crawl away, a slip-knot is placed round her abdomen, but this often excites her and she will sometimes not settle down so comfortably as is to be desired. Tethering in this manner is often practised when attracting *Amphidasys strataria* and other fairly large-bodied moths belonging to other groups. It must not be forgotten that the imagines of many species that are reared during this and the succeeding months, including many rare species, can be easily paired and ova successfully obtained in confinement by enclosing the females over potted plants of their food, *e.g.*, *Eupithecia extensaria* pairs freely and lays her eggs readily when enclosed over a potted plant of *Artemisia maritima*. At this time, too, the first broods of many of our partially double-brooded Notodontids are on the wing, imagines occurring again in some seasons at the end of July and beginning of August in our southern counties, among these are *Clostera reclusa*, *Leiocampa dictaea* (the second-brood examples of which are rather smaller and paler than those of the first brood), *L. dictaeoides*, *Notodonta ziczac*, *N. dromedarius*, and *Lophopteryx camelina*. The imagines of the second broods of these species lay eggs and produce larvæ which are rarely fullfed until the end of September. Owing to the double-broodedness being partial it must be borne in mind that we have individuals that are in the pupal stage from June or July until the following May, whilst others have a pupal period only from October to May. In Scotland this partial double-broodedness is almost unknown, the above-named species being purely single-brooded, whilst, strange to say, the period of emergence for this brood is much extended, lasting over the whole time that the two broods occupy in this stage in the south, *viz.*, from the end of May to July, or from six to eight weeks. Gravid female lepidoptera, to be used for breeding purposes are now frequently sent by one lepidopterist to another ; such brood females will travel well by post in metal glass-topped boxes. A piece of something fixed rigidly in the box to which they can cling is most desirable. Leaves or any part of the foodplant that will wither and shake about loosely in the box not only makes the insect restless but often injures it sufficiently to break its legs and wings or even to cause death. These boxes should, for postal purposes, be carefully and immovably packed in a cardboard box, which should be then safely fastened and a label attached for stamping. Anything that will reduce the stamping of the box containing insects is very desirable and necessary. Never put the stamps on a box containing ova, larvæ, pupæ, living imagines, or set imagines. Always place them on the label. Larva-hunting and larva-beating will, at this season, occupy a great share of the lepidopterist's time, but the subject has already been dealt with (pt. 1, p. 24). It is well to remember that, in collecting larvæ of *Agrotis agathina*, the larger, almost fullgrown, examples, collected in May and June are those most free from parasites, the small larvæ found at this time of the year are usually small because they are stung, and their chief parasitical enemy appears to leave its hosts when the latter are about two-thirds grown. Bring home, therefore, the largest larvæ you can find, when collecting this species in May and June. During May, two of our rarer Micro-lepidoptera want working for on the moors ; these are *Lithocolletis vacciniella* and *Nepticula weaveri*. The larva of the former mines the underside of the leaves, that of the latter, which requires a great deal of finding, mines the upperside. At about the same time the northern collectors should make their bags of the larvæ of *Retinia resinana*, the large, con-spicuous, resinous masses produced by which are abundant in some parts of Aberdeenshire and Morayshire, and possibly occur in other suitable places. On the Thames marshes the fullfed larvæ of *Phorodesma smaragdaria* should be obtained ; it is remarkable, considering the way in which the *Artemisia maritima* is buried in the spring under the high tides at Benfleet and else-where, how these larvæ get through their hybernation, for they must be buried under every high tide. Continuing our notes from last month on rearing larvæ *(anteà,* pp. 21—23) we would observe that the use of a bee-glass is strongly recommended by many lepidopterists. This is an ordinary horticultural bell-glass, but a hole replaces the ordinary solid knob by which it is lifted. They may be of various sizes, generally, however, those in common use are 12 inches in diameter. It may be used basin-like, the stems of the foodplant passing through the hole in the bee-glass into a receptacle below filled with water, and on which the bee-glass rests. The stems of the foodplant must be well padded at the hole with wadding or cotton-wool, as many species, especially when nearly fullfed, will otherwise crawl into the water below. A piece of muslin is tied over the open-end of the bee-glass, and the cage is complete, the food usually keeping well and the larvæ obtaining a fair supply of air and light, whilst the habits of the larvæ can readily be observed. Another way of using the bee-glass is to reverse it, in the direction opposite to that suggested above, over a large earthenware propagat-ing-pan entirely or partly filled with well-baked earth—a mixture of sand and peat is frequently used. In this case ventilation is not easily secured, but the difficulties can be met by having a perforated zinc cylinder, reaching to the bottom of the soil, and resting on the floor of the pan, to prevent escapes, made of such a size that the bee-glass exactly fits in it, the latter being held in position at the top by a zinc rim a short distance from or at the top inside; this allows a free passage of air through the zinc cylinder and the hole at the top. A glass bottle to hold the food may be buried in the earth in the pan. When the food wants changing, put in a fresh bottle with a fresh supply of food, so that the larvæ can crawl from the old to the new; the old can then be removed without any unnecessary handling of the larvæ. One may take it for granted that it is always bad to handle larvæ, very few will stand much of it. A cover of *Sphagnum* on the surface of the earth affords hiding-places for the larvæ that desire it. Above all things, do not allow frass to accumulate. Frequently, pieces of cork, short lengths of dried stem of large umbellifers, and similar materials, are necessary for some fullfed larvæ that require something of the kind in which to bore to pupate. Do not forget that a closed glass-top box, or a closed glass vessel without proper ventilation, and placed in the sun, provides the quickest possible means of killing larvæ and imagines. Tugwell notes that the best results in breeding will generally be secured by confining oneself to moderate numbers, when there will not only be fewer deaths, but the insects obtained will often be finer. Overcrowding, especially if combined with stale food, is almost sure to bring on disease. This often takes the form of diarrhœa, and carries off the entire brood. One of the first symptoms of this is found in the frass becoming soft and watery, the larvæ themselves getting frequently soiled with it, the anal segments particularly so, and it is not often that one can do much to arrest this disease when once it asserts itself. A plan tried on a brood of some 30 *Dimorpha versicolora* with fair success when the larvæ showed unmistakable signs of diarrhœa was to take each larva separately in the open hand, hold it under a tap of running cold water, and gently brush it the while by means of a camel-hair pencil. The larvæ were then placed on clean fresh food and stood out in the sunshine. As a result, most of them safely pupated, and produced' in due course fine moths. Sleeving is generally adopted by those lepidopterists who wish to save time, &c. Sleeves are best made out of black book-muslin, which is less conspicuous than white and looks clean longer. Ova may be put (very carefully) into a sleeve about a fortnight before they are due to hatch, and, according to the size of the species, the number may vary from a dozen to three or four hundred in a sleeve a foot long. The branch selected must be free from earwigs and green fly; a sharp shake will dislodge the former, but the latter are much more difficult to get rid of, and, if any persist in clinging to the leaves, a fresh and free branch should be chosen. Long branches or those without much foliage should have their tips tied back towards the base. The tying on of a sleeve is most important. Always pull the free end quite tight, and have a string long enough to catch the gathered muslin at about an inch from the end, tying it within the gathers. At the base fasten with two or three twists of the string before tying, to keep in the larvæ and keep out earwigs and other enemies. When a change is required, cut the sleeve right off and take it and its contents under cover, open the bottom and pull out the branch over a clean news-paper; many larvæ will be left in the muslin, these need not be disturbed, and when the others are transferred from the newspaper and twigs back to the sleeve, the latter can be tied on a fresh branch. The removal of larvæ from the twigs requires care. Commencing at the top of the branch, clip off (with scissors if necessary) the little twigs to which the larvæ are clinging, into a jam-pot; by passing a small piece at the time under review the whole is rapidly gone over and thoroughly cleared; larvæ that insist on clinging tightly to the larger stems are best removed by passing a sharp penknife beneath them, leaving them on the piece of detached bark. Larvæ that live in webs may best be dealt with by holding the branch horizontally in the left hand and tapping it smartly with a pair of scissors over a newspaper, when most will detach themselves, and either fall down or hang suspended by a thread, when they can be guided into the jam-pot; repeat this operation till the branch is cleared. It will usually be found advantageous, if the brood be a large one, or the larvæ be of large size, to transfer the larvæ in their last stadia to a large roomy box or breeding-cage and finish rearing them by hand. This is also necessary with all larvæ that require earth in which to burrow for pupation. A few branches cut off and inserted in damp soil in a small pot will last quite fresh for three or four days. On the other hand, if the brood be a small one and does not require soil, with ordinary attention one may successfully rear as many as sixty larvæ of the size of *Selenia bilunaria* in a sleeve little more than a foot and a half long and about 9 inches in diameter. For many species the folds and creases of the muslin seem well adapted as sites for the making of their puparia, the majority of species preferring these to the twigs and leaves. It must not be forgotten that some larvæ which may have shown a most docile disposition to remain within the sleeve whilst being reared, will, as soon as food begins to run short or the exigency of pupation warns them that the environment is no longer suitable, eat their way out of the sleeve and seek for themselves that food or position with which the careless breeder has neglected to supply them. Little pupa-digging can be done at this season, but bank-raking for larvæ, pupæ or imagines has been suggested as work worth being adopted when pupa-digging is impracticable; it is claimed by those who practise it that it can be indulged in throughout the greater part of the year, and they aver that it is unquestionably one of the most successful methods of obtaining good specimens of the imagines of certain species. Find an over-hanging bank of any sort, a hole where a tree has been blown down, round the edges of a quarry, a landslip on a mountain-face or slope, however small or large, never mind how barren the place is, banks or sandhills, and particularly banks in lanes, caused by cuttings where the soil or gravel has fallen away a little from the vegetation. Having found any such place, first look carefully under the overhanging grass, and, lifting it gently with the left hand, pin and kill the moths and butterflies you find there. Peer well into the crevices before using your walking-stick. When you have got all you can see, and also when you cannot see anything, draw the point of your stick across the grass-roots, &c., hanging under the bank, and the chances are greatly in your favour that one or more Noctuids will fly out, or what is more general, roll down the bank, sometimes perfectly quietly, at others fluttering very much as they fall; in any case have a small net ready. Repeat this several times, and if windy or wet shake the plants well. To follow "bank-raking" requires both patience and perseverance, for, like pupa-digging, you do not find nuggets in every hole, and it is no joke walking on sand-banks, where your progress is like that of lawyers journeying heavenwards, but it has the advantage of being most advantageously practised when pupa-hunting is rather out of season and you can get the insects as good as bred without further trouble. Where banks are scarce, as on pasture-lands, small bundles of grass or hay should be pegged down and shaken whenever the ground is visited. It is true that the process produces unlimited quantities of common moths, but sometimes one gets a prize (Gregson).

MICROPTERYGIDES.—In May and June *Micropteryx thunbergella* is very common in Chippenham Fen, where they are to be obtained swarming about the "bird-cherry" blossom and the young fir cones. One day they were observed so thickly on the cones that fifteen were boxed in one large chip-box off one cone (Farren).

Micropteryx calthella often swarms in late May and early June in the flowers of *Caltha palustris*, and imagines of *M. seppella*, in the flowers of *Veronica chamaedrys*, the insects feeding on the pollen; often as many as half-a-dozen golden-green atoms may be seen among the stamens in a single flower.

ADELIDES.—Towards the end of May, imagines of *Nemophora pilella* are to be found among *Vaccinium* on the moors at Green Thorn (above Stoneyhurst), Glen Tilt, Rannoch, &c. They fly in the hot sun among the *Vaccinium*, and when startled soar up into the fir-trees. They are active in the net, and want care in boxing, even after they are in the net they may be lost.

LITHOCOLLETIDES.—The larva of *Lithocolletis pyrivorella* makes a short mine on the underside of leaves of *Pyrus communis* and its cultivated varieties, and also on *Pyrus malus* and *P. aucuparia*, in May and June, and again from September–November, hybernating as a pupa.

The larva of *Lithocolletis concomitella* mines on the underside of leaves of *Pyrus malus* and its cultivated varieties, also on *Prunus communis* with larvæ of *L. spinicolella*, in May–June, and again in September and October. It hybernates as a pupa and not as a larva, as do *L. cerasicolella* and *L. spinicolella*.

ARGYRESTHIIDES.—The bright green larva of *Zelleria insignipennella* is to be beaten from juniper in mid-May (in Headley Lane) when fullfed. It is very like that of *Cerostoma costella*, spins up almost at once, the imago appearing in mid-June. Possibly feeds also on some other foodplant.

COLEOPHORIDES.—The larva of *Coleophora binotapennella* mines *Salicornia* and does not construct a case till nearly fullfed in May. When fullfed it crawls in its case down to the surface of the mud, in which it spins its cocoon at a little depth, leaving its case sticking up at the surface.

The larvæ of *Coleophora albitarsella* are plentiful on *Glechoma hederacea* in early May; the imagines may be obtained plentifully in middle June by smoking (Miller).

Larvæ of *Coleophora ibipennella* should be sought quite early in May on birch. They often eat right through the leaf, leaving only the veins, somewhat in the manner of the larva of *Thalera lactearia*.

Search hawthorn bushes, 6 or 7 feet high, for the larval cases of *Coleophora hemerobiella*, at the end of May.

Sloe bushes and hawthorn hedges will yield young larvæ of *Coleophora anatipennella* early in May, but in order to breed specimens for the cabinet it is better to take the spun-up cases at the end of June. They will be found on the upper surface of the leaves.

Small plants of *Genista anglica* often yield numbers of cases of *Coleophora genistae* early in May. It is much more difficult to find them later.

Birch in May and June will often yield larvæ of *Coleophora fuscedinella* in abundance.

Larvæ of *Coleophora nigricella* are easily found, early in May, on hawthorn hedges or on young shoots of apple in gardens.

The larvæ of *Coleophora gryphipennella* may be found on rose-trees in gardens, trained against walls or fences, or on wild roses in hedges. Pale patches on the leaves will betray their presence.

The larvæ of *Coleophora solitariella* are to be found along hedge-banks on which *Stellaria holostea* grows, at the end of May. Search any plants which have a white or variegated appearance; the green plants are no good for larvæ.

The easiest way to take *Coleophora albicosta* is to beat the imagines out of sheltered furze bushes, on warm afternoons in late May.

GLYPHIPTERYGIDES.—In May, the imagines of *Heliozela sericiella* fly in little swarms round the twigs of oak-bushes in the full sunshine.

TORTRICIDES.—The larva of *Sciaphila sinuana* lives in a loose silken web spun among the flowers of *Scilla nutans*, feeding on the flowers themselves and also on the green unripe seeds, and occasionally, in confinement, nibbling the stalk. The pupæ are enclosed in slight loose white silken cocoons, spun, in confinement, among the flowers and stems of the wild hyacinths.

The larvæ of *Sciaphila nubilana* should be collected in May from whitethorn and blackthorn. It is the only way really to get (1) good males, which lose colour almost directly after they have flown, and (2) females, which are rarely taken in the imaginal state, and are not at all free fliers.

The larvæ of *Sciaphila pascuana* have been bred from larvæ found in May in spun-together tops of milfoil; a very curious pale yellowish form is also bred from larvæ found in the Essex saltmarshes on *Aster tripolium*.

The larva of *Bactra lanceolana* may be found well on into May, in boggy places, feeding and afterwards pupating in the stems of *Juncus conglomeratus*.

The larvæ of *Bactra furfurana* feed in May in stems of

Eleocharis palustris, ejecting green frass, and pupating in the stem in slight silken cocoons; the imagines appear at the end of June and in early July.

The larva of *Euchromia purpurana* feeds on the roots of dandelion and allied composite plants in May; the species prefers rough uncultivated ground.

The imagines of *Grapholitha gemmiferana* occur at the end of May (May 27th—June 2nd, 1890) among plants belonging to the nat. ord. *Leguminosae.* The perfect insect is like *Grapholitha caecana*, but with more ample wings and with the gemmated markings on the costa more developed *(Ent. Rec.,* iv., pp. 112—113).

The larvæ of *Grapholitha naevana*, may be obtained abundantly on the shoots of holly in May and June, the imagines common on trunks and fences near a few weeks later.

The imagines of *Phloedes immundana* are locally abundant among alder in May, a partial second-brood occurring in August.

The imagines of *Coccyx argyrana* are sometimes very common in May and June, resting on oak trunks; they may also be bred from oak-apples gathered during the winter.

The imagines of *Stigmonota germarana* may be beaten from, or netted flying around, oak-bushes in May and June.

The imagines of *Stigmonota internana* occur among furze, in some places very abundantly, at the end of May and in early June.

In early May the imagines of *Stigmonota perlepidana* fly in great numbers about the flowers of purple vetch. The males are very white-looking on the wing, the females much darker and less conspicuous.

The imagines of *Carpocapsa juliana* are to be found in May and early June at rest on oak trunks.

The larvæ of *Catoptria hypericana* are to be obtained in the young shoots of *Hypericum* in early May.

The imagines of *Pyrodes rhediella* may be taken freely at the end of May flying in the sunshine round the tops of tall hawthorn bushes or hedges. The larvæ feed in the green berries.

The pupæ of *Eupoecilia udana* are to be obtained in the stems of *Alisma plantago* in early May; the larvæ in the same position throughout the winter and early spring.

The imagines of *Eupoecilia udana* are to be found at the end of May and June flying over the water plantain *(Alisma plantago)*, in the stems of which the larvæ feed throughout the winter and spring.

In May the imagines of *Eupoecilia maculosana* are to be taken in abundance during the day flying among bluebells; the males look quite white on the wing; the females are altogether less active and much less frequently captured.

PLUTELLIDES.—In mid-May the fullfed larvæ of *Eidophasia messingiella* are to be found on *Cardamine amara :* they are best obtained by sweeping, and will spin up within a week (Zeller).

The young, active, light green larva of *Plutella annulatella* feeds in June upon the seeds and pods of *Cochlearia ;* when older, it feeds upon the fleshy leaves, making round holes and blotches in the leaf, principally working from the underside (Gregson).

PYRALOIDES.—The larvæ of *Depressaria assimilella* are to be found between the matted twigs of broom in May.

The larvæ of *Depressaria pulcherimella* feed on the flowers and young seeds of *Conopodium denudatum (Bunium flexuosum)* under a slight silken web, at the end of May.

The imago of *Metzneria littorella* occurs somewhat plentifully from May 6th-29th (perhaps later) on the cliffs near Ventnor. Owing to its retiring habits it may be easily overlooked, unless dislodged by smoke or other means from the *Plantago coronopus* (Walsingham).

The larvæ of *Butalis grandipennis* are abundant near Carlisle during May, feeding gregariously in webs on furze (Day).

The larva of *Aristotelia tetragonella* is to be obtained from *Glaux maritima* during May.

In May and early June the imagines of *Pancalia lleuwenhoeckella* are exceedingly abundant among violets, also to be found resting on the blossoms of daisies and other flowers common in their habitat during the afternoon. They are more active in the evening, and are readily swept.

The larva of *Lita plantaginella* feeds on *Plantago coronopus* in May; it may be found burrowing in the root to the depth of nearly half an inch, feeding on the substance of the root, spinning together the central leaves of the plant to conceal itself from view, and changing to a pupa in its burrow.

The larva of *Lita suaedella* burrows among the fleshy leaves of the *Suaeda*, which are something like thick, short pine-needles, spinning them down to the stalk in May. At the end of May it leaves its burrow to spin up in the sand or wood under the plant on which it has fed.

CRAMBIDES.—The larvæ of *Rhodophaea marmorea* are to be collected from blackthorn in May; look for webs spun closely along the branches and twigs; the imagines emerge in July.

PYRALIDES.—The larvæ of *Scopula lutealis* occur very freely at the end of May, on the lower leaves of bramble, also found on wild strawberry, *Plantago lanceolata, Ranunculus* and several other low plants; it has also been found on thistle and dock, so that it is evidently quite as general a feeder as are the larvæ of *S. prunalis* and *S. olivalis* (Porritt).

Scopula decrepitalis occurs in May from Kilmun to Lochgoilhead, more often in damp places near the shore, usually seen sunning itself, its wings stretched fully out on the bracken.

ORNEODIDES.—In May the young larva of *Orneodes hexadactyla* eats its way into a flower of honeysuckle, leaving a little frass outside; its food consists of the stamens and the styles of the pistils; it moves from one flower to another, entering also by a small round hole which betrays its whereabouts.

In May and early June the hybernated females of *Orneodes hexadactyla* may be found busily laying their eggs on the flower-buds of the honeysuckle, on almost any part of the corolla or on the margin of the calyx. The larvæ feed inside the buds and flowers in June and July.

PTEROPHORIDES.—In mid-May and early August the fullfed larva of *Agdistis bennettii* crawls to the top of a leaf of *Statice limonium*, spins numerous silken threads across it, attaches itself by its anal claspers, and after two days becomes a long characteristic Pterophorid pupa with a marvellous ability to throw its head up, and away from its resting-position.

Agdistis bennettii larvæ are given as being common on *Statice auriculaefolia* at Portland by Richardson ; feeds on *S. limonium* on the Medway marshes and by the sides of the Yar, in the Isle of Wight.

At the end of May search for bladdery-looking leaves of *Artemisia vulgaris* in which the pretty little hairy larva of *Leioptilus lienigianus* finds its home.

The young larva of *Leioptilus lienigianus* is to be found in May and gnaws oblong blotches near the tips of the upper leaves of *Artemisia vulgaris*, leaving the cuticle of the upperside entire and nearly transparent, eating the parenchyma, and carefully rolling back the downy skin of the upperside to the edge of the blotch (as is done by the larva of *Aciptilia galactodactyla* on burdock). The blotches are seldom more than half-an-inch long, but generally there are two or three of them side by side (Barrett).

The fullfed larvæ of *Leioptilus tephradactyla* are to be found feeding on the leaves of *Solidago virgaurea* in middle May. [In April the larvæ are very small, and no facts are known as to its hybernation or whether the eggs do or do not hatch before winter.]

The imagines of *Leioptilus mircrodactylus* are to be obtained in plenty towards the end of May and again in early August. It is to be found almost anywhere among *Eupatorium cannabinum* and in all sorts of localities, *e.g.*, the chalk cliffs at Kingsdown and Wicken Fen. It can be disturbed from the foodplant during the day, but flies quite freely about sundown, and is often in considerable abundance.

The green larva of the common *Aciptilia pentadactyla* is to be found on *Convolvulus arvensis* in May and June, and is readily reared (Harding).

The larva of *Aciptilia pentadactyla* feeds on *Convolvulus (Calystegia) sepium* in May, eating the leaves and flowers, and resting on the underside of the leaves and on the stems. The pupa is attached by the anal segment to the foodplant, and is of a greyish-green colour (South).

About the middle or end of May the wellgrown larvæ of *Aciptilia tetradactyla* are to be found feeding exposed on the leaves of wild thyme, which they so closely resemble in colour that they are not easy to detect (Bankes).

The larva of *Aciptilia migadactyla (spilodactyla)* is to be found on *Marrubium vulgare* from the end of May until July, usually several are to be found on one plant. South observes that it feeds on the terminal leaves, rests on the upper surface of the leaf in dull or damp weather, but hides underneath when the sun shines.

The green larvæ of *Marasmarcha phaeodactyla* feed on the terminal leaves of *Ononis* in May and June, being excessively abundant in some localities—Cuxton, Dover, &c. The pupa is generally to be found spun on the foodplant in early July.

GEOMETRIDES.—In May search the honeysuckle for larvæ of

Pericallia syringaria; they get far inside the big branches and very near the ground on the trailing ones ; they readily drop, so one must be careful to select the right piece before disturbing them, and be sure not to touch the trailing branches. They vary much in colour, from brownish to deep red. They are generally to be found at least six inches from the place where they have been feeding, and are hard to see on the stem on which they rest. After searching the bush, beat it for overlooked larvæ.

The eggs of *Ennomos fuscantaria* go over the winter and hatch very irregularly between the end of May and early July ; as a result the larvæ are to be found all summer of different sizes and are fullfed from mid-July until early September ; imagines appear from July until mid-October. The larvæ feed up well if sleeved on ash in gardens (Hewett).

During May and early June search birch and larch trunks for *Tephrosia punctulata;* difficult to detect on the white birch bark, but more conspicuous on the larch.

The lichen-covered branches of the large oaks, thorns and beeches near Lyndhurst station should be beaten for larvæ of *Cleora lichenaria* in May. The larvæ fall readily into the tray (luckily the lichen doesn't) so that all the bits of lichen in your tray will probably be larvæ of *C. lichenaria.* You had better take home a box of lichen and keep it just moist for them.

In the New Forest the larvæ of *Cleora glabraria* are common on the long lichens growing on beech, hawthorn and oak, and the imago is to be obtained from the same trees in July and August.

At Lyndhurst, on the dwarf firs growing among the heather, in May, *Boarmia cinctaria* is abundant often on the very small trees, difficult to see at first, but easier once you get used to the work; they rest on other trees, birch, &c., the larvæ feeding later on heather.

Hemerophila abruptaria is a very easy species to breed, sleeved on lilac or privet ; a large brood had only to be re-sleeved once; on spinning up some of them choose the crevices in the muslin, forming a much slighter cocoon than those which spin on the twigs, for, when the muslin is stretched, nearly all the pupæ in the muslin folds tumble out. Those on the twigs often form curious objects, lying one after the other, as many as six in some cases, all joined end to end. What strikes one most is the curious manner in which the larvæ (when at rest) hang pendant from the twigs. Perhaps it is owing to their being so crowded, but they remind one more than anything else of the pictures drawn in seedsmen's catalogues of fabulous crops of peas (Alderson).

Zonosoma pendularia usually sits on a birch-stem anywhere under five feet from the ground ; the best plan on a hot day appears to be to tap the stems sharply with a stick and net them as they fly off, which they do instantly ; they will hardly ever allow you to box them sitting, except towards evening, when it gets cold, or on a very cold day ; they often rise even before you strike the stem. When *Cidaria corylata* begins to appear, the difficulty of catching them is increased, as, though the colour and flight are very different, it is difficult to keep the eye from following the *C. corylata,* which sometimes rise three or four at a time (Woodforde).

The alders want beating in early May for larvæ of *Geometar papilionaria,* which are difficult to see in May, although the leaves perhaps are only just showing, they look very like small stumpy alder twigs and are easily passed over even in the beating-tray; in their early spring coats they are reddish, with a green crest along the back, very like the alder leaf just breaking out of its sheath. Towards the end of May the almost fullgrown larvæ of *Geometra vernaria* may be beaten; insert the stick into a bush of *Clematis vitalba,* hurtle it about, and one will obtain a larger number of uninjured specimens than by beating the plants; the larva falls readily with a sudden movement of its foodplant, and does not require a heavy blow to dislodge it (Dollman).

In May (and again in July) the imagines of *Fidonia conspicuata* fly in the hot sunshine, whilst on wet and cloudy days they may be found resting with wings closed over the back on the underside of the broom bushes.

The pines about Lyndhurst give larvæ of *Thera firmata* in May; they are tedious to breed, they feed so slowly. For those who have a difficulty in getting Scotch fir, do not forget that a branch kept in damp sand keeps fresh a long while. Put the fir into the pot in small pieces, otherwise, when you come to change the food later, you will never find the larvæ. Be careful also not to throw away the green pupa of *T. variata,* lying along the needles near the base. *Ellopia prosapiaria (fasciaria)* also spins among the needles, but the pupa is red and easily seen. *T. firmata* will want sand to spin up in.

Lobophora hexapterata appears in the middle of May, sitting, when newly emerged, in the full rays of the sun, but is exceedingly wary if an attempt be made to box it; its colour assimilates well to the bark on which it rests, and it glides rapidly from one place to another. The species flies after sunset, the males searching up and down the stems of the aspen for the females, or flying at a short distance from the tree. The coldest evening does not prevent them being out, and they are sometimes over before the end of the month (Talbot).

The larva of *Eupithecia debiliata* (with a somewhat superficial resemblance to that of *Abraxas grossulariata)* feeds on bilberry, fastening two or three leaves together, in which it hides during the day (Kaye).

Eupithecia indigata is to be searched for on the boles of fir-trees at the end of May (Edleston).

The larvæ of *Cidaria prunata* should be fed on red-currant, not on black-currant, failure resulting when the latter is used as the foodplant. The ova of this species hatch in early May, and the larvæ spin up towards the end of June. The feeding larvæ require but little attention, and in spinning up appear to select a part of the stem of the foodplant just below the junction of the branches, several cocoons often being placed together so as to almost encircle the main stem (Adkin).

CYMATOPHORIDES.—During the first week of May dig for pupæ of *Cymatophora ocularis* at the foot of poplars. On May 2nd, 1872, I turned up no fewer than twelve pupæ of this species at a single poplar-tree at Malvern Link (Goodyear).

ÆGERIIDES.—In rearing *Sesia sphegiformis,* there is a tendency for the pupæ sometimes to drop out of their burrows when they are obtruded for emergence, and under such circumstances the imagines almost always fail to emerge without immediate manual assistance (Hall).

The larvæ of *Sesia chrysiditormis* are best collected when nearly or quite fullfed; the roots of sorrel and dock, in which they are feeding, should then be plunged into mould in boxes sufficiently deep to be covered with gauze or canvas (Tugwell).

PSYCHIDES.—The cases (larvæ) ot *Whittleia retiella* are to be obtained in early May on the saltmarshes at the mouth of the Thames. At Canvey, it is necessary to lie down and search very closely for the cases to obtain them in any numbers (Whittle).

The cases of *Whittleia retiella* are attached to a wiry grass (? *Poa maritima*) that grows in patches on the open marshes. The imagines also occur among this, more rarely among *Plantago* and *Atriplex*, the receptacle of a dead flower-head of *Aster tripolium* being strikingly like a ♂ *W. retiella* at rest.

The larvæ of *Luffia ferchaultella* can be collected in May and June on old fences, tree-trunks, &c. It is widely distributed— Broxbourne, Deal, Worthing, and many other localities giving the cases in great numbers.

The larval cases of *Fumea casta*, made of a few small straws arranged around a central cylinder of silk, are also to be found in abundance on trees, fences, palings, &c., in middle May, when the larvæ climb up for pupation; they are then readily seen although one never observes them during the winter and spring on the ground. The female remains within the case for fertilisation.

EUTRICHIDES.—In May the larvæ of *Eutricha quercifolia* vary from $1\frac{1}{2}$in. to $4\frac{1}{2}$in. in length, and are fairly easy to find when one is accustomed to their appearance, resting on buckthorn and sallow bushes quite near the ground, often, on Wicken Fen, amongst the grass and reeds, and preferring the small bushes to large ones.

At the end of May and beginning of June when the larvæ of *Trichiura crataegi* have separated, and also later when they are almost fullfed, they may be found sunning themselves on the small blackthorn or whitethorn bushes, stretched out lengthwise on a twig, on the outskirts of woods and thickets.

ATTACIDES.—The young larvæ of *Saturnia pavonia* live gregariously; their little colonies are to be found from the end of May into June, the larvæ separating after the third moult, and feeding more or less singly in July and August. When young they prefer thick cover and are to be found low down in the willow beds where the long grass and herbage grow around the small willows, whilst the larger larvæ are found higher up in the bushes, occasionally fully exposed, but usually even these are most abundant in the densest part of a bush.

SPHINGIDES.—To pair *Amorpha populi* leave them in a cage together, they pair during the night, preferably the second night after emergence; they remain paired for some hours.

In order to secure cross-pairing for purposes of hybridisation, place the individuals that you wish to cross in one cage, and a ♀ of the same species as the ♂ in an adjoining cage, near enough for the ♂ in the first cage to be affected by the scent diffused by the ♀ in the second.

In rearing *Mimas tiliae* the larvæ do well on fresh young shoots of lime till the third moult, they then want less succulent food, and feed best on the smaller dark green fleshy leaves from the top or upper branches of the tree (Bacot).

In their earlier stages the larvæ of *Mimas tiliae* usually rest on one of the veins on the underside of a leaf, but in the later stages choose a twig or leaf-stalk for the purpose. They seem to dislike the light, and, when their food is changed, get underneath the leaves as soon as possible (Bacot).

During May and early June, low marshy ground in which *Scabiosa arvensis* is growing should be worked for *Hemaris bombyliformis (tityus)*. *Pedicularis sylvatica* is, in such situations, a favourite flower to attract the species, so also is *Ajuga reptans*.

The imagines of *Hemaris tityus* are to be found hovering over flowers of *Pedicularis palustris*, *P. sylvatica*, &c., in June. Striking sideways is useless in capturing this species. Always strike down while the moth hovers over the flower, and then lift the bottom of the net, and it will flutter upwards (Carrington).

DELTOIDES.—*Madopa salicalis* occurs at the end of May in the woods near Haslemere; in a hot, dry year the specimens were all walked up out of the long grass and rushes edging the wood-paths; in another, a wet year, the specimens all occurred in low bushes of beech or oak. They generally rise up before one, settling again in the grass at the distance of a few yards. One example was met with at 10 a.m., but all the rest occurred between 2 p.m. and 6 p.m. (Barrett).

CHLŒOPHORIDES.—Larvæ of *Hylophila bicolorana*, beaten from oak in May and June, may be sleeved out on oak, when they invariably spin a boat-shaped cocoon on the underside of an oak leaf and emerge satisfactorily (Moberly); in confinement they may be kept in cardboard boxes, and usually spin up well on the sides of the boxes (Studd); shut up in the dark with their foodplant, they spin on the leaves and emerge without further trouble (Atmore).

LIPARIDES.—From the beginning of May till the first week in June is the best time to search for the larvæ of *Dasychira fascelina* in Scotland. During the early part of May the larvæ can be picked up in very varied stages of growth; from 2nd instar to fullfed. When moulting the larvæ spin a slight silken web on the top of the heather, and when small this catches the eye of the collector much quicker than does the caterpillar. Sometimes you will find the shed skin attached to the web and the larva gone; in this case it is always well to search the heather carefully near this spot and you will not infrequently be rewarded by finding the freshly moulted caterpillar (Haggart).

NOTODONTIDES.—The larvæ of *Asteroscopus sphinx* are to be beaten in numbers from apple and oak, principally the former, the leaves of which in confinement they prefer to any other food. From the middle to the end of May the imagines of *Leiocampa dictaeoides* should be sought at rest on birch.

Leiocampa dictaea has a most inconvenient habit ot emerging between 9 p.m. and 11 p.m. I have bred it now for several seasons successively, and find this an invariable rule. The males, too, must be killed as soon as possible after the wings are dry, or they will be worthless as specimens in the morning (Thornewill).

It may be well to note that *Leiocampa dictaea* and *L. dictaeoides*, bred from larvæ taken on Cannock Chase in 1891, had entirely different times of emergence, for whilst the former invariably emerged between 9 p.m. and 11 p.m. at night, the latter made their appearance in the afternoon (Thornewill).

Stauropus fagi is to be found at all hours of the day, from early morning until dark, on all sides of the tree and at all heights from the ground. Three out of four moths were found on small trees, but then, our beech woods, though of old standing, are cut severely, and there are twenty small trees to one of fair size, so that says nothing. *S. fagi* seems just to come out and sit on a tree in a very commonplace fashion. They were found over a period of two months from the middle of May till the middle of July (Holland).

A newly-emerged female *Cerura furcula* or *C. bifida* should be placed on a tree-trunk at dusk in garden (or elsewhere). By dark one will often find a male paired with her.

NOCTUIDES.—At the end of May or beginning of June, *Pharetra menyanthidis*, *Arctomyscis myricae*, and *Hadena glauca* come to sugar, but one cannot always depend on a good "sugaring" night. One often has an afternoon to spare, and (if on the Scotch moors) may get them all three at rest on the rocks at the lower elevations, especially where the heather has been burnt within the last year or two, and is still small.

In May, at night, the larvæ of *Leucania littoralis* are to be taken in the greatest profusion, and with them often the conspicuous and beautiful larva of *Actebia praecox*. It is of no use confining these larvæ by means of a muslin top, as they at once eat their way through it and escape.

The imagines of *Leucania littoralis* are a long time on the wing, often from late May until well into July, and fullfed larvæ are also obtainable from the beginning of May until well into June. Shake the sand-grass growing on the sandhills and keep the larvæ in a large tub with a glass covering ; a muslin covering is useless, as the larvæ bite holes through and escape.

The pupæ of *Tapinostola elymi* are to be found in late May and early June by working the roots and rootstocks of *Elymus arenarius;* the species is common at Cleethorpes, the Hornsea sandhills, &c.

The eggs of *Nonagria sparganii*, to be found in the folded edge of leaves of *Iris pseudacorus* all the winter, hatch in the middle of May, the young larvæ immediately commencing to mine the leaves of the foodplant.

I secured a good series of *Luperina cespitis* by sweeping grass from the middle to the end of May for larvæ, which were then not rare. The young larva is light green with white dorsal and subdorsal lines, but later on it becomes of a beautiful bronze colour, the lines being yellowish-white, and in this coat it exactly agrees with Guénée's description of the larva of *Neuronia popularis* (Newnham).

In early May the fullfed larvæ of *Apamea ophiogramma* are found at Castle Bellingham in the stems of *Arundo phragmites*, the topmost shoot of which is withered, and with frass coming from one or more holes in the stem lower down; the larvæ bore into the stem from the outside, and move from one plant to another; they feed up well on cut stems placed in water, and they pupate in any loose friable matter on the ground or in the withered stems of the preceding year's growth (Thornhill).

From the middle to the end of May on the moors about Burnley, *Hadena glauca* is to be obtained; almost all are found resting on the walls with *Cuspidia menyanthidis*; it was observed, however, that while all the *H. glauca* were low down near the base of the walls, the *C. menyanthidis* were near the top (among the bird-droppings) (Clutten).

In Scotland the sallows growing on the moors do not come into bloom until the beginning of May, and often last till quite the end of the month. This is a rare chance to take *Hadena glauca*, which come to the sallow flowers in abundance; they flit about from catkin to catkin, never resting long in one place, and can be easily netted. If the bush be shaken, they do not fall to the ground in the manner of the Tæniocampids, but fly off if disturbed in the least. The time of flight begins about 9 p.m. and lasts for about three-quarters of an hour. I have netted in this manner over two dozen fine *H. glauca* in an evening, and most exciting sport it is (Haggart).

The eggs of *Hadena glauca* are laid in batches on sallow in May, hatching in about a fortnight, the larvæ changing as they get older from green to deep brown, and fall to the ground if disturbed; they pupate just below the surface of the ground in a loose cocoon of silk and earth.

The larva of *Taeniocampa populeti* feeds on aspen (and other species on poplar), residing between two leaves united by a web; it is very transparent, always colourless (pale yellowish-white), with a black head; found on all tall aspens in the Epping district (Doubleday).

Larvæ of *Tiliacea citrago* are to be obtained in May from spun together leaves of lime. These larvæ are easily seen between the leaves against the sky. They seem to stick to their tents in the day, but come out at night to feed, and may then be beaten into "the Bignell" from lime and sometimes from nut. They are best found young, for when nearly fullfed they appear to be more independent, and wander about in the day (Holland).

The larvæ of *Tiliacea citrago* are nocturnal feeders, concealing themselves either between leaves, or under bark, among rubbish, in crevices of bark, etc., by day, and coming out at night to feed In captivity, the lime-leaves soon wither and curl up, and these curled leaves form most convenient hiding-places for the larvæ in the daytime, in which they spin thin, slight, cocoons.

The lime-trees in avenues, parks, and on roadsides usually have some of the lower branches removed, and, this being done annually, causes a woody excrescence to be formed on the trunks by the growth of numerous small twigs with abundant leafage. These spots I have found to be the favourite feeding-grounds of *T. citrago* larvæ if searched for after dark ; but except on lime bushes, where they have fewer opportunities for concealment, it is useless to attempt to beat them out by day (Fenn).

The larvæ of *Tiliacea citrago* spin up in May but remain in the cocoon some weeks before assuming the pupal state.

The larva of *Mellinia gilvago* is to be beaten from elm, frequently in abundance during the last week of May and beginning of June. Those of *Thecla w-album* are usually obtained at the same time and place.

The larvæ of *Cucullia chamomillae* feed on *Pyrethrum maritimum* during May and June ; when young they require to be carefully searched for, owing to their resemblance to the flower-buds and to their habit of curling themselves round the stem of the foodplant ; they prefer low-growing, flat plants, and feed up very rapidly, and one may find half-fed larvæ on plants which have been searched in vain a fortnight before (Still).

Plusia iota larvæ will feed and thrive well in confinement on hawthorn ; they are sometimes taken wild on *Stachys sylvatica* (South).

In the gardens of the Bexley district *Plusia moneta* is now quite common ; above 210 are recorded as captured in the larval and pupal stages as well as some 20 imagines netted of the first brood alone—June and July. The autumnal brood at present appears to be very partial.

The imagines of *Anarta melanopa* abound on the tops of many of the Scottish mountains in May and early June. When such an altitude has been reached that the heather grows thin and sparse and the grey lichen takes its place as a covering to the surface, there *Anarta melanopa* may be seen flying rapidly in the sunshine, or even on sunless days if the air be mild. I have never, in any locality, observed *A. melanopa* lower down than where the lichen begins to take the place of other plants (Fraser).

About May 15th, when the bearberry *(A. uvaursi)* is in bloom at the lower elevations, gather a large bundle of it and carry it to the hill-tops, which *Anarta melanopa* frequents. Lay a long line of small bunches of the flowers on the ground, and visit these bunches as you would patches of sugar. Sometimes the insect will be visible on the flowers, but more often it is well inside the bunch, and the best means of capture is to quickly place the net flat on the ground over the bunch of flowers. A bright sun is necessary to make the insect move freely (P. C. Reid).

Just about sunset search the borders of rock, especially where the heather has been recently burnt, and you will find *Anarta cordigera* at rest, sometimes singly, but generally *in cop.* The latter are easy to box, but the former is often extremely wide awake, and should be quickly netted. About May 15th is a good average date (P. C. Reid).

The larvæ of *Catocala sponsa* and *C. promissa* feed high up on the oaks, and are rarely beaten from the lower branches, but when there is a very great storm of wind one finds the larvæ occasionally climbing up the trunks afterwards.

ARCTIIDES.—The fullfed larvæ of *Callimorpha hera* will spin up as well among leaves and other rubbish, placed at the bottom of the breeding-cage, as in soil, and emerge quite as successfully.

Young larvæ of *Arctia villica* can be fed up on lettuce as long as it lasts and then on cabbage; they will spin up in March and appear as imagines in April (Riding)

In June the ova of *Spilosoma urticae* may be obtained; larvæ sleeved as soon as hatched on elder did fairly well but probably would have done better with a more varied diet (Whittle).

LITHOSIIDES.—In mid-May the imagines of *Gnophria rubricollis* are sometimes to be seen flying in profusion round the tops of the larches in the hottest sunshine from noon till 2 p.m., after which scarcely any more are to be seen on the wing (Reeks).

The larvæ of *Nudaria mundana* are to be found in May in colonies numbering thousands, feeding on lichens on beech trees and stone walls in the Cheltenham district (Robertson). They are also common at Portland, feeding in little companies on lichen under stones in May (Richardson).

PAPILIONIDES.—The larvæ of *Thecla w-album* are sometimes to be obtained in numbers at the end of May and in early June by beating elms. Dixon recommends *(Ent. Rec.,* x., p. 137) searching for them and gives an interesting account of his success by adopting this plan.

You must search the honeysuckle in May for larvæ of *Limenitis sibylla* ; look out for freshly-eaten leaves and then search the stem round and be careful not to overlook the trailing branches.

In May beat sallows in all sorts of positions for larvæ of *Apatura iris*—sallows that stand high and dry or in the middle of a marsh even furnish larvæ—for the ♀ wanders very far in search of sallows, and you never know on what stunted little bush may be feeding the horned head which is so dear a prize (Hewett).

In early May the ♀ *Gonepteryx rhamni* lays her eggs on *Rhamnus*, choosing the underside of a leaf, or the twig itself, or a bud. When a bud is chosen, only a single egg is laid as a rule on each ; a terminal bud is frequently chosen, and occasionally the upperside of a leaf (Grover).

JUNE.

We have already given *(anteà,* pp. 4 and 19) a few short notes on the mode of sugaring, and as June is, *par excellence,* the month in which an abundance of species are lured to this bait, a few further notes on the subject may not be out of place. No one has yet been able to tell us what makes a good sugaring season, or even what combination of circumstances makes a good sugaring night. Prolonged drought (especially with east wind) is stated to be bad, an early-rising and bright moon almost equally so, but rather in the direction

of making them more wary than that they are attracted in fewer
numbers; showers after drought (especially with a westerly or south-
westerly wind) bring about atmospheric conditions that seem to
tempt the moths to the feast, and we have had several experiences
in which swarms of moths have been attracted for hours in pouring
rain and in a high wind. Although Noctuids are most generally
attracted, certain species of Geometrids, Tortricids, Pterophorids,
Lithosiids, Pyraloids, Sphingids, Nolids, &c., are well known to
be attracted. Why does *Nola strigula* come to sugar whilst *Nola
cristulalis* will not? No one seems to know. Sugaring gives rise
to the question as to the "rights" of sugarers when several are
working in the same locality. This must always remain more or
less a matter of etiquette, and depend upon good-nature and good
manners (and the reverse). We have heard of lepidopterists pinning,
in the afternoon, a paper in one of the ridings of a wood (belonging to
someone else), announcing that they claimed it as against all
comers for the evening. We have known cases in which collectors
have sent their servants to sugar certain well-known grounds for
local species in the afternoon, so as to be the first in the field,
and in order to exclude other workers from the same town; we
remember once having tea at Chattenden with two other lepidopterists, one of whom left us as soon as tea was over, and sugared
every available tree between the keeper's house and the Cliffe
entrance gate, leaving us the distant parts of the wood, where
catching the last train was an impossibility. All one can say is
that bad manners and selfishness are not confined to any section
of humanity, and that among lepidopterists one finds occasional
examples of bad-mannered and selfish people. As to the manner
of taking insects from sugar. We box everything, take all our
specimens home alive, empty them into the large killing-tin on our
return, pour in the ammonia, and go to bed, leaving them until
the morning. Never leave the moths alive in the boxes all night.
Hewett does not know what a lepidopterist can do at sugar without the cyanide bottle and oxalic acid. He confesses the plan
is tedious, but the specimens are perfect. Stupefy in the bottle,
turn out at about every sixth insect, oxalic them carefully, and pin
sideways into the box, which should be very slightly damped. It
involves a small wait after every insect, even with a new bottle, and
some method in putting away the pen and oxalic carefully. As to the
habits of the moths at sugar, many species drop from the sugar
when the light is put on, but can be picked up. Many walk from
the sugar to another part of the tree near, and want looking for.
Some only stay for a short time, *e.g.*, *Thyatira batis*, *Heliothis
marginata;* others drink for ever, *e.g.*, *Triphaena pronuba*, *Noctua
festiva*. Some come early and disappear, *e.g.*, *Cymatophora or;* others
come late, *e.g.*, *Aplecta tincta*, which is rarely seen before 11 p.m.
The various methods of sugaring adopted in various districts is
interesting; in woods the tree-trunks are utilised, as already described;
on bare sandhills the fences, on moors the stones, on the fens knots are
made of the reed-tops, whilst on the bare chalkhills, cliffsides, sandhills, &c., thistles and other plants must be carefully sprinkled with
the sweets. At Freshwater, beet, knapweed, *Heracleum*, as well as
thistles, are utilised for sugaring for the capture of *Agrotis lunigera*, but

Onopordium acanthium, a sturdy thistle growing all along the cliffs, is the most remunerative to work, 35 examples of *A. lunigera* have been taken off a single isolated thistle of this species on the cliffs at one time; four-fifths of the captures of this species are said to be made within two feet of the edge of the cliffs, whilst those thistles growing upon the open turf are not nearly so attractive as those growing amongst the loose chalk rubble in the crevices and upon the face of some of the seaward slopes. For such cliffs as these Hodges states that it is preferable to have a smaller and less open lantern than the pattern ordinarily used for sugaring, owing to the difficulty of keeping the latter alight if it be at all windy. Sugaring on the northern moors is always interesting to the southerner, and in Rannoch, Morayshire and Aberdeenshire, in June, the sight of purely northern species such as *Arctomyscis myricae*, *Cuspidia menyanthidis*, *Hyppa rectilinea*, &c., always excites the lepidopterist new to the district, whilst other species, *e.g.*, *Apamea rurea*, &c., are obtainable in such variety as is unknown in the south. In the same localities, by day and in the early evening, larvæ of such species as haunt the heather, *e.g.*, *Noctua neglecta*, *Agrotis agathina*, *Lycophotia strigula*, *Plusia interrogationis*, *Larentia caesiata*, *Lasiocampa* var. *callunae*, can always be obtained, whilst cocoons of *Trichiura crataegi* var. *ariae*, *Dasychira fascelina* and *Plusia interrogationis* are also worth searching for. The mode of work being so different from that usually indulged in by the southerner makes it so much the more interesting. As to the great difference in the habitat of a local species in different areas of its distribution, one may instance *Arctomyscis myricae*, which, in Rannoch, is only found on the moors at some distance above the sea-level, whilst, in Aberdeenshire, it has never been recorded from or observed on the moors, but is found commonly in the larval stage on the coast, by nearly every roadside, on the edges of fields, in fact, almost everywhere where there is an abundance of sorrel and plantain, except on the moors; these are the chief foodplants of the larvæ, although they will eat ragwort, bramble, Scotch thistle, &c., but not *Myrica gale*, after which, oddly enough, the species has been named. We have already detailed *(anteà*, p. 20) a method of assembling *Amphidasys strataria*, a late-flying moth. During June, assembling *Sesia sphegiformis* may be successfully accomplished in its haunts. A newly-emerged, restless and "calling" female will attract many males during the whole of the morning, a quiescent one is absolutely valueless for attractive purposes. The pairing of this species is peculiar, the male taking hold of the female while on the wing, and immediately dropping immovably suspended from the female, reminding one of certain Hepialids. Smith gives *(Ent. Rec.*, i., p. 67) a most interesting account of a successful mode of assembling the males of *Stauropus fagi* in a Bucks beech wood. An open place on the ridge of a hill of some elevation, where three roads, passing through the wood meet, is pointed out as a favourable spot, a virgin female 24 hours old being hung up in a muslin cage on an oak, about 4 feet from the ground, just before 11 p.m., the males being on flight from just before 11 p.m. till a little after midnight. The males are netted and transferred at once to a cyanide bottle, and afterwards pinned, this method ensuring them being in fine condition. Wet seasons are said to be favourable to the larvæ of this species. Flowers are great this month, especially is this so in some coast districts, *e.g.*, Isle of Man, Howth, Sligo, &c. At Howth and in the Isle of Man the Dianthœcias are on the wing. They will occasionally come to light, but the only practical way to take them is to stand and patiently watch, gazing intently at a clump of *Silene maritima*, allowing such species as *Acidalia marginepunctata*, &c., to pass, so that the net may not be otherwise engaged when *Dianthoecia caesia*, *D. luteago* var. *barrettii* or *D. capsophila* does arrive. The Noctuids, in the dim light, appear, close to the flowers, almost invisible, and a rapid stroke is necessary to make sure of the prize. Often the flight of these does not last for more than a quarter of an hour on one evening, and sometimes does not come at all. At Sligo, the flowers of the common rocket are reported as most attractive both to Plusias and to *Theretra porcellus*, whilst at Horrabridge, at campion flowers at dusk, many good species have been taken. Light, too, in some seasons, pays well. We have already referred (pt. 1, p. 52) to the success sometimes attending this mode of capture in the Fens, but not only is this so, but large numbers are obtained in well-placed localities, near woods, moors, &c., even when quite close to towns and villages, *e.g.*, Still records that, on one evening in June, 1890, between 11 p.m. and 1.30 a.m., during rain, he captured, on the borders of Dartmoor, more than 100 insects, including *Plusia iota*, *P. chrysitis*, *Habrostola urticae*, *Dianthoecia carpophaga*, *D. cucubali*, *Xylophasia rurea*, *Sphinx ligustri*, *Dasychira pudibunda*, *Lophopteryx camelina*, *Selenia lunaria* and many others. From June 24th to July 26th, 1890, Highbury fields, and other areas in north London were closely worked for *Zeuzera pyrina (aesculi)*, the localities being visited once or twice each day. Some 120 examples were taken, a solitary ash-tree, with a diameter of only a trifle over 3 inches, at a height of four feet from the ground, yielding no fewer than 27 specimens. Warm morning sun seemed necessary to bring out the imagines. When the mornings were wet and the temperature low, none emerged. When the early morning was bright and warm, followed by rain before midday some emerged, but were all more or less crippled. In one locality ash was almost the exclusive home of the insect, one specimen, however, emerged from a privet bush, and there was evidence that the hawthorns were infected. In Highbury Fields, the trees attacked were, without exception, young elms of a diameter of from an inch and a half to three inches ; the young planes which alternated with these were entirely unaffected. The damage to the trees was very considerable ; in the case of the large trees whole branches were killed, whilst the stems of the young trees were so weakened as to snap across in a high wind, and the solitary ash tree was completely killed above a height of 4 feet from the ground. The normal hours of emergence may be set down as from 11 a.m. to 4 p.m. The moths began to appear on June 24th, but, till July 4th, they were only found on the small trees. It may be, that the heating effect of the sun's rays takes longer to penetrate the large trees. A lilac cut to pieces on July 7th produced several larvæ still feeding, and by no means fully grown. This would seem to point to the probability that *Z. pyrina*, like *Cossus ligniperda*, remains more than one year in the larval state.   Of late years many of the most local of our British butterflies have been reared in confinement.   A word or two more of rearing lepidoptera in confinement may be advisable, and particularly at this season of the year, as to the methods of obtaining ova from butterflies and moths.   Butterflies and many moths require feeding when ovipositing.   Montgomery says that this is best done by taking a piece of lump sugar, dipping it in water, and shaking it as dry as possible; he then recommends taking the insect by the wings in the left hand, holding it so that it can grasp the sugar, and stretching out its proboscis with a fine needle.   As a rule, as soon as the trunk is extended the insect begins to feed, and may be gently released and left to take its fill.   Refractory species can sometimes be induced to feed by placing the sugar close to the insect and gently blowing over the sugar towards it.   After feeding once or twice they become used to the treatment, and will extend their trunks as soon as one picks them up or even blows into the cage.   Montgomery observes that a specimen of *Miselia oxyacanthae* kept alive for some time would run zigzag towards the sugar from one end of the table to the other, waving its antennæ in great excitement.   Butterflies will rarely lay eggs except in the presence of the sunshine and the foodplant.   A few exceptions, chiefly among the Satyrids—*Melanargia galatea*, &c., are of course well known.   For the smaller species Montgomery uses a bottle-cage with great success.   A pint white glass circular bottle with the bottom cut off, and a slice of cork (about a quarter of an inch in thickness) thrust into the neck is placed neck foremost into a gallipot of suitable size.   A spray of the foodplant is plugged with cotton wool into a hole in the cork, so that the end reaches some water in the gallipot; put in two teaspoonfuls of sand, covering the cork and shoulders of the bottle and damp it slightly.   Next put in the butterfly and cover with mosquito-net or leno, secured with a black india-rubber band (12, Faber's).   (If a red or old black band be used the sun will melt it.)   Place the cage on a sunny window-ledge where it will be free from interference.   For the larger species a larva-cage may be used.   The base should be filled with sand, the food plugged into the well with cotton-wool and the perforated zinc lid replaced by mosquito-net.   It must be borne in mind that too much sun will kill the butterfly, so that if the weather be hot, the cage should be exposed for short intervals only; there should be plenty of ventilation, a free interchange of air, and the sand should be kept well damped.   Feed the butterfly, and in the evening the piece of foodplant on which eggs have been deposited may be removed, and a fresh spray inserted.   In the bottle-cage *Chrysophanus phlaeas* has been kept alive 26 days, and, in the larva-cage, *Dryas paphia* for a month.   The Satyrids will lay well on a tuft of grass growing in a large size flower-pot, and enclosed by a glass cylinder, covered with mosquito-net, or in a large glass jam jar in which a tuft of grass has been set in some damp sand, the top also covered with mosquito-net.   After the eggs have hatched, cover with a piece of linen, keep the jar in the light, and in an airy position, damp the sand regularly, and the larvæ will feed on well until they have attained a fair size.   To obtain eggs of small moths the bottle-cage, as described above, may be used, and for large restless moths, Catocalids, Sphingids, &c., a large larva-cage carefully fitted. For Acidaliids air-tight, white metal, glass-topped boxes, three inches or more in diameter, are to be much preferred to the generally used card- or chip-boxes. A spray of the food (never wet), carefully fitted into one of these boxes so that it cannot shift, and the females placed in with it are often all that is needed. Even a species like *Theretra porcellus* may be kept in such boxes for oviposition with success. The great objection to chip-boxes is the way in which eggs laid in them dry up and fail to hatch; with those species that pass many months in the egg-stage this usually entails a very serious loss, which the use of the metal boxes appears largely to obviate. The occasional dripping of a few drops of water into a chip-box in which eggs are laid on a collecting expedition is a great aid to success. Montgomery also suggests that ova taken from the foodplant should be placed on a piece of damp rag in these metal glass-topped boxes. He relates how a batch of *Theretra porcellus* ova, kept without special treatment, at first developed well, the young larvæ being seen moving about within, but nearly all failed to make an exit. Later, another batch was placed on the damp rag, 98 out of 110 hatched, and none of the 12 failures contained a larva.

UNCLASSIFIED.—When sugar fails one frequently gets many insects by visiting flowers with a lantern at night in June and July ; those of thistles and *Lychnis flos-cuculi* are often the most productive (Newnham).

To kill delicate imagines, without producing *rigor*, try the following method. Place in the cyanide killing-bottle, and directly the insect is motionless, transfer to a glass pickle-bottle which has a wide mouth, with a large hollow-headed glass stopper. In the hollow head of the stopper have packed a closely rammed filling of bruised young laurel shoots. This is an excellent way of killing all insects, as they never stiffen, and will keep relaxed for days. It is most useful at sugar (Dollman).

For setting small light-coloured Geometrid imagines, try a dark setting-board, preferably dark green. The correct placing of the wings can be arrived at much more easily than when a white ground is used, as their shape is sharply defined against the colour, even when seen through tracing paper.

If a breeding-cage in which pupæ are kept be of metal, like a wire-gauze meat safe, an excellent plan for keeping the pupæ just sufficiently damp will be found in having a double thickness of stout serge, or old flannel, which has been soaked in water and wrung half dry, tied round it. By this means the atmosphere is kept slightly damp, and provides as close an imitation of natural conditions as is possible. The damping and wringing of the cloth does not take more than a few moments, and can easily be done each day when the cage is examined. There is not the slightest chance of mildew occurring.

Working flowers of *Echium vulgare* in the evening in late June and July pays well in the south-eastern counties—*Sesia stellatarum* comes until dusk, *Theretra porcellus* after dusk, but ceases as soon as

H

it is time to light the lamp. *Thyatira batis*, *Gonophora derasa*, and *Heliothis marginata*, like the two Sphingids, want netting, but Noctuids, *Agrotis lucernea*, &c., stick well to the flowers and want knocking into the cyanide bottle.

From an elevation of 2500ft. and upwards *Pachnobia hyperborea (alpina)* and *Psodos coracina (trepidaria)* are flying from June 21st, onwards. *P. coracina (trepidaria)* flies freely if (which is not always the case) the sun is shining. *P. alpina* requires a night of hard cold work to secure. But all through June the pupæ of both can be easily obtained on the tops of hills of sufficient height, by turning over the moss and lichens which cover the ground, especially along the edges of foot- and sheep-paths, and round bare patches. *P. alpina* pupæ are hardly, however, to be got except in the "even" years, 1902, 1904, &c., while those of *P. coracina (trepidaria)*, though generally to be found, abound most in odd years, 1903, 1905, &c. (Reid).

ELACHISTIDES.—From the middle to the end of June the imagines of *Elachista scirpi* are to be found flying over reeds at Pitsea (Whittle).

In late June and early July *Cataplectica farreni* is to be swept, near Cambridge, by the side of the road, among plants of *Pastinaca sativa*, &c.

The larva of *Antispila pfeifferella* is abundant on *Cornus sanguinea* in June and July, fixing its case for pupation on the undersides of stones (Richardson).

LITHOCOLLETIDES.—The larva of *Lithocolletis sorbi* makes a long mine on the underside of the leaves of *Pyrus aucuparia* and *Prunus padus* in June-July and again in September-October. The species hybernates as a pupa.

The larva of *Lithocolletis spinicolella* mines on the underside of leaves of *Prunus communis*, *P. domestica* and their cultivated varieties, in June and July, and again from September to March. The imago is rather smaller than *L. cerasicolella* and of a shining golden-ochreous colour.

The larva of *Lithocolletis cerasicolella* mines on the underside of leaves of *Prunus avium*, *P. cerasus* and rarely on their cultivated varieties, also on *P. mahaleb*, *P. armeniaca*, *P. domestica*, and *Pyrus communis* in June, and again from September to March. The rufous-orange colour of the imagines separates them from those of *L. spinicolella*.

The larva of *Lithocolletis oxyacanthae* mines on the underside of leaves of *Crataegus oxyacantha* and also sparingly on *Pyrus aucuparia*, in June-July and September-October. This species hybernates as a pupa.

The larva of *Lithocolletis blancardella* mines on the underside of leaves of *Pyrus malus* and its cultivated varieties, also on *P. aria* and *P. communis* in June-July, and again in September-October. The species hybernates as pupa as does *L. concomitella*.

COLEOPHORIDES.—The larvæ of *Coleophora genistaecolella* should be collected from *Genista tinctoria* in June.

The larvæ of *Coleophora bicolorella* are rather late feeders; the cases, which are somewhat like those of *C. viminetella* (not pistol-shaped, as sometimes stated), may be found on alder and hazel in June.

The middle of June is a good time to search sloe bushes for larvæ of *Coleophora paripennella;* a honeycombed sloe leaf is a good guide.

The larvæ of *Coleophora murinipennella* are to be found in the heads of *Luzula campestris* in June. To take the imagines, visit a meadow in which *Luzula campestris* grows, between 6 p.m. and 7.30 p.m., on a warm evening early in May. The grey moths may be seen flying low down among the herbage.

CHRYSOCORIDIDES.— The green, hairy larvæ of *Chrysocorys festaliella,* feeds on the upper surface of bramble and raspberry leaves in June and again in October, making conspicuous blotches on the leaves. They spin their beautiful cocoons on the underside of the leaf.

TORTRICIDES.—The larvæ of *Tortrix diversana* is to be found from the commencement to the middle of June on birch and elm. Locally abundant in Essex and Kent.

The imagines of *Tortrix piceana* are to be found during the early part of the afternoon on the heather under the pines in the Esher district.

About 3 p.m. the imagines of *Tortrix piceana* commence their flight over and around the pine-trees, and continue until nearly 8 p.m. They fly high, must be obtained with a net attached to a long bamboo, and are not to be taken (except occasional specimens) with an ordinary net. Common in the Esher district.

Tortrix costana var. *latiorana* is not uncommon on saltmarshes bordering the Thames; the larvæ may be found on *Aster tripolium, Statice limonium,* &c., the first week in June (Thurnall).

The imagines of *Tortrix icterana* frequent grassy places (often damp) at the end of June and in early July; it comes out of the rank herbage freely after 4 p.m., and occurs freely on the wing at dusk.

Tortrix viburniana frequents moors where *Erica* and *Calluna* are the chief features of the herbage; the imagines fly in the sun in June, and in fine weather are on the wing freely from 10 a.m., being less abundant in the afternoon.

A *Tortrix,* usually referred to *viburniana,* is to be found in the larval stage on the saltings of the Kent and Essex marshes throughout the first half of June, feeding on *Aster tripolium, Statice limonium,* and spun-together tops of *Artemisia maritima.* Most probably quite distinct as a species from the moorland *T. viburniana.*

The imagines of *Tortrix corylana* are generally to be obtained by beating oaks in the Loughton and Ongar districts in late August. The larva is to be found on oak at the end of June (Thurnall).

The larvæ of *Peronea logiana (tristana)* are locally abundant on *Viburnum lantana* from June-August.

The larvæ of *Peronea hastiana* occur in Wicken Fen almost continuously from June until September in rolled leaves of sallow, without any suggestion of a break to divide the appearances into separate broods.

The larvæ of *Peronea aspersana* are to be found throughout June on *Spiraea filipendula, Poterium sanguisorba, Potentilla anserina, P. tormentilla* (Thurnall).

The larvæ of *Dictyopteryx holmiana* are to be found towards the end of June on bramble, particularly on those plants growing in hedges (Thurnall).

The larvæ of *Dictyopteryx forskaleana* are to be found abundantly in June on sycamore and maple, on which the larvæ feed. The imagines are to be beaten therefrom sometimes in large numbers a little later in the season.

The larvæ of *Sideria achatana* are to be found in early June, sometimes abundantly amongst hawthorn ; they spin two or three dead leaves to a twig of the foodplant and hide therein, coming out at dusk to feed (Thurnall).

The larvæ of *Ditula semifasciana* are to be found on *Salix caprea*, &c., throughout the latter half of June, especially abundant by the sides of ditches, &c.

From the middle to the end of June the imagines of *Ephippiphora nigricostana* fly freely around the bushes, among which *Stachys sylvatica* is growing.

In mid-June the local *Ephippiphora turbidana* is to be taken in some places freely, flying sluggishly among butter-bur.

The imagines of *Sericoris bifasciana* fly freely in the afternoon high around pine-trees at the end of June and in early July. Common in the Esher district.

Sericoris micana is very common at Orton in mid-July, flying amongst rough herbage in meadows with *Eupithecia pygmaeata* and *E. satyrata* (Wilkinson).

Mixodia palustrana flies in dozens round the trees in the Altyre Wood at Forres, also common in the Black Wood at Rannoch.

The larvæ of *Sciaphila chrysantheana* feed in early June on *Tussilago farfara*, turning down a lobe of the leaf or puckering the leaf by partly drawing two portions together with silk (Thurnall).

The dull-brown larva of *Sciaphila penziana* is to be found in May, feeding upon sea-pink, *Festuca ovina*, &c., spinning among its foodplants a fine white silky tubular web in which it lives and in which it pupates in June (Gregson).

The imagines of *Phtheochroa rugosana* are to be taken flying at dusk in early June along hedgerows wherever *Bryonia dioica* occurs at all commonly. The imago should be killed as quickly as possible as it is restless and soon injures itself.

A high wind at the end of June (or in early July) will disturb the larvæ of *Paedisca occultana*, and they may sometimes be seen in profusion swinging by their silken threads from all the fir-trees (Harrison).

Retinia duplana is best obtained by beating ; those that are captured on the wing damage themselves very much by plunging about in the net (Reid).

Retinia posticana is to be beaten out of small fir-trees by day ; they are also to be seen fluttering round the trees at dusk and on quiet afternoons (Reid).

The imagines of *Coccyx cosmophorana* occur near Carlisle in June, appearing to fly in the fir woods quite naturally in the morning sunshine.

The imagines of *Coccyx ochsenheimeriana* are to be found among *Abies cephalonica* in early June, flying about four o'clock in the afternoon in the sunshine, at the ends of the branches of this species of fir (Walsingham).

Early in June the imagines of *Bactra furfurana* may be disturbed in marshes from its foodplant, *Eleocharis palustris*, in the stems of which the larvæ feed and pupate.

The larva of *Grapholitha naevana* is to be found in June, drawing together the terminal shoots of holly *Sorbus aucuparia*, &c. It is of a greyish-green colour, with a dark brown or blackish head.

The imago of *Phoxopteryx upupana* flies very rapidly over the tops of oaks, birches and other trees in their immediate vicinity in June. It flies in the sunshine from 2 p.m., or a little earlier, until an hour before sunset. It frequents woods and appears very local. In flying it somewhat resembles *P. mitterpacheriana*, but the latter flies later in the day, and is heavier on the wing. The larvæ feed on birch in August and September.

The imagines of *Phoxopteryx derasana* are sometimes abundant about the middle of June amongst *Rhamnus catharticus* flying in the evening.

Patches of *Agraphis nutans* sometimes produce imagines of *Eupoecilia maculosana* in abundance during the first week in June. The species flies freely in the morning sunshine.

PYRALOIDES.—The larva of *Semioscopus anella* rolls the leaves of hornbeam into the form of a tube open at both ends in June. It is especially likely to be found in large woods.

The larvæ of *Gelechia albipalpella* are to be found freely on *ista tinctoria* in mid-June.

The larvæ of *Aristotelia ericinella* are to be found in June, spun between the stems of the ling.

When searching for pupæ of *Butalis grandipennis* amongst furze, one takes *Gelechia longicornis* flying past from one patch of heather to another.

The fullfed larva of *Gelechia morosa* should be collected in the fens (Wicken) about the middle of June ; the moths emerge usually towards the end of July.

The larva of *Depressaria douglasella* feeds on *Daucus carota* in June, spinning the terminal pinnæ of the leaf together so as to form a tube; later it often spins a web across the midrib, and, like some other carrot-feeders, leaves the plant to go into pupa. Occurs commonly at Howth and on the Lancashire coast (Gregson).

The larva of *Depressaria yeatiana* feeds in a tube formed by turning over the pinnæ of the leaves of *Daucus carota ;* it is fullfed during the last week of June and the first week of July, when it descends to the surface of the ground and pupates in a slight, white, silken cocoon.

Collect the stems of *Œnanthe crocata* for larvæ and pupæ of *Depressaria nervosa ;* hundreds may be sometimes obtained (Baxter).

The larvæ of *Depressaria rotundella* feed on the pinnæ of the leaves of *Daucus carota* in June and July.

The larvæ of *Nothris durdhamellus* are to be found in the latter half of June on *Origanum*, the imagines emerging the first half of July (Miller).

CRAMBIDES. — Although the males of *Platytes cerussellus* are frequently abundant in the morning sun the females are rarely to be taken in numbers until the afternoon when they get into the upper part of the grass tussocks, among which they live.   Occur from about the middle to the end of June.

The rare *Epischnia bankesiella* occurs at Portland and Lulworth and probably all along the rocky part of the coast towards Swanage ; taken on flowers at night, or flying at dusk in late June and July.

Honeycomb affected by larvæ of *Achroea grisella* should be isolated in early June, pupation taking place by the middle of the month among the destroyed cells, the imagines emerging by early July.

Imagines of *Crambus ericellus* are on the wing from the middle to the end of June ; they are easily disturbed in the daytime ; the species is abundant on Green Crag, Borrowdale, near Keswick.

PYRALIDES.—The larvæ of *Botys pandalis* will feed in confinement on golden-rod, marjoram and *Teucrium*, living exposed until the third moult, after which they spin cases of *Chenopodium, Origanum, Solidago*, beech, &c., open at both ends, and live therein, never protruding their heads to feed during the day unless in darkness and perfectly undisturbed ; they are fullfed in early September, pupating within their cases, which they carefully close.   The imagines emerge in the following June.   (In confinement they will sometimes emerge in November.)

The larva of *Scoparia lineolalis* feeds under *Parmelia parietina*, which grows upon the rocks at Howth, the Isle of Man, &c., and is fullfed in June ; it pupates in a slight web beneath its foodplant, and the imago appears in July.

The imagines of *Perinephele lancealis* may be beaten out of *Eupatorium cannabinum* during June.

Towards the end of June the imagines of *Ebulea stachydalis* can be disturbed among plants of *Stachys* growing by the side of wood-ridings ; it appears to be most easily obtained in early morning and at dusk.

The imagines of *Nascia cilialis* come freely to light in Wicken Fen from the third week in June onwards, when they are usually to be taken on the front of the lamp glass, although some are to be boxed off the sheet.

PTEROPHORIDES. — Imagines of *Agdistis bennettii* are to be found in June, almost throughout the whole month, on the tidal marshes of the southern and eastern coasts ; abundant at Hartlepool, banks of Thames and Medway, Portsmouth, banks of Yar in Isle of Wight, etc.

The pupa of *Cnaemidophorus rhododactylus* should be looked for in late June and July ; it is to be found attached by its anal segment to the foodplant, in close proximity to where it has fed, either on the peduncle, leafbud or leafstalk.

In the middle of June, marshy localities where *Senecio aquaticus* grows should be worked for the imagines of *Platyptilia isodactyla ;* the males are not difficult to disturb among the rough growth in such places, but the females are scarce.   They are most abundant just before dusk, when they fly somewhat freely. The moth appears again in late August and September.

In June the larva of *Leioptilus lienigianus* sews together a lower leaf of *Artemisia vulgaris*, so as to make a secure little tent, inside which it feeds as before, only making larger blotches side by side between the ribs of the leaf, until the greater part of the parenchyma is devoured, after which it deserts this habitation, makes another lower down, and so on, constructing four or five tents before becoming fullgrown.

The fullgrown larva of *Leioptilus lienigianus* leaves its tents in the leaves of *Artemisia vulgaris* in late June, and pupates on the lower part of the stem of its foodplant or some adjacent object. The imago emerges about the second week of July.

In June the larvæ of the summer brood of *Leioptilus microdactylus* which have all the appearance of internal feeders, feed in the flowering stems of *Eupatorium cannabinum*, generally at the axils of the leaves on such stems, but sometimes at some distance therefrom. The entrance-holes of the mines are not easily detected, being exceedingly small, lightly filled up from within, and with no loose frass making them conspicuous.

The green larva of *Marasmarcha phaeodactyla* feeds on the underside of the leaves of bird's foot trefoil in June (Harding) ; always found by us on *Ononis spinosa* (Tutt).

The green larva of *Œdematophorus lithodactylus* feeds on the underside of the leaves of common fleabane, *Inula dysenterica*, in June ; the imago appears in July (Harding).

The green larva of *Aciptilia baliodactyla* feeds on *Ononis reclinan* and may be found in June ; the imago appearing in July (Harding).

The larva of *Aciptilia baliodactyla* feeds on the terminal leaves of *Origanum vulgare* in June ; it is generally to be found at rest in the daytime on a plant, the top of which it has caused to droop by biting into the stem ; but it feeds in the evening (South).

The pupa of *Aciptilia migadactyla (spilodactyla)* is usually to be found on the upper side of the leaf of its foodplant, *Marrubium vulgare*, at the end of June or in July, generally along the midrib. The pupa is of the same dark green colour as the leaf, the longitudinal dorsal markings agree exactly with the elevated ridges of the leaves, and the hairy surface of the pupa corresponds exactly with the hirsute surface of the leaf. Best time usually in July, but may be found from June to August in different seasons.

AVENTIIDES.—In the middle of June the larvæ of *Aventia flexula* have been beaten from *Prunus spinosa* at Loughton.

DREPANIDES.—Some moths lay almost all their eggs in one or two batches and in one or two days. A ♀ *Drepana falcataria*, however, captured towards the end of June, lived for about a fortnight, laid about the same number of eggs every day (or night), till the total amounted to nearly one hundred (Merrifield).

Pairing of *Drepana harpagula* is not difficult. If the sexes be placed together in a roomy box, copulation takes place in the evening, and the moths separate usually in about 8 or 10 hours.

The eggs of *Drepana harpagula (sicula)* are laid by the parent moth on the very edges of the leaves of *Tilia parvifolia* in June. The larvæ are at first very restless, and are then best confined in an airtight jampot; later they settle down better to their food, and spin up in August and September. The larvæ will not thrive on *Tilia europaea*.

BREPHIDES.—The ♀ s of *Brephos parthenias* and *B. notha* lay freely in confinement, and eggs are readily obtained; the larvæ feed up without trouble, but unless some rotten wood, cork, or similar substance be provided into which they can enter for pupation, the full-fed larvæ will perish miserably.

GEOMETRIDES.—The larvæ of *Epione vespertaria* are obtained by sweeping the dwarf sallow in June, and are not difficult to rear. The female larvæ are larger and stouter than those of the male. Singularly enough, the proportion of female larvæ swept generally exceeds that of the male (Hewett).

The larvæ of *Eugonia fuscantaria* feed up well on ash, but they vary much in the rapidity of their growth, and they pupate very slowly, so that some of them will have pupated before others are more than an inch in length.

At the end of June and in early July, stand in the wider rides of the southern woods to net the quickly flying zigzagging males of *Angerona prunaria;* they commence to fly at sunset and continue flying till after dusk. The ♀ s are to be beaten from the trees.

The larvæ of *Amphidasys strataria* from the same brood will feed up at very different rates under identical conditions, some being quite fullfed by the beginning of June, others not till quite the end of the month or even later.

In confinement, emergence of *Amphidasys betularia* usually takes place between 7 p.m. and 11 p.m. in June; it frequently happens that the moths are not in a fit condition to kill before one retires to bed, as a result, those that are left in the cages all night are usually worthless in the morning. One rarely breeds really good specimens unless they recognise this fact.

The ♀ s of *Boarmia consortaria* will lay their eggs in June in a chip-box, hiding them under the films of wood; one can easily overlook the eggs, unless one looks carefully, when the green tint of the eggs often shows plainly through the thin layer of wood above them.

The larvæ of *Boarmia cinctaria* will feed freely on sallow in June as well as on the usually accredited foodplants—*Erica tetralix* and *E. cinerea*.

In June, *Nemoria viridata* is exceedingly plentiful on a piece of heathy ground adjoining the Upton estate near Poole (Spiller).

Mid-June is the time for many of the better Rannoch insects, and the higher mountains (Schehallion, &c.) give an abundance of *Psodos coracina (trepidaria)*.

The larvæ of *Pseudoterpna pruinata (cytisaria)* are to be obtained, sometimes freely, on *Genista tinctoria*, in June.

The eggs of *Geometra vernaria* are laid on a twig of *Clematis vitalba*, on the rind, never on a leaf; they are laid one upon the top of the other until a little pile of 12 or 14 stand out at right angles, with the twig like a small broken lateral twig or thorn.

The eggs of *Geometra vernaria* are laid in late June and early July, one on the other in little steeples of about ten or a dozen, on the stems of *Clematis*, each little batch looking like a leaf-petiole or tendril shortly broken off. The larva hatches in late July, stretches straight out from a leaf-stalk, and is easily beaten from its foodplant.

Whitethorn, blackthorn and oak should be beaten in early June for larvæ of *Hemithea strigaria (thymiaria)*, which spin up among the dried twigs of the foodplant, and produce such deeply-coloured imagines as are rarely taken wild.

On midsummer day, 1892, I was able to secure some thirty examples of *Nemoria viridata*, which were knocked out from the furze-bushes in Guernsey, all in the finest possible condition (Hodges).

With a stick, tap roadside hedges of stunted oak and maple for *Zonosoma porata* and *Z. annulata*. The ♀ s of both species lay freely in chip-boxes if furnished with a leaf of the foodplant.

A ♂ and ♀ of *Zonosoma orbicularia*, bred in June, were placed together in a large glass cylinder placed over sallow twigs standing in wet sand, the whole being put into the garden among growing sallows. Pairing took place in due course (? about 10 p.m., and lasted but a short time), and the ♀ laid fertile eggs. These hatched, and the imagines appeared as a second brood at the beginning of August (Merrifield).

The egg of *Zonosoma orbicularia* is laid almost uniformly on the edge of a leaf, generally two or three on a leaf, sometimes four or five; in rare instances the egg is laid on the stalk or midrib, or on the edge of one of the stipules. The eggs hatch in from ten days to three weeks, according to the temperature (Merrifield).

The imagines of *Acidalia dilutaria (holosericata)* are on the wing from the middle of June until the first week of July; they may be started during the day from grass, low bushes of privet, &c., and fly rather weakly at dusk (Durdham Down, Clifton Down, Leigh Down).

Imagines of *Acidalia emarginata* may be beaten out of the long grass in wood ridings in the day time, but they fly commonly at 11 p.m. and later; they may also often be found at rest at dusk on the long grass under bushes and hedges (Fenn).

In mid-June, among maple, the imagines of *Asthena luteata* sometimes fly in great abundance about 7 p.m. They can also be beaten during the day.

The middle of June is the time for *Asthena sylvata*. It is a shy and retiring insect and difficult to catch; it sometimes sits on the oaktrees, but unless it is a very cold day will not allow one to get near it, rising as soon as one approaches, and then requires to be netted. Most are obtained in alder swamps, where the imagines sit often quite low down on the stems of the trees.

When disturbed, *Asthena sylvata* at first goes straight away, and, if there be a low bush near, generally flies right through it and is lost to sight; if, however, the ground be open in the line it has taken, if often doubles back and gives a chance. Very occasionally it drops down and hides in the grass.

Imagines of *Venusia cambricaria* are to be obtained occasionally in early June sitting on the mountain-ash trunks; after this they are rarely to be seen till after the middle of July (Clutten).

I

The first week in June in the New Forest will usually give a large number of the larvæ of *Selidosema plumaria* by searching at night on *Calluna vulgaris*. (Those of *Noctua neglecta* are to be obtained in numbers at the same time.)

In late May and early June, the fullfed larvæ of *Scodiona belgiaria* are to be found on our Scotch moors. The best way to find them is to search the heather, in stony localities, where it grows clumpy. In fine sunny weather the caterpillars may then be found stretched at full length on the top of the heather. They are easily disturbed, and, if so, they curl up into a ring holding on by their claspers and much resemble a bird-dropping when in this attitude (Haggart).

In late June bare patches amongst heather ought to be well worked for imagines of *Scodiona belgiaria*. They rest flat on the ground and are not difficult to see. They occur on Bolton Fell, Bowness Hough, &c.

Towards the end of June, *Melanippe hastata* flies strongly and in fine condition at Bowness, in the wettest part, probably driven there by the strong wind which is frequently blowing.

In Scotland, towards the middle of June, those localities where the eyebright grows should be searched for the imagines of *Emmelesia blandiata;* they may be found at rest during the daytime on walls, trees, &c., and can be taken on the wing at dusk (Haggart).

The best time to capture *Melanthia bicolorata* is about an hour before sunset, when it is to be found on the wing among alders, &c. (Finlay); in Epping Forest it appears to fly at sunset, but may be beaten out of blackthorn in crowds during the afternoon in late June (Buckell).

The allied *Melanippe rivata* and *M. sociata* will both lay eggs freely in confinement if little pieces of *Galium mollugo* be introduced into the boxes in which they are kept; the eggs are laid singly on the underside of the leaves at the edge and generally near the tip.

During the last week of June the imagines of *Melanippe tristata* occur very commonly on the northern moors (Kildale, &c.), flying in the sun over heather (Hewett).

Melanippe tristata may be boxed from, or netted when it flies from, the stone-walls which form such a conspicuous feature in North Derbyshire scenery in June (Thornewill).

Larentia caesiata occur in great plenty at Kildale at the end of June and in early July; they rest on the trunks of the pine-trees and are to be obtained by beating the trunks and catching the moths as they fly off (Hewett).

During the first week of June the yellow-striped larvæ of *Oporabia filigrammaria* are rather conspicuous feeding on common ling (Clutten).

In June the larvæ of *Cidaria silaceata* (second brood) feed up well and rapidly on *Circaea lutetiana* and *Epilobium montanum;* only part of the pupæ gives imagines in August, the remaining pupæ going over until the following May.

Scotosia rhamnata nearly always emerges from pupa late at night, usually after 11 p.m.

The imagines of *Lithostege griseata* are to be disturbed from among the herbage at the edges of the fields at Tuddenham from the middle to the end of June.

On the Devon coast, plants of *Jasione montana* should be carefully worked for the larvæ of *Eupithecia jasioneata* in late June or early July.

ZEUZERIDES.—*Zeuzera pyrina* is said to be common throughout the London suburban districts, not on the lamps, but on the ground under them where they are often to be found in the early morning at rest.

Imagines of *Macrogaster arundinis* commence to come to light at Wicken and Chippenham during the first week of June, although not really abundant until towards the end of the month.

The best time in ordinary seasons for *Macrogaster arundinis* at light is during the last fortnight of June and the first week of July. In 1891, from July 3rd-6th, the captures at Chippenham Fen ran into three figures for a single lamp.

ÆGERIIDES.—The imagines of *Sesia ichneumoniformis* are to be obtained by sweeping the flower-heads of tall grasses, especially towards evening (Bankes). We have also obtained them in June by sweeping flowers of *Lotus corniculatus* in full sunshine.

The best way to obtain *Sesia sphegiformis* is to get on the ground between 7 a.m. and 8 a.m. and search the alder stems carefully, when the pupæ will be found protruding ready for emergence, or the insects actually emerged. Care should be taken to keep a virgin ♀ for assembling; the virgin ♀ s do not call till nearly midday, and pairs, *in cop.*, are almost always taken in the afternoon, sometimes as late as 5 p.m. (Hamm).

A virgin female of *Sesia sphegiformis*, captured near Basingstoke in the early morning, attracted no fewer than twelve males the first afternoon (Hamm).

Sesia sphegiformis emerges in the early morning, and it depends whether the morning be bright and sunny or the reverse how early the imagines come out. I have not the exact times, but I know that on sunny mornings I used to be up before 8 o'clock looking after them, as they are fidgety insects after they have dried their wings. They used to prefer the sunny mornings for emergence, and I used to expose the pupæ to the morning sun. On dull mornings they came out, if at all, much later, about 8 a.m. or so, and I remember one came out about 10 a.m. It was very interesting to watch the whole performance—the beak of the pupa first breaking the bark over its hole, then the pupa forcing its way out to three parts of its length, after which the pupa-case would burst, and "*sphegiformis*" would emerge. Sometimes when their wings were dry, and the imagines saw me, they took fright and went from the top of the cage to the bottom like a flash of lightning, so that I could not see where they had gone, and had to look about at the bottom to find where they were (Robinson).

Towards the end of June search the trunks of old apple-trees for *Sesia myopaeformis*, early in the morning before the sun falls on the trunk.

Trochilium bembeciformis emerges from pupa early in the morning, and flies directly the sun comes upon it (Griffiths).

Trochilium bembeciformis is best taken between 7 a.m. and 8 a.m., sitting on the poplars just after they have emerged, and *S. apiformis* is to be captured in the same manner. It is, I fancy, the best method of obtaining these insects. *T. bembeciformis* never comes out after 8 a.m. (Robinson).

The brown ova of *Trochilium bembeciformis* are laid on the under-side of willow or sallow leaves in early June, generally near the mid-rib, one to five eggs on a leaf.

CYMATOPHORIDES.—In late June a ♀ *Cymatophora ocularis* should be enclosed in a cardboard box with living sprays of its various foodplants—poplars and willows. An analysis of the eggs laid by a ♀ so enclosed resulted as follows :—On weeping willow, 5, all on uppersides of the leaves ; on black poplar, 7, 7, 3, 1, of which 8 (including a batch of 7) were on undersides of leaves ; on white poplar, 16, 10, 5, 1, 1, all on upperside except one ; on balsam poplar, 2, 2, all on upperside except one—total 70. The eggs sometimes touch one another (as in one batch of six), but are at other times laid singly. Five eggs were laid on one tiny white poplar leaf (Raynor).

In June, in some years, the larvæ of *Asphalia ridens* are in great numbers on oaks all over the New Forest. Sleeving is perhaps the best way of rearing them, and if moss be placed in the sleeve they will spin up in it without the slightest trouble. The pupæ should never be removed from the cocoons, and it must not be forgotten that a large number of pupæ usually go over two or three winters.

Beat the low boughs of oak for larvæ of *Asphalia diluta*, and, if one be found, keep industriously to the tree or trees. Also search leaves spun together. The larvæ are best fed separately, as they spin up among the foodplant. Separate rearing will be found the most success-ful mode of treatment with all larvæ which have this habit. It is trouble-some, but the results yield a harvest which will repay the exertion.

If any larvæ of *Asphalia flavicornis* be discovered, search the birch tree or bush carefully, as the larvæ are often present in batches on the same growth.

ANTHROCERIDES.—To obtain females of *Adscita geryon*, work for the species between 4.45 p.m. and 5.45 p.m. They can then be obtained without difficulty (Fox).

The males of *Adscita geryon* fly in swarms among the long grass in the morning sun, the females are usually difficult to obtain, hiding low down on or near the ground ; later in the day they affect the flowers of their district, resting helplessly thereon, and are easily boxed without the need of using a net.

The true *Anthrocera trifolii* is well out in early June, in a variety of situations—meadows, chalkhills, &c. ; the imagines are rather small, and frequently subject to modification in that the normal red spots of the forewings and the hindwings are of a yellow colour. The cocoons are usually spun low down near the ground.

During the first week of June the imagines of *Anthrocera hippocrepidis (stephensi)*, intermediate in superficial appearance between *Anthrocera trifolii* and *A. filipendulae*, may be found in meadows, on hillsides, and similar situations. They are of the small size of true *A. trifolii*, have the spotting almost like that of *A. filipendulae* except that the sixth (lower, outside) spot is usually very faintly developed or absent on the upper side.

At the end of June and in early July, *Anthrocera viciae* is to be obtained in its local habitat in the New Forest (for details see *Brit. Lepidoptera*, i., p. 465); the males fly freely in the sunshine or may be found at rest on flowers, the unfertilised ♀ s sitting on low plants where they await the males; pairs *in copulâ* may be sometimes swept in numbers in the afternoon.

Throughout June and rarely in early July, the imagines of *Anthrocera purpuralis* abound in the Burren district of Clare, a large, stony, highland extending into county Galway. At Abersoch the species is equally abundant rather earlier in June than it appears to occur in Ireland, whilst in early July it is very abundant in Argyllshire, south side of Oban, at the north of Loch Etive, at Taynuilt, and in the Island of Mull. When will the Scotch collectors re-discover the species at Ram Heugh, near Stonehaven, by the seaside?

PSYCHIDES.—*Epichnopteryx pulla* frequently swarms in June in meadows, grassy pastures and mossy banks in Kent and Essex; the males thread their way through the long grass from about 10 a.m. to 3 p.m., in full sunshine, like animated soot-flakes. The ♀ s, which do not leave their cases, remain low down and are rarely seen in nature.

LACHNEIDES.—The larvæ of *Trichiura crataegi* appear to choose, by preference, closely cut hawthorn hedges, fully exposed to the sun; often those bordering roadsides are selected, and, although the larva is somewhat conspicuously coloured, it is not very readily seen. In the evening it climbs to the tops of the young shoots in the hedges, and is then best found between 8 p.m. and 9 p.m.

The best way to find the larva of *Trichiura crataegi* is to search between 8 p.m. and 9 p.m., when it crawls up the tender shoots at the top of whitethorn and blackthorn hedges, especially those cut the previous year. It may also be found more rarely by day, especially in dull weather. Oak, ling, bilberry, apple, and *Salix cinerea* are also foodplants.

In confinement, the larvæ of *Trichiura crataegi* want rearing under the best possible conditions, for Holland notes that it is an irregular feeding species; only the quickest feeding ones appear to be successful, some batches feeding on slowly through most of the summer; these almost always die in the last instar.

The fullfed larva of *Poecilocampa populi* rests in a perfectly straight position on the trunk or on a branch of a tree, and is especially fond of the sun; affects oak in the Perth district, and is not easily beaten in the daytime; rests near the ground on trunks of black poplar in Bishop's Wood, and may be beaten from lime at Croydon. It appears to love crevices in the bark of oak-trunks in our southern woods, clings very tenaciously, is difficult to see on account of its colour, and equally difficult to beat.

Larvæ of *Malacosoma castrensis* are still very abundant on the marshes at the mouth of the Thames; these larvæ are not at all particular as to their foodplant; many more may be observed on *Statice limonium*, *Atriplex portulacoides*, *A. littoralis*, as well as on various coarse grasses, than on *Artemisia*. If placed in a cage with birch, rose and seawormwood, they show a marked preference for the rose and birch, particularly the former, for which the larvæ have a great liking (Whittle).

The young larvæ of *Malacosoma neustria* lie on the top of low hedges, sunning themselves in a group. Before making any attempt at capture, slip a net, cap, or sheet of stiff paper below the community, as, on being disturbed, they immediately wriggle from their position with the activity of lizards, and drop to the ground.

During the first fortnight in June assembling *Macrothylacia rubi* males with a virgin female is exciting and profitable sport. The best time to attract the males is from 7 p.m. till 9 p.m. Some evenings they arrive a little sooner, but seldom.

A mild evening and slight breeze are the most suitable weather conditions when assembling *M. rubi*. It appears that the males always come up against the wind. The female is sometimes attractive for three successive nights, but the first night is always the most seductive. The simplest method of procuring the pupæ of *M. rubi* is to search in bright weather in March the locality where the larvæ have been observed the previous autumn. They may then be found crawling about and they spin up without further trouble (Haggart).

Be careful not to pass by the nest of *Lachneis lanestris*, mistaking it for spider's web, which it closely resembles in the earlier stages of the larva.

ATTACIDES.—In Yorkshire the larvæ of *Saturnia pavonia* prefer the lower shoots of whitethorn bushes on or at the edges of heaths, whilst in the fens they are very abundant on the plants of *Spiraea ulmaria* by the sides of the ditches.

SPHINGIDES.—The larvæ of *Hemaris fuciformis* are very easy to detect upon honeysuckle leaves when young, as they rest on the underside of the leaves, and through which they eat small round holes, often in pairs on opposite sides of the midrib.

To find the larvæ of *Hemaris tityus* towards the end of June, look for leaves of the blue scabious with holes bitten in them; many leaves with holes will, of course, not produce larvæ, whilst at other times plant after plant will yield one.

The larvæ resulting from the early imagines of *Macroglossa stellatarum* are usually to be found on *Galium* from the middle of June until the end of the month, when they pupate on the surface of the earth in slight cocoons and give birth to the August imagines; larvæ, the produce of the latter, are again to be found abundantly in late August and September. The appearance of the imagines is, however, so irregular, that larvæ (like imagines) are almost always obtainable in their seasons from May to November. The larvæ are most readily found between 6 p.m. and dusk.

Theretra (Choerocampa) porcellus is attracted to various flowers the first fortnight in June; strangely, *Vinca major*, which is almost scentless, but has splendid nectaries, is one of the plants most attractive to this species; it will yield more specimens than early honeysuckle, white pinks, &c., although these plants have their share.

Walking through the woods in the afternoon I gathered a lot of honeysuckle bloom, and afterwards set it, in moss, in a field where *Galium verum* grew in abundance in Galway; at dusk it was visited by *Theretra porcellus* as well as by many species of Noctuids (Barrett). The first and second weeks in June are the time for early imagines of *Phryxus (Deilephila) livornica*, if there has been an immigration of the species. They appear at the flowers about this time, and are often taken with *Theretra porcellus* at rhododendrons.

The imagines of *Sphinx pinastri* are reported from Waldringfield, Leiston, Aldeburgh, Saxmundham, &c., from June 8th to August, on trunks of Scotch fir; they rest from about 4 to 14 feet from the ground in every aspect.

In their early stages the larvæ of *Dilina tiliae* usually rest on one of the veins on the underside of a leaf, but, in the later stages, choose a twig or leaf-stalk for the purpose. They seem to dislike the light in confinement, and, when their food is changed, get under the leaves as soon as possible.

Amorpha populi and *Smerinthus ocellata* males can be readily attracted by virgin females.

The young larva of *Smerinthus ocellata* usually rests on one of the larger veins of a leaf and has a most tenacious grip; when older, the larva rests on an upright twig, which it grasps with its anal and last pair of abdominal prolegs, assuming an upright attitude, with the fore-parts of the body raised and the head drawn back, resembling a leaf very perfectly.

The young larvæ of *Amorpha populi* rest similarly to those of *Smerinthus ocellata*, but when older they nearly always rest with the head downwards, and, although the fore-part of the body is raised, the head is curved inwards towards the leaf or twig; they often grasp the stalk of a leaf with the hind claspers only and hang down behind it, and it is quite remarkable how small a sallow leaf suffices to hide a fullfed larva.

The larvæ of *Amorpha populi* sometimes rest on half-eaten leaves of poplars, so as to represent the eaten portion of the leaf themselves, and are then so well protected that, with any wind, it is almost impossible to detect them. The larvæ are much easier to find on misty mornings, and before the sun is up, and the same is true of the larva of *S. ocellata*, the protective coloration being most perfect in sunlight or full daylight (Bacot).

In order to ensure the satisfactory pupation of the larger Sphingids, take a large flower-pot, put into it the usual crocks for drainage, and, upon these, no less than an inch of moist earth. On this place the larva, and then invert a smaller pot (or the upper part of one) over it, so as to enclose it in a chamber at least as large as the original cocoon would be. Then fill up the large pot with moist earth, taking care that this should entirely cover what may be called the artificial cocoon, and leave it so for at least three weeks. The plan proves very successful whilst larvæ covered with damp moss or earth often fail completely (Edgell).

NYCTEOLIDES (CHLOEPHORIDES).—In June the oak-trees in the Monkswood section of Epping Forest should be beaten for larvæ of *Hylophila bicolorana*, they may often be found crawling over the oak-trunks when nearly fullfed. Also common in some years in the New Forest.

The larvæ of *Hylophila bicolorana* are most frequently found in thickly-grown, green, oak foliage. It is best to rear each specimen separately, or the pupation of the earlier ones will be disturbed by those still feeding.

DELTOIDES. — *Herminia cribralis* can be obtained on boggy ground—marshes, fens, &c.—during June; it is easily disturbed at dusk ; also comes to light later in the evening.

LIPARIDES.—At the beginning of June, the batches of ova of *Notolophus antiqua* form conspicuous objects on the tops of the heather. Over 30 batches have been counted in one day on one of our northern moors.

The fullfed larvæ of *Leucoma salicis* always sp·n up in the leaves of the trees on which they have fed up, not on the trunk or on the ground, except in very exceptional cases.

NOTODONTIDES.—Eggs of *Stauropus fagi* laid June 2nd began to hatch June 18th. The little larvæ ate nothing for a day or so, then fed up well sleeved on apple. They often fought when they crossed each other's paths, and a number lost legs or portions of legs in these battles, but this loss did not always prevent their pupating (Holland).

A newly-emerged ♀ of *Lophopteryx camelina* attracts ♂ s readily between 11 p.m. and midnight, in June. The males do not appear to fly at all much before 11 p.m.

The fullfed larvæ of *Asteroscopus sphinx* may be beaten in early June from whitethorn, blackthorn, elm, &c., and require plenty of fresh food. When about to pupate, each larva should be isolated in a large flower-pot, when it usually pupates satisfactorily.

The conspicuous bluish-white larvæ of *Diloba caeruleocephala*, with yellow dorsal and lateral lines, are to be found in June, especially in the afternoon, resting lengthwise on long, newly-formed shoots growing out at the top of hawthorn hedges.

The top shoots of hawthorn hedges are often stripped by the somewhat conspicuous larvæ of *Diloba caeruleocephala* in early June ; the larvæ usually rest on the more or less bare shoots which stand upright at the top of the hedges.

NOCTUIDES. —It is a good thing to have a small net always at hand when sugaring. Many insects, such as *Thyatyra batis*, *Gonophora derasa*, &c., can be often taken while they are flitting round the trees.

On Bolton Fell, in June, moths come all night to sugar, and continue coming till day has fairly broken ; they are commonest just at dark and just before sunrise. Among the hosts of insects that usually come are *Crambus sylvellus*, *Lycophotia strigula*, *Hadena pisi*, *H. dentina*, *Leucania comma*, *Cuspidia menyanthidis*, *Thyatyra batis*, *Xylophasia rurea* and abs., *Apamea unanimis*, *Hadena thalassina*, *Agrotis exclamationis*, *A. segetum*, *Noctua plecta*, *Apamea basilinea*, *Pharetra rumicis*, *Triaena psi*, &c.

Arsilonche albovenosa comes to light throughout June, sometimes in considerable abundance ; a partial second brood also appears in August.

Female *Cuspidia menyanthidis* will lay freely in confinement and worn ♀ s should be kept for ova and not thrown away.

The larvæ of *Cuspidia menyanthidis* feed very rapidly on osier and willow and are fullfed by the end of July (Day).

The imagines of *Acronicta leporina* sometimes fly freely at dusk in the open places in woods.

It is advisable when sugaring for *Craniophora ligustri* in June, to have the treacle on early. I have made the bulk of my captures of this species on evenings when sugaring, before it was time for lighting up (Haggart).

The imagines of *Moma orion* come to sugar during the first fortnight in June from about 9 p.m. till 9.20 p.m., and settle with wings closed at the top of a sugar patch, looking remarkably like a piece of green lichen.

Leucania obsoleta always emerges between 7 p.m. and 8 p.m. The wings expand and dry very rapidly, and for a short time the insect sits head downwards with its wings closely appressed to the reed, looking like a node in the reed stem. It soon flies, however.

By tying up the clumps of marram grass on the sandy parts of the coast in July and sprinkling these with sugar, large numbers of *Leucania littoralis* are to be obtained ; after a time they get right into the middle of the clumps and are then easily overlooked.

Mamestra albicolon and *Agrotis ripae* are to be obtained at the same time and in the same places, but are as frequent on sugared posts, fences, &c.

The larvæ of *Tapinostola fulva* are to be found from mid-June to the end of July (or a little later) mining downward within the inner white lower part of the triquetrous flower-stem of *Carex paludosa*, a few inches more or less above the root while young, and nearer the root when fullgrown. No external trace of their presence can be seen, for though a slight blackish coloration does really exist, yet this is so completely masked by the closely investing leaves as not to be detected without very strict examination (Buckler).

At the beginning of June search should be made for *Chortodes arcuosa* in meadows at dusk. The males fly at dusk, but the females must be worked for later, as they then sit upon grass-stems, &c.

The first week in June is the time for *Cloantha perspicillaris*, one of our rarest Noctuid moths (see *Ent. Rec.*, iii., pp. 159–160).

The larvæ of *Apamea ophiogramma* in confinement feed up readily in short stalks or pieces of the stem of their food-plant, and will pupate therein, emerging well at the end of June, if the pieces be not allowed to get too dry. We have kept them on a damp piece of blotting-paper in a tin box with satisfactory results.

During June *Apamea unanimis* sometimes swarms in amazing numbers at sugar in Wicken Fen. It is generally out at about the time that *Hydrilla palustris* is being worked for at light.

The first fortnight in June is generally the time for *Hydrilla palustris* at light in Wicken Fen ; no doubt this species could also be obtained in greater numbers in the Carlisle district if the same means of attraction were adopted.

Phothedes expolita (captiuncula) is very abundant in Merlin Park, Galway, in June, and is sometimes very easy to secure whilst feeding on the flowers of wild thyme (Walker).

Eggs of *Neuronia reticulata (saponariae)* laid in June, hatch in

K

July; the larvæ will eat *Chenopodium* and broad-leaved knot-grass but seem to prefer willow (Day).

Imagines of *Pachetra leucophaea* are to be taken at sugar near Stanting; they are very shy and are apt to fly off as soon as the light is brought near them.

The exact locality for *Pachetra leucophaea* in the Canterbury district is known locally as the "Devil's Kneading Trough," about a mile and a half from Wye (Kent) S.E. Railway Station; it occurs on two very high banks, the farther and higher being the best (Parry).

Larvæ of *Pachetra leucophaea* received in June fed up on *Poa annua*, growing at first quickly, and then slowly, appearing to be fullfed at the end of October though still eating a little after that date; they went on like this till the end of January, still eating occasionally, when they began to spin up, the last in February; the moths emerged in early March the pupal stage lasting a month. The larvæ were throughout kept indoors in glass jars and fed on *Poa annua*, which was changed every second day until the larvæ became sluggish in December and January when it lasted about a week; but, throughout, the larvæ were regularly disturbed, and not allowed to rest for hybernation or otherwise. When they reached the fourth skin each larva had a separate glass, small, less than three inches high and two in diameter, covered by a glass plate; this kept the food fresh, whilst undue moisture was prevented by half an inch of clean dry sawdust at the bottom, changed with the food, and a sheet of blotting-paper under the glass cover, which also was dried at each change of food. The larvæ usually hid themselves in the sawdust during the day, and often made therein a smooth cocoon-like cavity, but without using any silk. I find that I rarely fail to rear anything to which I pay sufficient attention, especially if I individualise each larva in this manner. Dampness, stale food, and crowding are the great enemies of success in rearing larvæ in captivity, and they all result from trying to do more than the time and attention available justify (Chapman).

Among Noctuids that want to be killed as soon as they are caught, and that require especially careful handling, I would mention *Rusina tenebrosa* and *Apamea ophiogramma;* I kill these at once with oxalic acid, and set them with the least possible delay (Burrows).

A few docks, planted in a gauze-covered tub, will suffice for the rearing of the larvæ of *Dipterygia pinastri* without further trouble. A captured ♀ will nearly always give ova when boxed.

Aplecta tincta emerges from pupa about 5 p.m.; the imagines have to be killed very quickly, or they injure themselves so as to become almost useless (Alderson).

Eggs of *Aplecta prasina*, laid in June, produce larvæ that feed up freely on dock, plantain and lettuce, often becoming fullfed in the autumn in confinement, the imagines emerging later, if kept under suitable conditions, whole broods coming out in November and December.

The beautiful white aberrations of *Agrotis ripae* are abundant locally in June on the Chesil Beach, the larvæ feeding on *Chenopodium, Salsola kali,* &c., in September.

To rear *Agrotis agathina* larvæ successfully, they should be given sallow as well as heath, both of which should be slightly moistened every evening just before dusk; the same remarks apply to larvæ of *Plusia interrogationis* and *Noctua castanea (neglecta)* (Lawrance).

In Aberdeenshire, *Agrotis simulans (pyrophila)* is one of the best insects to be worked for. It is obtained at sugar, but more often at flowers—the yellow iris, thistles, dock, reed, ragwort, and, in gardens, sweetwilliam being particularly attractive. The imagines need thorough searching for, and, as the best localities are wild, rough places, and dark, windy nights are most productive, one has to be very careful. It is a most uncertain insect, some nights appearing in fair numbers, but generally singly, and with miles between them. The imagines have also a habit of crawling beneath bark and loose planks on palings and outhouses. At sugar, immediately the light is put upon them, they make a dive to hide, if they are not too busy with the sweets.

Place ♀ s of *Agrotis simulans* on growing plants of dock and knotgrass for ova; the latter are yellowish in colour when newly deposited, and about the size of and very similar to those of *Triphaena orbona*, but they get dark before the larvæ emerge. The latter are dark green, very like those of *Noctua augur*, but more sluggish.

The eggs of *Agrotis simulans* are laid on a withered grass culm, high up, and nearly always near the junction of a leaf with the stem, and, apparently, only one ovum upon each leaf. The ♀ keeps up a constant fluttering all the time she is depositing. The larvæ feed upon grasses, dandelion *(Taraxacum officinale)*, groundsel *(Senecio vulgaris)* and other low plants.

Searching tree-trunks, palings, &c., seems to be the only way to take *Hadena glauca* in some localities, as the species rarely comes freely to sugar; females lay pretty readily in confinement, and the larvæ feed up well if sleeved on sallow.

Females of *Hadena genistae* lay very freely in chip-boxes and the larvæ feed up very well on broom in confinement.

During the last fortnight of June dusking over the flowers of *Silene otites* at Tuddenham will give females of *Dianthoecia irregularis*.

The first week in June is a good average time for specimens of *Dianthoecia caesia* in good condition, on the coast of the Isle of Man; in early seasons, however, the species is out by the second week in May (Clarke).

The female *Dianthoecia capsophila* lays her eggs about dusk— from 9.30 p.m.-10 p.m.—in June, on the flowers of *Silene maritima* in the same way as does *D. caesia;* the larvæ hatch in a week or ten days, and at once eat their way into the young tender pods.

About the end of June or beginning of July collect the larvæ of *Dianthoecia capsophila* from pods of *Silene maritima*, place them in flower-pots, with loose mould, and cover with muslin; the larvæ feed well on the flowers and pods and pupate in the pots, the imagines emerging the following June.

The imagines of *Hecatera serena* are to be searched for in June on fences, palings, tree-trunks, etc.; they also come to sugar and flowers at dusk, and are occasionally abundant at light. The pine-trunks around Tuddenham give large numbers of this rather common species in favourable seasons.

The larvæ of *Taeniocampa miniosa* prefer the large juicy oak galls to oak leaves. They grow to a large size when fed upon this food, and, as the galls keep juicy longer than the leaves, they are very useful in satisfying the voracious appetites of the larvæ (Broughton).

The first week in June is the best time to beat the larvæ of *Cosmia paleacea (fulvago)*; they are in great abundance on oak and birch in some seasons in Sherwood Forest.

In June a walk down a row of old willows, searching beneath loose bark for larvæ of *Dyschorista upsilon* and *Catocala nupta*, is always productive and interesting.

The pupæ of *Tiliacea citrago* should be kept in a box in the shade all the summer; they will then emerge well at the end of August and in September.

Larvæ of *Mellinia gilvago* beaten from wych-elm, will feed up well on ash and sycamore (Robertson).

The larvæ of *Cucullia verbasci* are to be found feeding in June on *Scrophularia aquatica* as well as on *Verbascum*.

The taking of imagines of *Cucullia chamomillae* in June at Clevedon at *Lychnis dioica* (Mason) must be very unusual; the imagines generally appear in March and April and are found on fences, &c.

Heliothis dipsacea is out with *Acidalia rubricata* in the first fortnight of June, in its favourite spots on the Breck Sands, Tudden-ham, Brandon, &c.

The barren grass field at Torquay, *i.e.*, the first one you come to along the cliffs after passing Kilmore on the cliff path to Anstey's cove, is a favourable haunt for *Acontia luctuosa;* the species flies in the hot sunshine, among field-bindweed ; it also occurs in another field entered by passing through a little wood at the end of the first field ; also occurs at Lulworth, Portland, &c. (Fox).

Anarta myrtilli appears to pair at dusk or dark as some numbers are to be taken in June in the early morning, 2 a.m.—4 a.m., when searching for *Coenonympha typhon*, all in pairs, but none actually *in copulâ* at that time (Wilkinson).

Cocoons of *Plusia festucae* are to be obtained in mid-June spun up in leaves of *Iris* and other plants growing by ditch-sides ; the leaves are usually bent downwards somewhat sharply, although some of the stouter ones resist the contraction of the silk ; a leaf with a natural bend is frequently chosen.

Plusia chrysitis, P. pulchrina, P. gamma, P. interrogationis, and *Cucullia umbratica* were irresistibly attracted during June, in Glen Lochay, in the evening, by the flowers of the melancholy thistle *(Cnicus heterophyllus)* (Morton).

Search well the flowers of the red dead-nettle *(Lamium purpureum)* at dusk, and, in suitable places, *Plusia pulchrina (v- aureum), P. iota* and *P. chrysitis* will be common ; *P. interrogationis* is also to be taken in its own special localities in this way.

In the walled garden of a Scotch "lodge," where the raspberries were in full blossom and the flower-beds edged with blue pansies, moths came to these two attractions in astonishing numbers. The low-growing flowers were thronged with *Plusia chrysitis*, while almost every blossom on the fruit bushes had its *P. bractea, P. festucae, P. iota,* or *P. pulchrina.* Other common species were, of course, present (Arnold).

Plusia moneta pupæ should now be searched for, spun up on the underside of the leaves of its foodplants—*Aconitum napellus* and *Delphinium* (Lawrance).

During the first week of June, search the plants of *Delphinium* growing in gardens. On these the fullfed larvæ and rich, golden-coloured cocoons of *Plusia moneta* are to be found, the latter attached to the underside of the leaves.

The yellow cocoons of *Plusia moneta* are placed on the underside of a leaf of monkshood in June, without bending or warping it, but they may be detected without difficulty, and are sometimes quite conspicuous.

Larvæ of *Catocala promissa* are to be found in ·June in chinks of oak; they are hard to find, being so much like the lichen covering the tree. When taken, the larva tumbles about exactly like the larva of *Cucullia verbasci* (Holland).

ARCTIIDES.—Do not chase the imagines of *Nemeophila russula*, but mark them down and put the net over them, holding up the bag end of it. The specimens may thus be taken unbroken. The ♂ s will often start from the low growth by half-a-dozen at a time; on such occasions search for the ♀ .

LITHOSIIDES.—The imagines of *Lithosia sororcula* are sometimes to be beaten in numbers in the first week of June, from five until seven in the afternoon, on the outskirts of a wood, fluttering gently out from oak, maple, blackthorn and ash, sometimes settling immediately on a leaf, but more generally flying straight out from the wood into the open fields, at a height of from three to four feet from the ground. For the purpose of beating, a 9-foot pole is used, and a very hot day is necessary for good work (Raynor).

Eulepia cribrum is to be met on the heathy ground on the Wimborne Road, near Bournemouth. In places the ground is covered under the heath with a ground lichen, which is said to be the foodplant. The only time to find this insect is when the sun is out, when the imagines have to be beaten out of the heath, and as they ascend must be caught with a swift stroke of the net, as it appears impossible to see where they fly to, appearing to pass clean out of sight, something after the manner of *Macrothylacia rubi* (Hooker).

In June, in favourable seasons, the imagines of *Gnophria rubricollis* are to be seen in the New Forest, flying in the daytime around the tops of the trees almost everywhere, their manner of flight being unmistakable. When captured, like most of their congeners, they feign death.

The larvæ of *Lithosia caniola* are to be found in June, feeding on lichens on the face of the rocks at Bolthead, Dartmouth, &c. Birchall notes that, in confinement, the larvæ feed exclusively on leguminous plants. Jones bred the species on flowers of broom.

During the third and fourth weeks in June the open parts of Box Hill sometimes swarm with the imagines of *Setina irrorella*.

HESPERIIDES.—The eggs of *Syrichthus malvae* can be obtained by sleeving females out on bramble in the sun. They do not seem even then, however, to lay at all freely.

PAPILIONIDES.—At the end of June and the beginning of July search the underside of holly leaves for the larvæ of *Cyaniris argiolus*. The leaves affected have the appearance of being mined (Windham).

In June the females of *Lycaena arion* may be seen depositing their little white eggs near the base of a tiny blossom of thyme (Merrin).

Cupido minima may often be found in hundreds in June sitting in rows of half-a-dozen or more on the grass stems on the outskirts of a wood (especially if on chalk) in the afternoon; they are also to be found sitting in the hot sun in great numbers at puddles, if there be any in their immediate neighbourhood.

The most successful way to procure the imagines of *Polyommatus artaxerxes* is to visit their habitat after sunset. They may then be taken in numbers clinging to the rushes, grasses, marsh-thistle, &c., which generally occur in their localities (Haggart).

The larvæ of *Zephyrus betulae* always sit on the underside of a leaf of sloe along the midrib, and are most difficult to see. I have never been able to detect one in daytime in nature (Turner).

An umbrella is preferable to a tray in beating for *Zephyrus betulae*, as it can be fitted into the structural irregularities of the blackthorn more successfully.

The low branches of oak, with thin growth of foliage, on isolated trees, will often prove the best to try for larvæ of *Zephyrus quercûs*. Search the tray carefully as the half-grown examples imitate the fallen bud-sheaths exactly in colour.

The eggs of *Nemeobius lucina* may be obtained by collecting primrose leaves about the middle of June; the eggs are generally laid one on each leaf, but sometimes more (up to five) may be found; they hatch at the end of June and the larvæ are easily reared, pupating at the end of July.

The larvæ of *Pyrameis cardui* are, in years when an immigration has taken place in May or early June, most abundant in their little globular homes of spun-together thistle leaves in late June and early July.

The beautiful eggs of *Limenitis sibylla*, with deeply-set hexagonal basins, and sharp prominent spiny points giving rise to five gossamer-like hairs, are laid on the edge of the underside of a honeysuckle leaf in late June or during July.

Coenonympha typhon may be taken in some numbers on Bolton Fell in June at daybreak from 2 a.m. till 4 a.m. sitting on the tops of the heather. They are very quiet at this time and may be easily boxed. Females kept for ova will lay freely if put in a glass jam jar along with a few shoots of grass; they scatter the ova singly all over the jar.

One of the Irish localities for *Melampias epiphron* is described by Birchall as "a grassy hollow where a little hut is erected for the shelter of pilgrims," about half way up Croagh Patrick on the Westport side (the insect was captured here in June). It has also been recorded from the edge of a wood at Rockwood near Sligo at about 1000 feet elevation.

A ♀ *Colias edusa* placed under a bell glass with a sod of white clover will lay plenty of eggs on the upper surface of leaves; the eggs of a batch often hatch irregularly even when the whole is deposited within a few hours.

In late June, carefully search the seed-pods of cuckoo-flower, garden-rocket, etc., for larvæ of *Euchloë cardamines*. Examine those where the growth of the seed-pod seems irregular, which will be owing to the feeding of the larvæ, and the latter will be found closely imitating the growth there.

The larvæ of *Gonepteryx rhamni* are sometimes very abundant from the beginning to the end of June on *Rhamnus frangula*. They should be searched for early as most of the larvæ appear to wander away to pupate.

In examining buckthorn for larvæ of *Gonepteryx rhamni*, place yourself so that the sunlight falls across the leaf, showing the shadowed side of the larva, when it is at once discovered. Otherwise it so exactly resembles the midrib along the centre of the leaf (where it rests) that many will escape notice.

JULY.

The annual summer holiday of many lepidopterists takes place during this month and August. To the southerner, accustomed to woods, lanes, chalkhills, marshes, and coast districts, a change to the northern moors and mosses is in every way desirable. These have their own distinctive features. The surface is generally plentifully covered with heather and birch, whilst poplar, willow, alder, oak, and other bushy, rather than tree, growth studs the ground. Among the heather, sallow, *Myrica gale*, and cotton grass, one finds, in one stage or another, an abundance of *Saturnia pavonia*, *Macrothylacia rubi*, *Spilosoma fuliginosa*, *Celaena haworthii*, *Charaeas graminis*, *Hydroecia lucens*, *Pharetra menyanthidis*, *Anarta myrtilli*, *Larentia caesiata*, and many other species, whilst, on the bushes, the various prominents—*Leiocampa dictaea*, *L. dictaeoides*, *Lophopteryx camelina*, *Notodonta ziczac*, *N. dromedarius*—are usually abundant, as well as *Pygaera reclusa*, *Cerura vinula*, *C. furcula* and *C. bifida*, whilst *C. bicuspis* is not unknown. *Acronicta leporina* is usually not rare on the alders, and *Coenonympha davus* flies in the boggy places. Such localities prove, therefore, an excellent change to the southern worker, and whilst the lepidopterist from the north finds relaxation in the southern woods and forests, the southern one may find equal change of scene and variety of fauna on the mosses of Lancashire, Yorkshire, Cumberland, Westmorland and Scotland. July in Scotland is almost always successfully spent, and the professional collectors have made Moray apppear almost an El Dorado for lepidopterists. It is here that the marvellous aberrations of *Triphaena comes (orbona)* are captured, some absolutely black, whilst the usually rare *T. orbona (subsequa)* is not at all uncommon. Larvæ of *Eupithecia togata* may then be obtained in the fir-cones but are, at this time of the year, still very small. Among Geometrids, *Acidalia fumata*, *Ellopia prosapiaria*, *Larentia caesiata*, *Thera simulata*, a local grey race of *Boarmia repandata*, and, later, the bright Scotch forms of *Eupithecia sobrinata* are abundant. By beating the junipers, *Gelechia boreella* and *Hypsilophus juniperellus* are to be obtained, whilst larvæ of *Dimorpha versicolora* and *Asphalia flavicornis* are usually abundant on the birches. The resinous nodes of *Retinia resinana* are in numbers on the fir-trees, and if sugar does pay, some of the marvellous aberrations of such common species as *Xylophasia monoglypha (polyodon)*, *Noctua festiva*, *Noctua dahlii*, *Agrotis nigricans*, *Dyschorista suspecta*, &c., are in countless profusion, whilst *Noctua depuncta*,

Heliothis marginata and other species are always to be taken. In August, *Noctua castanea* and its aberrations, *Noctua glareosa*, *N. sobrina*, *Aplecta occulta*, *Lithomia solidaginis*, and *Epunda nigra* replace the commoner species whilst the splendid forms of *Agrotis cursoria* and *A. tritici* tempt many to become interested in the intricacies of the Agrotid variation. Reid observed that in 1890, a marvellous year for moths at flowers and sugar, they came in swarms to the former during the last half of July and the first half of August, even in the rain, a most unusual habit. The blossoms of the common heath, *Calluna vulgaris*, are to be reckoned among the most attractive of flowers. During July and August, *Noctua glareosa*, *Agrotis lucernea*, *Hepialus sylvinus*, *H. velleda*, *Crocallis elinguaria*, *Oporabia filigrammaria*, *Larentia caesiata*, *Eupithecia minutata*, *Triphosa dubitata*, *Cidaria immanata*, *C. populata*, *C. testata*, *Charaeas graminis*, *Celaena haworthii*, *Polia chi*, *Cleoceris viminalis*, &c., are to be taken on almost every northern heath of any size. We have already sufficiently stated the advisability of continued regular work at flowers. Abroad we have seen the large Salvias that grow in the lower mountainous regions swarming with Sphingids and Noctuids, and, in July, wild sage has proved one of the most attractive plants to Noctuids at Portland and in other districts where it is found. *Agrotis simulans (pyrophila)* rarely, if ever, attracted to sugar, sometimes comes commonly to wild sage flowers, whilst *A. lucernea* and *A. lunigera*, usually in prime condition during the first fortnight of July, are to be taken abundantly at the same plant. The Portland locality for *A. simulans* is exceedingly rough and rocky, and one must work solely for this species when after it. At Freshwater, where wild sage does not appear to grow, it is necessary to sugar flowers to attract these species. *Marrubium vulgare* and *Ballota nigra* thus treated are tremendously attractive, especially to *A. lunigera* and *A. lucernea*. We may also point out that ragged robin and other flowers are, during this month, especially attractive to Plusias—*Plusia festucae*, *P. pulchrina*, *P. iota*, *P. chrysitis*, &c. Prideaux advises lepidopterists to try sugaring sprays of bramble blossom, as well as tree-trunks, and observes that it can sometimes be especially well done in bushy districts where there are no trees. By this means, such species as *Thyatyra batis*, *Gonophora derasa*, *Cerigo matura*, *Xylophasia hepatica*, *Agrotis puta*, and *Luperina cespitis* are reported to have been taken in abundance, whilst Geometrids are more freely taken thus than at sugared trees—*Gnophos obscurata*, *Larentia olivata*, &c., being frequent visitors. Those sprays on which the young berries are forming will be found to hold the sugar best. Marshes in July are not to be neglected, and a note on the way in which larvæ of *Nonagria geminipuncta* are obtained, will suggest the way in which other Nonagriid larvæ may be taken. About the middle of July go carefully over a reed-bed, looking closely for a reed, the middle leaf of which looks dead, but which is otherwise sturdy. This appearance is due to the work of the larva inside the stem. Having found a locality, towards the end of the month search the reed-bed thoroughly from end to end, when reeds giving the above indications will be found to have, near their bases, a small oval or circular spot, over which only the external skin of the reed is left. This has been left by the larva, and is for the exit of the moth, the puparium being quite close to the hole. Cut below the node showing this mark, place the cut reeds in thoroughly wet sand to prevent contraction; careful cutting below the node is necessary, and continued damp is an essential. Excellent results have been obtained by cutting open the reed, taking out the pupæ and placing on damp flannel or damp moss. The moths emerge directly after the middle of August, usually between 6.30 p.m. and 8.30 p.m., but should be allowed to thoroughly dry their wings, which are exceedingly soft and flaccid for a time, before killing and setting. The imagines can, of course, be taken by means of a lantern at this time in their native haunts, suspended from the reeds, drying their wings, or sitting with their wings folded round the reed. Later in the evening they will be found paired in similar situations. Most interesting is Smith's note about the way to take and rear and interbreed *Jocheaera alni*. The larva is conspicuous and should be the object of search; sitting on the upperside of the leaf, it is more easily seen than most larvæ, and it is easily dislodged. When young it is black and white and more likely to be free from parasitic larvæ. The larva is to be found throughout July and pupates usually before the end of August, so that from July 20th to the end of August a special search may be made for it. As it likes sunshine and mounts up to the top of a bush or hedge to enjoy it, the sunny side of a hawthorn-hedge is a likely place to find it. The moth is capable of laying 360 ova, so where one larva is found another should be looked for near. These, when fullfed, should be provided with pieces of elder or raspberry-cane, ready bored and dried, to pupate in. When by these, or any other means at your disposal, you have got together about twenty pupæ of ♀. *alni*, you may wish to multiply your store by breeding. For this purpose you may divide your pupæ in spring, and keep half of them in a cooler place than the rest. Then the ♂ s of the earlier batch will come out first and may be kept as specimens; but the ♀ s of the first batch will come out at the same time as the ♂ s of the later; put these ♂ s and ♀ s in a roomy cage of muslin, and feed every night with honey and water. They should live for ten days or more and will lay their eggs before dying, more or less freely, on the muslin. If the cage can be kept at an open window with a north or east aspect, the prospect of success will be greater. The larvæ thus obtained can be fed in sleeves on growing trees, if at hand, and lime has been found most convenient. The capture of larvæ from the wild is still necessary to continue the brood in health ; and the protection of the larvæ from ichneumons and other foes, including birds, will give the collector exercise for his patience and ingenuity. A few more remarks upon ♀. *alni* and the young collector may then set to work, either in the way above indicated or some modification of it, and help to make this rarity more known. Smith submits with some diffidence that August is the best month to look for this larva, especially in the New Forest and the southern counties, but in the more northern localities, including Yorkshire, captures have been recorded in September. It seems partial to elevated spots, and this may account for some of the later appearances. The larva seems of a thirsty habit, and, in dry weather, cannibalism is apt to show itself. The use of the syringe in the evening, before the larvæ begin to feed for the night is obvious. The same has been remarked of other larvæ. Although instances are known of the larva of *J. alni* having been taken in gardens, the wilder woodlands are its usual haunts. Hawthorn has been indicated as a favourite foodplant, but the broad leaves of the lime and wych-elm have their attractions, and the nut no less, and the bramble on a sunny bank. In wet weather the larva lies under a leaf, using it as an umbrella. Beautiful and interesting in all its stages, *J. alni* is not difficult to rear, and, when once ensconced in a bit of raspberry cane, it is a pupa ready packed for travelling or storing, ready to place in the sun any morning early in June that you wish to bring out the imago. While the wings are still hanging down, the ♀ is readily distinguished by a glance at the ovipositor. If the ♀ be kept two nights regularly fed, before the ♂ is put into the cage, the result is more favourable, as a rule, and, if one's object be to obtain a brood the sacrifice of a few specimens must be made cheerfully. Warm nights and patience to wait are two conditions necessary for success. We have already stated *(antea*, p. 22) that, in rearing larvæ, many species should be transferred to regular breeding-cages when large or in numbers. Of these, that designed by Young, of Rotherham, and some slight modifications thereof are the best. Young's cage consists of a shallow, red, earthenware pan (7 inches in diameter, 3½ inches deep, with the drainage-hole enlarged and fitted with a zinc tube 1¼ inches in diameter, which comes up flush with the rim of the pan). Into the tube a zinc bottle drops, with a flange at the shoulder, which completely closes the tube. On the rim of the pan, and extending 2 inches above it, is a band of fine perforated zinc, closely fitted to the lower rim of a glass cylinder of the same diameter as the rim of the pan, and 6 inches in depth, which rests on three brackets soldered inside the perforated zinc band (½ inch from the top edge). A perforated zinc lid closes the cage. A modified form of this is described by Montgomery as consisting of a square seed-pan, 14 inches by 10 inches, to which is fixed a stout rim of angle zinc. The tube is soldered to a square of perforated zinc, which covers the drainage-holes. To the rim is fitted another rim of angle zinc, which bears a perforated zinc rim framed in wood. Between the two frames of angle zinc is shut a diaphragm, with a round hole in the centre to take the upper rim of the tube. The hole is bordered with a flange which is covered by the flange at the corner of the zinc well that drops into the tube. The top of the cage consists of a six-sided frame of Oregon pine, fitted on four sides with 20-ounce glass, and the top with perforated zinc, while the remaining side fits exactly on the frame containing the zinc rim and kept in its place with strips of oak. Any part of these cages being broken can be replaced without much trouble and expense in a few minutes, and are so strong that they can be piled three high without danger. Cages, on the same plan, but in which leno replaces the glass and zinc, the leno being glued into the rebates and protected with squares of ½ inch wire net, kept in place with wooden fillets, will also be found exceedingly useful. Hairy larvæ do well in these cages, but the leno soon gets dirty; it also rots quickly, and is somewhat troublesome to renew. As to the practical use of these cages, Montgomery says that "he uses the diaphragm until the

JULY. 83

larvæ are fullfed, and if they happen to be of a species which spins on the surface of the ground, it is not removed. It is covered with sand, damped if necessary, and greatly assists in keeping the cages in a good sanitary condition. When cleaning—place the top of the cage with the zinc downwards, and, in it, place food and larvæ, covering with a spare diaphragm, flange downwards; take off perforated zinc frame and brush it well; remove diaphragm, and knock off sand and loose leaves into the iron basin; fit parts together; add fresh sand and food, and it is done in less time than it takes to tell. With larvæ requiring earth, remove diaphragm when they are ready, fill with loam, and cover with a thin layer of sand. By this means, the earth does not become too dry before the larvæ use it, no mildewed frass is mixed with the earth, and the weight of the cage is reduced during the time it is most handled. For surface-spinning larvæ—place a layer (about $\frac{1}{2}$ inch) of scalded moss on the diaphragm, and cover it with sand. When living plants are required, the diaphragm and tube may be removed and the plants raised in the seed-pan. This has worked well with the Satyrids, a succession of pans of grass, raised from seed in greenhouse, being obtained; ordinary turfs of grass contain too many vermin for the purpose. It is also a saving of much time and trouble to go through all the cages every day and fill the wells. This is effected by the 'feeding bottle,' which consists of a long, narrow pickle-jar, three-quarters of a yard of elastic-tubing, and a piece of glass-tubing drawn out to a fine point. When filled with water this forms a syphon, the flow being regulated by pressing the elastic tube between the thumb and bottle. When not in use, both ends of the tube are stuck in the bottle, and it is ready for instant service. This apparatus can also be used for spraying the surface of the earth or moss in the cages when too dry, and for watering the food of those larvæ that require it."

ADELIDES.—During early July the flowers of scabious should be worked for the imagines of *Nemotois cupriacellus* (Miller).

OPOSTEGIDES.—In mid-July the imagines of *Opostega crepusculella* are common on Wanstead Flats, running up blades of grass at sunset and taking short flights (Miller).

LITHOCOLLETIDES.—The larva of *Lithocolletis mespilella* makes a short mine on the underside of leaves of *Pyrus aria*, *P. aucuparia*, *P. torminalis*, *P. cydonia*, *Cotoneaster*, morella cherry, and also, in confinement, on cultivated pear, in July, and again in September—October. The later brood hybernates in the pupal stage.
Imagines of *Lithocolletis kleemanella* are to be bred from mines in the underside of leaves of *Alnus glutinosa* in July and September (Atmore).

GLYPHIPTERYGIDES.—If you grow yellow stone-crop *(Sedum acre)* in your garden you will find in July an abundance of the beautiful *Glyphipteryx equitella* on the windows, walls and fences.

TORTRICIDES.—The larvæ of *Peronea cristana* are to be beaten

84 PRACTICAL HINTS FOR THE FIELD LEPIDOPTERIST.

from whitethorn in Epping Forest, &c., in mid-July (Thurnall).
At the end of July and beginning of August the larvæ of *Peronea autumnana* are to be obtained in rolled up leaves of dwarf sallow (near Morpeth).

The larvæ of *Peronea rufana* live in July and August on *Myrica gale*, drawing neatly together the terminal leaves and eating out the heart of the shoot. The imagines appear in September (Barrett).

The larva of *Phtheochroa rugosana* is to be taken in July and August feeding in the fruit and shoots of *Bryonia dioica*.

At the commencement of July the imagines of *Orthotaenia striana* are locally abundant; on favourable evenings, for about an hour before dark, both sexes are to be obtained on the wing in great numbers; otherwise the female is generally reputed to be rare.

The imagines of *Grapholitha nigricana* are to be found in mid-July among silver-fir *(Abies picea)*. They fly in the early afternoon, rather high, and must be worked for with a long-handled net.

In July the long silken tubes containing the pupæ of *Sciaphila* var. *colquhounana* are to be found in the lichens which cover the rocks in Unst. In the latter part of the month the imagines are to be found in great numbers on the rocks, to the colour of which they closely assimilate (King).

The black larvæ of *Sciaphila colquhounana* feed on sea-thrift in the Isle of Man; one looks for a web spun upon the cushion of the foodplant, and when this is noticed the larva should be found hiding among the roots where it usually pupates in a long silken tube ; also found under sea-plantain, although thrift appears to be the only foodplant (Partridge).

Larvæ of *Acroclita consequana*, more or less gregarious, are to be found spinning up plants of *Euphorbia portlandica*, and possibly other species of *Euphorbia*, in July. The imago emerges in late July or August, and is rather easily disturbed from amongst low plants, at Portland, &c. There is an early brood but the later is much the commoner.

In July or August an elm near Icklingham produced the rare little Tortricid, *Argyrolepia schreibersiana* (Wratislaw).

In early July the imagines of *Trycheris mediana* are sometimes abundant on heads of *Heracleum sphondylium* during the day.

Sericoris irriguana is common on all the mountains between Braemar and Glen Shee at a high elevation, frequenting slopes covered with *Vaccinium* and *Alchemilla*, the ♂ s flying in the sunshine, the ♀ s very rarely seen on the wing. A splendid locality for *S. irriguana* is the western slope of the hill at the back of the hotel near Glen Shee ; the species is also common on the mountains near Loch Laggan.

COSSIDES.—Where trees by the roadside, in residential localities, are infected with *Cossus ligniperda*, the imago is quite as likely to be found drying off on the garden palings as on the trees, for the larva, travelling to construct the pupal cocoon, will often enter the ground under the weather-boarding of a wooden fence.

PYRALOIDES.—In the middle of July the pupæ of *Orthotaelia sparganella* should be sought in plants of *Sparganium ramosum* growing by the sides of ditches; the infested plants will be known by the central leaves being withered. The plants should be cut off low down and carefully opened, a plant generally contains a single pupa, but sometimes two and even three will reward the collector's search. The pupæ should be laid on damp sand in a flower-pot, and slightly covered with finely chopped moss, the pot covered with a piece of muslin, and placed out of doors in the shade (Mason).

The larva of *Pancalia leuwenhoekella* feeds in the petioles of *Viola hirta*.

Anacampsis anthyllidella is common in May and August amongst *Anthyllis vulneraria;* the larva of the second brood feeds in July on the pods (hiding in the calyx-tubes) instead of mining the leaves like that of the first brood.

The larvæ of *Lita salicorniae* are to be obtained in July, spinning together two small plants of *Salicornia herbacea* growing near each other, or spinning up only one plant which it distorts considerably.

The imagines of *Œgegenia quadripunctata* are common in July among *Parietaria officinalis* at Portland.

The larvæ of *Hypsilophus schmidiellus (durdhamellus)* are common in leaves of *Origanum vulgare* in June, the imagines appearing in July.

Depressaria ciniflonella is to be taken on old stone-walls at Rannoch in July and August.

CRAMBIDES.—During the first week of July the imagines of *Crambus cerussellus* are to be found in great numbers resting on the marram-grass on the Deal sandhills, among the grass on the Cuxton chalkhills, &c.; they are readily disturbed in the afternoon sunshine, the white ♀ s being, however, much more retiring than the dark-coloured males.

Crambus contaminellus flies in great numbers after dark in July on the Deal sandhills; it comes to light freely on most evenings and can only be got in large numbers by netting as they come up to the lantern. On those evenings, however, when moths will not fly, but are to be taken in hundreds paired and at rest on the marram, this species may be obtained with swarms of *Lithosia* var. *pygmaeola* and various species of Agrotids.

Crambus dumetellus is generally considered nocturnal in its habits, but on the Culbin Sands at Forres, on quiet, warm days it flies in dozens in the sunshine; also common at Braemar, on Schiehallion, &c.

Crambus furcatellus will only fly in the sunshine on quiet, warm days; occurring on grassy slopes on several of the mountains near Braemar, &c., at about 3500ft. elevation.

Crambus myellus is very local, flies for about ten minutes on very quiet dark evenings, just as it is growing dark, whilst as many as 13 examples have been found in a single day by searching young fir-trees growing among the heather. Ova are best obtained by half stupefying a ♀ in the cyanide bottle, and then allowing her to recover (Reid).

Imagines of *Crambus myellus* hide during the day in fir-trees, dusk and early morning being the best time to obtain the species, the imagines being readily disturbed by beating the firs; a dozen obtained at 4 a.m. by this means on one occasion in a few minutes; open rides in woods and isolated trees most productive

—Glen Tilt, Rannoch, many localities in Aberdeenshire, &c. Although *Ilithyia semirubella* may be disturbed in July during the day from the long grass and herbage, which it appears to love, its true time of flight is for about an hour just before dusk, when large numbers may be obtained in some localities.

In July the imagines of *Rhodophaea tumidella* are regular visitors at sugar, large numbers sometimes being present on sugared oak-trunks.

Imagines of *Phycis abietella* are to be taken not uncommonly at dusk, flying on the outskirts of Scotch fir woods (Atmore).

At the same time the imagines of *Phycis ornatella* and *P. subornatella* are also on the move, a fair bag being readily made at Cuxton, Boxhill, and other places on the chalkhills any really fine evening in July.

PYRALIDES. — *Scoparia alpina* frequents the lichen-covered summits of the higher mountains, never coming apparently below 3000 feet, flying with a short jerky flight in the afternoon sunshine, rising from among the herbage at one's feet, and skimming away for a dozen yards or more, on all the mountains between Glen Shee and Braemar, very common on the highest point of Creag Leacach and Glas Maol.

Imagines of *Scoparia basistrigalis* are common at Edlington, on tree-trunks in July, also near Bexley, &c.

ORNEODIDES.—The larva of *Orneodes hexadactyla* feeds on the flowers and not in the buds of honeysuckle in July (Gregson).

GEOMETRIDES.—In July the local *Epione parallelaria (vespertaria)* is to be obtained in plenty in its well-known local haunts, near York, the males flying commonly from 6 a.m.—9 a.m., the females not to be found except by diligent searching.

The usual date for the imagines of *Epione parallelaria (vespertaria)* is from July 6th-10th. Its time of flight is just about dusk, and is continued until late into the night. It also flies (and much more commonly) on warm sunny mornings about 6 a.m., although its appearance is sometimes retarded until about 9 a.m., this of course depending upon the amount of sunshine. It appears particularly averse to cloudy and windy mornings, the passing of a large cloud across the sun being sufficient to cause it to cease flying. Its flight in the morning usually lasts about an hour, although odd specimens may be seen for a longer period. The specimens taken at night are usually finer and richer in colour than those taken in the morning. I have noticed that after it has been out some time, it generally flies earlier in the morning than when first emerged (Hewett).

The female of *Epione parallelaria (vespertaria)* is very rarely indeed taken in the morning, but almost always at night, at rest on the dwarf sallow, and requires careful search. It is very sluggish, and is seldom taken on the wing (Hewett).

Search the trunks of larch trees in Scotland in the afternoon for the imagines of *Cleora glabraria*. At the same time the imagines of *Ellopia prosapiaria (fasciaria)* can be obtained by searching the base of Scotch fir trees.

Ellopia prosapiaria may be commonly found early in July just emerged from the pupa, either on the ground at the roots of the pine-trees in the wood, or on the bark of these about half a foot above the grass (Carlier).

To give the "Emeralds" a favourable chance of preserving their colour while being killed, place them in the cyanide bottle until moribund, and then stab in the thorax between the second and third pairs of legs with a steel mapping-pen which has been dipped in a saturated solution of oxalic acid. The solution must be a saturated one, to kill. Drop the crystals into boiling water until no more will dissolve; pour off the liquid, and if, when cold, it precipitates any crystals, it is a saturated solution.

Geometra papilionaria appears to be most punctual in its time of flight, seldom occurring much before 11 p.m., from which time until midnight it may best be obtained (Hewett).

Geometra vernaria can be taken between 5 p.m. and 7 p.m. on *Clematis vitalba*. It is at this time that the species emerges, and can be readily seen if present.

Towards the end of July imagines of *Phorodesma smaragdaria* are to be taken in the evening in its well-known haunts on the Essex coast.

The larvæ of *Boarmia consortaria* feed up rapidly in July on birch, oak or sallow if sleeved, but will perish miserably unless they be supplied from the beginning to the middle of August with earth to bury in when fullfed.

Boarmia abietaria is to be captured on larch trunks in July, it is difficult to see, has a habit of flying off suddenly as you approach and dropping to the ground and resting there; fanning the trunks for them is a good method (Bayne).

Imagines of *Boarmia roboraria* are to be taken in early June, at rest on the tree-trunks (often high up), stretched out as big as they can make themselves and are easily seen (Holland).

In July the larvæ of *Nyssia lapponaria* are to be found in Perthshire on hawthorn and birch (Christy), on ling, bell-heather and bog-myrtle (Cockayne).

The larvæ of *Ematurga atomaria* feed freely, during June, on clover and dock, preferring the flowers of the latter and are fullfed in early July. Occasionally a partial second-brood emerges in late July and August.

The larvæ of *Corycia temerata* are to be obtained on blackthorn in July.

The larvæ of *Aleucis pictaria* are also to be obtained on blackthorn in July.

During the first fortnight of July the imagines of *Acidalia ochrata* are usually actively on the wing by about 3 p.m., and continue on the move till dark, flying about *Ononis*, quite close to the herbage; after dark they may be found with a lantern whilst resting on the plants.

The imagines of *Acidalia rusticata* rest in the afternoon with outspread wings (much like *A. virgularia*) on leaves at bottom of hedges, cliffs, &c. The larvæ feed on hawthorn, ivy, lilac, bramble, &c.

Acidalia straminata occurs in the Reading district in early July, and is best obtained by brushing the heath during the afternoon and evening (Holland). Its flight in Yorkshire is from 8.45 p.m. to

9.15 p.m. (Bower).

The imagines of *Acidalia degeneraria* are to be taken during the first few days of July in the daytime in the Isle of Portland, being easily disturbed by beating bushes, flying only a short distance, and then settling on the ground or leaves with the wings flat out; also fly at dusk, and come to light.

The larvæ of *Acidalia degeneraria* (almost always obtained from eggs laid by captured or inbred ♀ s) are to be reared on chickweed *(Cerastium)* and knotgrass *(Polygonum aviculare)*. The larva is inconspicuous, of a brown colour, and falls off its foodplant as soon as it is touched.

Acidalia emarginata may be found by searching bramble hedges after dark; they hang with outspread wings from leaves and twigs, and are very conspicuous by lamplight (Kimber).

The larvæ of *Acidalia emarginata* thrive on the withered leaves as well as the fresh ones of *Convolvulus*, and have such a great protective resemblance to the withered stems of their foodplant that it is difficult to find them amongst the dried *débris;* the pupal stage is a very short one, the imagines emerging early in July. The eggs are loosely scattered, and hatch within a week or ten days of being laid, given suitable weather.

The larvæ of *Acidalia dilutaria (holosericata)* emerge from the egg stage at the end of July, feed well upon *Helianthemum vulgare*, stripping the bottom of a shoot for some distance of the skin, and then feed on the withered leaves at the end of a shoot; they hybernate, commence feeding again in March, and finally pupate in May.

To obtain eggs of *Asthena blomeri* in confinement, place the moths in a large muslin-topped glass cylinder, with a sprig of wych-elm in water, and they will be found to deposit ova freely. These are flattish, oblong, and of a sienna-brown colour, and are laid close along the ribs on the underside of the leaves. Ova deposited between July 13th and 17th, began to hatch on the 21st (Ash).

Timandra amataria will nearly always yield ova if boxed with a piece of dock, and the larvæ can be readily reared (airtight) on that plant.

About the middle of July, *Melanippe unangulata* is to be taken commonly on bramble-flowers in the neighbourhood of Cromer (Nicholson).

The newly-hatched larvæ of *Melanthia ocellata* feed well on *Galium verum* and *G. sexatile ;* they spin up in October and remain throughout the winter in their cocoons, pupating in the spring.

Imagines of *Thera firmata* are to be found sitting on Scotch fir trunks towards night with the wings folded over the back.

At the end of July and during early August, *Larentia caesiata* is to be kicked up from the ling on the northern heaths; it occurs, however, much more abundantly at rest on small pieces of rock; the latter sometimes appear to be almost covered with them on the sheltered sides.

Capsules of *Lychnis diurnea*, collected the last week in July at Scarborough, gave a good supply of *Emmelesia affinitata* and *E. decolorata* the following June, the former continuing to emerge till the end of July (Clutten).

In July, imagines of *Eupithecia pumilata* simply swarm in some years at Aldeburgh, on the trunks of trees in the neighbourhood of the furze-clad common where it stretches inland; twenty specimens were counted on a trunk; they were of a very fine form, some of the ♀ s looking like small *Eupithecia togata* (Sheldon).

Capsules of *Silene maritima*, collected in July on the western coast of Scotland (Oban), give a supply of larvæ of *Eupithecia venosata* the resulting imagines, but slightly darker than those from the usual inland foodplant, *Silene inflata*.

The foxglove flowers, collected on the smaller islands of the Inner Hebrides in July, produce larvæ of the fine form of *Eupithecia pulchellata*, known as var. *hebudium*.

The larvæ of *Eupithecia irriguata* are to be beaten from oak (? and beech) in July (Bishop).

By searching juniper-bushes in July, with a lantern after dark, the imagines of *Eupithecia sobrinata* can be freely taken; the fullfed larvæ can be beaten in great numbers in early June.

In July, *Eupithecia debiliata* is abundant in woods where the undergrowth is bilberry. The place to search for this species is on the trunks of large holly-trees growing amongst the bilberry. They are exceedingly difficult to see, being almost exactly the colour of the bark on which they rest; but a tap with a stout stick disturbs them, and they are netted easily (Russ).

PTEROPHORIDES. — The larvæ of *Agdistis bennettii* feed at Portland on *Statice binervosa* which grows on the cliffs by the sea, and not on their usual foodplant, *S. limonium*.

Platyptilia pallidactyla is to be taken in numbers only after dusk. In a piece of swampy ground, hundreds of specimens were flying at night early in July, but hours of laborious beating in the daytime failed to disturb a single specimen from among the heather, rushes and yarrow (Kimber).

The female *Platyptilia ochrodactyla (dichrodactyla)* oviposits at night in late July and August, thrusting the ova down among the disc florets of the tansy flowers.

The larva of the second brood of *Platyptilia isodactyla* mines the stems of *Senecio aquaticus* in July and August; when young it mines one of the smaller shoots near the buds, but when older it crawls further down, enters one of the larger branches at the axil of the leaf, bores down the interior, feeds on the pith until nearly fullgrown, when it enters the thick main stem of the plant and there completes its feeding, hollowing out a cavity in which to pupate. The presence of the excrement thrown out of the burrow always shows the whereabouts of the larva.

The larvæ of the second brood of *Mimaeseoptilus bipunctidactyla* feed on the flowers of scabious in July; they are very sluggish, eat through the bases of several florets, are thus completely hidden, and, until the flower-head is pulled apart, there is no sign of the larva within.

In the last week of July, larvæ, pupæ and imagines (chiefly the last-named) of *Aciptilia migadactyla (spilodactyla)* are to be found on or among *Marrubium vulgare*, the pupæ always on the upper surface of a leaf near the mid-rib.

The imago of *Aciptilia migadactyla* is to be found in July and

M

early August, appears to prefer to rest on the half-withered bunches of flowers and on the flowering-stems of *Marrubium vulgare*, to the former of which the moth assimilates remarkably well in colour, and is thereby excellently protected.

The pupa of *Aciptilia baliodactyla* is to be found in July fastened by the anal segment to some portion of the foodplant, often to a leaf or stem belonging to the withered top (South).

The imago of *Aciptilia baliodactyla* abounds in July and early August in some of our chalk districts in the south-eastern counties, *e.g.*, Folkestone, Dover, Cuxton, &c. It is to be disturbed occasionally during the day, but flies quite freely in the late afternoon and early evening, directly before and after sundown.

In late July and early August the pupa of *Leioptilus microdactyla* is to be found in a small cleared space just above the middle of the mine in the flowering-stem of *Eupatorium cannabinum* (Buckler). [The larva feeds inside the stems of *E. cannabinum* and the species is double-brooded.]

During the first week of July the imagines of *Leioptilus lienigianus* may be disturbed from the mugwort by day, or found on the mugwort plants with a lantern by night.

ÆGERIIDES.—In mid-July, 1901, a specimen of *Sesia andreniformis* was captured as it was resting on a leaf of the wild cornel or dogwood, *Cornus sanguinea*, near Gravesend.

Old currant bushes should be searched during the last week of June and the first fortnight in July for paired examples of *Sesia tipuliformis;* they appear to be most readily found about 4 p.m.— 6 p.m.

At Salford the imagines of *Trochilium bembeciforme* are to be taken freely in July on the trunks of black poplar ; above 60 pupa-cases have been counted projecting out of one tree, and almost every tree in the district appears to be infested with the species (Chappell).

The imagines of *Trochilium bembeciforme* may be found in July in numbers, just emerged, from about 7 a.m.—8.30 a.m. on poplar-trunks, the empty pupa-cases sticking out of the trunks beside them (Porritt). In the south of England this species usually affects sallows and willows.

ANTHROCERIDES.—In early July *Anthrocera exulans* abounds on the "flats," the tops of a range of hills extending for several miles at an altitude of 2000ft.—3000ft. through the deer forest of Braemar. The species always prefers stony parts, is rarely seen where the heather grows freely, is sluggish in its habits, but much attracted by flowers.

Anthrocera palustris is exceedingly local and gregarious, appearing in July, and is almost entirely confined to a marshy habitat—Sandwich, Freshwater Bay, Waxham, Ipswich, Tuddenham, &c. The insect is much confused with the early June flying *A. trifolii*. The cocoons are to be found on tall rushes, sedges or grasses in late June, when the imagines of *A. trifolii* are almost over.

PSYCHIDES.—Imagines of *Diplodoma herminata (marginepunctella)* are to be found on the boles of oak and fir in July. The female covers her eggs with a thick coating of fur in a similar manner to that of *Porthesia similis* (Edleston).

LACHNEIDES.—Those who have tender skins should be wary of handling the cocoons of *Lasiocampa quercûs*. The stinging power of these can be beaten by no other cocoon, nor by the living larvæ.

DIMORPHIDES.—Among the dwarf bushes in the Altyre woods, larvæ of *Dimorpha versicolora* are to be obtained in moderate numbers by close searching in July.

ATTACIDES.—In July and August the larvæ of *Saturnia pavonia-minor* are often found upon the lower shoots of whitethorn bushes growing on or at the edges of heaths (Porritt).

SPHINGIDES.—In July, on Tuddenham fen, or "common," as it is generally called, look for leaves of the scabious with holes bitten in them; one is frequently disappointed, for many things bite holes in leaves besides larvæ of *Hemaris tityus (bombyliformis)*, but patience is generally rewarded, and at last one comes to a region where there has evidently been a considerable deposit of eggs, and plant after plant yields a light green larva with red markings on the sides, on the underside of one of the leaves (Wratislaw).

The larvæ of *Sphinx pinastri* are to be beaten from pine in August and September, and appear fairly easy to rear, although too succulent food is fatal. The pupæ are to be dug under pines just beneath the surface of the earth (Thellusson).

NOLIDES.—In the New Forest, *Nola strigula* comes freely to sugar in some years about the middle of July.

DELTOIDES.—*Hypenodes albistrigalis* is to be obtained in July by stirring the brambles in the beechwoods of the Reading district; it also comes to sugar (Holland).

LYMANTRIIDES.—The cocoons of *Leucoma salicis* are to be found spun up in the leaves of willow and poplar in early July.

NOTODONTIDES.—To pair *Notodonta ziczac*, leave a ♂ and ♀ in cage all night. They pair during the night, and usually remain paired all next day, separating in the early evening.

A female *Notodonta ziczac*, enclosed in a leno sleeve over a willow branch placed in water, laid 193 eggs the first night after that of pairing, and 50 the next night; most were laid on the leno either singly or in groups (in no case exceeding four); about a score were laid on the willow leaves. They hatched on the tenth day from laying (Merrifield).

By carefully searching maple in late July, I obtained a good many larvæ of *Lophopteryx cuculla*, always resting or feeding on the edge of the eaten leaf. In looking for these, I found a larva of *Jocheaera alni* sitting on the upperside of a maple leaf in a curved position, right in the middle of the leaf, and looking very conspicuous (Holland).

The larvæ of *Lophopteryx carmelita* are to be found on birch; they do not sit on the twigs as do the larvæ of *Dimorpha (Endromis) versicolora*, but on the underside of the leaves, along the stalk and midrib, making the leaves hang heavy with their weight (Holland).

The larvæ of *Notodonta trepida*, very much like Sphingid larvæ that have lost the caudal horn, owing to their oblique lateral stripes, are to be found in July ; in confinement they are amazingly sluggish.

Larvæ of *Notodonta trepida* are to be found crawling down oak trees to pupate, from the middle to end of July (Holland).

In hunting for the larvæ of *Stauropus fagi*, it pays better to search than to beat. Like the moths, they are to be found for a long time. They may be obtained fully grown at the beginning of August, and quite young ones may be taken in September, the latter feeding as long as the leaves remain good. They have been found in nearly all the Reading woods, mostly on beech, but some on birch and oak. These larvæ did not spin up in the green leaves on the tree, as Newman says, but, in every case, in dead leaves at the bottom of the sleeve, or on the side of the sleeve itself. In the woods I have found them crawling on the ground in search of a pupating place (Holland).

· At the end of July careful watch should be kept on early pupating *Leiocampa dictaea*, as a partial second brood of imagines often appears at this time.

NOCTUIDES.—In marshy districts, where there are no large trees or fences to sugar, swarms of Noctuids sometimes come to sugared blackthorn and reeds ; the reeds should be tied up in knots.

I find the flowers of the common rush a great attraction to the Noctuids in July, they are easily taken by searching with a light after dark.

The larvæ of *Apatela aceris* may be found by standing under a young sycamore or horsechestnut-tree, and looking up at the undersides of the leaves, when the larvæ can be seen at rest in a half-coiled position.

In July the young larva of *Apatela aceris* sits curled in a note of interrogation (?) form beneath a leaf of sycamore or horse-chestnut, and eats only the lower parenchyma between the veins in its first stadium, in the second it still leaves the veins and upper cuticle, but, by the time it gets into the third stadium, it eats the whole thickness of the leaf, and traces of its feeding are pretty evident.

Towards the end of July, the larvæ of *Craniophora ligustri* are to be found by standing underneath young ash-trees and looking up among the leaves, when the larvæ can be with care distinguished, stretched at full length along the midrib of the leaf ; they feed up quickly and are not difficult to rear.

The larva of *Craniophora ligustri* always rests underneath a leaf of the foodplant (usually ash), as soon as large enough along the midrib, and when full-grown along the central petiole ; the tapering to either extremity assists it in eluding observation, and it is difficult to see even when full-grown, whilst a half-grown one is very readily overlooked even in captivity.

Larvæ of *Craniophora ligustri* are sometimes common on privet ; when a privet-hedge is bordered by a stone wall the pupæ appear to prefer to spin their tough but not hard cocoons under the "toppers" of the wall (Todd).

The newly-hatched larvæ of *Moma alpium (orion)* may be successfully reared in a test-tube on oak until the first change of skin, which occurs in about six days from hatching. They keep together, skeletonising the leaves till all but the veins are consumed. They can then be transferred to an ordinary breeding-cage, and will feed up rapidly, being usually fullfed within a month.

In some years, towards the end of July and in early August, *Leucania brevilinea* is to be taken in abundance at light at Horning ; in other years, *e.g.*, 1891, not one will come to light, and then they have to be taken flying at dusk, or settled feeding on flowering grasses (Bowles).

In mid-July the pupæ of *Tapinostola elymi* are to be found in stems of *Elymus arenarius*.

Tapinostola elymi abounds all over the sandhills on the coast of Forfarshire. It is most abundant in later July and early August, occurring most freely about midnight, when it can be boxed readily from the heads of the marram grass, and travels well in ordinary chip boxes.

The imagines of *Chortodes morrisii (bondii)* fly somewhat rapidly between half-past eight and nine o'clock in the evening during the first week of July ; they thread their way among the grass, and at Folkestone prefer the slopes leading from the beach up to the Lees.

On the rough broken ground forming the slopes of the cliffs immediately to the west of the cement works at Lyme Regis, well into Devonshire, *Chortodes morrisii (bondii)* occurs in abundance in the beginning of July (Goss).

Towards the end of July, by sugaring the posts of wire-fences which cross moors, *Mamestra furva* is attracted, sometimes in con-siderable numbers, together with *Charaeas graminis, Noctua festiva, N. umbrosa*, &c.

The imagines of *Phothedes expolita (captiuncula)* are best taken from the middle to the end of July. They fly (near Hartlepool) in short low flights over grassy places near the sea ; for five minutes or so many specimens will be seen, then, possibly owing to a slight change in the temperature, none will be seen for perhaps half-an-hour, then the flight will be repeated, &c. (Maddison).

In July the imagines of *Phothedes expolita (captiuncula)* are abundant everywhere on the coast of county Galway, flying about in the daytime in hundreds, and much more strongly marked and marbled than English specimens (Harker).

In early July the imagines of *Apamea ophiogramma* are to be taken freely flying over the ribbon-grass in gardens. The eggs are deposited in rows in the fold of faded leaves (close to the point) of ribbon-grass.

Near Cambridge, in 1892, *Apamea ophiogramma* was on the wing a long time ; the first examples were taken on July 5th, and the species was still in first-rate condition from the 15th—26th, and then others up to August 5th. The species is a genuine dusk flyer, about half an hour being the time in which one can take them ; three or four were taken by walking about with a light, but dusk is undoubtedly the time. They fly quietly, look very light on the wing, settle on different flowers, and are very quiet and easy to box (Farren).

During the early part of July, searching rushes and thistle-heads in a damp wood, near Church Stretton, resulted in a good series of *Xylophasia scolopacina* (Newnham).

Agrotis lucernea is very common on flowers in Portland, but never (with one or two doubtful exceptions on sugared flowers) taken at sugar, although the other Agrotids are more indiscriminate in their tastes.

In July the flowers of wood-sage are a great attraction to the imagines of *Agrotis lucernea ;* the dusk of the evening is the best time to find them.

In early morning during July imagines of *Agrotis ashworthii* may be obtained resting in chinks and crevices, where they hide during the day ; especially frequent on lichen-coloured rocks approximating in colour to themselves (Gregson).

The imagines of *Agrotis ashworthii* come to sugar in July, and like those of *Mamestra furva*, do not appear to be attracted until late in the evening. These are rarely in fine condition. Good specimens are best obtained, sitting on grass-stems, during the night by searching with a lantern (Gregson).

The eggs of *Agrotis ashworthii* are laid during July, in batches of from 30 to 150 eggs together on flower- or grass-stems, which have grown on the rocky faces of the mountains ; they are particularly frequent on the old flower-stalks of *Scabiosa columbaria*. The young larvæ emerge in about three weeks, and will eat almost anything growing in the neighbourhood of their birthplace—*Helianthemum vulgare, Sanguisorba officinalis, Thymus serpyllum, Myosotis, Anthriscus vulgaris, A. sylvestris, Hieracia* species, *Galium verum, G. mollugo, Solidago virgaurea, Campanula rotundifolia, Aira caespitosa* and *A. caerulea* (Gregson).

In July, 1892, some 50 *Agrotis obscura (ravida)* were taken in the neighbourhood of Chinnor, only two, however, at sugar, the rest being obtained by searching outhouses, &c. (Spiller).

Noctua stigmatica may be taken freely by sugaring foliage and twigs outside a beech wood. The ♀ s lay freely, and the larvæ feed up well on chickweed, dandelion and narrow-leaved plantain, pupating in confinement about November, and coming out in the winter under suitable conditions.

The young larvæ of *Noctua stigmatica* feed freely on carrot and chickweed, and often attempt to go straight on with their metamorphosis indoors in confinement, pupating and producing moths in the winter or very earliest spring.

The first fortnight in July is the time to sugar for *Mamestra abjecta ;* the Kent coast (Gravesend, Cliffe, &c.), Essex coast (Wakering Stairs, Havengore, &c.), and Lancashire coast (St. Anne's-on-Sea, &c.) produce it very abundantly in some seasons.

Last July I received 50 larvæ, newly hatched, of *Hadena genistae*. From these I obtained 40 healthy pupæ ; the larvæ that I lost, died when quite small. I fed them on knotgrass, giving them fresh food every day, and kept them, as I do all my larvæ, in a large flower-pot, with muslin tied over the top (Farren).

Hadena contigua, said by the authorities to feed on oak and birch, thrives exceedingly well on *Polygonum aviculare*.

Hadena trifolii ♀ s will deposit ova without trouble in chip-boxes, and the young larvæ can be reared well and quickly on knotgrass.

The ova of *Hyppa rectilinea* are laid in July; the larvæ soon hatch, and are fullfed by the end of October. They then go down for hybernation, entering the pupal stage in the early spring without coming up to feed again (Wylie).

If there is nothing you want on a tree when sugaring do not disturb the moths that are on the sugar. Even *Noctua xanthographa* and *Triphaena pronuba*, in spite of their greediness, will give confidence to more timid insects to come.

When sugaring in woods in July a careful watch should be made for specimens of *Triphaena orbona (subsequa)*, which may be amongst the *T. comes* which come to sugar.

Not far from Forres is a wood with a river running through it, and in the woodland paths, and on the river banks, *Triphaena orbona (subsequa)* is sometimes taken in great abundance.

About the beginning of July the fullfed larvæ of *Panolis piniperda* are to be beaten from Scotch fir-trees, sometimes in great numbers.

The imagines of *Dyschorista upsilon* pair readily and lay rather freely in confinement. The eggs are said to be covered with a thin transparent coating, somewhat similar to that covering the eggs of *Leucoma salicis*, but of an orange-colour and not white.

The larvæ of *Hecatera serena* are to be swept during the after-noon and at dusk in late July or August from various species of *Picris;* they are common on railway banks in Kent, on the *Dianthoecia irregularis* ground at Tuddenham, &c. The larvæ are also sometimes abundant on the blossoms of *Crepis virens* (Norgate). May also be swept in morning (Bower).

Capsules of *Silene maritima*, collected in July on the western coast of Scotland (Oban), gave a supply of larvæ of *Dianthoecia conspersa*, the resulting imagines proving almost identical with the southern form.

In the Shetland Islands (Unst) the larvæ of *Dianthoecia conspersa* feed on *Silene maritima* in July and August; in the south of England the larvæ feed on *Silene inflata* (McArthur). Also found on *S. maritima* in Devon (Bower).

The larvæ of *Dianthoecia irregularis* are to be swept in middle and late July from *Silene otites*, at Tuddenham, &c., or, when nearly full-grown, they may be found just below the surface of the ground at the roots of their foodplant by day, or obtained by searching with a lantern at night whilst feeding They will feed freely on *Lychnis floscuculi* in confinement, and also on garden varieties of *Lychnis*.

Imagines of *Calymnia pyralina* are to be taken throughout July (sometimes more plentiful in the early, and at other times in the later, part of month) coming to sugar very early in the evening, almost before any other Noctuids are on the wing.

The last fortnight in July is the time to sugar for *Calymnia pyralina*. Swansea appears to be a better locality for it than Reading, fifteen specimens having been taken on one evening at sugar among elms (Holland). A bottle full in one night has been reported from Huntingdonshire (Bower).

Bryophila impar is almost confined to Cambridge, where it is to be obtained sparingly on most of the walls of the town itself. The earliest date of capture has been July 27th, the latest August 23rd (Farren).

By the first week in July the larvæ of *Cucullia verbasci* are often quite large enough to show very marked traces of their feeding on the mulleins.

Cucullia umbratica rests in such a way as to imitate a knot, or flaw, on a grey paling.

Eggs of *Plusia bractea* laid in July will hatch in a week; the larvæ can be kept in a cold frame, and fed on lettuce and groundsel till end of August; then, if one wishes to force them, they must be put into a cucumber-house with a temperature from 68° F.—80° F., and fed on dandelion; some thus treated were fullfed from September 13th till end of month; imagines emerged September 24th-October 15th.

During July, beds of *Aconitum* and larkspur should be well watched for *Plusia moneta* hovering over the flowers at dusk; light near such beds is also attractive to the species. They are also attracted by flowers of *Nicotiana affinis*.

Eggs obtained from *Plusia moneta* in early July will, if kept at fairly high temperature when newly laid, produce larvæ that will sometimes feed up rapidly and yield autumnal imagines.

The larvæ of *Plusia ni* are very like those of *P. gamma;* some were taken when small in July at Penzance and Portland, the imagines appearing in September.

Show no hesitation with the "Crimsons" at sugar, but get the bottle under them quickly and quietly, and keep the lamp from flashing on the bottle. *Catocala promissa* is sometimes particularly skittish, and it is best to let the light disk from the lamp only just include it, so that the bottle is in darkness up to the last instant.

Catocala promissa is far more wary on the sugar than *C. sponsa*, but if it rain the insect will sit with closed wings as soon as it arrives, and can then be easily captured.

The last fortnight of July and the first week of August are the time for imagines of *Toxocampa craccae* in its haunts on the north Devon coast. The moth, when disturbed in the daytime, flies rapidly for a short distance and buries itself among grass and low plants; the best time to take it is in the evening from 8.30 p.m. to 9.30 p.m., when it flies steadily and slowly, and spends most of its time in sucking the sweets (or bitters) of wild sage (its great favourite), the hemp agrimony, and other flowers. Three were taken at one sweep of the net at wild sage. When captured, it is quiet in the net, and does not knock itself about in a pill-box. It is rarely seen many yards from its foodplant, and as its habits are rather sluggish, and its foodplant *(Vicia sylvatica)* is not common, it is not surprising that the insect is rare and local in Britain (Horton).

The larvæ of *Euclidia mi* can be swept from long grass by day. It is a long tapering larva, greyish or drab in tint, with two fine white subdorsal stripes. It has but two pairs of abdominal prolegs, and moves with a pseudo-geometrid action. It can be reared on couch-grass, or trefoil, but is very liable to be stung by ichneumons.

ARCTIIDES.—In Guernsey and Sark *Callimorpha hera* seems to have a preference for the coast, and many specimens may be obtained from the ivy, hanging in thick masses over the rocky cliffs. The moths, if disturbed, fly out, and settle again on the side of the cliff, and may be marked down and easily captured.

Damaged females of *Callimorpha hera* have only to be kept in a roomy box to ensure an abundance of eggs, which they lay in batches of from 20 to more than 100.

Ova of *Euthemonia russula*, obtained early, will produce a second brood in captivity if the larvæ be fed up well upon living plants of broad-leaved plantain. Set the plants in earth in a good-sized box, and keep in the sun, with a cover of muslin stretched on a frame lying on the top. The ♀ oviposits freely in a chip-box.

LITHOSIIDES.—In early July the rough cliffs near Freshwater in the Isle of Wight abound locally with *Setina irrorella*. In the early morning, 6 a.m., in cloudy weather, they are to be found at rest on the grass-stalks, many pairs *in cop*. With sun, some of the males fly rather freely, but an abundance of the very finest newly-emerged specimens can be obtained from those settling on the grass. Near Dover, the local collectors used to find it hanging on the grass on the cliffs in the early morning in large numbers. We have taken it abundantly near St. Margaret's Bay (Dover), flying freely from about 3.30 p.m. to 5.30 p.m.

In the Norfolk Broads, *Lithosia muscerda* appears to be very local, no doubt owing to the distribution of its food ; it is attached to alders and sallows, whether in small patches or in thickets, and the imagines fly at dusk with a regular, directly forward, "footman" flight, but are inclined to soar out of reach. They come to light from 10 p.m.—12 p.m., and then sit on the lamp most obligingly. It is difficult to procure perfect specimens, they seem to chip and tear with the slightest provocation. Females lay freely ; the eggs are round and yellow (Bowles).

Œnistis quadra comes to sugar in July ; the males appear to come earlier in the evening than the females, *e.g.*, on July 21st, 1871, six ♂ s came before 10.30 p.m., and six ♀ s after that time, in the New Forest (Farn).

The imagines of *Lithosia sericea* appear about the middle of July, hide low down among herbage, grass, &c., where they occur (the locality, however, seems to be very restricted), the wings wrapped very closely around the body. A few take a short flight about 4 o'clock in the afternoon, but the majority fly at early dusk. Their flight is somewhat heavy and slow and in a direct course. They look of a creamy-yellow colour when flying, and can be readily distinguished by their colour and the character of the flight from the species of Crambids, that are generally on the wing at the same time. The males readily assemble to a virgin female perched on the top of a stem of grass.

The imagines of *Lithosia complana* are frequently found on flowers in July during the daytime. *Eupatorium cannabinum* is especially frequented. Sometimes they occur in numbers in the early evening, males and females flying together, as if assembling, but with no real evidence of its being the case.

Calligenia miniata flies at dusk in early July, and sometimes occurs in considerable numbers in a limited space, possibly ♂ s assembling. May also be beaten by day.

HESPERIIDES.—Towards the end of July the pupæ of *Pamphila comma*, enclosed in a very slight silken cocoon, may be found spun up

N

among the short herbage near the ground, but not actually on the ground, on the downs which they frequent.

Pamphila comma is restricted locally to particular spots on the South Downs. These are generally sunny and windy slopes. If one specimen be found, keep to the place for a close examination in sunlight. The insect sits at rest during cloudy intervals.

From the middle of July into August, the imagines of *Thymelicus actaeon* abound on the slopes of the cliffs at Swanage. July 15th is a good normal date.

PAPILIONIDES.—During July, in its more restricted haunts, eggs and larvæ (in all stages of growth) of *Papilio machaon* can be taken. The eggs, from one to a dozen, are usually laid on the underside of a leaf.

In taking *Dryas paphia* and *Limenitis sibylla* with the net, it will be found, upon trial, that an useful proceeding is to wait until the end of the afternoon, when the setting sun, shining through the tree-stems, lights up the bramble bushes along the borders of the rides. These insects are then to be seen gently hovering and flitting from leaf to leaf, making capture so easy that selection can be made.

Melampias epiphron swarms in early July all over Ben Cruachan, south of Cruachan Burn, from 1000 feet to 3000 feet elevation.

A favourable time to take the Lycænids is just before sunset on a fine day, when they can be boxed from the long grass or herbage upon which they have settled for the night. It is a good plan to pay particular attention to those patches of growth which catch the last rays of the departing sun.

If *Zephyrus quercûs* be required, make a small net six inches in diameter, which will take the place of the top joint on a fishing-rod. The matter then resolves into the question of selecting your tree. This insect is extremely fond of flying around and sunning itself on ash as well as oak. Wait until it settles, and sweep it off the leaf.

Plebeius aegon can be conveniently boxed from the heather at the close of day. The imagines are then often found congregated in large numbers in favourable positions.

In the late afternoon, during the first fortnight of July, the imagines of *Thecla w-album* leave the flowers and rest on the tops of the smaller elms and ash-trees, as well as the hazel bushes in their vicinity.

Worn ♀ s of *Thecla w-album* will oviposit freely if sleeved out in the sun on elm in early July.

AUGUST.

In August commences what is frequently known to lepidopterists as the "second" larva season of the year, and hedges, trees, &c., must be worked continuously with the beating-stick, especially after the middle of the month. A drooping and fading central shoot of thistle, reed, ragwort, &c., is almost always a sure sign of larval work within; cut deep into the root with a knife and split up the stem to discover the cause of the drooping. We have already *(anteâ,* pp.

80-81) shown how to work, in July, the reed-beds for *Nonagria geminipuncta*, in August one must work *Typha latifolia* on the Norfolk Broads and elsewhere for *Nonagria cannae*, and in later August and early September the same plant for *Nonagria arundinis*. In the Broads, at least in the Horning district, *Typha latifolia* is known as the "He Gladdon," the allied *T. angustifolia* being known as the "She Gladdon." Careful equipment for working *N. cannae* must be attended to. India-rubber waders or fishing-stockings are absolutely necessary, the former the more preferable. A strong and sharp knife is needed to cut off the *Typha* stems, a bag in which to place the cut stems, and a strong plank for crossing dykes and bogs is most desirable. Beds of strongly-growing plants by the edges of the rivers or broads are not, as a rule, much affected, straggling or over-grown patches in small ponds in the bogs are much more likely to produce the desired quarry. Although *T. latifolia* is usually chosen as food, *T. angustifolia*, especially if on the borders of small beds of *T. latifolia*, also produces a fair number of larvæ. The usual indication, a fading of the two inmost leaves to yellow, shows the whereabouts of the larva, although if a larva has recently effected an entrance this is scarcely noticeable and many apparently healthy-looking plants, especially of *T. latifolia*, produce larvæ, the entrance-hole in the stem indicating its presence within. In searching, it is best to take one of the central leaves of *T. latifolia* in each hand, and gently pull the two leaves apart, when the larva or pupa, if present, will be detected. The closely-wrapped leaves of *T. angustifolia* do not allow this method to be carried out with success, and the outer leaves must be unrolled till one be found marked with a semi-transparent spot, like a black bruise, which has been caused by the larva having eaten a hole almost through the leaf, before pupation, in order to facilitate the emergence of the imago later. The tip of the knife may be carefully inserted and the thin piece readily removed to make sure the tenant is within. The stem should be cut some six inches below this hole. A large number of larvæ of *Nonagria arundinis* may also be found at the same time, but they are readily distinguished, the larvæ of *N. cannae* being green and those of *N. arundinis* dirty pinkish-brown. The larva of *N. cannae* pupates head upwards, and the pupa has a very distinct beak pointing upwards, whilst the emergence-hole is necessarily above the pupa ; on the other hand, the larva of *N. arundinis* pupates head downwards, the beak is less large and stands out at right angles, and the emergence-hole is beneath the pupa. After having obtained a supply, place the lower ends of the cut stems containing the pupæ in wet sand, in tubs or large flower-pots, enclose in a large box with plenty of ventilation, and water freely every day. Above all things, there must be no contraction of the stems, otherwise the pupæ are inevitably crushed and fail to produce moths. Stems containing larvæ may be treated in the same manner, and the larvæ must be provided with fresh green stems, which they soon enter, should they crawl out of the old ones. The imagines emerge from about the middle of August to the middle of September, when, no doubt, they might be taken in numbers in their haunts, at night with a lantern, were the localities they affect more readily accessible for night-work. Many good Scotch insects are still to be worked for. Henderson gives

(Ent. Record, i., p. 70) a most interesting account of the capture of *Dasydia obfuscata* in the Western Highlands, in the neighbourhood of Garelochead. The species is extremely local but occurs commonly in some districts, whilst it is entirely absent in apparently precisely similar neighbouring spots. It frequents rough heath-covered ground at a moderate elevation, where there is an abundance of bare rocks cropping up here and there. On these the moths rest during the day, with wings spread fully out, in which position their strong protective resemblance to the stone cannot fail to strike any one seeing them *in situ,* the difficulty, indeed, being to see them at all. In colouring, the species varies very considerably from light granite-grey, in which the appearance of the rocks, where it is bare and clear, is very closely imitated, to a dark leaden form, almost indistinguishable from a lichen growing plentifully on the rocks, the resemblance being heightened by the rounded wavy lines on the wings corresponding to the outlines of the lichen. When disturbed from its resting-place, after the first fright (and flight) is over, the insect invariably makes its way to the nearest rock, not by flying, but by walking or running along, using its wings as aids, in the manner of an ostrich. It seems to have a very special aversion to being "blown upon" not only keeping to the leeside of the rock, but creeping in to the shelter of any over-hanging part, always provided there be a flat surface on which to spread out the wings. In the course of collecting in a special locality which the species frequents, one comes to know one or two corners, forming miniature caves, where, if there has been any wind at all during the previous night, one can almost certainly reckon on finding several ; on one occasion no less than nine examples were turned out from a retreat of this kind, of such limited space that there was no room for the whole without the extended wings overlapping. By dint of caution the lot was boxed, one by one, without the aid of the net, the moths being very averse to leaving such comfortable quarters. Gregson recommends August as a good month for "mothing," *i.e.,* for capturing moths on the wing, by hedge-sides, over flower-beds, in gardens, in woods, outside woods, and particularly in lanes ; the generality of moths caught at this time being Tortricids and Geometrids, although species of other families, Crambids, Noctuids, &c., are also to be taken, but, to obtain Noctuids at all abundantly during this month, flowers of all kinds, growing in a variety of places, should be visited regularly just before and after dark. In our opinion, June and July are, above all, the months for mothing, successful as its prose-cution may be in August, while sugar is the great means for enticing the Noctuids in numbers—*Mamestra abjecta, Helotropha leucostigma, Hydroecia paludis, Tapinostola hellmanni,* and numberless other good species, being attracted. Searching ragwort shortly after dusk, say from about 9 p.m. to 10.30 p.m., is, in suitable places, more productive even than sugar, particularly in certain Scotch and Irish coast localities. At Howth, for example, sugar is said to be quite useless whilst the ragwort is in flower, Geometrids, as well as Noctuids, being freely attracted. Sugaring flowers also will often pay well when sugar on tree-trunks is unproductive. The treacle thus applied should be thin, and without causing too many

drippings. Large heads of *Eupatorium* thus treated pay exceedingly well, whilst flowers of common heath, *Hypericum* and golden rod, are naturally very attractive, and *Gonoptera libatrix* is reported as being freely attracted by over-ripe blackberries, the imagines gorging them-elves with the juices until they drop like leeches into the boxes, which they stain badly with the juicy exudations from their bodies. *Triphaena ianthina* is similarly somewhat freely attracted. *Noctua dahlii* is often common on heath blossom as well as sugar. Although light is used all the year round, and many easily worked "light-houses" are to be obtained from the dealers, yet August is specially noted for its results in this direction, and a tolerably dark night, with south or south-west wind, is usually productive. Even a duplex lamp behind the window of a room that is well placed and overlooking a fair extent of wild open country will produce a very good supply of specimens on a favourable evening, and Robertson notes a mode of using ordinary French windows that open inwards as a means of capturing moths attracted to light. He places a table in the room, about 2 feet to 3 feet from the window, puts two lamps thereon a little apart, leaving the windows open till 12 p.m., when they are partially closed, about 10 inches or so being left open, and the catches tied together, and, to prevent the windows opening wider or closing altogether, a chair is placed between them. On the moths coming to the light and finding the window against them, they struggle along the incline formed by the slanting windows, and eventually find themselves in the room, which should be papered with a light wall paper, and have a white cloth on the table. Almost all the moths attracted ultimately find their way inside, and, once in, show no disposition to retreat. Frequently in the morning, the ceiling and walls are studded with moths. Two night's captures by this simple expedient produced — *Hydroecia nictitans*, *Noctua umbrosa*, *N. baia*, *N. triangulum*, *Leucania pallens*, *Triphaena ianthina*, *Hadena pisi*, *H. oleracea*, *Selenia bilunaria*, *S. lunaria*, *Leiocampa dictaea*, *Lophopteryx camelina*, *Cidaria russata*, *Acidalia aversata*, *Melanthia ocellata*, *Coremia unidentaria*, *C. ferrugata*, *Eupithecia pumilata*, *Melanippe galiata*, *Eupithecia tenuiata*, *E. subfulvata*, *Boarmia gemmaria*, *Metrocampa margaritaria*, and many other species, some in considerable numbers. We have already made several references to the various kinds of larva-cages *(anteà,* pp. 22, 82, &c.) used by lepidopterists for rearing larvæ. We may add that, when a lepidopterist breeds insects on a large scale, the larva-cages are usually kept in a larva-house. Some people utilise a conservatory or greenhouse, but Montgomery describes a larva-house specially designed for this purpose as follows: It much resembles a shelter for meteorological instruments. The back and two ends are louvred shutters, the front three doors covered with ½-inch wire net to exclude cats, &c., and to allow the sunshine to enter. An ordinary penthouse roof covers a flat roof, the tri-angular space at each end being left open. A shelf composed of bars with equal spaces between divides the interior. The ends of the house are north and south, the doors face east, and a thick hedge behind shelters it from the glare of the afternoon sun. For about a couple of hours every morning the sun shines right across the face of the house, and every north and east wind blows through it, effectually preventing hybernating larvæ from feeling hungry, and ova from hatching prematurely. Underneath are two bins, one containing sifted loam, which is occasionally watered, and the other silver sand. In one or other a bag full of scalded moss is kept, as well as an enamelled iron basin, cotton-wool, and spare wells for the larva-cages.

TINEIDES.—Imagines of *Tinea nigripunctella* should be sought in August, in outhouses, stables, &c., where they sit on the walls, reminding one strongly of a *Gracilaria* in their attitude, and they may be readily boxed.

In early August, the rough-headed *Phygas bisontella* flies abundantly in the morning sunshine among grass, rushes, etc., the pale ♀ s very easily seen, the dark males rather less conspicuous; the latter assemble freely to newly-emerged ♀ s.

ELACHISTIDES.—In August, in the leaves of reeds *(Arundo phragmites)* at the sides of ponds and ditches, the larvæ of the second brood of *Elachista cerussella* are to be found forming their long narrow mines, whilst those of the April brood feed in reed grass *(Phalaris arundinacea)* (Barrett).

During August small holes in the spun-up seeds of *Pastinaca sativa* suggest larvæ of *Cataplectica farreni;* the larvæ spin two seeds together, and are best obtained by collecting the heads where the seeds show signs of being eaten, and should then be kept in a linen bag. The larvæ appear to enter the seeds at the base, and, eating the contents, pass out at the side, slightly spinning the eaten seed to another, and so on.

LITHOCOLLETIDES.—At the end of August the mines of *Lithocolletis distentella* are locally abundant on oak—very common in Wyre Forest.

ARGYRESTHIIDES.—The larvæ of *Argyresthia aerariella* are to be collected in August at the Brushes, near Manchester, feeding on the berries of mountain-ash, the imagines emerging in late May and early June.

The larvæ of the autumnal brood of *Chauliodus chaerophyllellus*, on *Pastinaca sativa*, appear to straggle over a long period, fullfed larvæ being found from the last week of August to the end of September, the moths emerging from early October.

TORTRICIDES.—The imagines of *Peronea* var. *perplexana* are to be taken abundantly at Armagh in August and September, they are to be obtained with others of the same genus by beating the hedges in the afternoon. The moths, thus disturbed, fly out and are captured, but they need considerable quickness, both of eye and hand, as they dart down to the ground and hide at the roots of the herbage, or else make their way back into the hedge, from whence it is not always easy to dislodge them a second time. They are mostly found among hawthorn, but have been beaten out of hedges formed of a mixture of blackthorn and bramble.

Most of the books say that *Peronea hastiana* appears in August, and many a bag of dwarf-sallow and osier heads had we filled in

July in hope of breeding this species, and, much to our disgust, had never succeeded in getting any but odd specimens. One year, however, we collected, in August, a number of osier tops for larvæ of *Halias chlorana*, and were surprised to find a good many tenanted with apparently newly-hatched Tortricid larvæ, which produced in September and October a fine varied series of *P. hastiana.* Tops of dwarf sallow sent us by Baxter from St. Anne's-on-Sea during September of the same year contained quite small larvæ of this species, and the imagines from these appeared continuously from October up to December 1st. The larvæ and pupæ were kept under cover (in a greenhouse) and were probably thus hastened some days. The autumn, right up to the end of November, was exceptionally mild, and it would be interesting to know what would be the latest time of emergence in a fairly cold autumn, or whether the latest pupæ could go over the winter in that stage, and not emerge until the following spring? One can hardly suppose that a large number would naturally be killed off.

In mid-August, on the Wallasey sandhills, *Peronea permutana* is to be seen in the afternoon sunshine flitting about among the burnet rose ; the species is, however, much more abundant at dusk (Ellis).

The imagines of *Dictyopteryx holmiana* are to be beaten from hedges containing bramble in early August. They come to sugar in swarms at dusk on a favourable evening.

In early August (and July), *Paedisca occultana* abounds in woods of Scotch fir ; the imagines may be obtained in swarms in the morning, merely by tapping the branches, or sweeping among the half-dead twigs ; they soon settle down again and may be netted or boxed without difficulty. This species occurs in June-July in South of England.

In August, the terminal shoots of *Myrica gale* are done up in balloon-like bundles by the larvæ of *Penthina dimidiana.*

The imagines of *Phoxopteryx siculana* fly in great abundance in August at dusk, among the buckthorn bushes on Wicken Fen; by collecting on the edge of a large clump of these bushes, a very long series was obtained. *Phoxopteryx inornatana* was taken among sallow in smaller numbers at the same time.

The imagines of *Ephippiphora tetragonana* are to be obtained locally (Loughton, Hunstanton, &c.) by beating the wild rose bushes at the end of July and beginning of August. They fly freely in sunshine from 5 p.m. to 7 p.m.

The imagines of *Semasia spiniana* are to be taken flying in the afternoon in late August among whitethorn.

In clover fields, in August, tiny whitish-looking atoms are to be seen flying in the afternoon sunshine; these will prove to be *Stigmonota compositella.*

The larva of *Eupoecilia ciliella* feeds in August on the seeds of the cowslip, leaving the seed-vessels when fullgrown and spinning up in hollow sticks or dead stems, where it hybernates, pupating in the spring, the imago emerging in June (Barrett).

PYRALOIDES.—*Poecilia nivea* sits in the crevices of the bark of oak-trees in August, and great skill is required to capture the species, as the imagines are exceedingly active, and, when a pill-box is brought near them, they either fly away or dodge over the hand of the would-be captor.

In the first week of August the larvæ of *Plocheuusa inopella* are readily found on heads of fleabane; the tenanted flowers are easily seen, as the larvæ sever the bases of the florets, causing them to wither (Whittle).

CRAMBIDES.—The larva of *Homaeosoma cretacella (senecionis)* mines the top of the stems of ragwort whilst young, afterwards living within a compact web, which envelops the whole of a small flower-shoot. The central blossom may sometimes be seen thus surrounded, but only in small plants; when vigorous, the lateral flowering stems will alone be attacked. It is sometimes common at Chattenden, and the perfect insects from there are usually larger than those from other localities. *Homaeosoma nebulella* are to be bred from the dwarf thistle, *Cnicus acaulis*, in which the larvæ feed in August. [The larvæ of *H. senecionis* feed in the tops of ragwort under a web; the fat brown larva being quite at home among its frass.]

In the second week of August, *Crambus myellus* is on the wing during the evening; several captured in the neighbourhood of Mony-musk (Mutch).

In the first week of August, the grassy parts of Chippenham Fen give an abundance of *Crambus selasellus*, which are disturbed readily during the daytime and fly freely at dusk.

At the end of August and in September, the nests of *Bombus agrorum, B. venustus, B. sylvarum, B. lapidarius, B. latreillellus, B. derhamellus, Vespa vulgaris, V. rufa*, &c., should be collected for larvæ of *Aphomia sociella*. The larvæ spin the nest into a hard mass about the size of a cricket-ball and pupate therein, the imagines appearing in early June. The larvæ are scarcely ever found in underground nests of bees or wasps.

CYMATOPHORIDES.—In early August the larvæ of *Cymatophora or* are to be found on aspen (often small bushes preferred in Scotland and in moorland districts, where the species is generally distributed). These larvæ hide between two spun-up leaves during the day, very much in the same manner as those of *Asphalia flavicornis*, except that the larva of the latter, as a rule, spins one leaf folded in half, whereas that of *C. or* spins two leaves together, and comes out to feed at night (Hill).

If unacquainted with the larvæ of *Cymatophora duplaris* and *C. fluctuosa*, be careful that they are not tilted out of the tray under the impression that they are giant Tortricids by their behaviour and appearance.

GEOMETRIDES.—The larvæ of *Tephrosia extersaria* may be beaten from birch and hazel in August and September; considerable variation in the size of the dorsal tubercles is noticeable, dependent on the smoothness or roughness of the foodplant chosen.

The larvæ of *Phorodesma smaragdaria* are exceedingly abundant in some years on the Essex saltmarshes, in August and September. They can readily be seen, but one must work on hands and knees, and search even to the roots. Often by beating the plants, and then parting the stems, several will be found in a little colony on one bunch of the foodplant.

During the first week of August, a hedgerow adjoining a stubble-field where the dwarf *Convolvulus* runs riot is a good locality for *Acidalia emarginata*, which can easily be dislodged by tapping the hedge, and then capture the imagines on the wing (Dollman).

Young larvæ from eggs of *Zonosoma orbicularia*, hatching in beginning of August, can be sleeved on sallow, and, in about a fortnight, will possibly pupate, if the weather be really warm ; autumnal emergences are very frequent in confinement, and pupæ want watching closely in early September.

In early August, larvæ of *Emmelesia alchemillata* are common in flowers and seed-heads of *Galeopsis tetrahit* wherever these are to be found; they are rarely to be found wild on *G. ladanum*, although they eat it freely in confinement.

In and around Lochgoilhead (and elsewhere under similar conditions) one has only to find a hollow cavity with an overhanging ledge of rock, or matted roots of trees on the edges of the burns, or by the sides of the road, and *Larentia olivata* will start out at the slightest provocation.

I have taken the imagines of *Larentia olivata* commonly about the middle of August, by dusking along the sides of a wood where the heath bedstraw grows freely. The insect flies very low and can be quickly identified on the wing by this habit (Haggart).

Larentia caesiata is common in August, resting on rocks and stones, which they resemble most perfectly, chiefly on moors or in moorland districts ; the Shetland examples are particularly well marked, and have a much darker median band than those from Rannoch, &c.

The best way of obtaining *Larentia flavicinctata* at Rannoch is to search for them at rest on the limestone rocks on the mountain-sides. They are difficult to find, as their colour almost exactly resembles that of a yellow lichen growing on the rocks. They almost invariably choose the limestone rocks, and it is of hardly any use looking for them elsewhere (Hill).

In August, the seedheads of *Galium verum* should be most carefully searched for the larvæ of *Anticlea sinuata*—Tuddenham, Bury St. Edmunds, &c.

Cidaria testata, showing a great deal of brown in the colour of the wings, occurs in large numbers on Skiddaw; those taken on Ullock Moss are larger and tinged with pink (Beadle).

In some years, during the first week of August, the plants of *Thalictrum flavum* in Wicken and Burwell fens are covered with larvæ of *Cidaria sagittata*, hundreds being taken, and then for years they may not be found at all.

The alders are worth working during August, in Scotland, for *Melanthia bicolorata (rubiginata)* ab. *plumbata*. They can be beaten out during the day or can be more easily netted when the natural flight commences at dusk (Haggart).

Search the trunks of fir-trees during the afternoon for the imagines of *Thera firmata* (Haggart).

Phibalapteryx vitalbata may be beaten from *Clematis vitalba* by day, as well as be taken on flight in the evening. It is a free layer and easily bred.

Living ♀ s of *Eupithecia subciliata*, confined under a large bell-glass, with some sprigs of maple placed in a bottle of water, will deposit eggs at the junction of the footstalk of the leaf with the stem, or on the next year's bud, but all are pushed in and concealed from view, the only mode of discovering them being to pull off the old leaf. Although laid in August they do not hatch until the first week of the following April, and the larvæ are fullfed in May.

Elder is given as a foodplant for *Eupithecia albipunctata*, in Portland, by Richardson.

In the middle of August, the larvæ of *Eupithecia togata* feed in the fir-cones on trees standing from 20 feet to 40 feet high, in the Altyre Woods; an opera-glass is very useful for detecting the infected cones; the species is by no means easy to work, as many trees may be climbed without finding cones containing larvæ (Mutch).

Larvæ of *Collix sparsata* are, in the favoured localities of this species, to be found commonly on the underside of the leaves of *Lysimachia vulgaris* in August.

PTEROPHORIDES. — In August, stand in a patch of *Statice limonium*, where *Agdistis bennettii* occurs; until 7.45 p.m., not a moth will make its appearance; about this time many will appear simultaneously, fluttering up the stems, and by 8 p.m. they will be in large numbers. For about half-an-hour they are very active, then the increasing darkness makes them difficult to see, and one must, if he has not enough specimens, complete his bag by searching with a lantern, but this method is too slow as a rule to be profitable. Also to be taken similarly in June, but rather later in the evening.

The pupæ of the second brood of *Platyptilia gonodactyla* are to be found in August and September in silken cocoons which the larvæ spin on the underside of the leaves of *Tussilago farfara*, and in which the pupæ are suspended in true Pterophorid fashion.

In early August hunt in woods and beside hedges and ditches for the larvæ of the *Stachys*-feeding plumes. Both red and green larvæ are to be found, the former generally smaller, and to be met with before the purplish-red corolla has fallen off, the latter larger and on plants which are seeding. The larvæ of *Amblyptilia acanthodactyla (cosmodactyla)* and *A. cosmodactyla (acanthodactyla)* occur on *Stachys sylvatica* at the same time (Riding).

The little, fusiform, brown-striped larva of *Leioptilus osteodactyla* feeds in the flower-heads of golden-rod from August to October, remaining therein until the downy seeds fly away; they then hybernate till April, and appear to pupate in May without any further feeding, the imagines emerging in July.

HEPIALIDES.—In August the imagines of *Hepialus sylvinus* are abundant in the Monkswood section of Epping, flying over heather. *Eupithecia minutata* and *E. nanata* are also usually taken in abundance at the same time and in the same place.

LACHNEIDES.—The imagines of *Trichiura crataegi* emerge in the early evening, 5 p.m. to 7 p.m., and, as the males fly about swiftly almost as soon as the wings are dry, they should be carefully watched and quickly killed if wanted for the cabinet. The females rest quite still until they are fertilised.

The males of *Trichiura crataegi* are strongly attracted to light. Their natural flight takes place about 7 p.m., so that the lamps should be worked early in the evening for this species. The females do not appear to be so attracted.

Male *Malacosoma neustria* cross pretty freely with female *Malacosoma castrensis;* the resulting hybrid is known as *M.* hybr. *schaufussi.*

SPHINGIDES.—Lord Rendlesham and his sons captured eleven specimens of *Sphinx pinastri* at Woodbridge, Suffolk, during the daytime in the first part of August, and left several damaged specimens on the trees. Ova were obtained and larvæ were recorded as feeding up on September 4th.

The larvæ of *Eumorpha elpenor* appear to prefer *Galium palustre* before all other foodplants.

At first sight it may appear difficult to find a few larvæ of *Manduca atropos* in a large field of potatoes, but, in fact, it is fairly easy to do so. The larvæ, when nearly full grown, are very voracious, and quite strip the haulm of the leaves, so that you may walk down between the rows of potatoes and take ten yards on either side of you with ease and with the certainty of seeing any *M. atropos* larvæ that may be there, the bared stems showing up a long way off. You will soon see if the larva is still about by the abundance of fresh frass ; if the frass be dried up the larva has probably pupated. In this manner I found thirteen in two mornings' work at Deal, and bred them all by forcing. I much advise forcing for large Sphingids. Forcing is simple and safe. An ordinary biscuit box, with a partly glass lid, makes a good and simple forcing cage. Lay in two inches of clean sand, and on it place the pupæ. Do not bury them in the sand. Cover them over lightly with a layer of damp sphagnum, and with a few bushy twigs for the moths to crawl upon to expand their wings, and the cage is complete. The cage only requires to be placed in a warm room at from 75°F. to 80°F., and you will then breed your moths in midwinter with better results than by waiting, and with plenty of time to see to them in the dull season (Tugwell).

LYMANTRIIDES.—The occasional autumnal female imagines of *Demas coryli* lay eggs in August that hatch in early September ; the larvæ can be started on beech, but the leaves are usually all off before they are fullfed ; finish feeding on nut.

NOTODONTIDES.—Towards the middle of August the green larvæ of *Leiocampa dictaea* are to be taken on aspen. As they get older the colour changes to dull purple, and the red anal tubercle gets smaller.

Upon a single aspen on a Derbyshire heath I took nearly 80 eggs of *Leiocampa dictaea*, besides larvæ both of this species and *Pterostoma palpina ;* the birches in the same locality gave larvæ of *Notodonta dromedarius* and *Leiocampa dictaeoides* (Crewe).

Larvæ of *Drymonia trimacula* are to be beaten from oak at the end of August, fullfed. *Notodonta dromedarius* and, of course, *N. camelina* fall frequently enough by beating birch. *Cerura furcula* is to be obtained from sallows, *Acronicta leporina* on birch (Holland). The larvæ of *Notodonta dromedarius* are to be beaten more frequently from isolated and badly grown birch-trees than from large healthy ones.

Breeders of *Stauropus fagi* should always look out for autumnal emergences; individual examples frequently occur from September to October, especially when fairly large numbers are being reared.

The larvæ of *Cerura bicuspis* may be beaten from alders during September and October; the cocoons are made in a depression of the trunk, usually low down (from about six feet up the trunk to the surface of the ground), and are most difficult to discover.

NOCTUIDES.—At the end of August the larvæ of *Arsilonche albovenosa* sometimes swarm on the herbage in Wicken Fen, although a partial second brood of the imagines will, at the same time, be coming freely to light.

When one larva of *Acronicta leporina* is found, others will probably be discovered in the vicinity.

From the end of July to the middle of August the imagines of *Leucania brevilinea* are to be obtained freely at light in the Norfolk Broads, Horning being the most worked centre.

Ova of *Leucania albipuncta* laid on August 18th, 1899, hatched August 29th, the larvæ fed up rapidly on grass, and by October 5th many had pupated; they were kept in a room where was a fire every day, the pupæ being left undisturbed and the surface of the earth not damped; imagines commenced to emerge October 26th.

Nonagria neurica occurs in the neighbourhood of Lincoln. It is a species likely to be overlooked by any one not knowing its habits; it seems to have a period of about three weeks; *i.e.*, the greater part of the month of August, commencing flight soon after 8 p.m., threading its way rapidly low down among the reeds. The best way to net it is to stand by with a good lamp shining on the reeds, when the specimens can be easily taken on coming into the radius of light.

To obtain pupæ of *Hydroecia petasitis*, all you need is a good strong trowel and plenty of patience. I mention this last essential because one day I was out digging with a soft-handed collector, and by the time he had got eight he had had enough, at the same time showing me his blistered hands. Where we get them, they are so plentiful as to require no special method of working. We simply settle down and dig indiscriminately the ground before us, for *H. petasitis* larvæ leave the roots and effect a subterranean transformation. In places where they are unknown, and the foodplant occurs, search for affected plants, *i.e.*, plants with burrow at the crown of the root. Suspicious plants, with withered leaves break off when the burrow, made by the larva, is disclosed. Many plants, though, are slug-eaten, and the leaves withered from this cause. But those who undertake to get pupæ have a rough job on hand (Collins).

Towards the middle of August, ragwort blossom begins to be exceedingly productive in Scotland; *Hydroecia lucens* occurs almost everywhere, and, near Perth, a fine local form of *Agrotis obelisca* (var. *hastifera*) abounds thereon.

At dusk, in the middle of August, the females of *Charaeas graminis* are often to be taken in numbers, both in the Norfolk

Broads and the Cambridge fens, fluttering up the reeds and grass-culms, and paired couples are often to be found from 10 p.m.-11 p.m. It is remarkable that the males fly freely, and often in great numbers for about an hour in the early morning between 7.30 p.m. and 9.30 p.m., when no ♀ s are to be found.

During August the imagines of *Celaena haworthii* are to be obtained sitting on the flowers of the ragwort in the day-time (Booth). They are also to be taken off the bloom of *Erica* in the day-time, and have a habit of dropping off the flowers at the approach of the collector, long before he is sufficiently near to box them (Stott).

Towards the end of August and well on into September, a warm over-cast afternoon will give an abundance of *Celaena haworthii* flying over the mosses on the moors, and it appears that if the atmosphere be clear and the sun shining there are few specimens on the wing, whilst, if the afternoon be calm, warm and moist, the species flies pretty freely; the imagines may also be taken at night from the blossoms of *Calluna vulgaris* (Finlay).

The imagines of *Acosmetia· caliginosa* are to be caught in the en-closures between Lyndhurst and Brockenhurst during the first fortnight of August (also in July). They occur in the rides where the grass is long. To obtain them, walk slowly along the ride, holding the net by the ring and stir the grass in front by sweeping gently with the end of the stick as with a scythe. The longer the stick, and the slower you go, the better, otherwise you lose many moths that rise behind you. They can easily be mistaken for the common Crambids, remaining on the wing only a short time and then returning to the grass, where they are difficult to see. Sometimes they are to be disturbed freely on a rainy day.

If the larvæ of the Agrotids are kept in a fair-sized box or tub, with some inches of damp earth in the bottom for them to burrow in, and only sufficient food for the 24 hours given them at a time, they will probably thrive well.

The flowers of ragwort are especially attractive on the coast, *Actebia praecox*, *Agrotis cursoria*, *A. vestigialis*, *A. tritici*, *A. nigricans*, *Charaeas graminis*, *Luperina cespitis*, *Triphaena interjecta*, &c., being frequently abundant thereon in August (Gregson).

Imagines of *Agrotis agathina* are out from about August 18th; they only fly for about three-quarters of an hour just at dusk, and are then not difficult to net with the aid of a lantern, as they do not fly fast, and are very quiet in the net.

To find the larvæ of *Agrotis ripae* in August and September, dig round the prickly saltwort, found so plentifully at many places on the coast; at Hunstanton I found them very common, not less than 260 in four hours. I have at other places dug five hours for 50 (Farren).

In mid-August, on the Crosby and Wallasey sandhills, the flowers of ragwort are, during the daytime, a great attraction for *Agrotis vestigialis* (Walker).

On the coast sandhills in August, the imagines of *Agrotis vestigialis* and *Actebia praecox* are to be obtained by shaking the overhanging roots of the sand-rush (Robertson).

Quite the end of August is the earliest time that one can hope to secure *Agrotis obelisca;* in late years, the middle of September is most prolific, and if the weather remains suitable and good for sugaring, the species may be found until October.

The larvæ of *Agrotis ripae* (and its white var. *obotrictica*) are to be found towards the end of August, by searching at night with a lantern, on *Salsola kali, Cakile maritima* and *Atriplex maritimum,* and during the day by scraping away the sand immediately around the foodplants ; the larvæ are only to be obtained where the food-plants grow entirely in the sand, no larvæ being found on those plants growing where there is any humus mixed with the sand, however luxuriant the plants may be, and larvæ kept in confine-ment will not enter earth formed of sand mixed with humus, but die ; of those kept in large boxes filled with sand, the last examples disappear finally by the end of September ; most of a large batch of larvæ thus reared were found to have bored into the wooden bottom of the box (about 1½ inches in thickness) in which they were confined, and pupated therein, the imagines appearing the following June (Speyer).

To get eggs of *Agrotis lunigera,* place females in a bandbox with some honey on a small piece of sponge. Some thus treated were looked at every day, but no eggs could be seen, and, at the end of ten days (three of the moths having died meanwhile) the survivors were allowed to escape, and the experiment was supposed to have been a failure, until, just as the sponge used for the honey was about to be thrown away, some ova happened to be noticed therein, and, on careful examination, the cells of the sponge were found to be full of eggs, which had evidently been deposited by the moths as far inside as they could reach, and as much concealed as possible. There were no eggs on the outside of the sponge. No foodplant had been placed in the box, and, as the box was simply a card-board one with smooth sides, the sponge was the only place affording concealment (Brown).

The larvæ of *Agrotis lunigera* hatch in early August, and will feed on dandelion ; when this fails they will feed on slices of carrot, and thrive well on this diet till dandelion is again available in the spring. Larvæ kept in confinement continued to feed from August right through the winter up to the end of April— that is to say, some of them did, as the slices of carrot were always eaten more or less, even during the severest winter weather. They were kept in two large flower-pots under an open window facing the north, and led an entirely subterranean life, the four inches of earth in the pots being honeycombed by their burrows. The food was invariably eaten from the bottom, never on the top or edge, the slices of carrot being scooped out on the underside. On several occasions a dandelion leaf was noticed sticking straight up in the pot, having apparently been grasped in the centre and partially dragged into the hole occupied by the larva, just as if it had been bent and forcibly stuck into the ground (Brown).

Agrotis puta in the larval state feeds up rapidly on knot-grass.

Triphaena interjecta rarely comes to ragwort flowers till the commoner species have been boxed, or have retired for the night, usually not before 11 p.m. (Gregson).

Triphaena interjecta is peculiar in its habits on Wicken Fen ; instead of flying late in the afternoon at blossom, as it usually does in wooded districts, it comes freely to sugar in mid-August, several dozens sometimes being shaken out of the sugared "knots" in one evening. We have not noticed it as coming especially late.

Heather blossom is exceedingly attractive in August to *Noctua glareosa* and *N. dahlii;* the latter species is very uncertain in its appearance, but sometimes swarms in its well-known haunts—Aberdeen, Sutton Park, Sherwood, Morpeth, etc.

I find sugaring the posts of a wire fence crossing a tract of moorland profitable work in August. The imagines of *Noctua* var. *neglecta* are attracted, along with hosts of commoner Noctuids, such as *N. festiva* and *N.* var. *conflua*, in great variety, and I get charming varieties of *Charaeas graminis* in this manner (Haggart).

In August the imagines of *Noctua depuncta* come freely to sugar in the Galashiels district. I have frequently got a good bag of them when little else would come to the bait (Haggart).

The larvæ of *Hecatera serena* feed on *Crepis virens*, and in no instance have I found them on either of the sow-thistles, though they are very abundant in its locality. The larvæ feed at night on the inside of the flowers, but are easily taken during the day by shaking the plant into a net. They seemed difficult to rear, as many died when nearly fullfed (I tried change of diet to the sow-thistles, but it was refused), and a few were ichneumoned. Out of more than 100 larvæ, only 18 were reared to the imaginal state. Some dozen or so died in the pupal stage (Riding).

From the middle to the end of August, *Plusia festucae* and *Helotropha leucostigma (fibrosa)* occur by all the ditch-sides at Heysham Moss, near Morecambe (Porritt).

Three dozen larvæ of *Dianthoecia cucubali* were taken from the flowers of *Lychnis floscuculi* during the first week of August on Balerno Bog (Carlier).

The larvæ of *Hadena pisi* are exceedingly abundant on dwarf sallow on the Wallasey sandhills in mid-August (Walker).

In the Morpeth district, *Polia chi* ab. *olivacea* prefers to rest on the trunks of ash-trees in August.

The imagines of *Polia chi* ab. *olivacea* are to be obtained in numbers by searching the dry stone-dykes which divide fields, &c., in the Galashiels district (Haggart).

In localities where *Bryophila muralis* is found it may be sought for with increased prospect of success on "honey-dew" evenings, as the insect is strongly attracted by this condition of things.

During the first three weeks of August the imagines of *Bryophila muralis* emerge between 5.30 p.m. and 7.30 p.m. This is the best time to look for them as they are far more conspicuous when stretching their wings or when at rest on the spot where they have just closed them; when they have flown and chosen a resting-place for themselves they are exceedingly hard to find, so well do they harmonise with their surroundings (Woodforde).

In late August and September an occasional imago of *Tiliacea aurago* falls to the beating-stick, but a great majority appears to hide in the herbage and leaves below; at dusk the imagines will be found to have come to the top of the grasses and other plants, or feeding on the blackberries; they take flight at once on some evenings although sugared twigs will generally stop them.

Calymnia affinis is one of the least demonstrative of insects, in appearance, that is likely to be seen at sugar. It will often cling, closely appressed to the bark, some little distance from the sweets, and it is always best to give a glance around when the insect is likely to turn up, to make sure that it is not present.

Mellinia gilvago comes freely to light in August (Cambridge, Reading, &c.), and is sometimes taken from the lamps in and near towns in large numbers, probably only where avenues of elms exist in the suburbs of the towns.

Towards the end of August, at about 6 p.m., search should be made at the base of ash-trees and on the *débris* around for freshly-emerged imagines of *Cirrhoedia xerampelina.*

In the Church Stretton district, imagines of *Cirrhoedia xerampelina* are to be obtained at the foot of some large ash-trees, clinging to blades of grass at the end of August and early in September, the trunks rarely giving a single specimen.

In the middle of August the imagines of *Lithomia solidaginis* occur freely on a piece of boggy heath, near Wilsden. At rest, this species has a most remarkable resemblance to the excrement of grouse—the male particularly so. It folds its wings round its body, clasps a stone with its legs and raises its body to an angle of about 30°; its markings, colour, shape and mode of attachment make the imitation almost perfect.

Lithomia solidaginis abounds in Cannock Chase in late August. The imagines are very easy to find as they sit on the birch-tree trunks during the daytime, principally with their heads thrust into some crevice in the bark, so that their bodies stand out at right angles to the trunks of the trees, rendering them conspicuous. As many as 150 have been taken on one day, the greatest number on a single tree being 7.

The larvæ of *Cucullia lychnitis* will eat *Scrophularia aquatica* as well as *Verbascum*, preferring the buds and blossoms.

Everyone who has the opportunity should devote some days during the first fortnight of August to searching for larvæ of *Cucullia gnaphalii* in the Kent woods. It is easy to make a small, light beating-tray of black calico sewn to a frame of stout iron wire, and, having found an opening in the woods where the golden-rod is plentiful, each plant should be beaten gently with a light stick into the tray, which should be held against the plant-stalk, low down near the ground. Plenty of larvæ of *C. asteris* will be obtained, and of Eupitheciids too, but the larva of *C. gnaphalii* is unmistakable, with its dull green colouring and purple dorsal stripe. An average of one to a hard day's work is good (P. C. Reid).

The imagines of *Stilbia anomala* are best taken on Cannock Chase in early August, beginning to make their appearance about 7 p.m., and continue flying till about 8.30 p.m. The insect is very conspicuous on the wing, the ample lower wings making it appear almost white when flying; it gets up suddenly out of the heather and short grass, flies ten or twelve yards, and then drops down again, folding its dark upper wings closely over the lower ones, and thus, in a moment, becoming almost invisible, so that, unless one marks very exactly the spot where it falls, it is impossible to detect it. Frequently, however, it will fly up again almost directly, when, of course, it may be "snapped" with the net (Thornewill).

Whilst waiting on Cannock Chase in early August for *Stilbia anomala* to appear, the early evening can be well spent in searching for the larvæ of *Anarta myrtilli*, which are fairly abundant on the higher shoots of the heather, as also are those of *Eupithecia nanata* on the flowers of *Erica cinerea* (Thornewill).

Stilbia anomala is common in early August on the heathery slopes of hills, &c.; in many places flies freely just as it gets dark, and wants a sharp eye and hand to take it at that time; much more easily obtained between 9 p.m. and 10 p.m., when both sexes can be boxed at rest on the tops of reeds, grasses and heather in its habitat.

The weather undoubtedly exercises a considerable influence on the flight of *Stilbia anomala*, my largest number (48) being taken on a very still evening, when the flies were exceedingly troublesome and a thunderstorm was impending, while, on August 6th, when it was very cold, though fine, and a strong N.W. wind was blowing across the Chase, not a specimen was to be seen. Although the ground was repeatedly crossed during the daytime, no *S. anomala* were ever disturbed, though two specimens flew off palings in the early evening (Thornewill).

There is a great difference in the flight of the sexes of *Stilbia anomala*, that of the male being exactly like that of a Geometer. Indeed, when *Cidaria testata* and *populata* are about, you cannot distinguish them by their flight. The female, however, buzzes about like a Noctuid, keeps close to the heather, and only flies about two yards at a time (Freer).

ARCTIIDES.—The larvæ of *Spilosoma fuliginosa* are very abundant on dock and other low-growing plants on the Wallasey sandhills in mid-August (Walker).

Callimorpha hera is very fond of visiting by day, in August, blossoms of hemp agrimony *(Eupatorium cannabinum)* (P.C. Reid).

The larvæ of *Callimorpha hera* prefer the leaves of garden roses to dandelion. They want some well-sifted sand in which to pupate, and are exceedingly easy to rear (Moberly).

The larvæ of *Euchelia jacobaeae* will feed on coltsfoot and groundsel, as well as ragwort *(Senecio jacobaea)*.

LITHOSIIDES.—The larvæ of *Lithosia sericea* feed a little throughout the winter on chickweed and grass, although when newly hatched they will feed on *Polygonum aviculare*. In the spring they seemed to prefer *Spiraea ulmaria* (Day).

PAPILIONIDES.—About August 1st *Erebia aethiops* makes a first appearance at Pitlochrie; from then it literally swarms in several localities, more especially in a copse at the back of the Hydropathic, where it is a common occurrence to get four or five specimens in the net at one stroke. The species is also common in the Pass of Killiecrankie, Glen Tilt, and on the Dunkeld Road, but the males everywhere predominate.

The male imagines of *Erebia aethiops* are to be obtained in many parts of Scotland during the first week in August. They fly only during bright sunshine, and should be sought for in clearings in woods where *Aira caespitosa* (the wavy hair grass) grows freely. The females do not emerge in any number until about the third week in August, when the males are well over. Time of emergence is between 11 a.m. and 1 p.m. (Haggart).

Colias hyale, in its early days after emergence, often keeps to the restricted flight of a single field, and if one is netted others may be seen. Afterwards it takes to the downs, or open country, and then only solitary examples are met with.

Colias edusa, as well as *C. hyale*, may be waited for, down wind, at the border of a clover field, which will facilitate capture, as the insect has to turn against wind to try back.

Satyrus semele is best approached down hill, as it nearly always flits quickly up the gradient, and consequently can be seen to rise and be netted *en passant*.

SEPTEMBER.

There is much work for the collector to be done in September, both among the larger and the smaller species. Almost all methods of collecting can be successfully prosecuted, and sugaring is, especially during the first half of the month, often particularly successful. *Leucania albipuncta*, *L. vitellina* and *Caradrina ambigua* are among the better species to be especially worked for, the latter, only added to our fauna a comparatively short time ago, has been particularly abundant of late years, whilst *Leucania l-album* is also always a possible capture. *Agrotis obelisca*, so often confused with *A. tritici* by even fairly advanced lepidopterists, is at its best in the early days of the month, and may be taken in some seasons in large numbers at sugar in its chosen localities of which the Freshwater cliffs are the best known. One sometimes notices a complete absence of the beautiful Xanthiids at the sugared trees, even when other Noctuids are fairly numerous, and, at such times, it is well to search the dry-looking feathery heads of the long grasses in the neighbourhood of sallow-bushes, which, especially when covered with honey-dew, appear to be particularly attractive to these and other species. On the outskirts, and in the clearings, of woods, where such grasses are particularly abundant, scores of *Citria flavago* and *C. fulvago* may sometimes be taken, *Hydroecia micacea* being similarly attracted. But the honey-dewed leaves of the sallow-bushes themselves are also most attractive, and are said to be especially productive on dewy evenings, and on evenings following a fine hot day, *Helotropha leucostigma*, *Leucania lithargyria*, *Noctua baia* and other species frequently accompanying the "sallow-moths" at the feast. The present site of Westcombe Park railway station was, 20 years ago, partly an orchard and partly a market garden, the latter being allowed to go out of cultivation about 1883. During the next three or four years, before the land was taken over by the speculative builder, huge beds of dock-plants flourished amazingly. These plants were the home of thousands of *Hydroecia micacea* which, singly and in pairs, could be taken on the dock-plants in large numbers, as soon as it was dark, throughout the greater part of the month, with the aid of a lantern. Similar areas possibly would produce the same species in equal abundance. Larva-collecting is, however, the favourite work of the lepidopterist during this month, and the breeder of lepidoptera has to consider not only ways and means for those larvæ that will pupate in the course of a few weeks, but also for those that hybernate, and have to be carried through the winter. Such species as *Boarmia repandata*, *B. roboraria*, &c., are best wintered in sleeves on the growing foodplant, birch, &c., so also are those of *Phorodesma smaragdaria* on southernwood, but this method is not practicable for many species Hybernating larvæ divide up sharply into two very distinct groups (1) Those that hybernate completely, and (2) Those that do so partially. The former usually spin a silken pad on which they rest immovably the whole of the winter; the latter move about, nibble at their food in mild weather, but make little headway in their growth. The first group are best hybernated off the foodplant so as not to be exposed to mildew, &c., the second are best wintered on it. Tugwell states that it is a good practice to place in cages with hybernating larvæ some plant that will not readily get mouldy, *e.g.*, true ivy, or dry *Sphagnum ;* on these the larvæ can rest without much fear of injury. Keep in a cool greenhouse, avoid absolute drought and sunshine, as the former tends to dry up larvæ and the latter wakens them too early and before Nature has provided any growing foodplant on which to feed them. They must be watched carefully in March and April, and, as soon as they begin to move, fresh food must be provided, and they should then be moved to a warmer place. Montgomery states that small hairy larvæ may be carried through the winter in bottle-cages, arranged so that a glass phial is plugged into the neck of the bottle, the bottle filled up to the mouth of the phial with (1) a layer of sand and moss packed tightly round the phial and damped, and (2) a layer of loose moss, the larvæ secured with leno and an elastic band. The small quantity of food obtainable during the winter is thus made the most of; it is easily renewed, and the larvæ are readily cleansed by renewing the loose moss and putting in a little clean sand. Larger hairy larvæ will go through the winter in an ordinary cage, the base filled with sand, covered thickly with moss. Even larvæ of *Macrothylacia rubi* do well thus, but require to be placed on a sunny window-ledge, where the rain cannot reach them till they spin up. Species that cling to the leafless stems of trees and shrubs winter satisfactorily in the ordinary cages, the twigs being plugged into the wells, and the base of the cage filled with damp sand. Acidaliids do extremely well on damp sand covered with moss and dead leaves, on which is sprinkled such food as is available, and watered once a week. Many Noctuid larvæ will go through most successfully in flower-pots half-filled with sand and covered with moss, but it must not be forgotten that most of these require some food, passing the winter in a nibbling state, whilst others feed up entirely during the winter. In connection with this it must be remembered that sliced carrot is excellent winter food for many Agrotids and that larvæ of *Peridroma saucia* will feed up readily during the winter upon dock, &c. Eggs that pass the winter in that state should, as soon as possible after they are laid, be placed in a cool place and not brought into a warmer until the proper foodplant has advanced sufficiently to support the larvæ that will hatch from them. The

proper treatment of pupæ also has to be considered, and it may
be at once laid down as an axiom that excessive wetness, excessive
drought and an unusual temperature must be avoided. Subterranean
pupæ are, perhaps, best left *in situ* if they have gone down well
in large flower-pots, which can be sunk in a bed of light sandy
soil in the garden and taken indoors when the time for emergence is
near. Montgomery states that when his are due to emerge he
digs them up, separates the cocoons, which are often in such
close masses that a number of imagines would often be unable to
emerge if left in the larval cage. Cocoons spun up in the food
can be cut out and placed on the surface of the moss-covering
of the pupa-cage, whilst pupæ spun up on the surface of the ground
are best moved to the pupa-cage as soon as they are hard enough
to handle. But above all things a moderately dry and cool situation
should be chosen in which to keep them. Sphingid pupæ are usually
especially badly kept by young lepidopterists, and, because the larvæ
of these (and other superfamilies) have a subterranean habit, there
is a general opinion that soil is necessary to their proper treatment
through the winter, so that, as a rule, they are either placed
on or, still worse, in the soil. Treated thus, pupæ almost invariably
die, the particles of earth getting into the movable incisions, blocking
up the spiracles, &c. Such pupæ are best placed on *Sphagnum* or on
a bed of ordinary moss, kept only moderately damp, when one's
worries are much lessened, few moulding or drying up, and a
very large percentage of perfect specimens will usually be obtained.
Cocoa-nut fibre is now pretty generally used instead of soil by all
breeders of lepidoptera who still look upon soil as somewhat
necessary, and occasionally this is mixed with about one-fourth or one-
third of coarse clean sand. These substances allow free drainage, are
free from vermin, and rarely produce any fungoid growths. On this
account the materials make a good bed on which moss (well baked)
can be placed and the pupæ placed thereon. If kept in pots,
careful arrangements for thorough drainage are necessary, for most
hybernating pupæ suffer much more from an excess of moisture
than from an excess of dryness. Cover with perforated zinc to
exclude mice and other pupal enemies; the projecting edges of
the zinc cover should be turned down to prevent them being moved
accidentally. One other point may be noticed. For those who have room
in their gardens, a place should be found for a selection of oak, alder,
ash, birch, willow, etc., to be planted a little later in the year (November), so that sleeving can be carried out the next year. Some breeders
of lepidoptera " sleeve " larvæ on such a large scale that thousands
of imagines of various species are reared every year—*e.g.*, hundreds
of *Jocheaera alni* on alder at Bathampton, &c. One may here
remind lepidopterists that, when fullfed, Acronyctid larvæ should
have pieces of rotten wood, short pieces of pithy bramble, elder-
stems, or similar material placed in the sleeves, so that they may
pupate therein.

UNCLASSIFIED. — Throughout September, sallows, alders and
birches are the most productive bushes for larva-beating. Sweeping
for larvæ also pays well at this time, one should sweep every plant,
for larvæ are to be taken almost everywhere at this season (Gregson).

TINEIDES.—Larvæ of *Tinea (Meessia) richardsoni* are to be found in flattish, elongated cases, made of the greenish-grey microscopic lichen which covers the stones at Portland ; only found on the undersides (rarely on the sides) of stones, and, being small and of exactly the colour of the stones, are inconspicuous ; always scarce, and one may often turn 50 or more stones before finding a case ; any of the heaps of stones on the under-cliff may produce cases. The species occurs on the Swanage coast, and will probably be found in other limestone localities if looked for. The imago is glossy, nearly black, with silvery fasciæ and spots, and is hardly ever seen at large. It emerges early in June. When first found, it was supposed to be identical with the continental *T. vinculella.*

LITHOCOLLETIDES.—The larvæ of the second brood of *Lithocolletis pyrivorella* make short mines on the underside of the leaves of *Pyrus communis* and its cultivated varieties, and also on *Pyrus malus* and *P. aucuparia*, from September to November, hybernating as pupæ. Also feed in the same way in May and June.

The larvæ of the second brood of *Lithocolletis blancardella* mine in the underside of leaves of *Pyrus malus*, and its cultivated varieties, also in *P. aria* and *P. communis*, in September and October, hybernating as pupæ. They also feed in the same manner in June and July.

The larvæ of the second brood of *Lithocolletis oxyacanthae* mine in the underside of the leaves of *Crataegus oxyacantha*, and also sparingly in *Pyrus aucuparia* in September and October, hybernating as pupæ. Also feed in the same way in June and July.

The larvæ of the second brood of *Lithocolletis spinicolella* mine in the underside of leaves of *Pyrus communis*, *P. domestica* and other cultivated varieties, from September to March. They also feed in the same way in June and July.

The larvæ of the second brood of *Lithocolletis concomitella* mine in the underside of the leaves of *Pyrus malus* and its cultivated varieties, also in *Prunus communis*, in September and October, hybernating as pupæ. They also feed in the same manner in May and June.

The larvæ of the second brood of *Lithocolletis sorbi* make long mines on the underside of the leaves of *Pyrus aucuparia* and *Prunus padus* in September and October, hybernating as pupæ. Also feed in the same way in June-July.

The larvæ of the second brood of *Lithocolletis mespilella* make short mines on the underside of the leaves of *Pyrus aria*, *P. aucuparia*, *P. torminalis*, *P. cydonia*, *Cotoneaster*, Morella cherry, and also in confinement on cultivated pear in September and October, hybernating as pupæ. Also feed in the same way in July.

The larvæ of *Lithocolletis cerasicolella* are to be collected in September in the leaves of *Prunus avium ;* the mine is elongate, placed between the lateral veins of the leaf, reaching from near the midrib towards the margin ; the larva pale yellow, with dark brown or black head.

The larvæ of the second brood of *Lithocolletis cerasicolella* mine in the underside of the leaves of *Prunus avium*, *P. cerasus*, and rarely in their cultivated varieties, also in *P. malaheb*, *P. armeniaca*, *P.*

domestica and *Pyrus communis*, from September to March.   They also feed in the same way in June.

COLEOPHORIDES.—For the greater part of its life the larva of *Coleophora binotapennella*, Sta. *(salicorniae*, Wk.) mines in the fleshy stems of *Salicornia herbacea* which grows in much the same places as *Suaeda*. It then constructs an untidy rough-looking case composed of bits of plant which it attaches to the outside of its mine and shortly afterwards crawls down to the mud and burrows to the depth of an inch or so, leaving its case sticking up on the surface.   Owing to the short time during which it inhabits its case it is somewhat difficult to collect in the larval state and is not easy to breed, but, in compensation to the collector, the imago is sometimes to be found in abundance amongst the foodplant, in excellent condition in July.   It is pale ochreous with two small blackish dots on the forewing.   Portland, &c., on salt-marshes and muddy coasts.

With *Coleophora flavaginella*, as with others of this genus, it is best to collect the larvæ, as they are generally more easily obtained, and the imagines are often difficult to identify even when in fine condition which in caught specimens is unusual.   The larva feeds on *Suaeda maritima*, an annual, with small fleshy needle-like leaves, which grows about highwater-mark.   It has been rarely found on the perennial *S. fruticosa*, but its presence there is believed to be accidental.   The case is somewhat cylindrical with mouth turning to one side and triangular tip.   The colour is whitish, with darker brown stripes, generally of two shades.   The larva spins up on the stem of the foodplant, near the ground, and may be found there through the winter.   It does not seem to object at all to being covered by the tide.   It must be kept out of doors, fully exposed.   Stick the stems in a flower-pot, half full of earth, with muslin stretched over the top, and bring the pot indoors as soon as the imagines should appear, about June.   The imagines are not easy to collect in a wild state, but can sometimes be found amongst the herbage, at Portland, &c.   A muddy coast or estuary seems to suit the foodplant.

In September, gather *Solidago* for cases of *Coleophora virgaureae ;* the cases are difficult to distinguish among the flowers and seeds, so that it is best to pick the flower-heads and place in a box, when crowds of the larvæ will sometimes crawl up.

GLYPHIPTERYGIDES.—Larvæ of *Acrolepia autumnitella* are to be found in mines, which cause conspicuous greenish-white blotches in the leaves of *Solanum dulcamara*, especially in those parts of the plant which are most concealed from view.   They are generally fullfed about the middle of September [I have, however, found the larvæ in October], and spin beautiful spindle-shaped cocoons made of an open network of brownish silk, something like the cocoon of *Chrysocorys festaliella*.   The moths emerge in two or three weeks (Richardson).

Amongst the larvæ beaten from birch in late autumn a peculiarly pale, transparent larva may be noticed, rather sluggish, and of small size ; this is probably *Roeslerstammia erxlebella*, which will shortly spin a cylindrical cocoon within a rolled-over margin of birch leaf. It is said to be more usually attached to lime.

TORTRICIDES.—The larvæ of *Peronea logiana (tristana)* occur in September on *Viburnum lantana*, eating the undersurface of the leaves, making, in places where the larvæ are common, a tangled mass of drawn-together and dead leaf membrane on the lower part of the bushes. They spin up among the leaves, and the imagines appear in October and November.

During September working hedges, &c., will give *Peronea schalleriana* var. *latifasciana*, *P. variegana* var. *cinana*, *P. comparana*, and other interesting species; mixed hawthorn, sallow and nut hedges preferable.

In September, the leaves of *Arctostaphylos uvaursi* are mined by larvæ of *Coccyx nemorivaga*, the leaf usually chosen being one of those forming the rosette terminating the shorter shoots; the affected leaf is very obvious, being divided, about the middle, by a slightly oblique transverse line into a basal green healthy portion and a terminal part that is red, brown or black. The mine of the larva occupies the dividing line, and often has a slender branch or two into the terminal part of the leaf; in confinement *Arbutus unedo* forms a good substitute food.

The larvæ of *Phoxopteryx mitterpacheriana* feed in the folded leaves of oak and beech in the autumn, pupating therein in spring (Thurnall).

Eriopsela fractifasciana has been bred in May from larvæ taken at Box Hill in the preceding autumn, feeding underneath the radical leaves of *Scabiosa columbaria*.

The imagines of *Paedisca ophthalmicana* may be beaten freely from aspens towards the end of September. Equally common on the trunks.

The little pink larva of *Semasia ianthinana* is to be found in hawthorn berries in September and October with the greyish larva of *Laverna atra*. It leaves them when fullfed and spins up in the bark. In confinement, give the larvæ cork in which to pupate.

The larvæ of *Stigmonota weirana* are very abundant in Epping Forest at Brentwood among beech; they are to be found in September and October between two leaves spun together, pupating therein; very easy to breed (Thurnall).

The larvæ of *Stigmonota redimitana (nitidana)* feed between two leaves of oak, spun together, in September and October.

Acorns in September and October are sometimes much infested with the larvæ of *Carpocapsa splendana*. They should be collected, and cork should be given the larvæ for pupation.

The larva of *Grapholitha nigromaculana* feeds on the seeds of *Senecio jacobaea* in September, and is sometimes very common.

The larva of *Catoptria albersana* may be found in September in rolled-up leaves of honeysuckle and is not difficult to breed (Thurnall).

The larvæ of *Catoptria scopoliana* are to be obtained commonly in the heads of *Centaurea nigra*, those of *Catoptria fulvana* are confined to *Centaurea scabiosa*. The heads should be collected in September and October.

The larvæ of *Catoptria cana* are very common in the heads of thistles in September.

The larvæ of *Catoptria wimmerana* appear to be strictly confined to saltmarshes, occur freely along the marshes of the Thames in

September and October, when they are to be found spun up in the tops of *Artemisia maritima.*

The little pinkish larva of *Eupoecilia hybridellana* is sometimes to be taken in abundance, feeding in the seed-heads of *Helminthia echioides* in September. It is reputed to be difficult to breed.

The larvæ of *Eupoecilia dubitana* are to be obtained feeding on the young seeds of *Senecio virgaurea*, *S. jacobaea*, *Hieracium tridentatum* and *H. umbellatum* in September.

The larvæ of *Eupoecilia notulana* feed in the stems of *Mentha hirsuta* in September, also stated to feed in *Inula dysenterica.*

The larvæ of the larger spring form of *Eupoecilia angustana* are to be found feeding on the seeds of *Plantago lanceolata* in September and October.

PYRALOIDES. —The larva of *Metzneria littorella* feeds on the seeds of *Plantago coronopus*, among which it may be found in the months of September and October, forming a slight gallery between the seeds and stem, and always making an opening communicating with the interior of the stem in which the seeds are fixed ; into this it retires with the head upwards. It is presumable that it pupates within the stem, but I have not found it below the level of the seed-heads (Walsingham).

PTEROPHORIDES.—The autumnal and winter larvæ of *Leioptilus microdactyla* feed in the stems directly below the flower-bearing pedicels of *Eupatorium cannabinum*. Many of the affected plants can be at once distinguished from the dwarfing of the central head of blossoms, caused by the attack of the larva on the terminal portion of the stem being made when it was very tender, so that the side bunches of flowers over-reach it; when the stem is attacked lower down, where it is harder, the blossom is not affected (Bignell).

CRAMBIDES.—In mid-September, the larvæ of *Pempelia hostilis* are to be found on aspen. A brown, curled, dead or dying aspen leaf, spun to a living green one, is the home of the little grey larva, of which two or three may live in one group of leaves, the galleries, however, separate. At the end of the month the larvæ are nearly fullfed, and two half-dead yellow leaves with bands of silk stretching in all directions to the adjacent fresh ones, which are freely eaten, make a large and conspicuous object. In confinement, the fullfed larvæ leave their nests and spin up freely in rolls of paper, pupating at once, the imagines appearing in June.

The larvæ of *Homaeosoma nimbella* are common in September and October in the flower-heads of *Matricaria inodora* and other *Compositae.*

COSSIDES.—To take the fullfed larvæ of *Cossus cossus (ligniperda)* in abundance, it is necessary in early September, when they are wandering to find a suitable place for their winter cocoons, to search the stumps of willow trees that have been cut down ; in these they sometimes occur in considerable numbers. One must first rip off the bark from the stumps, under which will be found a number of cocoons ; then, if the wood is at all rotten, split it (the stump) in every possible place, and the cocoons will be found almost anywhere. I have taken numbers, I may say dozens, in this way. The stumps can be visited at intervals after the first time, and while there is any wood *C. cossus* will be there (Quail).

The fullfed larva of *Cossus cossus (ligniperda)* makes a cocoon, frequently on the ground, in September and October ; it remains therein till April, when it pupates, emerging in late June or in July.

Fullfed larvæ of *Cossus cossus* placed in a tin (the lid well perforated) half filled with sawdust, the tin placed on a shelf in the kitchen, will spin up, and rarely under this treatment fail to produce imagines in due course.

Cossus cossus larvæ are best hybernated with some chunks of bark and wood placed in a galvanised pail covered with perforated zinc. This is preferable to the flower-pot and glass-cover treatment, which is nearly sure to generate mildew. Cut a disc of the perforated zinc to fit inside the pail, an inch or two from the rim, and on this lay a ring of stout leaden pipe bent to fit neatly (Dollman).

DREPANIDES.—Larvæ of *Platypteryx harpagula (sicula)* are to be beaten in Leigh Woods in the first half of September ; difficult to obtain now as the trees on which they occur have grown large and therefore are not very practicable for working.

Give plenty of foodplant, and room also, to larvæ of *Drepana hamula*, as otherwise they may disappear gradually by the process of cannibalism.

The larvæ of *Drepana lacertula* should be searched for on quite young birch trees, which are more frequented by the larvæ than the larger ones, and can be thoroughly examined by simply standing over them for the purpose.

GEOMETRIDES.—Search carefully along hedgebottoms in September for female *Ennomos tiliaria*, where they may be found at rest, and when so found rarely fail to give a profusion of eggs (Gregson).

Tree trunks must be searched in September for imagines of *Ennomos erosaria*, which are sometimes abundant on oak, birch and beech in the New Forest, Epping Forest. &c. Eggs are readily obtained if a ♀ be taken, and the species is not difficult to rear.

Imagines of *Ennomos autumnaria (alniaria)* are to be obtained from gas-lamps in September and October (Deal is its chief centre, but Gosport, Chichester, and other localities produce the species); eggs are freely laid in confinement.

The larvæ of *Ellopia prosapiaria* pass the winter on needles of Scotch fir, commencing to feed early in the spring, and are fullfed from the middle of May to the middle of June.

In September a good supply of larvæ of *Phorodesma smaragdaria* may be obtained on the Essex salterns. Whittle states that the larvæ are sometimes curiously local, occurring on special clumps of *Artemisia maritima* to the exclusion of others. On searching over a mile of saltings, he found that, in 1894, all the larvæ obtained occurred on the same three clumps of the foodplant from which he obtained them in 1893, and that one of these yielded five-sixths or more of the larvæ taken, the proportion to each clump being much the same as the previous year.

The most likely spots for finding the hybernating larvæ of *Geometra vernaria* are on the shoots trailing on the ground or at a very slight elevation (Newman).

. The larvæ of *Numeria pulveraria* are to be found on hawthorn in September.

The larvæ of *Fidonia conspicuata* are to be beaten from broom from the second week to the end of September ; when nearly fullfed, careful brushing for the larvæ pays as well or better than beating.

The dimorphic larvæ (green and putty coloured) of *Zonosoma porata* can be beaten freely from scrubby oak in September and October ; they pupate in late October and emerge next spring.

The larva of *Macaria notata* will feed up well and quickly on birch, which is a foodplant generally more easily obtainable than either alder, sallow, or blackthorn. It is a reddish-brown larva, with lemon-yellow, diamond-shaped, lateral blotches, not unlike that of *Hybernia defoliaria*.

Cut off sprigs of the unripe capsules of *Bartsia odontites* towards the end of September ; turn these into band-boxes, with muslin over the top ; in a few days larvæ of *Emmelesia unifasciata* attach themselves to the muslin, when they should be removed to a cage with fresh sprays of foodplant, and the bottom of the cage covered with a layer of sand ; in this they make small oval cocoons ; the pupæ often go over two years before emergence ; the seed-heads occasionally require to be sprinkled with water.

There is a striking similarity between the larvæ of *Melanippe fluctuata* and *Coremia designata*, and as they are both to be taken on rape and cabbage, it is best not to mistake the latter for the former. The larva of *C. designata* may be known from the other by the tint of the dorsal surface which is not so olive-green as in *M. fluctuata*, but browner ; the head, prothoracic, and first three abdominal segments being very dark on the upper part. Also examined closely, there will be found four to six obscurely-marked, diamond-shaped blotches, with dark centres on the summit of the dorsal region.

In tapping bedstraw on heaths for the larvæ of *Larentia viridaria (pectinitaria)*, be watchful for the very young examples. They are exceedingly small, and coil themselves up tightly on being dislodged. As they are of a bright red colour, almost scarlet, they look like anything rather than larvæ.

Larvæ of *Lobophora sexalata* and *Eucosmia undulata* are to be found on sallow in September. The latter prefer, in the New Forest, the tufty, pony-trimmed bushes, most likely because they are more suitable for the making of their leafy tents (Fletcher).

Eggs of *Camptogramma fluviata* laid on September 13th, produced larvæ, which fed on knotgrass and dock, and pupated, the imagines emerging between November 1st and 18th (Mera).

In September, 1892, 27 larvæ of *Cidaria reticulata* were taken in one afternoon near Windermere ; the larvæ appear to feed almost exclusively on the seeds of wild balsam, entering the seed-pod about the middle ; in the daytime they are to be found resting at full length along the midrib on the underside of the leaves (Moss).

The imagines of *Chesias spartiata* can be beaten from broom in the daytime, when they flit quickly to the next bush, and can be marked down.

The best way to take the larvæ of *Eupithecia subfulvata* is to look over yarrow and tansy after dark with a lamp when they may be sometimes found in abundance. The ab. *oxydata* is bred rather freely from larvæ collected in the Morpeth district.

The larvæ of *Eupithecia succenturiata* are half grown about the end of the first week in September, on mugwort; they rest by. day among lower leaves of their foodplant and are only to be taken at night. The pupa is dark buff, the wing-cases olive-green.

The larvæ of *Eupithecia helveticata* are obtained at Milngavie by the Glasgow collectors in September; three hours' work beating junipers rarely results in more than from one to two dozen, so that they want working for.

The pupæ of *Eupithecia helveticata* are spun up in a slight cocoon in the thick part of junipers. They are frequently shaken from the junipers when beating for the larvæ at the end of September and on through October.

The larvæ of *Eupithecia extensaria* feed, upon the Norfolk coast, near King's Lynn, on *Artemisia maritima*. The larva has a habit of standing, apparently, upon its head, *i.e.*, catching hold with its true legs and extending its body stiffly so that its hinder extremity is in the air. Usually, however, it remains during the daytime close to the stem, twisting itself among the leaves and blossoms, but at night feeding voraciously on both. The species appears to be gregarious, and excessively local, frequenting sheltered clumps of the foodplant, but not extending its range very far, although the *Artemisia* is plentiful on the coast (Barrett).

In the latter part of September and in October the imagines of *Eupithecia stevensata* fly close to the short turf of the downs in the neighbourhood of Dover, or visit the flowers thereon. More than thirty males were netted in one evening, attracted by a newly-emerged ♀ resting on a plant of golden-rod (Webb).

The larvæ of *Eupithecia pimpinellata* feed on the seeds of *Pimpinella magna* and *P. saxifraga*. They seem to be more partial to the former, green larvæ assimilating in tint to green unripe seed-capsules and red larvæ to the purple ripe ones.

LACHNEIDES.—Pupæ of *Trichiura crataegi* not emerging at the normal time in August or September should be carefully put aside till the following year, as it is a very common habit, even in the southern counties, for this species to spend two (occasionally three) winters in the pupal stage.

The larvæ of *Macrothylacia rubi* feed on grass until three-parts grown, when they collect in small colonies on the nearest bramble, and are afterwards found rambling in search of winter quarters; I have a numerous batch of the larvæ feeding in the natural condition in a sunny position, being only confined to a certain space of ground ; these were picked off the grass culms when about an inch long. I believe it is better to collect them when about this size than to wait till they are three-parts grown, as when of this size they are more likely to be stung, which accounts for failures, to some extent, in breeding the imago. My larvæ at the present time are quite as fine as any to be found in the wild state, they have been fed on bramble, but have also eaten most of the grass in their enclosure (Mason).

DIMORPHIDES.—The pupæ of *Dimorpha versicolora* may be kept out of doors during the winter in a flower-pot, and covered with moss, the imagines emerge well if taken indoors in spring (Draper).

SPHINGIDES.– I have been very successful in rearing *Manduca (Acherontia) atropos*, the method adopted is as follows : I feed the larvæ in large flowerpots, half filled with light mould for them to effect their transformations in. After they have been underground for ten days, I take them out, and put them in pots partly filled with mould and sand and well drained. I keep them in a warm room and well saturate them with water once a week. To keep up the moisture I put damp moss over them every third day. In this way I have bred fourteen as fine specimens as ever were seen, some of them $5\frac{1}{2}$ inches across the wing. One of them was only three weeks from the larva going underground to the appearance of the perfect insect, but a month is about the average time (Rogers).

The larva of *Manduca (Acherontia) atropos* should never be fed airtight, as, unless it has plenty of air and space, it is very liable to sweat, and, being a very delicate creature, this condition is fatal to success in rearing it.

The pupæ of *Sphinx convolvuli* appear not to be able to survive our British winter. Placed in an incubator in November or December, the imagines emerge in from fifteen to twenty days.

LYMANTRIIDES.—The small hybernating larvæ of *Leucoma salicis* want some rough cork or bark on which to hybernate ; they will spin a silken web in September, collect into companies and remain quiescent all winter, beginning to feed again in April as scon as the poplar, willow and sallow leaves begin to show. The cocoons of their parents are favourite hybernacula.

During the first week of September, the larvæ of *Notolophus gonostigma* may be found by searching or beaten from oak in the woods at Bexley (Newman).

Dasychira pudibunda larvæ can be brought up well on birch, and will readily spin up in crumpled newspaper, which should be fastened round the side of the cage containing them.

NOTODONTIDES. — The dull yellow and- black lined larvæ of *Phalera bucephala* sometimes occur in amazing numbers in September, October and even in November, crawling on fences, paths, walls, &c., in search of suitable places for pupation.

To suceeed in rearing *Leiocampa dictaea* and *L. dictaeoides* in confinement, the greatest care is required with the pupæ, which should not be left in the earth. Prepare a large box, say 20 inches long by 15 inches wide, and 12 inches deep, half-fill it with earth and make it very damp. Cover the earth with a layer of dry moss, upon which place the pupæ, then put another layer of dry moss upon them. Keep the box in a cool airy place until May, at which time it may be removed into a warm spot to feel the heat of the sun. This treatment I find best for all underground pupæ (Chapman).

In September, in suitable localities, the larvæ of *Lophopteryx camelina* are to be found commonly on alder, oak, willow and sallow. They throw up their heads over their backs in a kind of defensive posture.

The pupæ of *Lophopteryx carmelita* are to be obtained near Derwentwater at the foot of small birches in September (Marshall).

The larvæ of *Stauropus fagi* do not appear to spin up among the green leaves, but almost always, in dead leaves, with which they should be supplied when fullfed ; the cocoon is very closely spun between the leaves (Newman).

There is no need to keep the pupæ of *Stauropus fagi* out of doors, but a moderately damp atmosphere is necessary ; if the imagines do not appear by the beginning of June, exposure to a shower of rain is often advisable. An outhouse, if the pupæ be kept on damp sand, seems to produce satisfactory results.

NOCTUIDES.—Larvæ of *Cuspidia megacephala* are to be found in plenty at the end of September on the trunks of the black poplars which have been planted so extensively in many parts of London. I have usually found them commonest on sunny days, half curled round in a crack of the bark, and basking in the sun (Battley).

The larva of *Apatela aceris* likes to get behind a loose chip of wood or bark, but will utilise a crack in soft wood where it can make use of the tiny pieces of surface material to spin amongst the silk of its cocoon, which is made pretty level with the surrounding surface. In confinement it likes a mixture of wood-chips and dead leaves, and will spin up in dead leaves, moss, &c. The larvæ are to be found commonly in most parts of Kent, in early September, on fences, tree-trunks, &c., searching for a suitable pupating place.

Examine every example of belated *Leucania lithargyria* that you see at sugar, for it is very easy to mistake *L. albipuncta* for this insect by the light of the lamp. *L. albipuncta* is rather the smaller, has squarer shaped forewings, which are redder, and the hindwings lighter than those of *L. lithargyria*. The principal distinction is not, however, seen to advantage when the insect is quivering on the sugar. This is the sharp and bright definition of the white spot on the fore-wings, which has no upward blur towards the costal edge as is found in *Leucania lithargyria*. It is safest to take every one on suspicion.

Captured females of *Leucania albipuncta* have no objection to laying their eggs on a growing sod of grass under a bell glass ; the young larvæ appear in about eleven days, make elongated holes right through the blade between two of the ribs. Kept in a warm atmosphere, they will occasionally feed up quickly, and then the imagines appear the same autumn.

Leucania l-album should be carefully worked for on our south coast by means of sugar in early September. It occurs rarely in Guernsey, and has been recently captured (September 8th, 1901) at Sandown, in the Isle of Wight.

Sugared sunflowers form a very attractive bait in September. *Leucania vitellina* is often attracted thereto (Robertson).

The eggs of *Nonagria sparganii* are laid in September on *Iris pseudacorus ;* they are glued down firmly (in single file) in a groove formed by the convolute edge of a leaf-blade; the cement visible, shining like gum arabic, the margin stuck down so that, as it withers and turns brown, the eggs are almost concealed; the eggs are small for the size of the moth and assimilate wonderfully well with the brown colour of the leaf during winter.

In late September, on the marshes along the estuary of the Thames, *Calamia lutosa* is very abundant; on some evenings (often moderately cold ones) the moths are to be seen flying freely, at other times nearly all the specimens are found to be resting on reeds and other plants.

One may search luxuriant beds of reeds in September and October for *Calamia lutosa* without success, and yet find them in scores on the poor isolated patches and scattered stunted plants near, resting between 7.30 p.m. and 8.30 p.m., both sexes, not paired, silent and solitary, generally on the underside of a drooping leaf and sometimes upon a stem; they want boxing at once, waving the light, or moving it, or delaying, is fatal. It is most abundant the less promising the reed-growth appears.

Hydroecia micacea and *H. nictitans* are often attracted in early September in large numbers by sunflowers, which they appear to prefer to sugar.

Towards the end of September, imagines of *Celaena haworthii* are plentiful (near Morpeth), flying over the mosses on the moors about 3 p.m., whilst *Tapinostola fulva* is also abundant flying freely about 5.30 p.m. (Finlay).

Females of *Celaena haworthii* can be swept; practically all the specimens taken on the wing in the afternoon, with the net, are males (Finlay).

Aporophyla australis comes to sugar freely in September, and is said to be more abundant between 11 p.m. and midnight than earlier in the evening.

Caradrina ambigua is evidently a very easy species to force. During December I bred above 140 imagines from a single batch of eggs laid by an Isle of Wight ♀. No more warmth than that of an ordinary kitchen was necessary and the larvæ were fed throughout on dandelion (Gardner).

In early September, *Neuronia popularis* and *Luperina testacea* are attracted by light in large numbers.

During September the south coast is most prolific of good things, among others, *Heliophobus hispidus* is always considered worth working for. The imago comes to light, flies straight to the lamp, and settles down quietly on the glass. The species does not seem to be attracted by sugar.

The striped ribbon grass is a well known lure for *Apamea ophiogramma*. Plant it now. The division and subdivision of what was at first one stunted plant gave a supply of the necessary foodplant; the first summer I got none, the next seven, then twelve, and so on in increasing numbers until the insect was exceedingly abundant (Burrows).

In late September and early October, females of *Peridroma saucia* lay freely in chip boxes, each batch usually consisting of some 300-400 small ova. They hatch in rather less than three weeks, feed ravenously on dock, rape, cabbage, spinach, &c., show no inclination to hybernate, hide under the leaves by day when young, feed only by night, and bury themselves under the earth in the flower-pots in which they may be kept after they are half-grown ; placed in a hot-house in November, they will pupate in December and emerge from December to February, but in an ordinary cool room or greenhouse will continue to feed all the winter, and pupate in March or April, emerging in May.

The larvæ of *Peridroma saucia*, hatch in September, feed up rapidly and continuously in a warm room—cabbage and cabbage-stalks, marigold, slices of carrot, potato, turnips, etc., being eagerly devoured. They pupate in autumn and the imagines emerge shortly afterwards.

In September and October the young larvæ of *Agrotis ashworthii* may be found, three to five together, sometimes more, in the old webs left by the larvæ of *Botys terrealis*, on the flowering spikes of golden-rod, and are then from a quarter to three-eighths of an inch in length, and light-green in colour (Gregson).

During the first week of September *Agrotis agathina* ♀ scatters her eggs freely on the stems and leaves of heather in confinement ; they hatch in about three weeks and the larvæ feed indifferently on *Calluna vulgaris* and *Erica cinerea*, preferring the latter later, when they devour the leaves regularly downwards, beginning at the upper part of each twig, which they completely clear. They will go on feeding all the winter in a cool conservatory, but usually die off in the spring ; success is much more probable if the larvæ be wintered out of doors.

Larvæ of *Aplecta occulta*, kept in a kitchen in a flower-pot, fed twice daily on dock, grow very rapidly, and may be removed to a breeding-cage, when fullgrown, about the end of October, for pupation. Kept indoors, under the same conditions, the imagines emerge towards the end of November. Some larvæ of every brood resist the forcing and insist on hybernating.

During September, the larvæ of *Hadena pisi* are to be obtained on heather on the moors, sometimes in large numbers (Finlay).

In confinement, the imagines of *Epunda lichenea* pair readily it placed together in a roomy glass-topped box with a small quantity of honey. Placed together on September 23rd, by the 26th the ♀ s had each deposited some 200 ova on the loose paper lining the sides and bottom of the boxes. The ova changed to green on the 18th and hatched on October 26th.

In the middle of September, imagines (both sexes) of *Polia nigrocincta* are attracted to sugar in the neighbourhood of Ilfracombe.

The eggs of *Anchocelis litura*, laid in September, hatch in October (in about 4 weeks), the larvæ feed slowly through the winter on grass, and can be pretty easily reared.

Mellinia ocellaris has been taken in many localities at sugar and light in September, usually with its near ally, *M. gilvago*.

Mellinia ocellaris occurs at sugar in September, pretty regularly in a locality not far from Wilmington, in Kent. *Mellinia gilvago* and *Tiliacea citrago* are usually taken freely in the same locality at the same time (Newman).

In earlier September, the females of *Tiliacea aurago* scatter their eggs on the stems and leaves of beech in confinement ; the eggs soon change colour, but the larvæ do not emerge till towards the end of April, when buds of beech and hornbeam want cutting open for them.

The cocoons of *Miselia oxyacanthae* frequently get hard and dried-up by contact with the surrounding earth. An occasional liberal shower-bath from the greenhouse watering-can in the period just preceding their emergence reduces the number of cripples perceptibly (Henderson).

All the known British examples of *Xylina lambda (zinckenii)* have been captured at rest (or at sugar) during the last day or two in September or in October.

The eggs of *Lithomia solidaginis* are laid during September, remaining in the egg-state throughout the winter and spring, until about April 26th, before hatching ; the larvæ readily take to hawthorn as a substitute for bilberry (Harrison).

The larvæ of *Cucullia lychnitis* in September cannot be confused with those of *C. verbasci*. Some are quite small still in September, whereas those of *C. verbasci* have been in the pupal state for quite a month, and they are of a green tinge as opposed to the white of *C. verbasci*. The markings correspond most exactly with Wilson's description. They are to be found on the flowers and seeds of the white mullein, and are very local (G. M. A. Hewett). Feeds on *Verbascum nigrum* in Surrey (Bower).

In September the larvæ of *Chariclea umbra* are sometimes found in great plenty on *Ononis ;* they appear to feed almost exclusively on the corollas of the flowers, and are much given to cannibalism if reduced to short rations (Prout). Newman states that they will feed greedily on knotgrass an easily-obtained substitute foodplant.

The larvæ of *Chariclea umbra* are sometimes obtained in hundreds on *Ononis* growing by roadsides, are sad cannibals in confinement owing to the difficulty of supplying them with full rations of their natural food—the young seeds of rest-harrow. Solitary confinement on plants of rest-harrow with foliage, blossoms and seeds is successful, but the best plan is to supply them with the green pods of scarlet-runner beans suspended among rest-harrow from the tops of the jars ; the larvæ fully appreciate the beans as well as the pods, and one can rear from 6 to 8 in a jar without fear of loss, but over-crowding will even then give rise to cannibalism. Green pods of peas are also eaten, but those of scarlet-runners are preferred (Norgate).

It is sometimes tempting to take *Catocala nupta* or *Mania maura* when they are found at sugar, though the killing-bottle in use at the time is too small to capture them comfortably. On such an occasion it will be found that if the mouth of the bottle be placed quickly over the insect, only allowing the top edge of its neck rim to touch the tree, the moth will dive into the bottle without getting its wings damaged.

The larvæ of *Plusia festucae* hybernate when very young, and begin to feed again in April, as soon as the *Iris* makes its appearance. Emergence goes on for some time when the imagines once begin to appear (Grime).

To force *Plusia bractea*, feed the young larvæ in a cold frame until they show signs of hybernation, then put them into a stove house where the temperature is from 65° to 80°, feed on groundsel or lettuce, pupation will take place in October, and the imagines emerge a fortnight later.

Sweep heather in early September for larvæ of *Anarta myrtilli*, *Eupithecia nanata* and *E. minutata*.

ARCTIIDES —Having failed in an attempt to winter larvæ of

Callimorpha hera on a growing plant of *Lamium purpureum* a later batch was put into the conservatory on cut branches of the same plant. The food being kept in water remained fresh during the winter months, until it could be replaced early in the year, the temperature never having been below freezing point. By this means eighty-five per cent. of the larvæ survived, while a friend of mine who put them on a growing plant lost over seventy per cent. (Jäger).

LITHOSIIDES.—The larvæ ot *Gnophria rubricollis* begin to appear in the beating-tray in September, and are fullfed towards the end of October; they feed on very minute lichens, mainly on oaks and beeches ; when fullfed the larvæ retire beneath the moss on the trunks, for pupation, and, as each takes a long time to pupate, pupæ should not be searched for until late in November; the white flimsy cocoons are hardly distinguishable from the spinning of certain spiders ; it is advisable to detach the moss from the top downwards, as the pupæ are not among the moss, but between moss and bark, and consequently often drop, whilst, not infrequently, after the moss is withdrawn, a pupa is found adhering to the trunk. Larvæ sometimes spin up low down near the ground, at others high up the trunk. Pupæ are best kept through winter on damp sand with a layer of moss, damp being very necessary to them.

PAPILIONIDES.—The eggs of *Colias hyale* may be obtained by confining the ♀ s with clover plants in the sun ; the young larvæ feed slowly until the end of November, when they hybernate ; in confinement they do this best by removal from the foodplant and being placed in a large chip-box covered with muslin, and kept at a temperature of about 40°F.–45°F. By the middle of February the larvæ are on the move, and should then be placed on a growing plant of clover. They feed on slowly through March and April, pupate in May and the imagines emerge in about a month.

Larvæ of *Pyrameis cardui* and *P. atalanta* found in September and October will pupate and emerge the same year ; they must, therefore, be kept under artificial conditions, and care taken of them, both as regards temperature and food, to ensure success. The species will not hybernate as larvæ or pupæ under any conditions.

OCTOBER.

The Eupitheciids are always objects of interest to the collector of Macro-lepidoptera. With few exceptions they form the only group that is really difficult for the beginner to name, and, as a result, one finds the average collector repeatedly asking for these in exchange instead of attempting to collect his own material. Throughout October certain plants yield the larvæ of many of these interesting species in abundance. A strong umbrella is needed, and, on the sides of this, the flowering-stalks of ragwort, golden-rod, *Angelica*, yarrow, &c., should be beaten for the larvæ, which are readily found and easily reared—*Eupithecia centaureata, E. absynthiata, E. satyrata*, and others in large numbers from ragwort, *Eupithecia albipunctata*, and its ab. *angelicata* from *Angelica*, *E. virgaureata* from

R

golden-rod, *E. subfulvata* from yarrow, &c. Many lepidopterists eschew
the umbrella and prefer to collect the heads of the various plants,
storing them away in large breeding-cages with a good supply of
silver sand and cocoanut fibre in which the larvæ can pupate. During
this month, too, the larvæ of the Drepanids should be carefully
worked for, that of *Drepana binaria* being much more common in our
oak woods than is usually supposed, and, possibly, is collected in
fewer numbers than one would expect, owing to its feeding high
up on the trees out of reach of the beating-stick. By the end of
October one is on the look out for the autumnal rarities, of which
possibly, next to *Orrhodia erythrocephala, Dasycampa rubiginea* is
the most sought after. In some districts, Clevedon, &c., the latter
is systematically taken when worked for at sugar and light, although
in some years much more abundantly than others. From the end
of October right on into December, mild evenings are likely to
produce it, and continuous sugaring on such occasions is the only
way to bid for success with this species. Frosty nights send the
individuals very quickly into hybernation, and a few successive nights
with frost will rapidly force all our hybernating species into their
winter-quarters ; when once this has happened, a later spell of
fine weather will not tempt them to move again until the following March, however abundantly it may bring out the autumnal-
emerging species that do not hybernate in the imaginal stage.
D. rubiginea is reputed to be a shy insect, occurring on the sugar or
at ivy in out-of-the-way places, sucking the sweets whilst partly hidden,
and requiring to be well searched for on ivy bushes that are not amen-
able to the beating-stick ; it also shows a partiality for small, isolated
ivy bushes with few heads of bloom. It is a sluggish insect, which
possibly accounts for its being usually in good condition when it is
captured. We have already *(anteâ,* p. 116) ventured some remarks as to
the preparation that should be made for carrying pupæ successfully
through the winter. Fenn suggests a cage or box with a leno covering,
the floor of the box supplied with stones for drainage ; above this about
six inches of earth, and this, in turn, covered with moss, the pupæ
and cocoons to be placed on the moss ; above the pupæ about four
inches of loosely-laid fern fronds to protect them from frost. Leave
the cage exposed (except for the leno covering) to the weather,
and, as a rule, there will be no reason to regret the result. Moisture
must never be allowed to accumulate, the appearance of fungi
must be followed by immediate action, and vermin rigorously ex-
cluded. Riding, for dug pupæ, suggests knocking the bottom out
of a box some seven or eight inches deep, the bottom to be
replaced with coarsely perforated zinc, or, better still, with wire
gauze. Level the edges at the top, and cut a piece of glass to
fit accurately. Place on the zinc or gauze, inside the box, a layer
of *Sphagnum* moss pressed down till it is at least one inch thick.
On this place a thin layer of cocoanut fibre, and on it the pupæ,
and cover them with about half-an-inch of the fibre. Between this and
the glass lid place some sticks crosswise from which the insects
can suspend themselves when drying their wings. The box itself
should be kept in position over a flat pan of water. The *Sphagnum*
should be first plunged in boiling water, then dried and slightly
damped before using. Studd winters his pupæ in wooden boxes,

18ins. long, 12ins. wide, and 15ins. deep, with hinged cover of glass, sloped from back to front; in the front and back are pieces of wire gauze, 15ins. by 5ins., fixed for ventilation; the bottom also made of wire gauze. The box is stood over a zinc tray of water the size of the box. In the bottom of the box is placed a layer of moss, 2ins. to 3ins. deep; above this a thin layer of cocoanut fibre is spread, and on this the pupæ and cocoons are laid. The cocoanut fibre prevents the most active pupæ wriggling down through the moss to the gauze. Bowles places his pupæ on silver-sand in a cage with a layer of chopped moss over them, and pins cocoons to the side of the cage; whilst Moberly also lays them on sand under boiled moss, damping the latter two or three times a week, except in frosty weather. Corbett recommends a large meat safe with perforated zinc sides, and about 2ins. of peat at bottom, as an excellent cage for wintering pupæ, and suggests that it should be placed out of doors. Many of our most successful breeders of lepidoptera never move the pupæ from the large pots in which the larvæ have gone down, but cover the soil with a layer of moss and the pot with a sheet of glass, leaving them under an out-of-doors shed till February and March, when they take them indoors and damp the moss occasionally. Mason keeps a large emerging-cage—really a pupa-house—fixed to the north side of his dwelling and covered on three sides with garden-netting, and a substantial roof to keep out the rain. The pupæ are never disturbed, and the soil is never damped until the spring. Experience is, however, the best guide, and one must experiment, even if one pays somewhat dearly, before one can hope to reach success in a most difficult matter.

NEPTICULIDES.—The larva and mine of *Nepticula auromarginella* are very like those of *N. aurella* in bramble. The larva is chiefly found in the autumn, about October, and the imago begins to emerge early in the year. The imago is distinguished from all other British species by having a golden tip to its wing as well as a golden fascia. It appears not yet to have been found outside the parish of Chickerell, where it was discovered in 1888. There are one or more continental Nepticulids with a golden tip, but they are quite distinct. This moth must surely have a wider distribution than one parish, especially as its foodplant is everywhere !

Edleston observes that the larvæ of *Nepticula subbimaculella* are somewhat gregarious ; in one blotch-mine he found five larvæ, occasionally four and three, and repeatedly two. It is further to be noticed that the conspicuous green patches of leaves affected by this and other Nepticulid larvæ in late autumn are also to be observed when they are tenanted by Lithocolletid larvæ ; *e.g.*, the leaves of *Sorbus aucuparia* show this peculiarity conspicuously when tenanted by larvæ of *Lithocolletis aucupariella.*

The larvæ of *Nepticula lapponica* feed in broad serpentine mines in birch, are light yellow in colour when fullfed, and are to be found at the same time as those of *N. betulicola*, *viz.*, October 1st to 20th (Threlfall).

In mid-October, plants of *Hypericum* are sometimes to be found containing large numbers of the larvæ of *Nepticula septembrella*. The larvæ are at that time in all stages of growth, the mines varying in size from the merest indication to the great discoloured final blotch.

COLEOPHORIDES.—Boyd found larvæ of a Coleophorid making very white blotches in October, 1900, at Danbury, on blackthorn ; an imago bred in the spring of 1901 has been identified by Barrett as the birch-feeding *Coleophora milvipennis.*

The fullfed larvæ of *Coleophora albicans* are to be obtained in October among the flowers and seeds of *Artemisia vulgaris* (Gregson). The rush-feeding Coleophorids want very careful separation. We have among the insects that used to be united as *Coleophora murinipennella* and *C. caespititiella,* six species—(1) *C. sylvaticella,* larva on *Luzula sylvatica,* imago flies in May. (2) *C. alticolella,* larva on *Juncus lamprocarpus,* imago flies in July. (3) *C. murinipennella,* larva on *Luzula campestris* and *L. multiflora,* imago flies in May. (4) *C. caespititiella,* larva on many kinds of *Juncus,* imago flies in June. (5) *C. glaucicolella,* larva on many kinds of *Juncus,* particularly *J. glaucus,* imago flies in July. (6) *C. agrammella,* chiefly on *Juncus conglomeratus,* but also on *J. effusus* and *J. lamprocarpus.*

TORTRICIDES.—The larva of *Halonota cirsiana* feeds quite at the bottom of the old flower-stalk of *Inula dysenterica* and *Centaurea nigra,* sometimes almost in the root of the plant. The stems should be kept out of doors during the winter, in a flower-pot, until the larvæ pupate in late spring ; they should then be brought indoors, cut for convenience to a length of 6 or 8 inches and placed in bottom of a box. The pupa wriggles up the stem a day or two before the emergence of the imago, which takes place at the end of June and in early July.

The larvæ of *Stigmonota coniferana* feed in the resinous matter exuding from young and old Scotch firs in Moray (Salvage).

The larva of *Argyrolepia zephyrana* feeds in the autumn and winter in the stems of *Daucus carota,* eating out the pith and filling the space with frass, still feeding on the dead stems or working back through the frass as late as April and spinning a very slight brownish cocoon in the tightly packed mass of frass in the stem, the moth emerging in June.

The larvæ of *Catoptria aemulana* are to be obtained somewhat freely on the Kentish side of the Thames, feeding on flowers and seeds of golden-rod in October.

The larvæ of *Catoptria tripoliana* are to be found wherever *Aster tripolium* grows in the saltings of the Kent and Essex coasts ; the heads of this plant should be collected in October when the larvæ are fullfed.

The larvæ of *Conchylis smeathmanniana* may be obtained in October feeding in a silken gallery in the seedheads of *Achillea millefolium* and *Anthemis cotula.*

The larvæ of *Conchylis stramineana* may be found in October feeding on the heads of *Centaurea nigra.*

The larvæ of *Chrosis tesserana* are to be obtained in October, feeding on the roots of *Helminthia echioides* and *Picris* growing on rough dry waste ground.

The larvæ of *Eupoecilia implicitana* are to be found in October feeding on the seeds of *Anthemis cotula, Matricaria inodora, Senecio virgaurea,* &c.

The larvæ of *Eupoecilia notulana* feed in October on stems of *Mentha hirsuta,* apparently entering at a joint and working upwards, feeding on the pith and leaving the lower part of the burrow tightly packed with excrement. They hybernate within the stems and usually spin up therein, the moths appearing in June (Barrett). Also in stems of *Lycopus europaeus* in Wicken Fen, the moths appearing in July (Walsingham).

The larvæ of *Eupoecilia notulana* hybernate in the stems of *Mentha hirsuta*, pupate therein and emerge through a small hole in the side hidden by a thin layer of skin, all that the larva has left at that point. The affected stems should be collected in the autumn months, kept out of doors during the winter; the imagines emerge in June.

COSSIDES.—Some two dozen larvæ of *Cossus cossus*, most of which were nearly, if not quite, fullfed, were placed in a large tin, into which I put a lot of old corks, thinking that they would spin up more easily in them than among elm chips and sawdust. I was surprised to find that they at once commenced to voraciously devour and to tunnel the large corks in all directions, apparently enjoying the new material as food. On removing several wine corks, nothing was left but an outer shell, and, in several instances, a fullfed larva was comfortably coiled up in the interior, where one would have thought it was almost impossible for such a large larva to find accommodation. Another remarkable thing was, that I could find no excrement, unless the cork was passed in the same, or a similar, state as when eaten; that it was eaten appears to be evident from the fact that most of the larvæ attained a much larger size (Mutch).

ORNEODIDES.—When a stack or thatch-covered outhouse is near the habitat of *Orneodes hexadactyla*, the species hybernates therein and may be disturbed by beating the thatch during the autumn months.

CRAMBIDES.—The larvæ of *Euzophera cinerosella* are common at Portland in the stems of *Artemisia absinthium* throughout the winter, and should be collected in early spring when nearly fullfed.

The larvæ of *Phycis abietella* are to be obtained in old fir-cones, in galls, and the dead wood of old Scotch firs in Moray (Salvage).

The larvæ of *Cryptoblabes bistriga* should be searched for on oak, from the middle till the end of October; at the same time one is sure to find the small hybernating larvæ of *Rhodophaea consociella*.

DREPANIDES.—In October, beating oaks (in the Gloucester district) gives plenty of larvæ of *Drepana binaria*, which fall freely into the umbrella, although imagines are rarely seen in the district (Merrin).

Grigg kept his pupæ of *Drepana harpagula* out of doors through the winter in an exposed situation, and the imagines came out well in May and early June. They can, however, be kept just as well indoors, the greatest chance of failure arising from their becoming too dry just before emergence.

CYMATOPHORIDES.—In the first week of October the larvæ of *Gonophora derasa* may be beaten from *Spiraea ulmaria*. The larvæ always hide amongst the leaves in the daytime. I have noticed the larvæ of many Noctuids that feed at night go to the same place day after day to hide (Cross).

GEOMETRIDES.—In October, *Himera pennaria* is frequently attracted in numbers to light. It is also to be found by searching bushes in woods, with a lantern, but is rather readily disturbed by the light.

Phorodesma smaragdaria occurs along the Essex coast line from Benfleet to St. Osyth. For the larvæ, examine closely every patch of *Artemisia maritima*. The caterpillar is a fluffy-looking spider-like creature, clinging to the twigs ; clothing itself as it does with little pieces of its foodplant, it is very difficult to detect.

To winter larvæ of *Phorodesma smaragdaria*, they should be sleeved out of doors on southernwood. Kept indoors the larvæ almost invariably die. Turned out of doors, exposed on southernwood (or other foodplant) they almost all disappear before spring. *Artemisia maritima* and *A. absinthium* die down during winter, and do not afford the stems on which to tie the bags. Southernwood has an abundance of suitable branches, and, protected by the bag, one often gets every larva through without loss (Burrows).

The larvæ of *Abraxas sylvata (ulmata)* swarm in Thorpe wood, near Worksop, preferring the elms scattered through the beech-wood ; in some seasons the foliage of the elms is literally eaten up (Hall).

The eggs of *Oporabia filigrammaria*, laid in autumn, go over the winter, and hatch in March ; the larvæ feed up very sharply as they are fullfed and out of sight by the middle of May (Harrison).

In October, *Phibalapteryx lapidata* occurs on grassy moorlands all over the west of Scotland at an elevation of about 1000 feet ; it flies freely between 5.30 p.m. and 6.0 p.m ; is also said to fly later in the evening at Kinloch-Rannoch, and to be obtained by searching with a lamp.

In October, *Phibalapteryx lapidata* occurs in the Black Wood at Rannoch, and is to be taken sometimes very freely flying along the sides of the paths after dusk.

From the beginning to the middle of October, *Eubolia cervinata* comes freely to light in those localities in which it occurs.

Up to the end of October the larvæ of *Eupithecia subfulvata* are still to be taken exceedingly commonly on railway banks, waste places, &c.

To obtain eggs of *Eupithecia subfulvata*, take a fairly large glass jug. place therein a tolerably large head of ragwort, also a good-sized piece of mugwort, and sprinkle the flowers with treacle. Enclose the ♀ s, and it will be found that they will lay most of their eggs on the mugwort. The larvæ will feed freely in confinement on chrysanthemum.

The larvæ of *Eupithecia subfulvata* are about half-grown in the middle of October, they rest stretched out at length, exposed on the stems and leaves of their foodplants, and can be easily obtained by searching during the day ; the pupa is rich red in colour with lighter (almost buff) wing-cases.

In hybernating tree-feeding larvæ, sleeving on growing trees will usually be successful if fallen leaves are placed in the sleeve to help to protect the larvæ from severe frost ; larvæ of *Pericallia syringaria*, *Acidalia inornata*, &c., winter well when thus treated (Robertson).

COCHLIDIDES.—Larvæ of *Heterogenea cruciata* must be searched for in October on beech ; cannot be beaten in numbers ; larvæ pupate readily on leaves or twigs ; very uncertain in appearance, common in

1884 at Lyndhurst, then scarce till 1892, when it was found in great numbers; the cocoons have never being obtained in the wilds (Hewett).

ANTHROCERIDES.—To hybernate the larvæ of Anthrocerids, remove them from their food altogether and place in a dry receptacle among bits of cork, &c. (Christy).

LACHNEIDES.—Larvæ of *Lasiocampa quercûs* var. *callunae*, about 1¼ inches long in September, should be hybernated in an airy cage out of doors; they change their skins about the beginning of April, after which they commence to feed again pretty freely (Christy).

Collect larvæ of *Macrothylacia rubi* in the first or second week of October; place in a large empty wine-case about 30 inches long, 14 inches broad, 20 inches high at back, 15 inches in front, with space in the back for ventilation, 15 inches by 5 inches, covered with perforated zinc; cover the whole with a close-fitting glazed pane and make secure with hooks. Inside at each end place a layer of *Sphagnum* moss about 6 inches deep, none in the centre where the jelly-jar stands with food for larvæ so long as wanted; the latter when full-fed roll in a close ring in the *Sphagnum*, remain there till spring when they come up and spin at the top of the *Sphagnum*. The whole should be kept out of doors (Finlay).

Fullfed larvæ of *Macrothylacia rubi* collected in October, placed in a wooden box half filled with peat dust, the lid perforated with air-holes, the box placed in a cupboard abutting on the kitchen-flue produced imagines between Christmas and the end of January.

SPHINGIDES.—Pupæ of *Manduca (Acherontia) atropos* obtained first week in October, placed in a small breeding-cage and laid on about 2 inches of very damp moss and covered with about an inch of the same material, the cage then placed in a shady corner of a small humid plant stove, in a temperature ranging from 65°F.—80°F., the moss covering the pupæ sprinkled with water every other day; imagines emerged at end of third week in splendid condition (Mason).

LYMANTRIIDES.—To hybernate *Dasychira fascelina* keep under cover, in bag, among dead leaves; sleeve out again on foodplant in March (Christy).

NOTODONTIDES.—A freshly-spun cocoon of *Cerura bicuspis* has a different tone from the surrounding bark, and would, one supposes, be easily seen; by midwinter, however, the tone differs very slightly and afterwards there is little or no difference between the cocoon and the surrounding bark, the lichens, &c., which the larva raises to the surface of the cocoon in spinning up, probably growing a little in the damp autumn weather.

The cocoons of *Cerura bicuspis* are much more easily seen on birch than on alder, as, on the latter, growths of lichen are usually worked into the cocoon which make it homogeneous with the bark; the cocoons occur at all aspects and heights from close to the ground to 20ft., but rarely on trees whose bases are close to water.

The imagines of *Diloba caeruleocephala* emerge from pupæ from 5 p.m.-7 p.m. in the first week of October. They pair readily in confinement, but a female will sit still for several days if she has not been fertilised, scarcely moving from the spot where she expanded her wings.

NOCTUIDES.—Males of *Dasypolia templi* are to be captured at light in the haunts of this species from the last week of October onward. Females appear not to be thus attracted.

In October, *Dasypolia templi* is to be taken freely in the Huddersfield district, but only by dint of hard work ; on an average a ton of stones must be turned over for every specimen. The best localities are old stone quarries in elevated situations, where the old rubbish must be turned over, when this and other species are almost sure to be taken. If there are no quarries at hand, the dross from the iron-forges, used for road repairing, is a good place for them, and they prefer that which is newly laid on. Females must be kept until the spring for eggs, the larvæ boring into the stems of *Angelica*, cow-parsnip, &c. (Varley).

In the neighbourhood of Halifax the imagines of *Dasypolia templi* may be obtained at night sparingly on the lamps in the outskirts of the town. They are much better obtained, however, in the early morning on the lamp-posts or on the ground near them, sometimes from ten to twelve may be taken upon or in the immediate neighbourhood of a single lamp-post (Halliday).

Sometimes when the evenings are cool and misty the imagines of *Calamia lutosa* fly freely, but when it is warm and clear the specimens are to be found resting on reeds and other plants.

The eggs of *Calamia lutosa* are laid in September and October, like those of *Apamea ophiogramma* in rows, in the folded edges of partly-dead leaves of reeds towards the tip.

Any one who wants *Apamea ophiogramma* has only to increase the quantity of ribbon-grass in his garden. In a year or two he will find extensive traces of larvæ that clear out the main shoots, leaving the outside shells full of greenish frass, the shoots being, however, so entirely hidden by the unaffected culms that they are difficult to find.

The young larvæ of *Agrotis* var. *ashworthii* feed up well to hybernating stage on willow ; if they insist on hybernating, place in roomy cage out of doors, but not exposed to rain ; fill bottom of cage with dry leaves for them to hide among, but with a supply of suitable low-growing plants for them to nibble at during winter if they choose to do so. By keeping the young larvæ in glass jars in a warm place, however, the greater part will feed up rapidly in the autumn, and pupate, the imagines emerging in November and December.

To hybernate larvæ of *Agrotis ashworthii* (and other Noctuid larvæ), procure a large deep and strong wooden box; place in it 6 or 8 inches of light sifted earth, plunge a few small flower-pots with the foodplant growing therein, into the earth (knot-grass is very useful for most species). Place the box in November into a cold frame, or leave it outside, covered with a sheet of glass, till early spring ; do not disturb it until foodplant is again available ; then take into a greenhouse or warm room, and the survivors will soon be observed feeding again.

I would recommend those lepidopterists who have not tried to hybernate Noctuid larvæ on the common marigold, to give it a trial.

It is far easier to grow in a pot indoors than most foodplants; larvæ of *Aplecta herbida*, &c., thrive on it (Norgate).

Young larvæ of *Epunda lichenea* hatch in October and feed up well throughout the winter upon stonecrop and chickweed (Day).

Ova of *Epunda lichenea* hatched nearly at end of October and the larvæ fed freely on stonecrop in a bottle for six weeks; they were then about an inch long and were placed in a large breeding-cage with muslin sides and top, three inches of soil, and supplied with a plant of stone-crop, about the size of one's fist, every three or four days; they fed up very rapidly, hid by day in the soil or under the plants, crawled up at dusk, fed at night, and had all gone down for pupation by the end of January (Woodforde).

The foodplant for many winter-feeding larvæ may be grown in large flower-pots, leaving them in the open with a piece of perforated zinc on top, which helps to exclude earwigs and prevents larvæ escaping; larvæ of *Caradrina taraxaci, C. alsines, Cerigo matura* feed well all through the winter in this manner.

The only known British example of *Mesogona acetosellae* was taken at sugar in a garden at Arlington in October, 1895.

Imagines of *Dasycampa rubiginea* bred or captured in autumn, can be kept through the winter if supplied with thin syrup, on which they will occasionally feast, remaining generally quiet, however, under curled-up withered apple leaves, with which they should be supplied; they pair at the end of February and in early March, oviposit throughout March and April and on into May, the young larvæ feeding up freely on apple and dandelion. The larvæ are all usually fullfed by the commencement of June. The imagines appear throughout September, October and on into November.

In October and November, tree-trunks in woods should be searched for imagines of *Xylina rhizolitha*. In some districts they show a marked partiality for resting on the trunks of Scotch fir-trees.

Tiliacea aurago often appears at ivy in October, the females lay pretty freely in confinement on the stems of the branches of beech, especially in the axils of the leaves and shoots, and rarely scatter their eggs on the muslin or the walls of the receptacle in which they are confined.

The larvæ of *Toxocampa pastinum* are to be obtained in much greater numbers in autumn than in spring on *Vicia cracca*. Holland notes gathering them at the rate of 150 per hour in autumn, while he could only get some 50 in two mornings on the same ground the next spring

ARCTIIDES.—The small larvæ of *Callimorpha dominula* may be found in October and November on *Symphytum officinale*, more abundantly, however, after hybernation in the spring. They will eat many things, but comfrey appears to be their chief food in nature in the New Forest district.

LITHOSIIDES.—At the end of October, in some years, the fullfed larvæ of *Gnophria rubricollis* are exceedingly abundant throughout the New Forest area. In such seasons the larvæ may be found rambling in every direction over the walls of houses, outhouses, logs, posts, fences. The cocoons are generally to be found in damp places.

S

138   PRACTICAL HINTS FOR THE FIELD LEPIDOPTERIST.

PAPILIONIDES.—The larvæ of *Colias hyale* become dormant in October, and should then be placed in a thoroughly dry position (not left on foodplant) and protected from frost, temperature not lower than 45°F. if possible; in middle January place them on a fresh plant of clover with plenty of young leaves when they will commence to feed again, pupate in April, and the imagines will emerge in May.

NOVEMBER AND DECEMBER.

In November the winter moths proper commence to appear. The males of *Hybernia aurantiaria* sit on the twigs of oaks and birches after dark, whilst those of the variable *H. defoliaria* affect a variety of trees and bushes, being also attracted with *Oporabia nebulata (dilutata)* in large numbers to the lamps, where such exist, more particularly in the vicinity of woods and hedgerows. The paired examples of *Hybernia aurantiaria* are to be found after about 9.30 p.m. At the same time the males of *Cheimatobia boreata* sit on the leafless twigs of the birches, whilst the females may also be taken in the same way, generally *in copulâ*. Fenn reports having seen the males of *C. brumata* paired with females of *C. boreata* and *vice versa*, the progeny not to be distinguished from *C. brumata*. At Oxton, near Exeter, Studd has, during recent years, taken large numbers of *Asteroscopus sphinx* at light. Why certain species are, and others are not, attracted by light is a problem that still awaits solution, as also is an explanation as to why the males of such species as *Asteroscopus sphinx*, *Poecilocampa populi*, *Heliophobus hispidus*, *Epunda lichenea*, *Neuronia popularis*, &c., are attracted freely, whilst their females are rarely, if ever, so attracted, and yet both sexes of many other species, *Ennomos fuscantaria*, &c., are lured to their doom. Moonlight nights are generally fatal to success at light, and a continued east wind appears to act as a deterrent; one often finds that a showery or stormy evening after a protracted spell of dry weather is exceptionally favourable, but almost all warm nights after a dry interval are good for work at light; and the absence of wind is a favourable determining factor. The bulk of the attractions are made from about an hour after dark until perhaps 2 a.m. in summer, and from about an hour after dark until midnight in the autumn. As to working light for *Poecilocampa populi* and *Asteroscopus sphinx*, besides the approved methods of working lamps in the districts in which they occur, and fixing proper attracting lamps, as supplied by Watkins & Doncaster, to one's house, or in other suitable positions, Holland describes *(Ent Rec.*, i., p. 20) how he works for them. As he points out, so many examples taken at the lamps are spoiled, for, when inside, of course, a moth never rests till it has been through the light. If it cannot get inside, it rests not on the glass, but on the dark supports or framework of the lamp, where its capture is easy, and, in this case, is almost invariably a good specimen. Acting on this observation, Holland adopted the plan of using two large lamps about 10 feet from the ground, and at the same distance apart, placed in front of a large dark sheet hung up about 15 feet from the lamps. He selected a dark sheet because the moths prefer a dark surface to rest upon. Visiting the sheet several times during the evening, he boxed 20 *Poecilocampa populi* and 3 *Asteroscopus sphinx* in perfect condition the first night, all quietly at rest on the sheet. No wonder that he decided that after this experience he would do no more two hours' march round the gas lamps for scorched specimens. Dadd gives a most interesting account *(Ent. Rec.,* xiii., p. 159) of the winter habits of the half-grown larvæ of *Senta maritima*, which, he states, live in the galleries that have been formed in the reeds by the larvæ of *Nonagria geminipuncta* or in those of *Typha latifolia* that have been made by the larvæ of *Nonagria sparganii* or *N. cannae.* The beds of reeds or *Typha* fringing lakes and rivers are the chosen haunts, and the larvæ are to be obtained by opening the reed-stems, which show that they were tenanted the previous summer by the larvæ of *N. geminipuncta;* in some localities every reed will have its tenant. It is suggested that the species is best worked when the water is frozen, otherwise the plants are most difficult to reach. The larvæ are said to feed on the more or less dormant creatures inhabiting these galleries, and not on the reed itself, and, when the reeds are collected, they require no other attention than a good supply of affected reeds kept more than moderately moist, in fact they should be well watered every day. The larva pupates in May in the old galleries in which it has lived. There are still many districts in which thatched roofs, straw stacks, stacks of hop-haulm and of bracken, and great piles of faggot brushwood are to be met with, and these are the haunts of many hybernating species of micro-lepidoptera. The heaps of faggots or brushwood are said to be a favourite haunt for *Coriscium brongniardellum* and *C. cuculipennellum* during the winter, as they certainly are the favoured hiding-places of most of our hybernating species of *Depressaria.* Barrett says that out of chip-thatches, *i.e.,* the chips sliced off in making rough hoops, he has obtained many specimens of *Laverna decorella, Depressaria ciliella, D. chaerophyllella, D. albipunctella, Gelechia humeralis, Gracilaria stigmatella, Coriscium brongniardellum,* &c., and states that he is not sure that this kind of thatch is not the best for collecting from, as the perfect shelter, with larger spaces for creeping into, seems to suit the moths. Still, straw-thatch is by no means to be despised, and, in a good locality, furnishes plenty of moths ; whilst many specimens may be obtained from the sheltered sides of ricks, although they are not generally found to be so productive as regular thatches. In Kent, stacks of hop-bine placed on the outskirts of a wood are exceedingly productive. We have, before now, moved, bundle by bundle, a whole stack of bine, and been rewarded with some thousands of specimens of Depressarias, as well as many other interesting species. As a rule, the moths are difficult to net ; one cannot easily manipulate a net and shake and throw a bundle of hop-bine at the same time, although this appears to be largely what one has to aim at. Barrett's original advice is as good as any we can offer. He says : The way to work thatch is to beat the edge with a stick, when the moths will fall out, and, if the weather be cool, they will probably drop to the ground ; if hot and at all windy they show great activity in getting away. Perhaps the easiest plan of collecting is to catch the moths as they fly away or flutter down, but certainly the most profitable way is to hold the net close under the place beaten, so as to catch everything that falls. The net, of course, soon contains a large collection of dirt, moss, straw or chips, and other rubbish, with sundry spiders, beetles, diptera, &c., among which, by careful scrutiny, in all probability, many moths may be found lurking, though some of the Depressarias do not wait to be looked for, but come running up the side of the net. Fortunately the commonest species appear to have this habit. *Depressaria applana*, for instance, in the autumn, will come running up the net in abundance, leaving all the better things at the bottom, and, in the spring, *D. arenella*, as well as *D. ciliella*, has the same habit. By shaking the rubbish, however, the other things may generally be induced to show themselves, and can then be boxed, and any others which will not move for that, can be disturbed by blowing sharply amongst the rubbish, a sort of treatment that many Tineina, and especially the Gelechiids, cannot endure. As in every other kind of collecting, much depends on the weather; wind, provided it comes from a mild quarter, is no disadvantage, indeed, it helps to drive everything into the net, except the dust, which it usually contrives to deposit pretty liberally in one's face and neck. Taking insects out of the net in windy weather is no easy task when the bottom is full of rubbish. In cold weather, however, with north or east wind, hardly a moth can be obtained, either they creep further in or hide in more protected places, while, on the other hand, a very hot sun makes the thatch too hot to hold them. Cloudy, moderately warm, and even stormy weather is favourable. Several Depressariid species, which are plentiful in thatch in the autumn, appear to desert it in the early spring. Such is the case with *Depressaria applana*, *D. alstroemeriana*, and *D. nervosa*, all of which are common before hybernation, but hardly occur afterwards. *D. applana*, we know, hides among its foodplant. *D. carduella* and *D. subpropinquella* have only been found in autumn, but, as they were scarce then, it may not be the rule. On the other hand, *D. arenella* and *D. propinquella*, which are sometimes scarce before the winter, appear very commonly after, and the two species of *Coriscium*—*brongniardellum* and *cuculipennellum*, *Gracilaria stigmatella* and *Laverna decorella* are decidedly commoner in the spring. The same appears to be the case with *D. heracleana; D. ciliella*, *D. chaerophylli* and *D. albipunctella* are equally common in autumn and spring, and so is *D. purpurea*, but it has a habit of flying briskly among hedges and by the sides of woods all day long in sunny weather in April, and, consequently, is not always to be found at home. *D. ocellana* and *D. umbellana* do not appear to affect thatch, though the former hides among the herbage and grass roots overhanging rivulets, and the latter may be disturbed from among ivy and dead leaves on hedge-banks as well as from furze. The number of other species belonging to other superfamilies to be obtained from thatch appears very inconsiderable, but one may obtain, in spring, specimens of *Anticlea badiata*, as well as *A. derivata*, *Cidaria miata*, *C. psittacata*, *Xylina socia (petrificata)*, *Xylocampa lithorhiza* and plenty of *Alucita hexadactyla (polydactyla)*.

UNCLASSIFIED.—Any air-tight box, in which set specimens of *Acidalia emarginata* are stored, should be closed and opened gently, otherwise the strong air currents, caused by the operation, will spring and break the wings. This remark applies to several other delicate Geometrids, and it is safest to pin them on a substratum of cotton wool.

The following preparation will be found useful in getting rid of mites. It consists of equal parts of oil of thyme, oil of aniseed and spirits of wine. I find this efficacious, both for destroying them in imagines and also for prevention. I went through a large collection for a friend of mine three years ago, which was swarming with them, and although not touched since, there is not a mite in one of the 40 drawers (Baxter).

COLEOPHORIDES.—In December, a visit to Benfleet (or any other similar coast district) will usually give an abundance of the cases of *Coleophora artemisiella* on sea-wormwood *(Artemisia maritima)* (Whittle).

During the winter months the cases of *Coleophora maritimella* are to be obtained from the seed-heads of *Juncus maritimus*. The larvæ form their cases in the seed capsules and are difficult to discover unless these are rubbed off the plant over paper, when those tenanted by larvæ will be seen walking away (Moncreaff). [There is some doubt in our mind as to the species indicated in these last two "hints."]

Cases containing the larvæ of *Coleophora badiipennella* may be found throughout the winter on the lower twigs of elm-trees and the upper twigs of elm-bushes. The particular elms may be noted and the larvæ gathered in the spring.

TORTRICIDES.—In early November, *Peronea mixtana* may be obtained by smoking overhanging banks where *Erica tetralix* grows. It is necessary to have a friend to assist when smoking for insects, one to stand net-in-hand to capture the insects, whilst the other smokes them out of their hiding-places.

In mid-November, the heaths want working for *Peronea autumnana* (Finlay).

The larvæ of *Stigmonota regiana* are to be found spun up under the bark of sycamores throughout the winter and the spring.

The larvæ of *Ephippiphora pflugiana* are common in thistle-stems throughout the winter months.

The larvæ of *Ephippiphora brunnichiana* are to be found in the roots of *Tussilago farfara* in the autumn and early winter.

The larvæ of *Ephippiphora trigeminana* are to be found feeding in the roots of *Senecio jacobaea* in late autumn, and are particularly partial to waste places and railway banks.

The larvæ of *Ephippiphora inopiana* may be found in roots of *Inula dysenterica* throughout the winter months.

The spun-up larvæ of *Carpocapsa nimbana* are to be found under rough bark on beech-trunks in Epping Forest during the winter months (Thurnall).

The spun-up larvæ of *Carpocapsa pomonella* may sometimes be found in the winter and spring under loose pieces of bark or moss on the trunks of apple-trees.

The larva of *Argyrolepia maritimana* is confined to *Eryngium maritimum*, and may be found throughout the winter far down in the sand in the roots of this plant.

The larvæ of *Argyrolepia zephyrana* may be found not uncommonly throughout the winter feeding in the root and lower part
of the stem of *Daucus carota.*

The larvæ of *Eupoecilia atricapitana* are to be found through
the winter in dead stems of *Senecio.*

ŒCOPHORIDES.—Males of *Lemnatophila phryganella* fly in early
November in woods, &c., especially among bracken, the semiapterous
females, with whitish wings and black markings, being found at rest
on the tree-trunks near.

CRAMBIDES. — Larvæ of *Euzophera cinerosella* are to be found
through the winter in stems of *Artemisia absinthium,* at Portland.
The larva works its way up the stem and pupates usually
near the end of a shoot. The moth emerges in July and can
often be disturbed among the wormwood or found resting on the
stems, generally near the ground. It is not difficult to breed,
but with this, as with many other species which hybernate as larvæ,
the stems should not be gathered too early in the winter. They are
better kept out of doors, fully exposed until pupation has taken place.

GEOMETRIDES.—In mid-November, *Cheimatobia boreata* is to be
taken, with a lantern, *in cop.,* in large numbers hanging on the stems
of rushes or on the small leafless branches of the birches, among
which the rushes grow; *Hybernia defoliaria, H. aurantiaria,* and
Himera pennaria are to be taken in fewer numbers at the same time.

In November and early December *Hybernia defoliaria* and *H.
aurantiaria* are both very common in Epping; the ♀ s are to be
found freely, after dark, on tree-trunks, with a lantern.

PSYCHIDES.—Between November 12th and 26th I found near
Southend 58 cases of *Proutia betulina,* all apparently containing adult
larvæ (Whittle).

LACHNEIDES.—The males of *Poecilocampa populi* come freely to
light during November; there is then often a break in December with
a fresh lot of imagines appearing towards the end of the month and
during the first week in January (Mason).

The larvæ of *Macrothylacia rubi* may be found on heaths and
commons, and particularly downs, throughout the winter, hybernated
on or just below the surface of the earth about furze bushes, the roots
of heather, &c.

Keep larvæ of *Macrothylacia rubi* out-of-doors until December,
then obtain a large bunch of heather, place in a jar of water in cage
with larvæ, and remove cage near to the kitchen fire; the larvæ soon
get lively, crawl about on the heather, and spin up in a week or two;
emergence commences in the middle of February (Butler).

To hybernate the larvæ of *Macrothylacia rubi,* put them into large
boxes out of doors. At bottom of box place five or six inches of *Sphagnum* moss; protect from excessive wet, as if the *Sphagnum* gets wet and
then there is frost many are killed (Finlay). Half-fill a large box with
light earth, lay in a quantity of bramble leaves, &c., cover with net and
keep in a cold frame in garden (Brady).

NOTODONTIDES.—Search for the eggs of *Ptilophora plumigera*. The sight of a spray of maple with a dozen eggs of *P. plumigera*, sprinkled up and down it, is attractive to other eyes besides tomtits' (Bernard-Smith). During November the imagines (chiefly males) of *Asteroscopus sphinx* are in some districts to be taken very abundantly at light. At Oxton, in Devonshire, 50 in a single evening have been recorded.

NOCTUIDES.—In November and December, search for cocoons of *Arctomyscis myricae*. In some years they are quite abundant in the Pitcaple district, and one takes as many as two or three on one stone. The species is distributed over all lowland districts of Aberdeen and Kincardine (I have seen the larvæ below high-water mark at Muchalls), but in Perthshire it is almost confined to the mountains. It has the reputation of being difficult to rear from ova, but, if fed entirely on birch, there is no difficulty. The ♀ deposits ova naturally upon clods and stones, more rarely on the foodplant, and if one ovum be got, more can always be obtained as they are generally laid in batches of threes or fours, and they are very easily seen owing to their bright red colour (Reid).

The first fortnight of November is generally a good time to sugar for *Calocampa exoleta*, which is very abundant thereat in some seasons.

LITHOSIIDES.—The larvæ of *Gnophria rubricollis* are to be obtained in November at Pokesdown, spinning their cocoons in the crevices of the bark and among the lichens covering the branches of trees ; the cocoons (with pupæ) are also to be found on the surface of the ground, underneath leaves, or among the fragments of lichen that have been washed down by the rain, the favourite position being where the earth lies high against a smooth part of the trunk and not under the arches of the roots. As many as 4 or 5 larvæ sometimes spin their flimsy cocoons together (Harvey).

The pupæ of *Gnophria rubricollis* appear to be very fastidious, sometimes occurring plentifully on a few particular trees and entirely absent on all the neighbouring ones, although seemingly equally suitable.

The pupæ of *Gnophria rubricollis* are to be obtained in a slight cocoon, surrounded by an outer covering of spider-like web. They have been found under the "topper" of a stone wall which surrounded a fir plantation (larch, spruce and Scotch), the larvæ feeding on the lichens on these trees and probably on the lichens on the hawthorns which hung over the wall (Todd).

PAPILIONIDES.—The eggs of *Thecla w-album* are to be found above or directly below an aborted leaf-bud, and harmonise so exactly with the colour of the bark of the elm-twig on which they are placed that only an entomologist could possibly detect them.

The yellowish milk-white eggs of *Zephyrus quercûs*, covered with a rough raised reticulation, are laid upon oak-twigs where they may be found during the winter months ; in spite of its colour an egg is not at all easy to see, looking like a small inconspicuous fungoid growth.

The larvæ of *Hipparchia semele* hybernate small, remaining on the grass all the winter, and show no tendency to burrow or hide ; they feed a little all the winter in suitable weather, but do not grow perceptibly till spring.

THE ENTOMOLOGIST'S LIBRARY.

Books written by J. W. TUTT, F.E.S.

These books are written by an Entomologist for Entomologists—Up-to-date information ; up-to-date synonymy ; entomology treated on lines of modern science.

THE NATURAL HISTORY OF THE BRITISH LEPIDOPTERA.

(A text-book for Students and Collectors),

Vols. I, II and III.   (Price 20s. each volume net.   54s. for the three volumes).

This work is the most advanced scientific text book ever issued on the British Lepidoptera. Besides chapters on the general subject, each volume contains in the systematic part a detailed account of the species, each of which is treated under a variety of headings, e.g., Synonymy, Original description, Imago, Sexual Dimorphism, Gynandromorphism, Variation, Egglaying, Ovum, Habits of Larva, Larva, Variation of Larva, Pupation, Cocoon, Pupa, Pupal Habits, Dehiscence, Variation of duration of pupal stage, Foodplants, Parasites, Habits, Habitat, Time of appearance, Localities, Distribution. The entomologist has, here, a revision of the superfamilies treated in a modern scientific manner, and the work is of first importance to workers at these groups in all parts of the world. The systematic part deals particularly with the species found in Britain and affords such a mass of detail concerning the British species as has never before been brought together. To the general biologist the discussion and details relating to the hybridism, gynandromorphism, variation and life-histories of the species dealt with, afford a mass of material not to be obtained elsewhere, whilst to the lepidopterist pure and simple, the mass of information will enable him to study his subject from many different standpoints ; to the collector the information concerning the habits, food-plants, habitat, and localities, is as full as it can possibly be in the present state of our knowledge, whilst it would take the phenologist and student of distribution years to collect anything like the number of facts bearing on their own special work that is here ready for their digestion. Each volume contains a great deal of original matter not only from the observations of the author, but also from Dr. T. A. Chapman, Messrs. A. W. Bacot and L. B. Prout, who have collaborated with the author in their own branches of study for the work. Besides these some 200 other lepidopterists have helped in different ways and in various degrees. The volumes contain a vast amount of absolutely new material relating to all the species treated, and, at the same time, the whole of the information to be obtained from the long series of volumes of *The Entomologist's Monthly Magazine*, *The Entomologist*, *The Entomologist's Record*, *The Entomologist's Weekly Intelligencer*, *The Zoologist*, *The Transactions of the Ent. Society of London*, as well as that contained in the works of Stainton, Newman, Meyrick, Barrett and others, also in the leading continental *Transactions* and Magazines, has been carefully summarised and noted. The works of all the leading continental authorities have also been carefully overhauled and the important facts gleaned therefrom. So much labour has been expended in making the volume worthy of acceptance to all lepidopterists, and the cost of production of so large a book is so heavy, that the support of every lepidopterist is earnestly solicited. To those lepidopterists who have become interested in the scientific study of the subject they profess, these volumes will open up a new world. The amount of labour expended in producing them has been enormous, and, expensive as the books may appear, are really cheaper than any other published work on the subject, for they represent a whole library of information that is otherwise practically unobtainable.

BRITISH BUTTERFLIES

(Illustrated.   Crown 8vo., Cloth, Gilt.   Price 5/-).

This book consists of 476 pages, contains 10 full-page illustrations, and 45 wood-cuts. There are figures of every British butterfly. Sometimes three or four figures of the same butterfly to illustrate the two sexes, underside and variation are given. The full-page illustrations and most of the wood-cuts have been drawn by the well-known entomological artist, Mr. W. A. Pearce.

Each British butterfly is described under the following heads :—(1) Synonymy. (2) Imago. (3) Variation, with summarised diagnoses of all described forms, British and Continental. (4) Egg. (5) Larva. (6) Pupa. (7) Time of appearance. (8) Habitat and Distribution. Besides these, there are extended remarks on each of the Tribes, Subfamilies, Families, Divisions and Superfamilies. The descriptions of the " Larvæ " and " Pupæ " are mostly original. There are 282 aberrations and varieties diagnosed, of which 111 are described for the first time.

At the end of each chapter is a brief summary giving the following information, in tabular form, for each species :—I. Dates for finding (1) the ovum, (2) the larva, (3) the pupa, (4) the imago. II. The Method of Pupation. III. Food-plants.

The preliminary chapters consist of a series on the structure of the Egg, Larva, Pupa, &c. ; also others on practical work—Collecting, Pinning, Setting, Storing, Labelling, &c.

PRACTICAL HINTS FOR THE FIELD LEPIDOPTERIST.
PART I.
Price 6/- Net.
(INTERLEAVED FOR COLLECTOR'S OWN NOTES).

This is one of the most useful books ever offered to the field-lepidopterist, and will save him time, trouble and expense in prosecuting his work. *One thousand two hundred and fifty* practical hints are included, telling the lepidopterist how, when and where to work for the more desirable species. No lepidopterist can afford to be without a copy of this book. The older collectors will not only find many hints that they do not know, but will find many facts that they may wish to remember presented in a compact form. To the younger collector it offers a mass of information that he could not hope to accumulate by himself in very many years of field work. To all it must remain one of the most necessary books ever published for the use of field-lepidopterists. The contents are divided into the following chapters : (1) January, February, and early March; (2) late March and April; (3) May; (4) June; (5) July; (6) August; (7) September; (8) October; (9) November and December. Each chapter opens with a general review of the field work than can be done in the period indicated, and this is followed by a classified list of the " Hints " available for the period. Now that the whole of the material is grouped, it offers a great mass of exact, useful and reliable information, bearing on the work of the lepidopterist in the field, telling him exactly what to do and how to do it in the fewest possible words and in the least possible space. Such information could only be gathered by the individual worker as the result of many years' observation and by reference to many books in which the facts are buried amongst a mass of other entomological detail. Lepidopterists, experienced and inexperienced, will find in this book much information that will suggest quite new lines of work in their collecting, and enable them to find, in close proximity to their homes, species which they had never suspected to be in their vicinity, and the saving of time and trouble will thus be enormous.

The published records of many of our most observant field-lepidopterists have been largely drawn upon in the compilation of this work. Among many others whose work has been laid under contribution are—Messrs. Alderson, Bankes, C. G. Barrett, Birchall, Bignell, Bower, Buckler, Burrows, Butterfield, Chapman, Corbett, Coverdale, Harpur-Crewe, Elisha, Farren, Fenn, Finlay, W. H. B. Fletcher, J. E. Gardner, Greene, Gregson, Hamm, Hellins, G. M. A. Hewett, Hodges, Hodgkinson, Holland, Horne, James, Jäger, Kane, McArthur, Machin, Mason, Merrin, Moberly, Morres, Newman, Norgate, Norman, Prout, Porritt, Raynor, W. Reid, Richardson, Riding, Robson, Robertson, Sheldon, Bernard Smith, Stainton, Stott, Threlfall, Tunaley, Tugwell, Warren, Whittle, Wratislaw, Lord Walsingham, &c.

The book has been interleaved, so that collectors can add therein their own notes, dates, &c. Reference has been made easy, the notes for each month being classed under the superfamily heads to which they belong. Those for June come under—Tineina (unclassified), Tineides, Adelides, Plutellides, Elachistides, Gracillariides, Argyresthiides, Coleophorides, Lithocolletides, Nepticulides, Tortricides, Pyraloides, Crambides, Pyralides, Drepanulides, Cymatophorides, Brephides, Geometrides, Pterophorides, Sesiides, Zeuzerides, Cochlidides, Psychides, Anthrocerides, Lachneides, Sphingides, Deltoides, Lymantriides, Nycteolides, Notodontides, Noctuides, Arctiides and Papilionides. These also will give an idea of the range covered by the notes.

STRAY NOTES ON THE NOCTUÆ.
(Demy 8vo. Price 1/-).

This contribution to our knowledge of the British NOCTUIDES should be read by every British entomologist. It contains detailed information, among others, of the following points :—VARIETIES and ABERRATIONS—The local races peculiar to Britain—True distinction between Varieties and Aberrations—Types of species—Scientific usage of the term and its general application. Full notes on the Orrhodias, *Leucania straminea* and other species in the British Museum—Identical North American and British species of Noctuæ—Representative North American species—The genitalia of NOCTUIDES—Identical Japanese and British Noctuæ—Classification of the Noctuæ—Arrangement of Genera—Criticism of the various methods of classification which have been introduced into England—Want of relationship between *Cymatophoridae* and *Bryophilidae*—Separation of *Leucania* and *Nonagria*—Position of the Plusiids as exhibited by our species—The position of the *Deltoides* among the Noctuæ, and many other matters of interest with which British entomologists should be conversant.

THE BRITISH NOCTUÆ AND THEIR VARIETIES.

(Complete in 4 volumes. Price 7/- per vol.).

The four volumes comprise the most complete text-book ever issued on the NOCTUIDES. It contains critical notes on the synonymy, the original type descriptions (or descriptions of the original figures) of every British species, the type descriptions of all known varieties of each British species, tabulated diagnoses and short descriptions of the various phases of variation of the more polymorphic species; all the data known concerning the rare and reputed British species. Complete notes on the lines of development of the general variation observed in the various families and genera. The geographical range of the various species and their varieties, as well as special notes by lepidopterists who have paid particular attention to certain species.

Each volume has an extended introduction. That to Vol. I deals with "General variation and its causes"—with a detailed account of the action of natural selection in producing melanism, albinism, &c. That to Vol. II deals with "The evolution and genetic sequence of insect colours," the most complete review of the subject published. That to Vol. III deals with "Secondary Sexual Characters in Lepidoptera, explaining, so far as is known, a consideration of the organs (and their functions) included in the term. That to Vol. IV deals with "The classification of the Noctuæ," with a comparison of the Nearctic and Palæarctic Noctuides.

The first subscription list comprised some 200 of our leading British lepidopterists, and up to the present time some 500 complete sets of the work have been sold. The treatise is invaluable to all working collectors who want the latest information on this group, and contains large quantities of material collected from foreign magazines and the works of old British authors' arranged in connection with each species, and not to be found in any other published work.

MELANISM & MELANOCHROISM IN BRITISH LEPIDOPTERA.

(Demy 8vo., bound in Cloth. Price 5/-).

Deals exhaustively with all the views brought forward by scientists to account for the forms of melanism and melanochroism ; contains full data respecting the distribution of melanic forms in Britain, and theories to account for their origin ; the special value of "natural selection," "environment," heredity," "disease," "temperature," &c., in particular cases. Lord Walsingham, in his Presidential address to the Fellows of the Entomological Society of London, says, "An especially interesting line of enquiry, as connected with the use and value of colour in insects, is that which has been followed up in Mr. TUTT's series of papers on 'Melanism and Melanochroism.' "

MONOGRAPH OF THE PTEROPHORINA.

(Demy 8vo., 161 pp., bound in Cloth. Price 5/-).

This book contains an introductory chapter on "Collecting," "Killing," and "Setting" the Pterophorina, a table giving details of each species—Times of appearance of larva, of pupa and of imago, food-plants, mode of pupation, and a complete account (so far as is known) of every British species under the headings of "Synonymy," "Imago," "Variation," "Ovum," "Larva," "Foodplants," "Pupa," "Habitat," and "Distribution." It is much the most complete and trustworthy account of this interesting group of Lepidoptera that has ever been published.

INSECTS AND SPIDERS.

(Crown 8vo. Illustrated. Price 1/-).

A really good introductory text-book to the study of general entomology. It contains 15 chapters, giving structural and characteristic details of the various orders of insects. These are entitled—" General external characters of insects," " Internal organs of insects and their functions," " Metamorphosis in insects," "The earwig." "Locusts and Grasshoppers," " Dragonflies," " Caddisflies and moths," " Beetles," " Flies," " Social Insects—bees, wasps and ants," The " Honey-bee," " Wasps," " Ants," and " Spiders."

THE "RECORD" LABEL LIST OF BRITISH BUTTERFLIES.

Arranged after the most recent systems suggested. Printed on one side of the paper only. For labelling cabinet.

(Copies 7 for 6d., 3 for 3d., not less than 3 sent. Postage ½d.).

To be obtained from **Mr. H. E. Page**, "Bertrose," **Gellatly Road, Hatcham, S.E.**

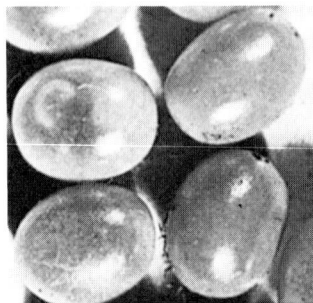

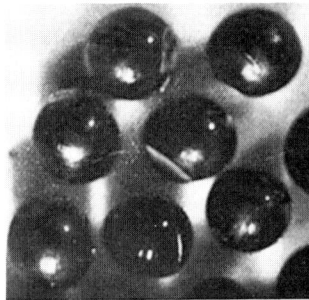

PLATE I.

EGGS OF LEPIDOPTERA.
(Photographed by A. E. Tonge.)

Practical Hints, etc., 1905.

PRACTICAL HINTS

FOR THE

FIELD LEPIDOPTERIST

BY

J. W. TUTT, F.E.S.

Editor of *The Entomologist's Record and Journal of Variation* ;
Author of *A Natural History of the British Lepidoptera* ; *British Noctuæ
and their Varieties* ; *British Butterflies, &c., &c.*

III

Price 6s. net (interleaved).

January 1905.

LONDON:
ELLIOT STOCK, 62, PATERNOSTER Row, E.C.

BERLIN:
R. FRIEDLANDER & SOHN, 11, CARLSTRASSE, N.W.

PREFACE.

The accumulation of a large number of additional Hints, the kind offer of Mr. H. J. Turner to compile an index to the three parts, and the wish to ask lepidopterists using the Hints to utilise their work for the scientific purpose of clearing up and publishing unknown details of the life-histories and habits of our lepidoptera, have led us to acquiesce in the idea of publishing a third part of *Practical Hints*, which is now submitted to the kind consideration of the entomological public.

It has been often stated that the main reason for collectors not turning their attention to the scientific description of eggs, larvæ, and pupæ has been the want of something sufficiently elementary from which to study the subject. In order to meet this objection we have written simple chapters on the egg, larva, and pupa of lepidoptera, which include just as much detail as should enable collectors to obtain a preliminary knowledge of the subject, and explain as simply as possible the structural points required nowadays to make such descriptions of any value to science. To make these subjects more intelligible we have given certain illustrations. For the photographs of the eggs which we have reproduced we are indebted to Messrs. F. N. Clark and A. E. Tonge, for the drawings of larval tubercles to Mr. A. W. Bacot, and for those relating to pupal structure to Dr. T. A. Chapman. We have also to thank Mr. Tonge for his chapter on the photography of eggs, and Mr. Clark for that on their preservation and mounting.

If this Part III should attract some serious workers to our study, we shall be well repaid. Any youngster can collect and learn to name the greater part of our British lepidopterological fauna in a few years of arduous and careful work, but this must always be only a preliminary to scientific study. Progress is most hindered by those who, having reached this stage, try to persuade themselves and others that there is nothing more to do, and who argue that they have taken up the study as a hobby, and not as a science, *i.e.*, that they have killed thousands of insects, not for study, but for a mere whim. The young collector has to remember that all the most progressive work is done by the young and by those who remain mentally young, and we can only reiterate that, as " There is no royal road to mathematics," so there is no royal road, but work, time, devotion, and study, by which to progress along the path of entomological science, that the best-informed lepidopterists of to-day are most aware of their own ignorance, that the beginner of to-day will set the pace to-morrow, and not the older collector who, already beaten in the race, croaks because he is permanently left behind and finds his mental equals in the beginners he affects to despise.

In the different parts, we have often given two or three hints relating to the same phase of a subject when different lepidopterists have made similar observations thereon from slightly different standpoints, or have drawn somewhat different conclusions from their observations. We also find that we have unwittingly repeated two or three hints that we had overlooked, had been already published. This, however, is apparently inseparable from handling so large an amount of material .(some 4,000 hints), at various times spread over a period of five or six years. For these blemishes we most humbly apologise. For carefully reading the proofs to see that no very bad blunders had been included, we are greatly indebted to Mr. B. A. Bower, and, as we have said before, for the Index we owe our very best thanks to Mr. H. J. Turner.

For the great success that has attended the publication of Part I (now practically out of print) and Part II (rapidly following it), we thank everyone who has in any way helped us. We trust Part III will find at least as satisfactory a reception. To those of our friends who, taking us on trust, have already promised their financial aid, we offer our most hearty thanks.

PRACTICAL HINTS

FOR THE FIELD LEPIDOPTERIST.

CHAPTER I

COLLECTORS, COLLECTING, COLLECTIONS.

Collectors of lepidoptera probably outnumber those of all other orders of insects, nor is this to be wondered at, for the lepidoptera are at once the most attractive of all insects and the most conspicuous of them are, perhaps, more readily distinguished than the species of any other order. There is no need to urge that collectors and collecting have a very definite and legitimate position in relation to science ; and since every collector must have a beginning, it follows that any hints that will put him on the right track, lead him to adopt good methods, and suggest lines of work that will tend to make his observations as a collector of service to science, must be useful in advancing the study of lepidopterology.

The legitimate position of the collector is easily stated. He obtains material on which scientific observations are based ; he should himself make observations on the living individuals he collects ; he should preserve well the insects he captures ; he should note exactly the date of capture and the locality where captured of each specimen ; and he should record carefully and with clearness the observations he makes. He should also be prepared to make deductions from his observations, for true science correlates facts, and suggests logical deductions from the observations made. Only the collector in the field can know the relationship of an organism to its environment, the fundamental basis of much of the modern science of natural history, and, hence, from the men who have started as " mere " collectors, attracted first of all by the beauty of some striking butterfly or moth, have arisen all our foremost scientific lepidopterists, both of the present and of past days, the only difference between these and those who, starting with them, have lagged behind, being the difference in the power to observe, or to record their observations, or to draw obvious conclusions from their observations. Without the collector no really scientific work on certain branches of lepidopterological study can be written, and the man who collects his own insects, makes observations, and records such, is a most valuable addition to the ranks of those who study lepidopterology. For the mercenary collector who collects insects, like a man collects old " pots," in an auction-room, one can only feel the heartiest contempt.

The popular idea of a lepidopterist, even now, is one who collects, and not one who studies, butterflies and moths. In 1855, Lubbock regretfully notes (*Ent. Ann.*, 1856, pp. 115 *et seq.*) the want of attention on the part of lepidopterists to the habits, anatomy, and physiology of

insects. This, he says, is the more to be regretted, because he fears
that we must confess that to make collections the end instead of the
means, *i.e.*, to collect merely for the sake of collecting, has a direct
tendency to narrow the mind. To aspire only to be able to say that
one has in one's cabinet a certain number of species, or some rare
sorts which nobody else possesses, is surely an ambition quite unworthy
of a true entomologist.

The legitimate function of collections is readily summed up:—
Collections are means, not ends. This embodies the whole *raison d'être*
of forming collections, and determines whether the making of the
collection is of real advantage or not to the maker. To the ignorant
and uneducated man, the collection, setting, and correct arrangement
of specimens, when obtained, is an advantage to the maker, and such
an one needs little defending, even though he go no further. To the
educated man, the making of collections with no other end in view
does not appear to be justifiable. To capture large numbers of specimens to gratify a feeling of possession is altogether insufficient ground
for forming a collection, and where this is the end, the collections
cannot be of any use to the collector. To every collector a collection
should be a record of observations :—

(1) Of observations made in the field by the collector himself.
(2) Of observations tending to comparisons of species which he has
    obtained from others.

Nothing is more disappointing than to find a collector with no ideas
of comparison, no appreciation of the development of species as exhibited by variable forms, no wish to have specimens of a species that
he knows well, from an outlying locality. The true collector will
attempt to learn all he can of the habits and lifehistories of the species
he collects, and will not be satisfied that his series is full if composed
of a few finely set specimens from any one locality. His collection
will show wherever possible, useful data—locality, date of capture,
&c.—relating to the specimens that he has, maybe, so carefully
brought together.

There is a stage in all collector's lives when the love of possession
and the beauty of the collection is perhaps the ruling passion. This
is quite natural, but when a man has been collecting some six or eight
years and finds that he has no more wish to know the how, when,
where, and why of his specimens than he did at first, he may consider
very fairly that his collection is doing him but little good. If he be
making no observations, his collection is certainly the end and not the
means, as it ought to be, to the end. A collection of insects which is
not studied is of as little use as books which are not read.

One first piece of advice to collectors of lepidoptera—collect all
lepidoptera. The large lepidoptera used to be called macro-lepidoptera,
and the small, micro-lepidoptera. Modern scientific methods have
taught us that this subdivision of the lepidoptera is, on the whole,
unsound and unnatural, and gives no idea of real relationship, the
actual goal attempted to be reached by all systems of classification.
The use of old-fashioned and antiquated books that have been rendered
largely useless by their own contained errors, and by the recent progress
in our knowledge of the lepidoptera, will perpetuate, for a time at least,
these obsolete terms, but no work on modern lines recognises any such
division, and we find among the smaller moths both highly specialised

and generalised groups, or, as they are more correctly termed, superfamilies, just as we find very highly specialised, and more generalised (*i.e.*, less specialised), superfamilies among the larger moths. The best collectors are those who master the manipulation of the small, as well as of the large, moths in the very earliest part of their careers as collectors, and do not restrict themselves to the collection of the larger species, the life-histories of most of which, at least so far as relates to our British species, are very fairly well known. The ardent collector will find much virgin ground yet unbroken among the smaller lepidoptera, and a careful specialisation of his studies to one of these groups, when he has obtained a good general knowledge of the whole, will give him the greater reward.

CHAPTER II.

EGGS AND EGG-STAGE OF LEPIDOPTERA.

STRUCTURE.—The eggs of many lepidoptera are among the most beautiful, as they certainly are among the most interesting, of all microscopic objects. A very short acquaintance with them is sufficient to teach one that they are exceedingly variable in size, shape, sculpture, and colour, but that a general similarity is observable among the eggs of species known to be really closely allied. Broadly speaking, the eggs of all lepidoptera are divisible into two groups, which are known technically as " upright " and " flat " eggs. Every egg has a tiny little rosette of minute cells, which contains a number of microscopic canals leading into the interior of the egg, and by means of which the fertilisation of the egg is effected. This rosette is the micropyle, and if the egg be laid so that the micropylar axis is vertical to the surface on which it is laid, it is called an " upright " egg, and if laid so that the micropylar axis is horizontal to the surface on which it is laid, it is called a " flat " one. Eggs of butterflies, Arctiids, Lymantriids, Noctuids, Notodontids*, &c., are upright eggs, the greater number of those of other moths are flat eggs, *e.g.*, those of Sphingids, Lachneids, Anthrocerids, Geometrids, Cymatophorids, Pyralids, Alucitids, and almost all the smaller moths. The eggs of Tortricids and Cochlidids are exceedingly flat and scale-like, whilst those of the Psychids, packed tightly inside the empty pupal case, are so delicate, that it would be impossible to separate them without destroying them were it not for the particles of wool that the ♀ distributes among them when egg-laying. It will be found, on closer examination, that the upright eggs are, as a rule, eggs with only two axes of measurement, the vertical and the horizontal, the latter being the same in all directions, a horizontal section being circular. In the flat egg there

---

* The synonymy used throughout this work agrees generally with that of (1) *British Butterflies*, J. W. Tutt, 1896 [Gill & Sons, 32, Warwick Lane, E.C.]. *British Noctuae and their Varieties*, J. W. Tutt, 1890-1892; *Natural History of the British Lepidoptera*, J. W. Tutt, 1899-1904 [A. Holder, 41, Wisteria Road, Lewisham, S.E.]

are three different measurements, length, width, and height, the two horizontal axes being different and the horizontal section roughly oval in outline.   Reference to pl. i., figs. 1, 5 ; pl. ii., figs. 1, 2, 3, 4, 6, will give a good idea of certain types of upright eggs ; whilst pl. i., figs. 2, 3, 4, 6 ; pl. ii., fig. 5, illustrate various types of flat eggs.

GENERAL APPEARANCE OF CERTAIN EGGS.—Eggs, as we have just said, may vary greatly in appearance.   The following general hints may prove useful :

1. Upright, ribbed eggs, the micropyle at top—
   (a) Hemispherical or rather more than a hemisphere, ribs coarse, shell more or less transparent, e.g., most Noctuids (pl. ii., fig. 3), Deltoids (pl. ii., fig. 4), some butterflies (pl. i., fig. 1), some Coleophorids (pl. ii., fig. 6).
   (b) Hemispherical, sculpture fine, shell more or less transparent, e.g., Endrosids, some Lithosiids (pl. i., fig. 5). Arctiids, Hesperiids, &c.
   (c) Hemispherical, ribs usually poorly defined or absent, shell opaque, e.g., Notodontids (pl. ii., fig. 2), Lymantriids (pl. ii., fig. 1), &c.
   (d) Tall, nine-pin shaped, ribs well-defined, e.g., Pierids, some Lithosiids, &c.
   (e) Flattened from apex, ribs almost restricted to edges, shell opaque, e.g., Lycænids (pl. iii., all figures).

2. Flat, comparatively smooth eggs, the micropyle at one end—
   (a) Green, smooth, with scarcely any trace of markings or surface reticulation, shell transparent, e.g., Alucitids, Sphingids (pl. i., fig. 2), Dimorphids.
   (b) Yellow, smooth, very delicate, one end glassy, shell wrinkled, very transparent, e.g., Anthrocerids (pl. i., fig. 3).
   (c) Green or yellow, changing sometimes to red-brown or leaden, with ribs (variable development) running lengthwise along sides, shell transparent, e.g., some Geometrids (pl. ii., fig. 5), Crambids, &c.
   (d) Green or yellow, almost smooth, very delicate, flat, and scale-like, shell transparent, e.g., Cochlidids, Tortricids, Pyralids.
   (e) Cream-colour, drab or mottled, shell opaque, surface covered with many minute black points, the crossing points of a fine surface reticulation, e.g., Lachneids (pl. i., fig. 6), Attacids, &c.
   (f) Black (shortly after being laid), shiny, surface finely pitted, shell opaque, e.g., Hepialids (pl. i., fig. 4).

This rough table will perhaps be sufficient to give the beginner some idea as to the position of any of the larger eggs he may obtain. Ability to name the eggs of even a fair number of lepidoptera at sight, however, is only to be obtained by years of study.   In recent years Chapman has revolutionised our ideas concerning them,. and has given us a mass of general and detailed information relating to them that has had a marvellous effect on all the more modern schemes of classification.

THE FIXITY OF THE EGG-LAYING HABIT.—The particular habit observed in the egg-laying of a female of a certain species will be found to be generally that common to all females of the same species, i.e., the females of a given species have a particular egg-laying habit, and, once the habit is known, the mode of obtaining the eggs is usually not difficult.   Further, somewhat similar egg-laying habits are usually noticeable throughout a long series of allied species, or even throughout a whole genus.   The Arctiid habit of laying a large batch of eggs near where emergence takes place, and then flying and laying sundry other batches later at different points, is very general in the group. Again, the Arctiid eggs are laid side by side with considerable regu-

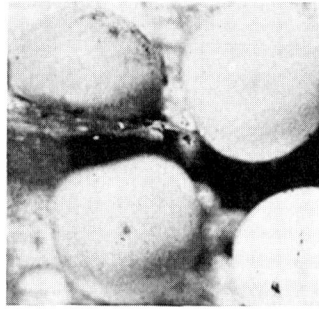

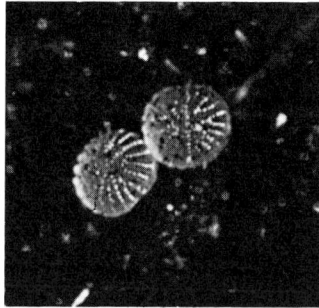

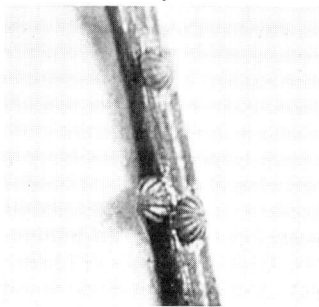

PLATE II.

EGGS OF LEPIDOPTERA.

(*Photographed by A. E. Tonge.*)

Practical Hints, etc., 1905.

larity, and the mode is found in a number of allied superfamilies. On the other hand, in superfamilies in which the habits of different genera vary greatly, great diversity in egg-laying may be observed, *e.g.*, in the Sphingides, &c., whilst Bacot further points out (*Ent. Rec.*, x., p. 31) how remarkably different the egg-laying habit may be in very closely-allied species. Of the common British Lymantriids he observes that the most closely-allied forms differ widely from each other, *e.g.*, *Dasychira pudibunda* eggs laid in close patches uncovered, whilst those of *D. fascelina* are covered with down ; those of *Lymantria monacha* uncovered and squeezed into crevices and cracks of tree-trunks, while those of the allied *Porthetria dispar* are laid in exposed patches covered with down ; the eggs of *Notolophus antiqua* are laid closely on the outside of the cocoon, those of *N. gonostigma*, between the inner and outer cocoon, in large loose masses, mixed with down from the body of the female. On the other hand, throughout whole superfamilies, *e.g.*, Eriocraniides, Nepticulides, Adelides, &c., a very similar method of egg-laying is followed by most of the species.

COLLECTING EGGS.—Egg-searching is a mode of work which, from the small size of many of the objects for which one makes quest, is attended with difficulties that cannot possibly occur to the seeker for larvæ, pupæ. or imagines. It is, of all forms of collecting lepidoptera, the most difficult, the similarity of many eggs to their environment being often such that they are absolutely not to be detected by the keenest searchers. Others, on the contrary, are not so difficult to find, and, with practice, one may obtain a considerable number of species in the egg-stage. Amongst those that are the most easily obtained may be mentioned the eggs of certain butterflies, Sphingids, Arctiids, &c., and those of many Noctuids are also not difficult to find. The eggs of day-flying species (butterflies and moths) may be obtained by carefully following up the females when on the wing, watching their habits, and so obtaining their eggs. Thus the eggs of *Papilio machaon, Colias hyale, Gonepteryx rhamni, Euchloë cardamines, Dryas paphia, Argynnis aglaia, Brenthis euphrosyne, Polyommatus bellargus, P. corydon*, and many other species are readily obtained, as also are the yellow egg-masses of the Anthrocerids (burnets) and Adscitids (foresters). The eggs of the Nepticulids, in spite of their small size, are readily discovered, with practice, with a hand lens, and those of the Psychids are most readily found by seeking for empty cases, from which the imagines have not long emerged. Among fairly conspicuous eggs, if one be in the right locality, one can hardly overlook the eggs of the Amorphid species, nor those of the Ceruras, and those of the ground-feeding Sphingids offer no serious difficulties.

IMPORTANCE OF KNOWLEDGE OF EGG-LAYING HABIT.—Almost every species, as we have noticed, has its own particular egg-laying habit, and this habit, strongly fixed and specialised in some species, is much more variable in others; still, on the whole, it may be assumed that when one has found an egg (or batch of eggs) of a particular species, careful observation of the exact position, mode of attachment, and other details, will enable any observant collector to find more of this particular species. Seeking for eggs of *Melanargia galathea* or any of the Hepialids in nature would be an absurdity, the eggs being scattered broadcast on the ground or among the roots of the plants on which they feed ; so, also, are those of so large a moth as *Lasio-campa quercùs*, but, usually, the eggs are attached to the foodplant in a particular way and at a more or less particular place.

POSITIONS CHOSEN FOR EGG-LAYING.—The most frequently chosen positions for eggs to be laid are on the upper- or underside of leaves, on or at the base of a flower-bud, on or at the base of a leaf-bud, and on the stems of herbaceous plants. Many are placed on the stems of more woody plants, a few are thrust into the chinks of rough bark, but generally they are placed in close proximity to the leaves and flowers on which the larvæ will feed. Exceptional cases like those of the honey moths— *Galleria cereana, Achroia grisella*, &c., the larvæ of which are most destructive to honeycomb, or those of *Aphomia sociella*, which feed gregariously in wasp and bee nests—usually placed on the surface of, or in a shallow hollow in, the ground—must, of course, be considered separately.

SOME EGG-LAYING HABITS.—Of those lepidoptera that lay their eggs on their respective foodplants, many lay the eggs solitarily, others side by side in clusters, whilst the Anthrocerids often heap their eggs in two or three layers. *Malacosoma castrensis, M. neustria, Lachneis lanestris, Saturnia pavonia, Anisopteryx aescularia*, &c., lay their eggs round the twigs of their various foodplants, forming a sort of necklace round them, in some cases being covered with long silky hairs, in others with a thick varnish, those of *L. lanestris* in particular making a large and conspicuous bundle on a hawthorn or blackthorn twig. The female Amphidasyds, Boarmiids, and Tephrosiids—*Amphidasys betularia, A. strataria, Tephrosia bistortata, T. crepuscularia*, &c.—like *Zeuzera pyrina*, are provided with long ovipositors to enable them to lay their eggs deeply in the crevices of the bark of the trees on which the larvæ feed. The eggs of *Trochilium bembeciforme* are laid on the underside of the leaves of osier, although the larva is a borer, and feeds on the solid wood of the stem and branches. The females of *Leucania littoralis* fold over the edge of a grass leaf, and lay their eggs in a string in the fold ; the ♀ *Apamea ophiogramma* lays her eggs similarly on leaves of reed or striped riband-grass, and the eggs of *Orrhodia vaccinii, O. ligula*, and others, are pushed into crannies when laid, so as to lose almost all the usual characteristics of a typical Noctuid egg. The eggs of *Geometra vernaria* and *Polygonia c-album* are laid one upon the other in rouleaux on the stems of their respective foodplants, those of the former resembling a slender twig or tendril of *Clematis*, on which plant the eggs are laid. The females of the Eriocraniids and Adelids are provided with a complex cutting apparatus, with which they cut out pockets in a leaf of the foodplant, and lay their eggs in the soft cellular tissue of the leaf.

RELATION BETWEEN POSITION OF EGG AND DURATION OF THE EGG-STAGE. —When eggs are laid naturally upon the leaves of deciduous trees or of herbaceous plants, it generally follows that the egg-stage is a short one. When, on the other hand, the eggs are laid upon the stems, leaf-buds, and the more persistent parts of plants, the egg-stage may last a considerable time. Although one could cite numberless exceptions, it happens that the Sphingids, Geometrids, Tortricids,

Pyralids, and their close allies usually adopt the former method, and, as a general rule, it may be asserted that a comparatively few species choose the latter, in other words the egg-stage is not, in many cases, the long one. The Catocalids, Xanthiids, Ennomids, Theclids, and many Lachneids and Lymantriids are to be noted among those with an egg-stage that extends to many months, and the eggs are always laid by these on a very permanent part of the plant. The eggs that are naturally laid on grass-stems, leaves of plants, &c., are usually white, yellow, or greenish in tint when laid, changing to a dull leaden colour as maturity proceeds, those laid on the twigs of bushes or trees are of a dirty whitish or grey tint, assuming frequently a purplish or red-brown tint, e.g., *Dichonia aprilina*, *Cirrhoedia xerampelina*, *Tiliacea aurago*, *Dimorpha versicolora*, *Ennomos autumnaria*, *Zephyrus quercûs*, *Thecla w-album*, *T. pruni*, &c., and it may be taken for granted that, as a general rule, those species which hybernate in the egg-stage, have eggs which rapidly change to some dark hue that corresponds well with the colour of the stem or twig on which the egg is frequently deposited. Those that are scattered on the ground are usually of a dirt-colour, or have a shiny pearly apppearance; in fact, with a few apparent exceptions, the colour of the eggs of lepidoptera soon become such as to make them difficult of detection by the various predaceous creatures that prey upon them, and equally difficult for the collector to discover them.

SIZE OF BATCHES OF EGGS.—The number of eggs laid by ♀ s of different species varies exceedingly, although for a certain species a very fair average is maintained. We have come slowly to look on the laying of a large number of eggs of exceedingly small size as a specialised condition intended to meet the immense destruction that occurs in the early stages in many species. Among the Noctuids, a ♀ of *Triphaena fimbria* will lay as many as 1200 eggs, a ♀ of *T. pronuba* from 800-1000 eggs, and *T. comes* 400-500, *Peridroma saucia* lays about 2000 eggs, whilst the ♀ s of many Noctuids, especially those that have larger eggs, rarely lay more than 200, the number accredited to a ♀ *Epunda lichenea*. Among the Arctiids a ♀ *Phragmatobia fuliginosa* may lay as many as 610 eggs, whilst *Spilosoma mendica* lays between 300 and 400. A ♀ *Zeuzera pyrina (aesculi)* lays above 1000 eggs on an average, and many others lay almost as many.

STRANGE LAYING PLACES.—The eggs of lepidoptera are sometimes laid in very unexpected and very remarkable situations, far removed from the foodplants of the larvæ. Riding records that *Triphaena pronuba* frequently chooses a piece of wire (in a fence) or a cord hanging loosely in a garden, for the purpose, and also reports batches of eggs of this species in two successive years (1895, 1896), on the meshes of a lawn-tennis net, and the same observer records the finding of eggs of *Macrothylacia rubi* on the trunk of a pine at a height of nearly six feet from the ground, whilst a couple of the lower leaves of a Weymouth pine were girdled by 70 or 80 rows (10 eggs in each) of the eggs of *T. pronuba*. Many ground-feeding Noctuids lay on the stems of dead plants, leaves of trees, &c., and so also do *Arctia caia*, *Spilosoma menthastri*, and other Arctiids. We have often seen large batches of the eggs of *Porthetria dispar* on the perpendicular face of a rock, some distance from the ground, and many feet from the nearest

8 PRACTICAL HINTS FOR THE FIELD LEPIDOPTERIST.

herbage. The butterflies, *Chrysophanus phlaeas* and *Polyommatus icarus*, frequently deposit eggs on objects adjacent to their foodplants, and so, more rarely, do *Pararge egeria*, *P. megaera*, and *Pieris napi*. Some of the Acidaliids, like the Hepialids, and certain Satyrids, *Melanargia galathea*, &c., sprinkle their eggs on the ground.

SEARCHING ON LEAVES.—Here, again, if success is to be obtained, something of the egg-laying habit of the species desiderated should be known. We know no work except our own *Natural History of the British Lepidoptera* where this information is to be obtained first hand. If, however, the collector watch a ♀ insect deposit one egg. others may most certainly be found in similar or even identical positions. It is rarely, indeed, that the female of a species chooses both the upper- and underside of a leaf for the purpose. Some select the upperside of leaves, others the underside, but usually the same species keeps to the same side, and frequently to almost exactly the same position on the leaf. When one egg of *Cerura vinula* has been found on the upperside of a poplar leaf, others are sure to follow by continuing the search, similarly with *C. furcula* and *C. bifida*. Or, if the egg of *Hemaris fuciformis* be found on the underside of a leaf of honeysuckle near the margin, it may be relied upon that others will be found in an almost exactly similar position ; similarly with the eggs of *H. tityus* on the underside of the leaves of *Scabiosa arvensis* and *S. succisa*. In other cases both the upper- and underside may be selected, as in the case of *Amorpha populi* and *Smerinthus ocellata*, although the latter rarely utilises the upperside. The underside of the leaves of the various foodplants of the Anthrocerids will usually give us large batches of the eggs of the burnet moths in their various localities. The undersides of the leaves of low plants give batches of the eggs of Arctiids, ground-feeding Noctuids, and some Geometrids. Haphazard searching may be successful, but we would again urge collectors to make themselves quite conversant with the egg-laying habit of a species, and then this knowledge may be followed up with every chance of success.

SEARCHING ON STEMS.—Eggs laid upon stems possibly offer more trouble in finding than those laid in almost any other position, because, as already stated, they are usually so similar in tint to their surroundings that their detection is most difficult. It is, however, to be noted that, in spite of this difficulty, the egg-laying habit is so fixed that, in a locality where a species is known to exist, if the habit be known, the eggs can usually be discovered by a keen collector. The eggs of *Trichiura crataegi*, laid in little rows of eight to ten, touching each other, like a little ribbon, along the twigs of hawthorn and sloe ; the eggs of *Poecilocampa populi*, laid singly or in little batches on the twigs of oak, poplar, hazel, apple, hawthorn, &c. ; of *Citria flavago* and *C. fulvago*, near the leaf-buds of sallow ; of *Cirrhoedia xerampelina*, of the Theclids, of the Ennomids, &c., laid as we have described elsewhere in this book, will not prove difficult to find. In fact, so specialised are the egg-laying habits of *Epione apiciaria*, *Himera pennaria*, *Chesias spartiata*, *Scotosia vetulata*, *Ptilophora plumigera*, *Tiliacea aurago*, &c., that, given the locality where it is known the species occur, there seems to be no reason why any smart collector should not find almost as many eggs as he may require.

OBTAINING EGGS IN CONFINEMENT.—1. *Pairing moths:*—First an impregnated ♀ is required. She may be obtained in one of two ways : (1) By capture in the field. (2) By rearing and pairing a bred ♀ with a ♂. This latter is not always to be done without thought and care; some species are practically easy to pair, others most difficult. Butterflies are possibly among the most difficult to pair in captivity. The following notes may be useful :—

ANTHROCERIDES : Pair freely, and lay their eggs quite readily if left in each other's society, *in the sun*, they will not stir in dull weather.

PSYCHIDES : Pair freely; one can hold the case containing a newly-emerged ♀ in one's hand, and the ♂ will extend its abdomen deep down into the case, and pair with the female whilst one holds it.

LACHNEIDES : Must be left together in the evening, they pair usually between 6 p.m. and 9 p.m. without trouble. A few dayfliers pair in late afternoon.

DIMORPHIDES AND ATTACIDES : Will pair freely in confinement during the day, if they be kept in a warm room, in a light, well-ventilated box or breeding-cage.

SPHINGIDES : Must be left together at night when pairing will possibly take place ; much more certain result with Amorphids than with the swifter-flying Sphingids. [A ♀ *Smerinthus ocellata* attached by a slip-knot round the upper part of abdomen to a tree in a London suburban garden is nearly sure to attract a ♂ , and one will find them paired next morning, nor will they separate till the evening.]

NOTODONTIDES, NOCTUIDES, ARCTIIDES : Will usually pair freely if left shut up together for a night or two. [Some species, *e.g.*, *Spilosoma mendica*, will not pair unless a current of air be able to pass through the cage in which they are confined. In confinement some Noctuids, *e.g.*, *Triphaena comes*, appear to want feeding before pairing takes place.]

GEOMETRIDES : Sometimes pair readily if confined together in a roomy cage. They usually pair in the evening, some, however, late at night.

2. *Procuring the eggs:*—Having obtained a fertile ♀ , the question arises how to persuade her to lay her eggs. The species of butterflies and moths are as different as possible in the ease or difficulty with which they may be persuaded to lay their eggs. In nature, each species has its own particular egg-laying habit, and the nearer one permits a moth to approach natural conditions in confinement, the more certain is one to succeed. Thus :—

BUTTERFLIES may be kept in a large, airy, well-ventilated fern-case with the growing foodplant (the case to be in the sun), and many will then deposit their eggs, *e.g.*, *Colias edusa*, *C. hyale*, *Leucophasia sinapis*, *Nemeobius lucina*, *Brenthis euphrosyne ;* and most readily of all the Satyrids—*Coenonympha pamphilus*, *Erebia aethiops*, *Enodia hyperanthus*, *Pararge egeria*, *P. megaera*, &c. Others want a large leno sleeve covered over a considerable portion of the food-plant. In this way, *Apatura iris*, *Limenitis sibylla*, *Thecla w-album*, *Zephyrus quercûs*, and others have been persuaded to lay. With the exception, however, of *Melanargia galathea*, which often lays freely in a chip-box, *Erebia aethiops*, and a few others, very few of our butterflies really lay freely in confinement. Feeding species should be given the opportunity of feeding before pairing and during laying.

SPHINGIDES : The Amorphids, sleeved on their foodplant, usually lay freely, sometimes also in a large enclosed box. The Sphingids proper, as a rule, are more difficult, the day-flying Hemarids particularly so. We have, however, known a ♀ *Sphinx ligustri* cover the sides of a large cardboard box with its eggs when it has been confined therein.

DIMORPHIDES and ATTACIDES : Our British species, *Dimorpha versicolora* and *Saturnia pavonia*, lay freely, preferably on branches of their foodplants, but on the sides of an enclosed box if nothing better be forthcoming.

LACHNEIDES : Lay more freely, perhaps, than any other moths. Some— *Cosmotriche potatoria*, *Lasiocampa quercûs*, &c.—will lay in one's hand if they be held when the egg-laying instinct is on ; others, as *Lachneis lanestris*, *Eutricha quercifolia*, want sleeving under satisfactory conditions to ensure eggs.

ANTHROCERIDES : The foodplant growing in a large, well-ventilated fern-case and put in the sun is sufficient to ensure the ♀ s being tempted to lay. They will lay freely sometimes in a closed chip-box provided they get plenty of light. [If

10 PRACTICAL HINTS FOR THE FIELD LEPIDOPTERIST.

the moths be confined in a closed glass-topped box, and put in the direct rays of the sun, they will die in a few minutes of sunstroke.]

NOTODONTIDES, NOCTUIDES, ARCTIIDES, and GEOMETRIDES : Vary much, individually, but enclosing these in a roomy box, or sleeving them on the food-plant, is usually quite sufficient to ensure obtaining eggs of many (? most) of the species. It must not be forgotten that most of the species belonging to these superfamilies require feeding on honey and water, or moistened sugar, as they obtain considerable food in nature.

For species that require crannies, &c., in which to push their eggs, a large chip-box, carefully scored so that the ♀ can get its ovipositor in the crack, or a piece of virgin cork, is a great advantage. *Dicycla oo*, *Anchocelis litura*, *A. pistacina*, *Tephrosia bistortata*, *T. crepuscularia,* and many others can thus be persuaded to lay, the ♀ s often failing to deposit any eggs at all under less favourable conditions. Folded paper, too, is most convenient for the use of these moths, and a ring of folded paper placed round the interior of a box obviates the annoyance caused by finding the eggs pushed between the lid and box, and being crushed in opening.

COLOUR CHANGES IN EGGS.—A series of colour-changes in eggs, that have a moderately transparent shell, is very general, and betokens the development of the embryo within, although some eggs undergo the first stages of colour-change even when infertile. That the colour is, in these cases, due to the contents is evident, for after the young larvæ have left the egg the shell is often quite clear, pearly, and transparent, *e.g.*, the eggs of many butterflies, Anthrocerids, Sphingids, Geometrids, Ægeriids, Pyralids, Noctuids, &c., but some eggs have dense opaque shells, and in these the colour is usually retained. The changes are generally very uniform for the same species; in some they are simple, in others complex. The egg of *Euchloë cardamines* is yellow when first laid, becomes orange within 24 hours, and with the exception of a slight change just before hatching, remains of this tint. The egg of *Leucania littoralis* is at first pale yellow, then orange, then mottled with reddish, and at last purplish, at which stage the embryo-larva can be seen coiled up in the shell. All field-naturalists should make themselves acquainted with the changes that eggs undergo, as it is otherwise impossible to tell what many eggs may be when they have assumed a coloration with which the collector is not familiar.

FERTILE AND INFERTILE EGGS.—The difference between fertile and infertile eggs is said by a high authority to " be known by the changes which take place in their colour, density, shape, &c." This may be perfectly true in some cases, but not in all. The infertile eggs of *Dimorpha versicolora*, of *Bombyx mori*, and a few others remain yellow, whilst the fertile ones rapidly go through the various colour changes to purple-brown or leaden-colour respectively. The infertile eggs of Sphingids quickly collapse, whilst those that are fertile remain plump except for a slight depression on the upper surface, and the transparent shell allows the observation of the development of the embryo in these and allied eggs. But whether the eggs of many species be fertile or infertile, the primary colour-change is often undergone, and one cannot tell until later, sometimes only by the eggs going over the normal period, that the eggs are infertile at all, *e.g.*, certain Hepialids, in which the change of colour, from white to black, takes place in a few hours from being laid, and remains permanently so, the eggshell itself really becoming black, and remaining so even after the young larvæ have hatched from the fertile eggs. The opaque eggs of the Lachneids, the Attacids, the Notodontids, &c., exhibit very little change, and fertile and infertile eggs remain practically alike. But, generally, a series of changes, to be followed out with interest by the lepidopterist, is soon observable in most fertile eggs, and these are more or less the result of the development of the embryo. Eggs laid on leaves and on grass stems are, as we have already noted, as a rule, of a white, yellow, or greenish hue, whilst those laid on the twigs of bushes and trees are frequently of a dirty white or grey, and rapidly assume a purplish or red-brown tint, e.g., *Dichonia aprilina*, *Cirrhoedia xerampelina*, *Dimorpha versicolora*, *Ennomos autumnaria*, &c., and become of such a hue that each corresponds well with the colour of the stem or twig on which it is normally laid in nature. Experience, however, and the free use of a good lens will soon teach the young lepidopterist whether his eggs are fertile, and he will soon learn which are sufficiently transparent for him to watch the actual development of the embryo, e.g., eggs of *Nemeobius lucina*, Anthrocerid eggs, Tortricid eggs, &c., may be readily mounted under the microscope, and the development of the young larvæ can be easily observed. [NOTE.—Do not forget that some soft-shelled eggs collapse considerably, even when fertile and when development is proceeding satisfactorily.]

PROTECTION OF EGGS.—Eggs are protected in a variety of ways, the first and simplest of which is frequently the rapid change of colour already mentioned, by means of which they assimilate to the surface on which in nature they are laid. This is especially the case with those that choose the bark of twigs, or stems of trees, on which to lay, e.g., the eggs of *Dimorpha versicolora* are yellow when first laid, on twigs of alder or birch, they rapidly become red-brown, and later purple-brown, assimilating most perfectly to the surface on which they are laid. The peculiar resemblance of a rouleau of the eggs of *Geometra vernaria* to a broken tendril of *Clematis vitalba*, the plant on which the eggs are laid ; the similarity of a ring of eggs of *Saturnia pavonia* to a bunch of dead ling flowers; the packing away of the eggs of *Tephrosia bistortata*, *T. crepuscularia*, *Biston hirtaria*, *Amphidasys strataria*, *Orrhodia vaccinii*, *Dicycla oo*, etc., into the deep cracks and crannies of the bark of the trees on the leaves of which the larvæ feed ; the covering of the eggs of *Anisopteryx aescularia*, *Lachneis lanestris*, *Porthesia similis*, *P. chrysorrhoea*, and *Porthetria dispar*, with silk from the extremity of the abdomen, may all be looked upon as modes of preservation adopted under different conditions. The female *Leucoma salicis* covers its eggs with a salivary-looking substance, but which is quite hard, and egg-masses so covered are not difficult to find on the undersides of poplar and willow leaves in August. The thick gum in which eggs of *Malacosoma neustria* and *M. castrensis* are embedded, must be an excellent protection, and when one considers that the latter are often under water, and that the eggs are to be found on the shores of the Thames and Medway marshes directly under the seawall, where the egg-rings have been washed up by the winter flood, the utility of the varnish is obvious. The eggs of Psychids are packed tightly in the pupa-case (which in turn is inside the case made by the larva), and the remains of the dead mother block up the entrance and prevent ingress to wandering ants, &c. Female Adelids, like the Eriocraniids, cut pockets in the leaves and place their eggs in the soft cellular structure of the leaf. Many other modes of protection in this stage will soon be discovered by the field lepidopterist.

EGG ENEMIES.—One authority states that " The great advantage of egg-hunting is the escaping those odious ichneumons which, in the larval and pupal states, so often blight the collector's legitimate hopes. It is stated, however, that the eggs themselves are sometimes stung. If this be true, it is, indeed, 'nipping one's hopes in the egg (bud).' Upon the whole, egg-hunting, perhaps, is not very productive, but it may still while away an hour or two." This is a strange sentence to be written by a naturalist for naturalists. Whole genera of minute microscopic Hymenoptera are comprised of species that lay their eggs within the eggs of other insects, whose larvæ feed on the contents of the eggs, pupate within, and emerge as imagines in due course to continue the work of destruction. Among the most destructive to the eggs of lepidoptera are the genera *Trichogramma* and *Telenomus*, *e.g.*, Nicholson records the rearing of 30 *Telenomus phalaenarum* from eight eggs of *Macrothylacia rubi*. Bacot records the destruction of a whole batch of *Arctia caia* eggs by the same species, and Bignell states that he bred 2100 imagines of the same parasite from 200 eggs of *M. rubi*, an average of more than 10 to each egg. Dimmock bred 30 hymenopterous parasites from a single egg of *Smerinthus excaecatus*, and there are dozens of similar records. Among other enemies that destroy the eggs of lepidoptera, spiders, ants, and mites hold the foremost place, but their combined destructive efforts possibly fall much below those of the true egg-parasites mentioned above.

DURATION OF EGG-STAGE.—The duration of the egg-stage varies greatly in different species. The shortest recorded periods are 2 days for *Acidalia virgularia*, and 4 days for *Timandra amataria* and other species. On the other hand, many species whose eggs hatch the same year have a much longer oval period, *e.g.*, *Lasiocampa quercûs*, 30 days; *Amphidasys strataria*, 30 days; *Hybernia leucophaearia*, 38 days ; and the Selenias, whose eggs hatch the same year (whilst those of the Ennomids, usually grouped with them by systematists, go over the winter), require from 20-30 days; *Selenia tetralunaria*, 23 days; *Selenia bilunaria*, first brood, 28 days, &c. But the length varies in different seasons. *Biston hirtaria* has taken from 17 to 37 days, *Hemerophila abruptaria* from 14 to 26 days, and *Selenia lunaria* 7 days in 1865, 12 days in 1861, 15 days in 1886 —all of the first brood. But different broods of the same species may vary in the same year. Thus, in 1865, one batch of *Camptogramma fluviata* took 5 days, another 10 days, and a third 21 days. Naturally those that go over the winter in the egg-stage pass a comparatively long period in this, and a correspondingly short one in the larval, pupal, and imaginal states. Thus the egg-stage of *Epione apiciaria* lasts as long as $9\frac{3}{4}$ months, of *Ennomos autumnaria* $7\frac{3}{4}$ to 10 months, of *Himera pennaria* 5 months, of *Oporabia filigrammaria* $4\frac{3}{4}$ months, of *Cidaria testata* 8 months, of *Chesias spartiata* $4\frac{1}{2}$ months. The egg-stage of *Thecla w-album* and *Zephyrus betulae* lasts from July until late April or early May, that of *T. pruni* from June until late April, of *Plebeius aegon* from July to April, of *Trichiura crataegi* from September to April, of the beautiful

Catocalids from July and August to April, and so on. The most
likely faults to be committed by the young collectors are (1) to overlook eggs that will quickly hatch in summer and produce larvæ that
intend to go on to a second brood ; (2) to put away hybernating
eggs with the comforting idea that they will hatch in spring, and
not look at them till March or April, when the larvæ have long since
emerged and died. Eggs about which any doubt occurs should be
kept in a glass-tube or other receptacle in which they can be easily
seen, and placed in such a position that they can be frequently
examined without any trouble.

MANIPULATING EGGS.—The eggs of lepidoptera are delicate structures, some so exceedingly fragile that the slightest touch destroys
them (e.g., Psychids) ; others, on the other hand, will stand a
certain amount of handling without much damage. As a rule the shells
are softer when newly laid than later, and hence more liable to injury.
Whether the eggs be fragile or not it may be taken for granted that
it is best to leave them in the position in which they are laid, and to
handle them as little as possible. If they be laid on a leaf, snip
off the leaf, if on a twig, cut off the twig, if on a trunk, slice off
the piece of bark, and if on cardboard, cut off the surface carrying
the eggs. If at all possible leave the eggs exactly in the position
that the parent laid them. Nothing shows better the delicate
structure of some eggs when first laid than those which, like *Apamea
ophiogramma, Leucania littoralis, Orrhodia raccinii, O. ligula,* &c.,
laid in the fold of a leaf or in crevices of bark, are twisted by the
folding leaf or the pressure of the rough bark into almost unrecognis-
able little masses (and appear quite different from the beautiful, upright,
ribbed, typical Noctuid egg), and yet in no wise injure the developing
embryo by their irregular form. The irregularity is, of course, formed
when the eggs are first laid, they soon get harder, and the irregular
shape becomes more or less fixed. In confinement, all sorts of plans
may be contrived to obviate too much handling, e.g., the growing foodplant in fern-cases is an excellent receptacle on which to tempt
captured females to lay their eggs, but ventilation must be seen to or
the fern-case may prove too damp; sleeving captured females on a
branch of the foodplant in a large muslin or leno sleeve ; pinning the
box, or piece of cardboard, or leaf on which a batch of eggs has been
laid, upon a branch of the foodplant, are all good methods that reduce
handling the eggs to a minimum.

KEEPING EGGS THROUGH THE WINTER.—The eggs of many species of
lepidoptera, that are laid in late summer or autumn, go through the
winter as eggs, the young larvæ not emerging until the spring. Among
the butterflies are those of *Pamphila comma, Plebeius aegon,* the
Theclids—*Thecla pruni, T. w-album, Zephyrus quercûs—Argynnis adippe,*
&c. The eggs of many Geometrids—*Scotosia vetulata, Ennomos
autumnaria, E. alniaria (tiliaria), E. angularia, E. erosaria, E. fuscantaria, Himera pennaria, Epione apiciaria, Chesias spartiata, Oporabia
filigrammaria, Cidaria testata,* &c.—also pass the winter in this stage.
Many Noctuids—*Polia nigrocincta, P. flavicincta, P. chi, Tethea retusa,
T. subtusa, Citria fulvago (cerago), C. flavago, Tiliacea citrago, T. aurago,
Cirrhoedia xerampelina, Catocala fraxini, C. nupta, C. sponsa, C.
promissa, Dichonia aprilina, Anchocelis pistacina—Ptilophora plumi-*

14   PRACTICAL HINTS FOR THE FIELD LEPIDOPTERIST.

gera, Poecilocampa populi, Trichiura crataegi, Pachygastria trifolii, Lymantria monacha, Notolophus antiqua, &c., and many other species. Eggs laid in nature are generally deposited under such conditions as are sure to be best suited for their preservation and safety, and will not be liable to destruction by being allowed to get too wet or too dry, states that frequently happen under artificial conditions. Those laid on paper, in a chip-box, or attached to some similar object, may usually be kept without injury in a cool greenhouse, or covered shed, exposed to the air. Damping is usually fatal, and closed boxes kept in a warm room are not the best of resting-places for eggs in winter, for even if they do not dry up or die off they will sometimes hatch prematurely when food is difficult to get, or, perhaps, quite unobtainable. In cases like this, buds will often suffice the young larvæ; thus early-hatching larvæ of *Cirrhoedia xerampelina* may be fed on the cut buds and early catkins of ash, &c., whilst those of *Tiliacea aurago* feed equally well on buds of sycamore, and early-hatching larvæ of *Asteroscopus sphinx* on cut buds and catkins of birch. Buckler suggested that the keeping of many kinds of eggs on a growing pad of the velvety moss so abundant on old walls, prevented them damping off or drying up, and maintained them in a healthy condition during the winter. Frequent examination of eggs during the winter is most necessary, as the uncertainty of their hatching under artificial conditions cannot be too often repeated. Care must be taken that the larvæ of *Œcophora pseudospretella* do not get at the eggs, as they will enter a box when very small and clear the entire batch.

CONDITION IN WHICH EGGS PASS THE WINTER.—The condition of eggs during the hybernating period is very interesting. The fully-formed embryonic caterpillar is to be found coiled up within the egg-shell throughout the winter in the eggs of *Argynnis adippe, Pamphila comma, Pachygastria trifolii, Malacosoma neustria,* and many other species. On the other hand, Buckler records that eggs of *Bombyx mori, Trichiura crataegi, Ennomos alniaria (tiliaria), E. quercinaria (angularia), Cheimatobia brumata, C. boreata, Scotosia vetulata, Ptilophora plumigera,* and *Polia chi,* have been examined from time to time, until the middle of January, with nothing but the faintest trace of the future larva to be discovered among the still fluid contents by a careful microscopic examination. It appears that the eggs of such species as *A. adippe,* &c., which normally hybernate as young larvæ inside the eggshell, are not so liable to hatch at early and unexpected times as those like *Tiliacea aurago, Cirrhoedia xerampelina,* &c., which, passing the winter normally in the fluid state, often develop their larvæ in January or February (or earlier) under favourable atmospheric conditions, and then appear unable to remain in the eggshell, but hatch at what seems to be both an unreasonable and unseasonable time, and are liable to starve unless proper precaution be taken. Great care should be taken to avoid sudden changes of temperature and moisture.

TREATMENT OF EGGS NEAR HATCHING.—When one has watched his eggs gradually change colour, and knows that they may shortly be expected to hatch, careful watch must be kept on them, for the young larvæ not only soon travel a considerable distance, but, owing to their small size, are frequently overlooked and lost. Eggs laid on cardboard, piece of chip-box, leaves, &c., should, in confinement, as the time for hatching approaches, be placed in a glass-topped box or a small wide-mouthed bottle, where they may be watched more readily, and the young larvæ may be left here, too, for a time after they have hatched, until at least you know how many larvæ you have, and preferably until after the first moult. They can then be removed to an ordinary rearing-cage, or sleeved out.

IRREGULARITY OF HATCHING.—We ought, perhaps, also to note that, in some cases, the hatching of a batch of eggs takes place very irregularly. This is particularly the case in those eggs that have a long oval period, and specially in eggs that hybernate, possibly as a safeguard against sudden changes of weather. Eggs of *Trichiura crataegi*, *Pachygastria trifolii*, *Notolophus antiqua*, *Ennomos quercinaria*, *E. tiliaria*, *Epione apiciaria*, *Polia nigrocincta*, *Cirrhoedia xerampelina*, and other species may have a hatching period extending over four or even six weeks—in fact, the hatching period of *C. xerampelina* has been noted as extending from November to March. Few reliable recorded observations on this point, however, are at hand. The difference is usually much more noticeable in batches of eggs kept under artificial, than in those kept under natural, conditions. The Catocalids, too, vary in the length of time in which the eggs remain as such, some hatching more quickly than others. The influence of temperature, not only on the hatching period and the vitality of eggs, but also on the rapidity with which the resultant larvæ feed up, is well known.

SENDING EGGS BY POST.—There are many methods of sending eggs through the post, and the facilities now offered for sending small packets cheaply make the actual space occupied of less importance than hitherto. The usual modes of sending eggs are:—(1) *Quills:* Cut off a piece of stout quill about 1 in. in length, plug up one end with cotton wool, place the eggs in the quill, and carefully stop up the other end with cotton wool. Eggs sent thus will usually carry well in an ordinary envelope. (2) *Cork:* Cut a small square or oblong hole in a piece of sheet cork; gum a piece of stamp-paper over the hole on one side, place the eggs in the hole, gum another piece of paper over the top, and the eggs will also carry well in an ordinary envelope. (3) *Cardboard or wood:* Treat as in 2, and the result is the same. (4) *Tin boxes:* Eggs laid on a piece of leaf, stem, or inside a chip-box must be left *in situ*, as removing them injures or destroys them. Such, obviously, may want more space than a quill, &c., offers. A small, but strong, tin-box is now the best receptacle in which to place the eggs, or, better still, put them in a small chip-box, and pack this carefully in cotton wool inside a slightly larger tin-box. The latter wrapped in a piece of brown paper, and provided with a label for the address and stamps, will usually travel quite safely. [N.B.—Always use labels for the stamps, as the Post Office authorities then usually stamp the label and not the parcel, and the latter naturally runs less risk.]

CHAPTER III.

PRESERVATION, PHOTOGRAPHING, AND DESCRIPTION OF EGGS.

PRESERVATION OF LEPIDOPTEROUS EGGS.—Various methods have been devised for the preservation of lepidopterous eggs. The following has been suggested : A batch of unfertilised eggs of *Notolophus antiqua*, laid on paper, was divided into two parts, one was dropped into methylated spirit and the other into turpentine. They were allowed to macerate for a fortnight, when they were removed, still firmly fixed to the paper and to each other. They were then placed in a glass-topped box for another fortnight, so that they might thoroughly dry. At the end of the second week, those in the turpentine remained absolutely unaltered, fresh, plump, and fully distended, but the spirit-soaked batch had nearly every egg more or less shrivelled, with the upper face depressed or pulled down, as it were, with the periphery more or less flattened and the two sides drawn together. The turpentine group, three months later, looked as though quite recently laid, making a beautiful top-light object for the 2in. binocular.

Clark, one of our most successful manipulators in this direction, writes : " Some efficient method of treatment for the preservation of lepidopterous ova as permanent specimens is very much needed ; unfortunately there are few which will retain their shape, let alone colour, indefinitely. It is obvious that the contained larva must be killed by some means or other, and if it were possible always to do this without injury to the shell, our object might be achieved. Amongst some of the methods which have been tried, are—*Pricking*, with a very fine needle, this might answer with the larger eggs, but would be useless with the smaller or more delicate. *Heat*, either dry or with boiling water, has been recommended, but I cannot say much in its favour. *Carbolic acid, benzine, naphthaline*, and strong *ammonia* I have tried, but without success. On the other hand, soaking for four or five hours in paraffin and drying on blotting-paper has been found to answer with such fragile eggs as *Pieris brassicae* and *Pyrameis atalanta*. Possibly a five per cent. solution of formalin might be effectual, but I have not yet experimented with it. There is no doubt but that the harder eggs, such as *Notolophus antiqua*, *Phalera bucephala*, &c., will keep almost any length of time without previous treatment, provided they are infertile. Whether an infertile egg has identically the same structure as a fertile one, is perhaps a matter for investigation. I have by me, at present, eggs of *Zephyrus quercûs* and *Nomiades semiargus*, which are as perfect as when I received them in 1900, and they have undergone no treatment whatever. It is probable that the Lycænid eggs generally will keep equally well if infertile. So far as my experience goes there is no known method for preserving the form and colour of the softer eggs, such as the Geometrids, although in the case of some Noctuids, *Heliophobus popularis*, *Agrotis agathina*, &c., in my possesssion, the larva has dried up, and the structure of the egg is as perfect as it was two or three years ago. When it is desired to study only the size, form, and surface structure cf an egg, I would particularly recommend mounting the empty shells. As good examples of these I may mention eggs of *Ennomos fuscantaria*, *E. tiliaria*, *Geometra vernaria*, and many other Geometrids. The external structure of the egg may be very well

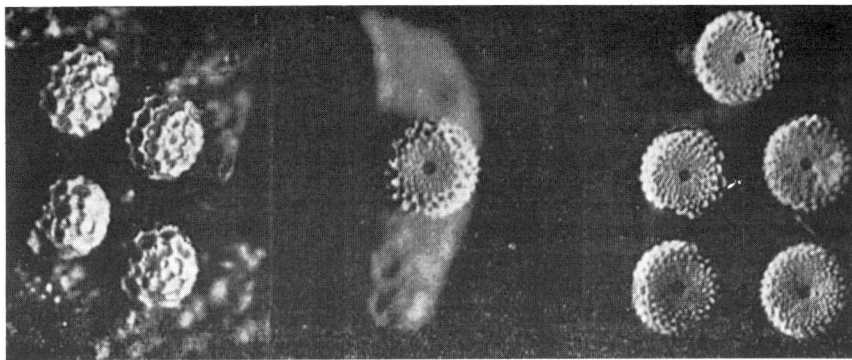

1. Chrysophanus phlæas.   2. Polyommatus corydon.   3. Polyommatus bellargus.

4. Nomiades semiargus.   5. Plebeius ægon.   6. Polyommatus icarus.

7. Cupido minima.   8. Cyaniris argiolus.   9. Callophrys rubi.

PLATE III.

EGGS OF LYCÆNID BUTTERFLIES.
(Photographed by F. Noad Clark.)

actical Hints, etc., 1905.

studied in the empty shells of these, as also the mode of emergence of the larva. They form very beautiful objects for the microscope, on account of their iridescent appearance."

MOUNTING OVA.—For mounting ova as permanent specimens, they should, in the case of the lighter coloured ones, be placed on a dark background. For this purpose make a disc of asphalt varnish on a glass slip, allow to dry and fix the eggs thereon with a thin film of gum. Make a cell either by ringing on successive layers of gold size with the turntable, or by using ready made cells of tin or vulcanite. The cell and contents may now be closed in the usual way with a thin glass cover, the edges of which are cemented with some suitable varnish, such as gold size or shellac. Eggs should be mounted in various positions, so as to present a lateral and an upright aspect, and whenever possible, should be shown *in situ* on their natural support or foodplant (Clark).

MEASUREMENTS OF EGGS.—Measurements may be made as in photo-micrography, or by means of the eye-piece micrometer, and a 2in. objective will be found to be the most suitable lens for general work, whilst for the study of the micropyle and finer details of structure a ¾in. or ½in. objective will be necessary (Clark).

PRESERVATION TO SHOW VARIOUS STAGES OF DEVELOPMENT.—To preserve the eggs so as to study the various stages of development in the embryo, the following methods have been suggested : (1) Distribute the eggs as follows in phials—one phial to be filled with carbolic acid, an egg put in, and the phial stoppered, the day after laying ; similarly each successive day, till the last phial contains the newly-hatched larva. (2) Kill one each day by heating in water at 80°C., then puncture the eggs with a fine needle, and stain with " Grenachar's borax carmine " or " Czochar's cochineal."

PHOTOGRAPHING EGGS.*—Lepidopterous ova being of considerable size from a microscopist's point of view, it follows that to obtain a satisfactorily sharp photograph of the entire ovum, a low power objective must be used. For this reason it is quite unnecessary to spend large sums in the purchase of expensive apparatus, such as would be required for high power photomicrography.

Anyone possessing a fairly good modern microscope and a " stand " camera, which racks out backwards, can with a very small amount of ingenuity fit up all the other apparatus he will require in order to obtain successful and valuable photographs.

Some of the best results which the writer has obtained, have been taken with a rectilinear photographic lens of 5in. equivalent focus, adapted to the body tube of a microscope in the place of the usual objective, and used without any eye-piece. This, however, necessitates the use of a special camera with a very long extension, 5ft. or more being required between the lens and the sensitive plate for an amplifi-cation of 10 diameters ; and the only reason for using a microscope at all is to obtain the necessary accuracy in focussing.

We will suppose we have an ordinary student's microscope, such as can be obtained from any of the leading makers for about £5, a 3in.

---

* We are indebted to the kindness of Mr. A. E. Tonge for this section.

microscopic objective, and a quarter plate camera with a backward extension, and that we wish to obtain photographs of lepidopterous ova magnified 20 diameters.

We must first obtain a baseboard to afford us a solid, flat, and smooth surface upon which to erect the completed apparatus. This will need to be at least three feet long by six inches wide, while a thickness of one inch will probably provide sufficient stability.

Extend the camera to its full length, and bore a hole for the camera screw exactly in the centre of the baseboard, and at such a distance from one end that the focussing screen is flush with the end of the baseboard ; screw the camera firmly down and then take the microscope, and, having lowered the body tube to a horizontal position, place it on the baseboard in order to ascertain the amount of adjustment required to bring the centre of the eye-piece into exact alignment with the centre of the lens flange on the front of the camera.

Should the microscope require raising, we must make a wooden bed of the necessary height, which can be firmly screwed down in the proper position, and on which we must fix two or three small wooden blocks to prevent the microscope from slipping when it has been placed in due alignment with the camera. If, however, it is the camera which is too low, we must apply the same principle by putting a block under the base, not forgetting to bore the necessary hole for the camera screw, and fix the wooden stops for the microscope direct on to the baseboard.

Having got over this difficulty we must next obtain a piece of black velvet, and make a light tight tube to be fixed to the inside of the camera front extending out through the lens flange, and long enough to be slipped over the draw tube of the microscope, where it may be held in place by an elastic band, the object aimed at being to exclude all light from entering the camera excepting that which passes through the microscope, and we must not omit in fixing it on to allow a little extra length for the play of the draw-tube when focussing.

Our apparatus being now ready for use, we remove the camera, place an object on the stage of the microscope, and, having illuminated it to our satisfaction, we look through the instrument and focus it. Now replace the camera, slip the velvet tube over the eyepiece of the microscope, and, putting the focussing cloth over our head, we rack up the camera until we obtain a satisfactory image on the focussing screen. We shall probably find that this is not enlarged quite as much as we desire, so we rack the camera out again, say a couple of inches, and then, still keeping our head under the focussing cloth, extend a " long arm " to the coarse adjustment of the microscope, and gently turn it towards the object until the image on the ground glass is sharp again. Having in this way adjusted matters to our satisfaction we have but to place the dark slide containing a sensitive plate in position, slip a piece of black card between the objective and the object, to act as a lens cap, draw the flap of the dark slide and make our exposure by removing the card for the necessary number of seconds, and replacing it before we close the slide.

Development may be proceeded with on any of the usual lines to which we may happen to be accustomed. The writer recommends pyro-soda as giving very excellent and uniform results.

The illumination of the object will probably cause us some trouble at first, as all the light we get through the camera is reflected from the object, owing to the ovum being an "opaque subject." This necessitates a fairly powerful illumination to enable us to see the image on the ground glass with sufficient clearness to focus it. Oil lamps are of very little use, as they throw out a great amount of heat, but incandescent gas, or better still, electric light, with a single bullseye condenser to concentrate the rays upon the object, will give very good results.

The writer uses a ·25 Ampière Nernz electric lamp about four inches away from the object, and, assuming this to give a light of 25 to 30 candle power, an exposure of from one to three minutes, according to the colour of the egg, will be found approximately correct, with such a plate as " Edwards' snap-shot isochromatic," for a magnification of 20 diameters.

Isochromatic plates should be used preferably to obtain the truest rendering possible in monochrome of the various colours presented by the object, and these should always be " backed " to avoid halation.

In mounting ova for use as photomicrographic objects, it will be found best to fix them, or the leaf, twig, etc., on which they may have been laid, firmly on glass slips, such as are used for microscopic slides, always bearing in mind that they must lie absolutely flat and at right angles to the axis of the microscope lenses if we wish to get our photographs as sharp as possible all over. Any strongly adhesive matter, such as "seccotine," will do for this, but must not, of course, be used so liberally as to show in the picture. A better plan for mounting loose ova is, perhaps, to mount them on gummed paper, of a colour which will, by contrast, show up the outline of the ova strongly. This can be done by cutting a piece of the paper about ¼in. square and laying it face upwards in the centre of the glass slip. Then cut another piece, say ½in. square, and punch out a round hole in the centre ¼in. in diameter, wet it, and press it down firmly over the first piece. The ova can then, under a hand lens, be placed in position on the gum with a fine camel-hair brush moistened between the lips, and by breathing gently upon them the gum will be softened sufficiently to hold them in place without injury.

The illuminant should invariably be placed to one side of the object, and a reflector or mirror used on the opposite side to reduce the shadows. By this means an appearance of rotundity will be given to the ova in the photograph, which would be lost if they were evenly illuminated from both sides.

In conclusion, always stick to one extension of camera for all ova photographed, as this will give uniformity in magnification unless a different lens is used, and facilitate comparison of size in differing species.

Ascertain your *exact* magnification by photographing a finely ruled scale in place of the ova and comparing the photo with the original scale.

If possible, mount at least one ovum in the field of view on its *side* as a guide to the shape and so that dimensions from two points of view may be obtainable.

Always pencil the name of the species on one corner of the plate as soon as possible after you have taken the photograph, as this will be quite readable after development and fixation, and will enable you to identify it at any future time in the event of your negatives getting "mixed."

DESCRIPTION OF EGGS.—The description of eggs has undergone a remarkable advance during the last few years. As illustrating the various phases of what authors have supposed to be suitable to the requirements of collectors of their own time, we may note the following :

1. *Goneptery.x rhamni*—
   (1) Stainton, 1857, *Manual*, i., p. 16. No description.
   (2) Newman, 1869, *British Butterflies*, p. 147. "The eggs are elongated and of a bright yellow colour."
   (3) Tutt, 1896, *British Butterflies*, p. 264. "Somewhat ninepin-shaped, wider in the middle than at the base, but tapering off towards the apex. The shell is pale, shining, silvery-green, changing afterwards to yellow, whilst twelve main longitudinal ribs have other less developed ones placed between them. Just before hatching, the egg of *G. rhamni* turns to a dull, brownish-grey colour, whilst the markings on the embryo inside appear, under a one-sixth lens, as a black spiral, running up the egg from the base. The base of the egg is distinctly flanged outside, owing to the weight of the egg, in its fluid condition, pressing upon the base when newly laid."

2. *Polyommatus icarus :*
   (1) Stainton, 1857, *Manual*, i., p. 61. No description.
   (2) Newman, 1869, *British Butterflies*, p. 129. "The eggs are laid on restharrow."
   (3) Tutt, 1896, *British Butterflies*, p. 177. "The egg is circular, rather flat, of a dull, greenish-white ground-colour, with shiny white raised reticulations. It has the ordinary *Echinus*-like appearance, and is of a delicate green colour, covered with small white elevations, joined by slender white reticulations. The base of the egg is more finely reticulated. The upper surface of the egg is depressed, leading gradually down from the convex outer margin to the micropyle, which is placed at the bottom of the depression ; the sloping sides of the depression reticulated in the same way as the sides, but the micropyle, which is very distinct, is comparatively smooth and composed of smaller cells. It differs from the egg of *P. bellargus* (1) in having the sides of the depression reticulated like the outside ; (2) in the depression gradually sloping to the micropyle, which is at the lowest point of the depression, and not raised again in the centre to form a platform, as it were, to contain the micropyle, as is the case in the egg of *P. bellargus*."

These illustrate fairly the progress that had been made up to 10 years ago in our knowledge of, and our ideas of what was required in descriptions of, the eggs of butterflies. Our own descriptions are, of course, wofully incomplete — without measurements and other important details—but they show how much room there is for good workers in this field. Following this up, we wished to illustrate the same advance in the description of the eggs of moths, but Stainton's moths, like his butterflies, appear to have laid no eggs, and Newman's moths must have belonged to the same category ; we fail to find a single egg described in Stainton's *Manual* or Newman's *British Moths*. It is to be assumed, therefore, that the description of the eggs of moths for scientific purposes is of quite recent growth, and we can only give two or three examples from those published in the *Natural History of the British Lepidoptera*, which, although by no means complete, will act as guides as to the minimum required :—

    1. *Anthrocera trifolii* (*Nat. Hist. Brit. Lep.*, i., p. 491).—Attached by its long side to the object on which it is laid ; also slightly attached to each other. Of a bright yellow colour ; the shell shiny ; roughly cylindrical in shape, the outline being somewhat oval, with a long oval depression on the upper side ; roughly, the length : breadth : height : : 10 : 7 : 6, the actual measurements being about

1mm. in length, ·7mm. in width. The two ends vary in different eggs; in some the micropylar end is broader than its nadir, in others there is practically no difference between them. The micropylar end is, however, flattened, and, in the centre of this, is a distinct crater, somewhat shallow, with a very simple stellate structure at the bottom of it. The egg is finely, and very faintly, ribbed longitudinally, the space between the ribs being reticulated irregularly. These ribs look almost like parallel striations, and 10 were counted on the upper surface of one egg, some of which crossed the central depression. At the micropylar end, this irregular reticulation becomes roughly hexagonal, or the longitudinal ribs fail; they are also absent at the opposite end, where, however, the irregular reticulation is less distinctly polygonal in form. There are some depressions in the egg, caused apparently by pressure, and the whole character of the egg suggests an exceedingly delicate structure. [Eggs laid June 8th, 1897; described June 10th, 1897.]

2. *Dimorpha versicolora* (*Nat. Hist. Brit. Lep.*, iii., p. 242).—When first laid yellow, soon assuming a dirty tint, the colour getting deeper till it is of a deep purple or red-brown hue, with a strong tint of bluish. Before hatching, air penetrates between the larva and the shell, the colour being then a dirty white, *i.e.*, a white semi-transparent shell over a black (brown-haired) larva. The eggs are plump and round when laid, but soon, by desiccation, develop a deep dent on one side, and this may assume really large proportions without detriment to the final exclusion of a healthy larva. The egg is nearly cylindrical, with hemispherical ends; it does, however, deviate a little from a strictly cylindrical form. The length is barely 2·00mm., and the greatest diameter 1·15mm., the least being 1·00mm. This is the extent of the flattening in a plump egg, but the flattening from shrinkage soon makes this flattening much more pronounced. The micropyle is represented by a minute cell, surrounded by a rosette of eight rather elongated cells, the whole rosette being about 0·03mm. across, and outside this is an area about 0·5mm. in diameter, with a well-marked network of cells, apparently separated by slightly raised lines; these cells are of very irregular form, usually lengthened in a direction radial from the micropyle, often four-sided, twice as long as broad, and about 0·025mm. in length, but longer as they recede from the micropyle; outside this area they become less distinct, but can often be made out, although they usually present the appearance of a minutely undulated surface. There is another appearance that is very marked, but of the exact nature of which it is more difficult to be sure. This may be described as beginning outside a radius of ·25mm. from the micropyle, and affecting more or less the rest of the egg, and presenting the appearance of a number of fine spicular hairs arising at a considerable number of the points of intersection of the mesh of network. These are very fine, and have a silvery sheen. Sometimes they appear to rise from a globular base. In some aspects, however, they look like small raised splashes, or marks of exclamation (!), laid flatly on the egg-surface with a minute free point standing off. They probably are portions of the network specially developed, and are not free spicules, except, perchance, at their extreme ends. The eggshell being almost transparent, and being on a dark ground, its inequalities and markings probably produce these puzzling effects in various lights. [Described April, 1901.]

3. *Sesia stellatarum* (*Nat. Hist. Brit. Lep.*, iv., p. 7).—Green in colour, of a somewhat darker green than the egg of *Theretra porcellus*, and more strongly sculptured; almost a sphere in form, but with a longer (micropylar) axis, and a circular outline at right angles to this; the length is almost exactly 1mm., and the transverse diameter 0·9mm. Some eggs are a little smaller than this, but the diameters have the same ratio. The surface sculpturing is in the form of a very shallow set of pits, the lines of the netting being little raised above the general surface, the reticulation being of 5- and 6-sided pits of a diameter of 0·02mm. A well-marked rosette of cells around micropyle; surface smooth and pearly, but not highly varnished. Some eggs have a fairly large, irregularly oval, hollow depression on the upper surface (opposite point of attachment); others without, and apparently quite filled with contents. Micropylar end somewhat flattened. Colour at first bright pea-green, changing to yellowish with green areas as the embryo matures. Before hatching, the young larva may be very distinctly seen through the transparent shell, and the structure of the young larva and its striking bifid hairs may be clearly distinguished. [Described July 10th, 1901.]

These are simply samples of comparatively recently-made descriptions, and suggest roughly the lines on which collectors should work in making their descriptions. No doubt in a few years these will be

considered as inadequate as are those of the earlier authors to-day.
To those who wish to know more of the eggs and the egg-stage of
lepidoptera we would urge the study of the chapters relating to the
egg and embryology of lepidoptera in *A Natural History of the British
Lepidoptera*, vol. i., pp. 1-30, besides which every student should
attempt to thoroughly grasp the general characters of the egg of each
superfamily by a careful study of the plates published in this part of
Practical Hints.

CHAPTER IV.

Larvæ and larval stage of Lepidoptera.

General structure of larva.—The larvæ of lepidoptera have, for
the last couple of centuries, at least, been of the greatest interest to
many naturalists. Their varied habits, the changes that they undergo
from the time they leave the egg until they become pupæ, the widely
different appearances that they themselves present, have been an un-
failing source of interest and pleasure to all those naturalists who have
interested themselves in the study of the order we are considering.
These and many other details will have to be considered by the lepidopterist who wishes to undertake the collecting and study of larvæ.
The merest beginner will have observed that a larva is roughly cylindrical in shape, that its head is very distinct from the remainder of the
body, that the latter is constricted at intervals along its length, so as to
divide the whole body into thirteen segments, of which the first three
belong to the thorax, and the last ten to the abdomen. The thoracic
segments are known as the prothorax, mesothorax, and metathorax,
and these bear the true legs; the last of the abdominal segments is
known as the anal segment; the 3rd, 4th, 5th, 6th, and 10th abdominal segments (or some of them) bear prolegs. The spiracles are
situated on the prothorax and on the 1st-8th abdominal segments, and
their position is often of considerable importance in a description. The
segments themselves again are subdivided, more or less distinctly, into
what are known as subsegments, and the number of these, usually
different on the thoracic and abdominal segments, and particularly
variable on the prothorax and 8th, 9th, and 10th abdominal segments,
is also often of great importance. Of still greater importance, however,
are the position, number, and character of the primary tubercles that
bear setæ or hairs, and are found on certain well-defined areas of the
body. The peculiarities of these have proved of great value to modern
systematists who have applied a special nomenclature to them, *viz.*,
tubercles i (=anterior trapezoidals, the two front tubercles on the
dorsum or back of each abdominal segment); tubercles ii (=posterior
trapezoidals, the two hinder tubercles on the dorsum of each abdominal
segment); tubercles iii (=supraspiracular tubercles, the two tubercles
of which one is above each of the spiracles on either side of each segment); tubercles iv and v (=subspiracular tubercles, two tubercles on
either side of each segment, which, in their normal position, are below
the spiracles, but are very liable to movement, in some superfamilies,

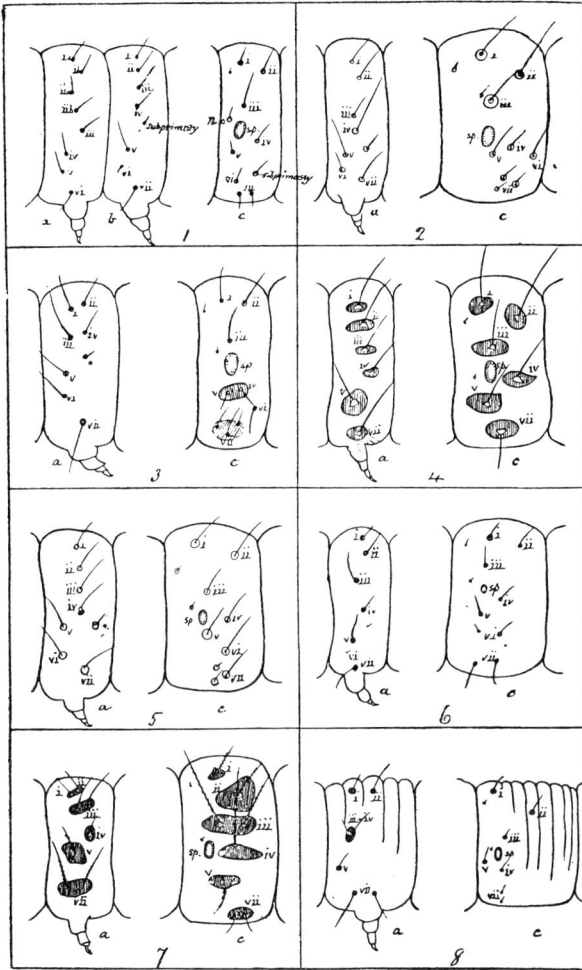

PLATE V.

DIAGRAMMATIC REPRESENTATION OF ARRANGEMENT OF
LARVAL TUBERCLES AND TUBERCULAR SETÆ.

Practical Hints, etc., 1905.

e.g., Sphingides, v, the front one, is moved up in front of the spiracle and becomes prespiracular, iv remaining as a subspiracular; in other superfamilies, *e.g.*, Noctuides, &c., iv, the hinder one, is moved up behind the spiracle, and so becomes postspiracular, whilst v remains subspiracular); tubercles vi (=lateral tubercles, a tubercle on either side of each segment between iv and v, and vii); tubercles vii (=marginal tubercles, one on either side of each segment, on the upper outside margin of the legs and prolegs).   These tubercles may bear only a single primary hair, or may consist of a flat plate bearing two or three other hairs, or they may be slightly raised and bear besides this primary hair other shorter secondary hairs, in which case the tubercle becomes modified into a wart.   There is usually a marked difference between the character, position, &c., of the tubercles on the thoracic and abdominal segments, and the most modified segments in this respect are the prothorax and the 8th, 9th, and 10th abdominal segments.   It frequently happens that there is considerable difference also between the character of the tubercles in the newly-hatched larva, and the character of the same tubercles after each successive moult, and no description of a larva can have real scientific value that does not take these facts into account.   This may at first seem a little complicated and difficult, but in reality, is not so much so as may, at first sight, appear.   At any rate, modern science demands that these details should be known, and any lepidopterist who aspires to push forward our knowledge of the larval stages of the lepidoptera must attempt his descriptions on the lines that biologists have found to be useful, otherwise their work will be useless and their time wasted. [Those who wish to study the structure of the lepidopterous larva in detail, we would refer to *Nat. Hist. of Brit. Lepidoptera*, vol. i., pp. 31-102.]

HINTS FOR DESCRIBING LARVÆ.—In describing a larva then—besides the preliminary useful statements as to colour and markings, the detailed measurements and appearance of head, thoracic, and abdominal segments, the position of a lateral flange, swellings, &c., the number and position of the prolegs, facts which used to be considered as ample even in the best descriptions, and which render such descriptions of no real practical use whatever to the present-day biologist in his researches on the relationship and classification of the lepidoptera—the following details should at least be included for each stage, *viz.* :—

(1) The number of subsegments to each segment, and the variation in their character on different segments.
(2) The position of the spiracles with regard to these subsegments.
(3) The character of the hooks on the prolegs; how arranged, whether in complete ring, or only on a longitudinal flange, &c.
(4) The position of the primary tubercles i-vii, and their variation in position on the different segments.
(5) The structure of the primary tubercles i-vii, and the position of the primary seta (or hair) with regard to secondary setæ (if any of the latter are present), and the variation in structure according to the different segments on which they are placed.
(6) The presence of any secondary tubercles other than those already noted as primary i-vii.
(7) The character of the skin, and the presence or absence of secondary hairs not connected with definite tubercular structures.

If these characters be noted in the newly-hatched larva, and in every stadium (or stage) after each successive moult, we shall have not only a superficial account of each instar, *i.e.*, the plumage of each stage, but a collection of scientific detail of the utmost value to our scientific biologists. It behoves every collector, therefore, who desires to push forward the scientific side of our study to perfect himself in the art of describing larvæ in the most useful manner scientifically. There is scarcely need to add that information relating to the escape of the young larva from the egg, details of the duration of each moulting period, observations on the actual process of moulting, habits of the larva which may become much changed as it gets older, general observations as to suitability of plumage to help in its protection, are exceedingly valuable, and should never be overlooked or observed without mention.

AID TO THE DESCRIPTION OF LARVÆ.—It has often been urged by those breeders of lepidoptera, who have been criticised for their poverty of description, that there is no easily accessible text-book from which the details relating to the tubercles here referred to may be studied, in order that comparison may be made with the larva about to be described. In order to meet this requirement we have obtained from Mr. Bacot drawings illustrating the position of the tubercles in larvæ belonging to widely different families. These we have reproduced in two plates published herewith, and reference to these should enable any student who is interested in the matter to work up his subject. We do not suggest that these cover all the forms that will be met with, but the study of these should enable the collector to deal with most of the larvæ he may meet.

IMPORTANCE OF DESCRIPTION OF NEWLY-HATCHED LARVA.—TREATMENT OF EGGS NEAR HATCHING STAGE.—It has already been stated that a detailed description of the newly-hatched larva is of the greatest importance, and that it is not too much to say that a description of this stadium is frequently of more importance (as an aid to accurate classification) than that of any later stage. This being so it is clear that the collector must obtain the greater number of his newly-hatched larvæ from eggs in his possession. In order not to overlook eggs near the hatching-stage and thus lose the larvæ when they hatch, it is best to keep them in a glass tube, glass-topped box, or similar receptacle, where they can be frequently and easily examined. It will also be found advisable oftentimes to rear them in glass tubes whilst they are of small size, taking care, however, that the escaping moisture from the foodplant does not condense upon the inside of the glass and so drown the living atoms of which one is taking so much care. At the same time the young larvæ are, in tubes, kept confined pretty closely to their foodplant, from which some species have, in confinement, a tendency to wander. Most larvæ, however, soon grow beyond the limits of a glass tube, although, for isolated examples, which are frequently required for observation, this mode of feeding can still be maintained with success.

TREATMENT OF NEWLY-HATCHED LARVÆ.—DATA WANTED.—It is well, as soon as a batch of eggs has hatched, to carefully count the number of larvæ. Nothing is easier than to overlook or lose newly-hatched caterpillars, and the only check one has upon one's self is to count them whenever any removal of food, &c., takes place, when any missing

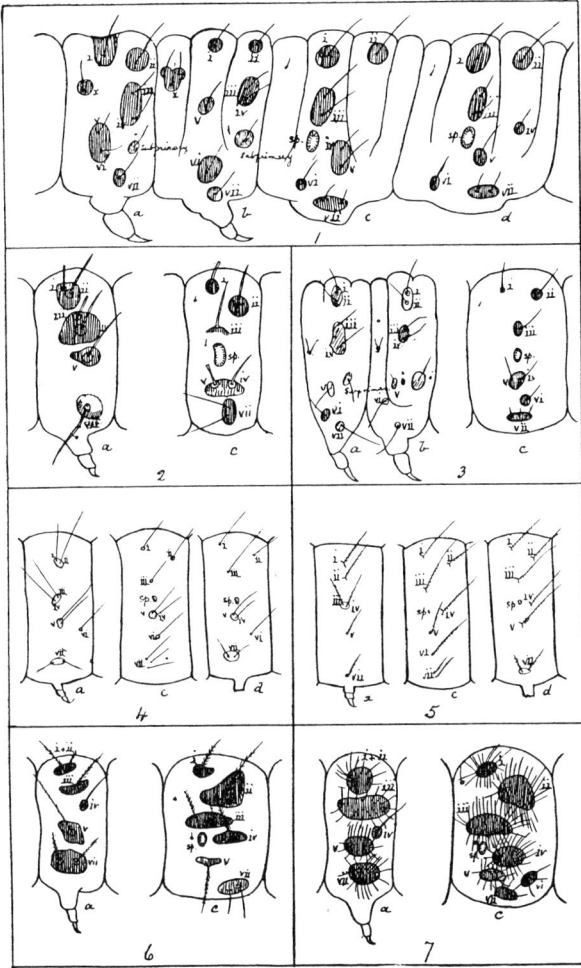

PLATE VI.

DIAGRAMMATIC REPRESENTATION OF ARRANGEMENT OF
LARVAL TUBERCLES AND TUBERCULAR SETÆ.

Practical Hints, etc., 1905.

ones are readily detected. To move such tiny larvæ the edge of a sheet of paper is frequently used, but much better than any other means is a camel-hair brush, a supply of which we look upon as a positive necessity. The date of hatching on a label attached to the tube is very useful as a check against entries made in note-books later concerning the larvæ. It may be added that at least one newly-hatched larva of a batch ought to be put into a tube in glycerine, and properly labelled for future reference and study, or, better still, in a weak solution (2 or 3 per cent.) of formalin.

EXAMINATION OF YOUNG LARVÆ.—Some young larvæ are exceedingly active and refuse to submit to examination under a microscope or lens without an anæsthetic or some other quieting influence. Subjected to chloroform for a few seconds a larva is usually quiet long enough to make the necessary examination, recovering soon, however, and none the worse apparently for the dosing it has received. A much simpler method, though, is to draw a film of clear water across a glass slide, in which the larva quickly gets helpless, and remains quiet enough under observation till the moisture has dried up. After a very little experience, there is no doubt that each individual collector will be able to devise methods of his own which will far surpass the rough-and-ready ones we have always adopted, and which we suggest here simply as an aid to the beginner and not as anything of value to the trained microscopist.

COLLECTING LARVÆ. — APPARATUS. — FEEDING-HABITS OF LARVÆ. — Apart, however, from obtaining larvæ by rearing from the egg, the collector is able to find larvæ, as our hints abundantly show, in a variety of ways, e.g., by searching, beating, sweeping, &c., for which operations he will require a supply of simple apparatus which is, for the most part, inexpensive, and can mostly be made at home. Among the most useful of the things needed we would suggest—a knife, metal glass-topped boxes, small tins of various sizes, a few perforated zinc boxes, a good supply of small linen or calico bags, and a larger bag. The use of the knife is obvious, and it is frequently required. The metal glass-topped boxes or small tins are exceedingly valuable for small larvæ, being airtight; larger larvæ will, however, sweat in them and undermine their constitution (even if they are not killed outright); the perforated zinc boxes prevent this untoward condition arising. The small canvas or calico bags are much more suitable for carrying such larvæ, especially if a small portion of the foodplant be included to keep the bag extended. The large bag is needed to carry a supply of the food-plant. Chip-boxes are an abomination, one always loses their best things if one uses them, they either fit badly, break, or get damp and unroll.

INDICATIONS OF PRESENCE OF LARVÆ IN FIELD.—As for the indications of larvæ when one is searching, this has already been so fully dealt with in the various hints that repetition would be largely waste of time. Suffice it to say that—eaten leaves (especially if cleanly eaten at edges), folded leaves, leaves spun together, leaves spun to a branch, rolled leaves, flowers and flower-buds spun together, leaves spun to flowering-shoot, are all, of course, primary indications of the near presence of leaf-eating larvæ. Similarly, frass pushed out from buds and seedpods, or extruded from stems, twigs, &c., the sickly appearance of plants, swellings on stems, &c., are the most prominent signs of internal-feeding larvæ. The presence of frass is also a sure sign, and as valuable an indication in exposed places where the large Sphingids feed, as in fields, lanes, and woods, where the lower leaves are often covered with the frass of larvæ feeding on the leaves of the upper branches. The veriest beginner has seen the frass-pellets on the herbage under an oak-tree when the tree above has been almost defoliated by an army of larvæ, of which *Tortrix viridana*, *Phigalia pedaria*, and *Hybernia defoliaria* are amongst the chief offenders. Some larvæ certainly feed exposed, they revel in the sunshine, an entomologist, without searching, is sure to see them, if he only finds himself in their locality. Such are the common Sphingids—*Smerinthus ocellata, Amorpha populi*; the Arctiids—*Arctia caia*: the Lachneids—*Macrothylacia rubi, Cosmotriche potatoria*, and so on. But the greater number of larvæ adopt certain methods of concealment, the habits of no two species probably being absolutely identical, and then these habits have to be learned and search made, and it must be noted that there can be no general law observed even for the species of the same superfamily in most instances. We might for example try to formulate general rules, but they would be open to many exceptions, *e.g.* :—

(1) SPHINGIDES.—Larvæ feed openly, rest exposed on foodplant, holding firmly to the foodplant when not eating—*Amorpha populi, Mimas tiliae, Smerinthus ocellata, Sphinx ligustri, Agrius convolvuli, Manduca atropos*, &c. (yet one has to search low down on the ground among the herbage, or even in the cracks just below the surface, for the larvæ of many species—*Theretra porcellus, Sesia stellatarum*, &c.). [The remark made by Knaggs and others that the larvæ of *Agrius convolvuli* and *Manduca atropos* hide away under sods is entirely inaccurate.]

(2) ARCTIIDES.—Larvæ feed openly in daytime, rest hidden on undersides of leaves—*Arctia caia, Spilosoma menthastri, Phragmatobia fuliginosa, Callimorpha dominula*, &c. (yet many Arctiid larvæ feed entirely by night, and hide so completely that they can only be obtained by the closest searching even when their habits are known).

(3) GEOMETRIDES.—Larvæ feed usually by night, remain stick-like, fixed, and rigid, standing out from a twig or resting along a stem, branch, twig, or leaf (yet many make hiding-places of leaves spun together—*Melanippe hastata, Hypsipetes elutata*; or conceal themselves in flowers—*Eupithecia pulchellata*; or hide in seedpods—*Eupithecia venosata, Emmelesia unifasciata*, &c.).

(4) TORTRICIDES.—Larvæ roll up leaves, within which they eat and live, wriggle out at one end when disturbed (yet the larva of *Penthina gentiana* lives on the pith of teasel heads, *Eupoecilia roseana* on the seeds of teasel, *Carpocapsa pomonana* bores into apples, *C. funebrana* into plums, and *C. splendana* and *C. juliana* into acorns and chestnuts, whilst *Endopisa pisana* feeds in pea-pods on peas, etc.).

It is quite true that the Nepticulids live in leaves, mining between the upper and under epidermis, that the Lithocolletids blotch leaves, living also on the cellular tissue, that Elachistids largely mine grass leaves, and so on, yet the great variation of habit in many closely-allied species leads us to suggest that the individual hints relating to individual species want to be followed up, rather than that the young collector should be set to consider general statements open to an endless amount of criticism due to the large number of exceptions.

LOCALITIES.—Searching by night, as well as day, has already been repeatedly recommended, beating has also been fully explained, as also has sweeping, and that many species may be obtained at night that are quite ungetatable in the day, is too well-known, even to beginners, to make it worth while to deal at length with it, still, some knowledge is wanted in the direction of what localities pay best at particular times,

Fig I. Wire gauze meat safe. Fig 2. Wire frame.

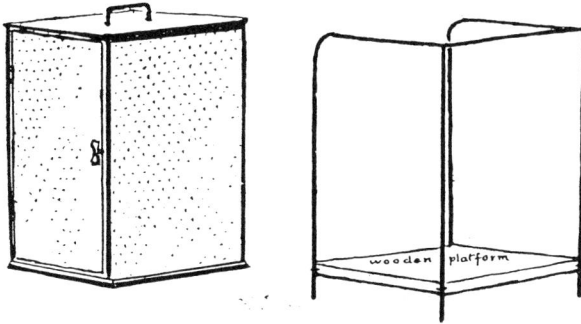

Fig 3. Tin baking dish to hold water.

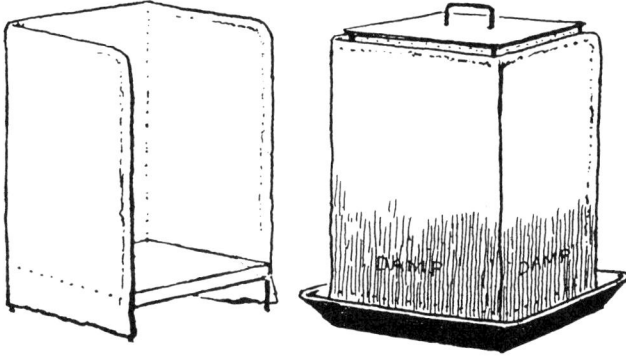

Fig. 4. Wire frame covered. Fig. 5. Breeding cage complete.

PLATE IV.

DOLLMAN'S BREEDING-CAGE.

Practical Hints, etc., 1905.

such as hedgesides, woodsides, and similar places, on mild evenings in February and March ; searching and beating sallow-bushes, on which the catkins are ripening or falling in April ; sweeping on moorlands and heathlands in May ; general larva-beating by the sides of woodland rides, or on the outskirts of woods in early June, and so on.

BREEDING-CAGES.—We have already in the previous parts (that can readily be discovered by reference to the index in this part) dealt at length with the subject of breeding-cages. The essentials are (1) To keep the food fresh. (2) To prevent an accumulation of moisture. (3) To prevent the formation of mildew, &c. Air and proper ventilation are, of course, prime necessities with larger larvæ, but with small ones they matter little. From those who breed insects in numbers as one breeds rabbits or fowls, giving them an abundance of food under healthy conditions, and caring nothing except that they live and produce imagines, to the scientist with his few examples, carefully arranged and placed readily at hand for continual observation and reference, each and every lepidopterist adopts (or uses) some form or other of breeding-cage in which to keep his larvæ. These are of all sorts and sizes, expensive, and inexpensive, and vary immensely according to the tastes and personal predilections of the individual. Our own are of the simplest possible description, and have always consisted of readily available materials, e.g., for small larvæ—metal glass-topped boxes, glass tubes, glasses with a piece of glass on top, jam-jars with a glass cover, bell-glass and plate, and similar objects, form the staple, but we confess to a weakness for the "Young" breeding cage (described pt. 2, pp. 21, 82) and to a rooted objection to any into which wood enters as a large part of the construction. Breeding-cages that do not offer ready facilities for observation are of very little use to the scientific collector and breeder of lepidoptera, and, except for breeding larvæ in quantity, for hybernating larvæ in the open on their foodplant, and for similar special purposes, "sleeving" cannot be recommended as of any real value to the scientific lepidopterist, owing to the difficulties of observation, &c. We have, however, considered all the best modes of breeding-cages, both on a large and small scale, and details, sufficient for the most exacting, will possibly be found by reference to the index. One breeding-cage, not hitherto noted in *Practical Hints*, is Dollman's (described *Ent. Rec.*, xv., p. 7) ; of this we give a sketch (*see* plate, figs. 1-5). Referring to the plate it will be noted that fig. 1 is an ordinary wire gauze meat-safe with tin back, top and bottom ; the front and sides being covered with gauze. The hook provided inside can be removed. This item costs about 6s. for a safe about 24in. in height. The back, top, and inside should be rubbed with earth, or have some fabric stuck to them to facilitate the hold of the insects when climbing to develop their wings. Fig. 2 represents a framework of stout wire rod about the substance of a slate-pencil. This is made 1in. larger in width and depth than the cage, and the legs should extend 2in. below the wooden platform. This platform on which the cage is to stand, can be firmly held in position by wire staples round the legs, being well hammered home. Fig. 3 is a common tin baking-dish, and should be large enough for the four legs of the frame to stand comfortably upon its flat bottom. Fig. 4 shows the frame covered on the front and sides. The covering may be either old flannel or a double thickness of stout serge, or any material which will readily absorb, and hold, water. The material employed should be allowed to come to the full length of the legs, and to hang below the platform, so that it will rest upon the bottom of the tin dish. Fig. 5 gives the apparatus in position. Fill the baking-dish two-thirds full of water, and stand the covered frame in it, seeing that the bottom edge of its coverings is well down in the water to the bottom of the dish. The breeding-cage can now be placed on the platform, and should have an inch of space left all round between it and the covering of the frame. The strong recommendation which this apparatus has is this—the water absorbed by the flannel to about one-third of its height keeps the atmosphere damp in and round the cage, while the open space allowed by the extra inch permits the air to circulate freely and freshly. The result of this is, that, while the atmosphere of the cage is damp, there is no possibility of producing mildew. There is no wetting of earth, moss, sand, or whatever material the pupæ may be stored in or upon, and the conditions are possibly as near an approach to natural ones as can be arrived at. The percentage of cripples emerging is almost nil, and there is no anxiety as to whether this or that species wants damping, as one glance at the flannel covering will tell one whether the dish requires more water or not. The entire concern being very compact it can be easily removed from one room to another, according as the question of temperature has to be considered. Reference to the index, published with this part, will be sufficient for the reader to find readily all details connected with the manipulation, breeding and keeping of larvæ, the structure of larva-houses, &c.

FOODPLANTS.—One of the most amazing things is the general ignorance displayed even by some experienced lepidopterists of the range of foodplants that are utilised by some species. Our own personal experience is usually so limited, and we are so apt to conclude that what we have learned is all that there is to learn, that it is a revelation to us to find that what we know is only a small portion of the whole. We had been collecting some ten years before we learned that the larva of *Cosmotriche potatoria* was not entirely a grass feeder, that the larva of *Eutricha quercifolia* was not confined to sloe, the larva of *Manduca atropos* to potato, and that the larva of *Eumorpha elpenor* lived on other plants than bedstraw. These were our early experiences. We knew where and how to find the larvæ of these species on these respective foodplants, but that others found them in entirely different localities on widely different plants, did not dawn on us till we began to look into the matter. Now, when one reads the various aids and hints to collectors about substitute foods, one is apt to wonder whether the substitute is not just as much the ordinary foodplant as the one for which it may be substituted, *e.g.*, one authority on this matter gives the following :—

Smerinthus ocellatus.—Foodplant : Apple.   Substitute foodplant : Poplar.
Choerocampa elpenor.—   ,,   Bedstraw.   ,,   Epilobium.
Saturnia carpini (pavonia).—Foodplant : Bramble.   Substitute foodplant : Heath.
Zeuzera aesculi (pyrina).—Foodplant : Apple.   Substitute foodplant : Ash.

and so on. As an illustration of the extent to which species can, and do, adapt themselves to various foodplants in different districts, &c., the following list dealing only with the few British species of four superfamilies, and taken from our own work *The Natural History of the British Lepidoptera*, may prove instructive :—

LACHNEIDES.

PŒCILOCAMPA POPULI : Poplar, aspen (*Populus*), hazel (*Corylus*), apple, crab-apple, pear (*Pyrus*), whitethorn (*Crataegus*), plum, cherry (*Prunus*), oak (*Quercus*), lime (*Tilia*), birch (*Betula*), alder (*Alnus*), wild rose (*Rosa*), sycamore, maple (*Acer*), elm (*Ulmus*), ash (*Fraxinus*), willow, sallow (*Salix*), beech (*Fagus*), horsechestnut (*Esculus*), larch (*Larix*), &c.

TRICHIURA CRATÆGI : Apple, pear, crab-apple (*Pyrus*), whitethorn (*Crataegus*), sloe, plum, cherry (*Prunus*), willow, sallow (*Salix*), oak (*Quercus*), birch (*Betula*), poplar, aspen (*Populus*), hazel (*Corylus*), ling (*Calluna*), beech (*Fagus*), bramble (*Rubus*), alder (*Alnus*), Cotoneaster, *Escallonia serrata*.

LACHNEIS LANESTRIS : Lime (*Tilia*), blackthorn (*Prunus*), sallow, dwarf-sallow, willow (*Salix*), whitethorn (*Crataegus*), bramble (*Rubus*), plum, cherry, apricot (*Prunus*), elm (*Ulmus*), ling (*Calluna*), birch (*Betula*), alder (*Alnus*), hazel (*Corylus*), pear (*Pyrus*), mountain-ash (*Sorbus*), buckthorn (*Hippophaë*), *Vaccinium uliginosum*, &c.

MALACOSOMA CASTRENSIS : Coarse salt-marsh grasses, sea-lavender (*Statice limonium*), southernwood, mugworts (*Artemisia maritima*, *A. campestris*), plantain (*Plantago maritima*, *P. lanceolata*), *Armeria maritima*, *Silene maritima*, *Atriplex portulacoides*, *A. littoralis*, *Suaeda maritima*, *Daucus carota*, *Centaurea jacea*, *Euphorbia esula*, *E. cyparissias*, *Hieracium pilosella*, *Athamanta oreoselinum*, *Helianthemum vulgare*, *Inula crithmoides*, *Erodium*, *Alchemilla*, *Geranium*, Campanula, rose (*Rosa*), birch (*Betula*), chrysanthemum, knotgrass (*Polygonum*), cherry, plum, blackthorn (*Prunus*), heath (*Erica*), ling (*Calluna*), dandelion (*Taraxacum*), apple, pear (*Pyrus*), poplar (*Populus*), oak (*Quercus*), sallow (*Salix*).

MALACOSOMA NEUSTRIA : Fruit-trees—pear, apple, crab-apple (*Pyrus*), plum (*Prunus*), almond (*Amygdalus*), cherry, apricot (*Prunus*), raspberry (*Rubus*), currant (*Ribes*), quince (*Pyrus*), &c.—rose (*Rosa*), osier, willow (*Salix*), birch (*Betula*), elm (*Ulmus*), bramble (*Rubus*), oak (*Quercus*), sloe (*Prunus*), laurel (*Laurus*), poplar (*Populus*), hawthorn (*Crataegus*), hazel (*Corylus*), beech (*Fagus*), maple (*Acer*), privet (*Ligustrum*), white and black alder (*Alnus*), juniper (*Juniperus*), &c.

PACHYGASTRIA TRIFOLII : Various species of grass—star-grass, etc., elm (*Ulmus*), hornbeam (*Carpinus*), bramble (*Rubus*), Artemisia, Spartium, Genista, Ononis spinosa, *Cytisus laburnum*, *Retama monosperma*, *Anthyllis vulneraria*, *Lotus corniculatus*, *Trifolium pratense*, *Medicago falcata*, *M. lupulina*, *Genista cinerea*, *Ornithopus perpusillus*, Ulex, *Calluna vulgaris*, *Plantago minor*, hawthorn (*Crataegus*), sallow, willow (*Salix*), oak (*Quercus*), blackthorn, plum (*Prunus*), walnut (*Juglans*), *Statice armeria*, beech (*Fagus*), ash (*Fraxinus*), poplar (*Populus*), raspberry (*Rubus*), sea-buckthorn (*Hippophaë rhamnoides*), Virginia creeper (*Ampelopsis*).

LASIOCAMPA QUERCÛS : Oak (*Quercus*), birch (*Betula*), sallow, dwarf-sallow (*Salix*), Erica, *Calluna vulgaris*, bramble (*Rubus*), plum, blackthorn (*Prunus*), bird-cherry (*Prunus padus*), whitethorn (*Crataegus*), maple (*Acer*), *Rosa canina*, sand-rose (*Rosa*), *Betula alba*, *Cornus sanguinea*, pear (*Pyrus*), hazel (*Corylus*), gorse (*Ulex*), broom (*Sarothamnus*), *Cytisus hirsutus*, *Rubus saxatilis*, Portugal laurel (*Cerasus lusitanicus*), ivy (*Hedera*), meadow-sweet (*Spiraea ulmaria*), ash (*Fraxinus*), elm (*Ulmus*), sea-buckthorn (*Hippophaë rhamnoides*), apple (*Pyrus*), bilberry (*Vaccinium myrtillus*), *V. uliginosum*, strawberry-tree (*Arbutus*), pine (*Pinus*), aspen (*Populus*), guelder-rose (*Viburnum*), mountain-ash (*Sorbus*), *Populus alnus*, strawberry (*Fragaria*), privet (*Ligustrum*), bog-myrtle (*Myrica gale*).

MACROTHYLACIA RUBI : Grasses, *Poa aquatica*, rush (*Juncus*), sedge (*Carex*), dandelion (*Taraxacum officinale*), *Potentilla reptans*, heath (*Erica*), heather (*Calluna vulgaris*), beech (*Fagus*), oak (*Quercus*), *Rosa spinosissima*, *Geranium sanguineum*, willow, sallow, dwarf-sallow, white osier (*Salix*), raspberry (*Rubus*), bramble (*Rubus*), *R. caesius*, rose (*Rosa*), clover (*Trifolium*), *T. repens*, bird's-foot trefoil (*Lotus corniculatus*), yellow vetchling (*Lathyrus pratensis*), birch (*Betula alba*), alder (*Alnus glutinosa*), ash (*Fraxinus elatior*), knotgrass (*Polygonum aviculare*), plum (*Prunus*), lesser burnet (*Sanguisorba*), hazel (*Corylus*), whortleberry (*Vaccinium vitis-idaea*), stork's bill (*Erodium cicutarium*), strawberry (*Fragaria*).

COSMOTRICHE POTATORIA : Grasses, rushes, striped riband-grass and reeds (*Phragmites communis*), *Digraphis arundinacea*, woodrush (*Luzula*), grasses (*Bromus sterilis*, *Alopecurus agrestis*, *A. pratensis*, *Holcus lanatus*), sedges (*Carex paniculata*, *C. riparia*, *C. caespitosa*), couch grass (*Triticum repens*), heather (*Erica*), bilberry (*Vaccinium myrtillus*), bramble (*Rubus*), dandelion (*Taraxacum officinale*), &c.

GASTROPACHA ILICIFOLIA : Willow (*Salix capraea*), poplar (*Populus tremula*), birch (*Betula alba*), hazel (*Corylus mandschurica*), bilberry (*Vaccinium myrtillus*), laburnum (*Cytisus*), apple (*Pyrus*), &c.

EUTRICHA QUERCIFOLIA : Fruit-trees generally, sloe (*Prunus spinosa*), hawthorn (*Crataegus oxyacantha*), peach, cherry, plum (*Prunus*), pear (*Pyrus*), Sorbus aucuparia, hazel (*Corylus*), buckthorn (*Rhamnus catharticus*), sallow (*Salix capraea, S. aurita, S. cinerea*), laurel (*Laurus*), barberry (*Berberis*), oak (*Quercus*).

DIMORPHIDES.

DIMORPHA VERSICOLORA : Birch (*Betula alba*), alder (*Alnus glutinosa*), sallow (*Salix capraea*), hazel (*Corylus avellana*), hornbeam (*Carpinus betulus*), lime (*Tilia europaea*), &c.

ATTACIDES.

SATURNIA PAVONIA : Birch (*Betula alba*), alder (*Alnus glutinosa*), white poplar (*Populus alba*), willow, sallow (*Salix capraea*), bramble (*Rubus*), apple, pear (*Pyrus*), hawthorn (*Crataegus*), meadow-sweet (*Spiraea ulmaria*), Tormentilla. Sorbus, raspberry (*Rubus*), rockrose (*Helianthemum halimifolium*), plum (*Prunus*), Potentilla, sloe (*Prunus spinosa*), dog-rose (*Rosa canina*), strawberry (*Fragaria vesca*), elm (*Ulmus*), hazel (*Corylus*), purple loosestrife (*Lythrum salicaria*), crossleaved heath (*Erica tetralix*), heather (*Calluna vulgaris*), sweet-gale (*Myrica gale*), restharrow (*Ononis arvensis*), dock (*Rumex crispus*), lilac (*Syringa*), whortleberry or blaeberry (*Vaccinium vitis-idaea*), elder (*Sambucus*), hornbeam (*Carpinus betulus*), oak (*Quercus robur*), sea-buckthorn (*Hippophaë rhamnoides*), walnut (*Juglans*), &c.

SPHINGIDES.

MIMAS TILIÆ : Lime (*Tilia platyphyllus, T. microphylla, T. ulmifolia*), all varieties of elm (*Ulmus*), hazel (*Corylus*), sallow, willow (*Salix*), honeysuckle (*Lonicera*), birch (*Betula verrucosa*), &c.; alder (*Alnus glutinosa*), ash (*Fraxinus excelsior*), oak (*Quercus robur*), sweet chestnut (*Castanea vesca*), crab-apple (*Pyrus communis*), bird-cherry (*Prunus padus*), walnut (*Juglans*), once.

SMERINTHUS OCELLATA : Sallow and willow (*Salix capraea, S. viminalis, S. triandra, S. repens*, &c.), poplars (*Populus nigra, P. italica, P. tremula, P. alba*), birch (*Betula*), rarely ; sloe (*Prunus spinosa*), bird-cherry (*P. padus*), apple (*Pyrus malus*), quince (*Cydonia vulgaris*), almond (*Amygdalus communis*), wild crab (*Pyrus communis*), wild plum (*Prunus*), bramble (*Rubus*), privet (*Ligustrum*), &c.

AMORPHA POPULI : Willow, sallow, osier, moorland willow (*Salix*), poplars (*Populus dilatata, P. balsamifera, P. nigra*), aspen (*P. tremula*), &c. ; plants of order Rosaceae—rose (*Rosa*), hawthorn (*Crataegus*), apple (*Pyrus*), Cotoneaster, &c.; ash (*Fraxinus excelsior*), common laurel (*Laurus*), laurustinus (*Viburnum tinus*), birch (*Betula*), &c.

HEMARIS FUCIFORMIS : Honeysuckle (*Lonicera xylosteum, L. caprifolium*, &c.), snowberry-tree (*Symphoricarpus racemosus*).

HEMARIS TITYUS : Scabious (*Scabiosa succisa*), field scabious (*Knautia arvensis*). lychnis (*Lychnis dioica, L. sylvestris*), Symphoricarpus lineocarpa.

SESIA STELLATARUM : Bedstraws (*Galium verum, G. mollugo, G. aparine, G. palustre, G. saxatile, G. verum*), madder (*Rubia tinctorum*), stitchwort (*Stellaria*), willow-herb (*Epilobium*).

EUMORPHA ELPENOR : Willow-herb (*Epilobium angustifolium, E. hirsutum, E. palustre, E. parviflorum*), bedstraw (*Galium palustre, G. mollugo, G. verum*), madder (*Rubia tinctorum*), buck-bean (*Menyanthes trifoliata*), enchanter's nightshade (*Circaea lutetiana, C. intermedia*), fuchsia (*Fuchsia fulgens*), wild balsam (*Impatiens noli-me-tangere, I. parviflora*), Balsamina repens, purple loosestrife (*Lythrum salicaria, L. purpureum*), virginia creeper (*Ampelopsis hederacea, A. quinquefolia*), vine (*Vitis vinifera*), apple (*Pyrus*), lettuce (*Lactuca*), honeysuckle (*Lonicera*), variegated holly (*Ilex*), bindweed (*Convolvulus*).

THERETRA PORCELLUS : Bedstraws (*Galium verum, G. mollugo, G. palustre, G. saxatile, G. aparine*), willow-herbs (*Epilobium angustifolium, E. hirsutum*), purple loosestrife (*Lythrum salicaria, L. purpureum*), vine (*Vitis vinifera*).

HIPPOTION CELERIO : Vine (*Vitis vinifera, V. hederacea*), Fuchsia, virginia creeper (*Ampelopsis*), Morinda citrifolia, Arum costatum, small willow-herb (*Epilobium*), white bedstraw (*Galium*).

PHRYXUS LIVORNICA : Vine (*Vitis vinifera*), bedstraw (*Galium verum, G. mollugo*), centaury (*Erythraea maritima*), garden centaury, dock (*Rumex*), knot-grass (*Poly-*

gonum), *Fuchsia*, marigold (*Calendula*), beet (*Beta*), scabious (*Scabiosa*), broad-leaved plantain (*Plantago*), sow-thistle (*Sonchus arvensis*), sorrel (*Rumex acetosella*), toadflax (*Linaria*), Euphorbiaceous plants, &c.

CELERIO GALLII : Bedstraws (*Galium verum*, *G. mollugo*, *G. saxatile*, *G. sylvaticum*), *Fuchsia*, *Clarkia*, madder (*Rubia tinctorum*), woodruff (*Asperula odorata*), willow-herbs (*Epilobium palustre*, *E. hirsutum*, *E. angustifolium*), wild balsam (*Impatiens noli-me-tangere*), *Escallonia*, vine (*Vitis vinifera*), *Tithymalus* = *Euphorbia*.

HYLES EUPHORBIÆ : Spurges (*Euphorbia cyparissias*, *E. amygdaloides*, *E. peplus*, *E. portlandica*, *E. paralias*, *E. esula*, *E. helioscopia*, *E. gerardina*, *E. piscatoria*, *E. guyoniana*, *E. dendroides*, *E. pinea*, *E. wulfenii*, *E. exigua*), fuchsia, vine (*Vitis vinifera*), dandelion (*Taraxacum*), lettuce (*Lactuca*), knot-grass (*Polygonum aviculare*).

DAPHNIS NERII : Oleanders (*Nerium oleander*, *N. odoratum*), *Cinchona*, periwinkle (*Vinca major*, *V. minor*, *Apocynum venetum*), milkweed (*Asclepias syriaca*), *Tabernaemontana coronaria*, potato (*Solanum*).

HYLOICUS PINASTRI : *Pinus sylvestris*, *P. laricio*, *P. excelsa*, *P. strobus*, *P. maritima*, *P. pinaster*, *P. abies*, *Abies alba*, *A. picea*, *A. pectinata*, *Larix decidua*, *Cedrus libani*, *C. deodara*.

SPHINX LIGUSTRI : Privet (*Ligustrum vulgare*), lilac (*Syringa vulgaris*, *S. persica*), mountain-ash (*Sorbus aucuparia*), guelder-rose (*Viburnum opulus*), laurustinus (*V. tinus*, *V. lantana*), Spiraeae—*Spiraea salicifolia*, *S. filipendula*, *S. ulmaria*; snow-berry-tree (*Symphoricarpus racemosus*, *S. parvifolia*), *Nerium oleander*, holly (*Ilex aquifolium*), &c.—said to prefer variegated varieties of holly—*Celtis australis*, spurge-laurel (*Daphne laureola*), *Ornus europaea*, elder (*Sambucus nigra*), honeysuckle (*Lonicera caprifolium*, *L. xylosteum*, *L. tatarica*), teasel (*Dipsacus fullonum*), horn-beam (*Carpinus betulus*), *Cytisus purpureus*, *Phillyrea angustifolium*, ash (*Fraxinus*), willow (*Salix*), sycamore (*Acer*), aspen (*Populus*), oak, evergreen-oak (*Quercus*), plum (*Prunus*), rose (*Rosa*), apple, pear (*Pyrus*), hop (*Cannabis*), Spanish lilac (*Euonymus europaeus*), laurel (*Laurus*), Portugal laurel (*Cerasus*), passion-flower (*Passiflora*), *Fuchsia*, &c.

AGRIUS CONVOLVULI : Bindweed and convolvulus (*Convolvulus arvensis*, *C. sepium*, *C. tricolor*, *C. soldanella*, *C. purpureus*), bell-vine (*Ipomoea coccinea*, *I. scandens*, *I. batatas*), *Zygophyllum fabago*, scarlet-runner (*Phaseolus maximus*), *Mirabilis jalapa*, *Batata edulis*, *Colia*, wild balsam (*Impatiens noli-me-tangere*), endive.

MANDUCA ATROPOS.—SOLANACEÆ : *Datura stramonium*, *D. tatula*, *D. metel*, *Solanum tuberosum*, *S. nigrum*, *S. dulcamara*, *S. lycopersicum*, *S. melongena*, *S. trilobum*, *S. persicum*, *S. candens*, *S. esculentum*, *S. sodomaeum*, *Nicotiana tabacum*, *N. glauca*, *N. rustica*, *N. affinis*, *Atropa belladonna*, *Lycium europaeum*, *L. barbarum*, *L. afrum*, *Lycopersicum esculentum*, *Physalis alkekengi*, *P. somnifera*. BIGNONIACEÆ : *Catalpa bignonioides*, *C. syringaefolia*. VERBENACEÆ : *Vitex agnus-castus*, *Stachytarpheta indica*. OLEACEÆ : *Syringa vulgaris*, *S. persica*, *Fraxinus excelsior*, *Ligustrum vulgare*, *L. japonicum*, *Olea europaea*, *Nyctanthes sambac*. JASMINACEÆ : *Jasminum officinale*. RUBIACEÆ : *Rubia tinctorum*. CAPRIFOLIACEÆ : *Sambucus nigra*, *Lonicera caprifolium*, *Symphoricarpus racemosus*. COMPOSITÆ : *Erigeron canadense*. URTICACEÆ : *Urtica*, sp. MORACEÆ : *Morus*, sp. CANNABINEÆ : *Cannabis sativa*. CHENOPODIACEÆ : *Beta vulgaris*. POLYGONACEÆ : dock (*Rumex*). CRUCIFERÆ : *Isatis tinctoria*. PLANTAGINACEÆ : plantain (*Plantago*). ZYGOPHYLLACEÆ : *Zygophyllum fabago*. RUTACEÆ : *Ruta graveolens*. CELASTRACEÆ : *Euonymus europaeus*. CORNACEÆ : *Cornus sanguinea*, *C. mascula*. UMBELLIFERÆ : *Daucus carota*, *Anethum graveolens*. PHILADELPHACEÆ : *Philadelphus coronarius*. AMYGDALEÆ : *Prunus domestica*. POMACEÆ : *Pyrus communis*, *P. malus*. ROSACEÆ : *Fragaria vesca*, *Spiraea trilobata*. CUCURBITACEÆ : *Coccinia indica*. LEGUMINOSEÆ : *Vicia faba-vulgaris*. LILIACEÆ : *Convallaria majalis*.

Most of these food-plants have been satisfactorily authenticated by well-known lepidopterists, the lists having been taken from the account of these superfamilies in *The Natural History of the British Lepidoptera*, where the names of the authorities are also given. Contradiction of some has been made, and doubt has been thrown in the case of others, whilst there may be a few errors due to wrong determinations, but, on the whole, the lists are no doubt very accurate, and are of the greatest value as showing what a wide range of food-plants some species, usually supposed to be confined to one or two food-plants, have.

GASTRIC EDUCATION.—The lepidopterist, however, is warned against supposing that all larvæ take readily, at all times, to any one of their known foodplants.   They may and they may not.   Collectors have again and again reported cases in which larvæ, brought up from the time of hatching on a certain foodplant, have refused another well-known food, and starved in preference to eating it, and it is supposed that, in such cases, the larva is unable to recognise or digest the strange food, in other words, that the alimentary canal is only able to deal with those foods to which it has become accustomed, although, had the larva been fed on any of the foods now refused from its earliest larval existence, or had it been given a variety of plants from among those known to be among its possible foods, it would be able to eat and digest them quite satisfactorily in its later stages.   It is obvious that detailed knowledge relating to this point is as interesting to the biologist as to the collector.

CHAPTER   V.

PUPÆ AND PUPAL STAGE OF LEPIDOPTERA.

THE PUPAL STAGE.—The lepidoptera constitute one of the groups of insects that has what is known as a complete metamorphosis, *i.e.*, a distinct change of form from the larva when it enters upon the quiescent or pupal stage.   There is no doubt that the advantage of a pupal stage of this kind is great, for it allows the insect to hide away from its enemies during the most critical period of its life, when the imago is being formed, and, as it is a more or less quiescent form, and the enemies of insects are most quickly attracted by movement, it follows that this also aids in its protection.   To take on this form, the full-grown larvæ adopt various devices, of which the formation of a puparium of some sort or other is the commonest.

THE PUPARIUM.—The puparia made by different larvæ vary greatly. It may consist merely of two or three leaves drawn together, or a hammock of silk suspended under a leaf or twig, or a chamber hollowed out in the earth, or a cavity formed in rotten wood, or hollowed out in or under bark, and carefully covered with silk into which gnawed pieces of the material in which the chamber is made are deftly interwoven, or a few stems of the foodplant drawn together on the surface of the ground, or a complicated silken cocoon of peculiar shape and outline, or one of many other structures at once hiding the larva from its enemies, protecting it from outside injury, and forming a suitable retreat in which the pupa, into which the larva turns, may undergo development into the imaginal state.   Some fullfed larvæ take much less precaution, and simply suspend themselves from a silken pad, either by means of the cremastral hooks on the anal segment alone, or by these combined with a silken loop round the body, when, as a rule, they are so excellently protected by their similarity to their surroundings, or by some one or other prominent objects frequent in their surroundings, that they are but rarely discovered, their form, colour, shape, and position, all aiding to deceive the unwary as to the real nature of the exposed pupa. Some fullfed larvæ merely throw off the last larval skin and become pupæ on the surface of the ground, resembling the little stones at the roots of the plant on which they have fed, and which they closely approach in form, tint, and general appearance, *e.g.*, *Cupido minima*, *Melanargia galathea*, &c.

THE EXTERNAL APPEARANCE OF THE PUPA—RELATION OF PUPAL AND IMAGINAL ORGANS.—At the moment of pupation the newly-formed pupal structures are externally coated with a liquid, and it is the hardening of this liquid, brought about apparently by its exposure to the air, which fastens the pupal organs into their places, and welds them into a whole. After the pupal stage has been assumed, Weismann supposes (and other observers agree with his view) that the pupal tissues undergo a certain amount of breaking up (or rather breaking down) by histolysis into "nutrient fluids and lowly differentiated units," this degraded material (if it may be so termed) being re-differentiated into the imaginal structures. The pupal shell, therefore, would appear to give us a distinct impression of the organs as they are at the time of pupation, the differentiation which occurs inside the pupa, giving us imaginal organs not always corresponding with the outer case of the pupa, in size, shape, or degree of development. Thus the pupæ of many microlepidoptera have highly developed palpi, although none exist in the imago stage, the organs evidently becoming entirely atrophied during the progress of development in the pupal stage. There is no trace of the cremaster in the imago, and the terminal abdominal segment also becomes practically obsolete in this stage, as also do other organs. On the other hand, increased development in strictly imaginal structures is very marked, especially with regard to the eyes, the replacement of a sucking-mouth for biting jaws, wing development, extension of tracheal passages (although the tracheal passage from the spiracle on the 8th abdominal segment becomes atrophied), development of scales, and final maturity of sexual organs.

VARIATION IN APPEARANCE OF PUPÆ.—With such a variety of habits in forming the puparia as those already mentioned, it is to be expected that pupæ will show considerable variation in appearance, form, colour, and structure, and so it is. Particularly do exposed pupæ, without puparia, vary in colour, form, and general appearance; hidden pupæ show comparatively little variation in these respects, and their variation is generally less superficial, and occurs in more fundamental and important structural details.

GENERAL SIMILARITY BETWEEN PUPÆ OF ALLIED SPECIES.—Apart from all superficial differences, the broad structural features of the pupæ of the species belonging to the same superfamily are generally similar, and there is frequently some particular character by means of which the pupæ belonging to a certain superfamily may be distinguished. It, therefore, becomes necessary for the lepidopterist to get some idea of the peculiarities distinguishing the various kinds of pupæ, and, when once the discrimination of those of the various large groups becomes moderately easy, rapid advance will readily be made by one who really devotes time to the work in distinguishing those of the various species. Still, it must be confessed that the difficulties of discriminating most pupæ is much greater than that of distinguishing larvæ, and, in some instances, even than those of eggs, and, although this may be in part due to the fact that none of our old-fashioned elementary text-books, and scarcely any of the modern ones, give us any useful description of individual pupæ, or even any really workable general description of the pupa belonging to a particular superfamily, it does not lessen the fact that the actual separation of closely allied pupæ is often very difficult, and it is not always easy to discriminate with certainty between some pupæ of superfamilies that are really not at all closely related, e.g., Noctuids and Geometrids, Notodontids and Lachneids. It may be well, therefore, if the collector make himself conversant with the general structure and appearance of some common Tortricid, Crambid, Pyralid, and Geometrid pupa, of an Anthrocerid, Lachneid, Attacid, and Sphingid pupa, also of some easily-obtained Lymantriid, Notodontid, Noctuid, and Arctiid pupa. If he do this he will soon get in his mind's eye the general contour and peculiarities attached to each, and will then be able, without much trouble, to say with few errors whether a pupa is to be referred to one of these main groups, or falls outside them. In addition to these he must then carefully study the, to many, more difficult pupæ, e.g., the Alucitid, Hepialid, Cossid, Nolid, Adelid, Lithocolletid, Elachistid, Pyraloid, Coleophorid, &c. If he will make sure of one well-authenticated type of each group, he will soon be able to refer others to these groups, and to deal with any peculiar ones that come his way by elimination, and reference to other groups not mentioned above.

MOVEMENTS OF LEPIDOPTEROUS PUPÆ.—Although the pupal is the quiescent stage of lepidoptera, and many are incapable of any but the least possible movement, yet others are able to move considerable distances within the limits of their puparia. This movement is not carried on by means of the limbs, but by means of hooks, which are really outgrowths of the chitinous coating of the abdominal segments, and the arrangement and working of which are subjects always to be carefully considered when a pupa is being described. The passage of the pupæ of Hepialids along their silken puparia, of Phragmatoecia arundinis along the reed-stem, of Eumorpha elpenor and Dimorpha versicolora from the puparia altogether, the remarkable somersault movement of the Alucitid or plume pupæ, as well as the protrusion of the thoracic and front abdominal segments, just before the emergence of the imago, of the large groups of pupæ-incompletæ, are all matters which require careful observation, and of which, at present, we know very little.

EXTERNAL STRUCTURE OF PUPA.—In a general sort of way the collector soon observes that pupæ are structurally similar to the imago, and he has little trouble in distinguishing the head, thorax, and abdomen. Attached to the head he notices the antennæ, legs, and maxillæ, as well as the glazed eye, a conspicuous lunular piece near the base of the antenna on either side. The mesothorax and metathorax he observes carry the wings, of which, however, the hindwings exhibit only a very narrow strip, whilst the abdominal segments, separated by incisions, he notices vary in size, and he recognises particularly the fact

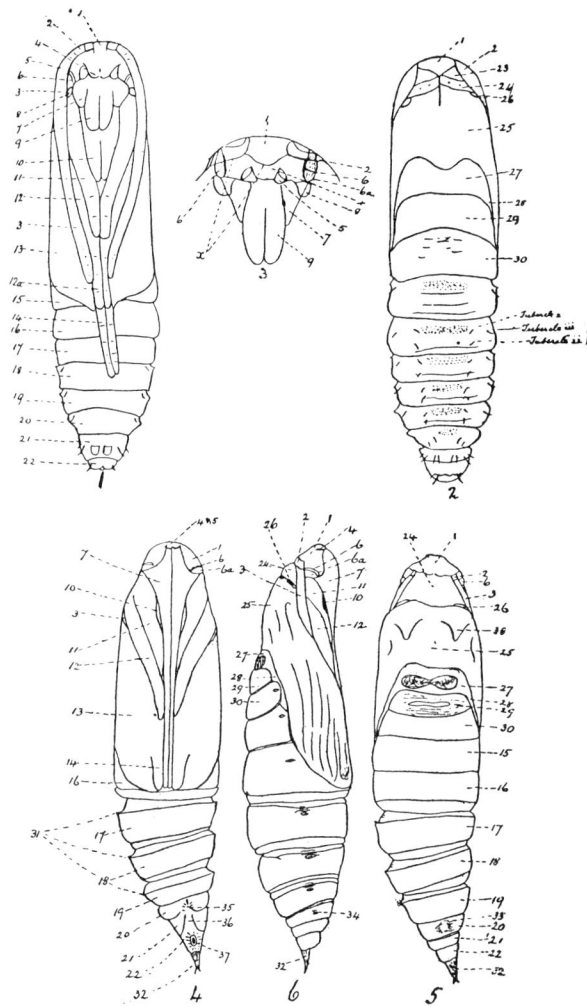

PLATE VII.

STRUCTURE OF LEPIDOPTEROUS PUPA-INCOMPLETA AND PUPA-OBTECTA.

Practical Hints, etc., 1905.

that, owing to certain incisions not being soldered, some of the segments are capable of movement, and that the number of these movable segments varies in different superfamilies. These movable segments constitute a most important feature in pupal structure. He will further note that the anal segment is considerably modified, often supplied with a projecting spike and many hooks for attaching itself to the silk. The anchoring structure of the anal segment (its armature), is known as the cremaster. The field naturalist further notices very early in his collecting career that, when the imago is fully formed and ready to leave the pupa, the latter acts very differently in certain groups. He will observe in certain superfamilies that the pupa-case remains within the puparium when the imago emerges from the latter. On the other hand, the pupæ of other superfamilies emerge more or less from the puparium, and the imago emerges from the protruding pupa. The segments and appendages of the pupæ of the former group (those that remain within the puparia) are much more solidly soldered together than those of the latter, and the pupæ are sometimes absolutely solid, in that they have no movable segments, rarely indeed, do they have more than three movable incisions, thus giving two movable abdominal segments. On the other hand, those pupæ that have considerable power of movement, have a greater number of movable incisions, consequently a greater number of movable abdominal segments, whilst the appendages are more or less incompletely soldered to the body segments. These two groups of pupæ are respectively known as obtect and incomplete. The same sort of pupa, *i.e.*, obtect or incomplete, is always found throughout the same superfamily.

DESCRIPTION OF LEPIDOPTEROUS PUPÆ.—If the collector be tempted to examine thus far the pupæ that he collects, he is almost sure to go a step further and make scientific use of his knowledge. As we have already noted, few pupæ have been properly described, and the field open here is practically new. There is no doubt that the difficulties of description have been the stumbling-block with regard to the pupa, and Newman's classical description of the pupa of *Smerinthus ocellata*, as " red-brown and glossy," is characteristic of all our British entomological works until Buckler and Hellins took a step forward, and Chapman taught us how to do the thing properly. It is with the idea of enticing collectors into so fertile a field for original work, that we give detailed drawings of two pupæ with the chief parts named, from the dorsal, lateral and ventral views. One is that of a pupa-incompleta (*Diplodoma herminata*, described in detail in *British Lepidoptera*, ii., pp. 151-2); the other is that of a pupa-obtecta (*Manduca atropos*, described in detail, *op. cit.*, iv., pp. 425-431). In describing a pupa our advice is to give at least the following particulars:—

(1) The general appearance, especially noting any particular and striking features of the coloration, &c.
(2) The general structure, especially any particularly striking points of development.
(3) The exact measurements.
(4) The general characters of the head, thorax, and abdomen.
(5) The number of movable abdominal segments.
(6) Peculiarities of the cremaster.
(7) Detailed description of the dorsal view—head (if visible), pro-, meso-, and metathorax (and wings), abdominal segments, traces of tubercular scars, subsegmentation, &c.

(8) Detailed description of the lateral view—head, antennæ, thoracic segments, and wings, abdominal segments, spiracles, traces of tubercular scars, lateral flanges, &c.

(9) Detailed description of the ventral view—mouthparts, antennæ, legs, maxillæ, wings (with comparative lengths, &c., of these three last-named parts), abdominal segments, proleg scars, genital organs, &c.

EXTENDED DURATION OF THE PUPAL STAGE.—To collectors, one of the most interesting points relating to the pupal stage is the fact that the pupæ of many species, in our latitude, occasionally extend their normal pupal stage, and many, whose ordinary habit it is to go over one winter as a pupa, will sometimes pass more than one winter in this stage. Pupæ that go over the winter may be grouped as :—(1) Those that mature in the autumn, the fully-developed imago really hybernating within the pupal shell, e.g., *Valeria oleagina*, the Tæniocampids—*Panolis piniperda*, &c. (2) Those that undergo but little development until shortly before the time at which the imago is due to emerge, and then mature rapidly, the imagines coming out without further delay. We doubt whether any of the first section go over the winter a second year, and it is one of the problems which collectors still have to solve. One of the species whose pupæ frequently go over more than one winter, is *Lachneis lanestris*, which has records for seven and five years, and then emerging successfully, whilst records for two, three and four winters are not infrequent. Often, in the same batch, a few imagines will emerge during each spring for several years in succession. Closely following on these are *Saturnia paronia* and *Dimorpha versicolora*, which are frequently reported to go over the usual pupal period. Among the Acronyctids, *Cuspidia megacephala* and *Acronicta leporina* are the species that most frequently go over two or more winters ; more than half a brood of the former has been known to go over to the second year, and, in some cases, imagines emerge quite satisfactorily from pupæ that have gone over as many as four, or even five, winters. *Petasia nubeculosa* is recorded as going over two, three, four and five winters ; *Sphinx ligustri* two winters; *Mimas tiliae*, two winters, although the habit is of rare occurrence among the Sphingids. Among the Geometrids, *Eupithecia togata, E. venosata, E. pygmaeata, E. linariata, E. expallidata, Emmelesia albulata, E. unifasciata, Nyssia zonaria*, &c., have been reported as emerging satisfactorily after having been two years in the pupal stage. The Notodontids appear to have largely adopted the habit. Besides *Petasia nubeculosa*, already mentioned, *Notodonta dromedarius, Lophopteryx camelina, Peridea trepida, Cerura vinula*, and *Phalera bucephala* have been recorded, whilst of the Noctuids, *Cucullia verbasci, Dianthoecia capsincola*, and one of the Cymatophorids—*Asphalia ridens*—have a great reputation in this direction. It is difficult to explain the causes that lead to these retarded developments of the lepidopterous imago. Among certain northern species—*Petasia nubeculosa*, &c.—the retardation of over two or more winters appears to be usual. Of others that have a wide distribution in the British Islands, e.g., from the north of Scotland to the south of England, such as *Eupithecia togata, E. venosata, Emmelesia albulata, Dimorpha versicolora, Saturnia pavonia*, &c., one finds the going over of southern pupæ to be of rare occurrence, whilst northern pupæ frequently do so, and, strangely, the latter are more prone to exhibit this peculiarity when brought south, than when reared in their own latitude under their normal conditions, in which cases it would appear that the retardation is due rather to an excess, than a defect, of temperature, a fact which seems amply proved when one considers how frequently pupæ, forced before their normal time, will resist the forcing, and not only not come out then, but will pass over the natural time for their appearance, and emerge twelve months later. It would appear as if the pupæ felt that the proper season for emergence having arrived and passed without their having made a move, it had become necessary for them to wait till another season. On the other hand, if the time required for the early development of the organs has been allowed to proceed normally, then a gentle forcing, judiciously increased after a time, generally hastens the final development and produces imagines. There must be, however, a great deal in the individual constitution, for, of pupæ from the same batch of ova, identically treated throughout their oval, larval, and pupal existence, some imagines will emerge at the normal time, whilst others will last on as pupæ for another year, and then will only emerge at the proper season, when the insects should take on the imaginal form. Details bearing on this and similar peculiarities involving exact data of temperature, &c., are greatly needed from all parts of the country.

DOUBLE-BROODEDNESS AND PARTIAL DOUBLE-BROODEDNESS.—Closely allied to the pupal conditions that we have just considered, is that of partial double-broodedness. It is, in normal seasons, the usual habit for some species in the south of England to be more or less double-brooded, the same species being normally single-brooded in the north of Great Britain, with, at most, only occasionally, a few odd second-brood specimens. Again, other species that are absolutely single-brooded in the north are occasionally partially double-brooded in the south, whilst many of these species in southern France become more or less completely double-brooded. The fact that these partially double-brooded species produce a much greater percentage of autumnal imagines in hot summers, makes it pretty certain that temperature plays an important part in their rapid development, yet, the fact that, reared under identical conditions, some individuals of the brood will emerge as autumnal imagines, whilst the remainder of the brood will go on as over-wintering larvæ or pupæ (according to the stage in which they normally hybernate), makes it certain that temperature is not everything, and that there is some hereditary influence at work, making a difference in the insects themselves, and, to these points, if they are to be elucidated, the observations of our collectors must be directed. As to our ignorance of these matters, we may note that, in the abnormally hot spring of 1893, many species, normally hybernating in the egg-stage had hatched, also those hybernating in the larval stage had pupated, whilst those that passed the winter as pupæ had emerged, long before their usual time in early spring, many species getting in an abnormal second-brood in July, August, or September, some even as early as June; yet that same year, whilst some of the over-wintering pupæ of *Pieris rapae*, *P. brassicae*, and *Euchloë cardamines* had produced imagines in March, so that fullfed larvæ and pupæ were everywhere abundant in May and June, other · over-wintering pupæ of the same species kept under identical conditions, still remained as pupæ in June, and we were at this time in possession of over-wintering pupæ, eggs. larvæ, and pupæ of the year, all at the same time, of these particular species. We still await an explanation of this varying effect of the same conditions on different pupæ of the same species. Of species which are naturally single-brooded, but which produce a partial double-brood occasionally, or even frequently, may be mentioned —*Papilio machaon*, *Leptosia* (*Leucophasia*) *sinapis*, *Nisoniades tages*, *Smerinthus ocellata*, *Amorpha populi*, *Stauropus fagi*, *Eumorpha elpenor*, *Euthemonia russula*, *Clostera pigra*, *Plusia chrysitis*, *P. festucae*, *Metrocampa margaritaria*, *Acidalia inornata*, *A. virgularia*, *Ligdia adustata*, *Pericallia syringaria*, and many others. In confinement, the regularity of appearance soon gets unsettled, and such species as *Notolophus gonostigma*, *Phragmatobia fuliginosa*, and other species, will produce a third brood, in fact, such insects as these, and many Noctuids— *Peridroma saucia*, *Aplecta occulta*, *A. herbida*, &c.— in domestication, evidently try to become continuously-brooded. It would be interesting to trace out in detail the exact range of these species to the south, and their habits in their more southern localities. One suspects this partial double-brood habit to be a remnant of a regular habit in these parts of its range, the double-broodedness having been largely bred out of the northern races by the extermination of the individuals showing it in cold, wet, and generally bad, seasons, and yet leaving sufficient trace of it in certain individuals to allow them to take advantage of it in specially warm and favourable seasons, sometimes, maybe, to their own undoing, if the season changes badly and the temperature is much lowered, after they have commenced the attempt.

DOUBLE-BROODEDNESS AND FIXITY OF HYBERNATING STAGE.—It is well for the collector to remember that, when partial double-broods of this kind occur, he must see to it that the larvæ, if the eggs hatch, have sufficient food to carry them through to the proper hybernating stage, for, with few exceptions, the habit is a fixed one, and brooks no change. For example, *Tephrosia bistortata* eggs hatch in May, the larvæ pupate in June and July, and, in later July, a partial second-brood of imagines emerges, whilst the rest of the pupæ go over the winter to the following March or early April before disclosing the perfect insect. The second-brood imagines pair in late July and early August and lay eggs, the latter soon hatch and the larvæ feed up in September and October, often when food becomes increasingly difficult to obtain. Still it must be obtained, for, the hybernating stage of the species being the pupal, the larvæ must reach the pupal stage or die, the insect cannot hybernate as larva, such late larvæ as fail to become mature, being killed off by starvation or cold. These autumnal-formed pupæ and summer-formed pupæ both go over the winter, and the imagines of all emerge in March and April, the nephews and nieces thus catching up their uncles and aunts, and mixing with them.

MODIFICATION OF PUPAL PERIOD BY HYBRIDITY.—Another point on which the observations of our collectors is needed is that relating to the pupal period of hybrids. Many of our experimenting lepidopterists have of late years crossed several allied species, and it has been repeatedly observed that hybridity has a tendency to unsettle the regular pupal habits of the parents as to the time of emergence, and appears to tend in the direction of an attempt to produce, as it were, continuously-brooded progeny. Thus *Smerinthus* hybr. *hybridus* (= *ocellata ♂* × *populi ♀*), *Tephrosia* hybr. *ridingi* (*bistortata ♂* × *crepuscularia ♀*), *Amphidasys* hybr. *herefordi* (*strataria ♂* × *betularia ♀*), &c., which should, in the ordinary course of things, have emerged the following year, produced imagines the same autumn. Exact information on the point is greatly needed.

FORCING PUPÆ.—What we have already noted with regard to the more or less complete or partial double-broodedness found in some lepidoptera possibly lends considerable aid in explaining why some pupæ are easily forced by a judicious application of moisture and temperature. For example the three British species of *Pieris*—*P. brassicae*, *P. rapae*, and *P. napi*—*Pararge egeria* (often with three broods), *P. megaera*, *Polyommatus bellargus*, *P. icarus*, and others among the butterflies are regularly, and possibly wholly double-brooded species, *i.e.*, probably all eggs laid in spring produce imagines the same year. So also among the Lithocolletids and other superfamilies of the smaller moths, many are undoubtedly double-brooded, whilst examples from each of the other large groups might be taken. But it is doubtful how far many so-called double-brooded species really are completely so. Many species emerge regularly in spring, and the eggs then laid produce larvæ, part of which, as we have already shown, may feed up rapidly, and produce pupæ that give a summer (or autumnal) brood of imagines, whilst the remainder feed slowly, pupate, and the imagines appear with the progeny of the summer (or autumnally) disclosed specimens the following spring—*Tephrosia bistortata, Acidalia virgularia, Timandra amataria*, and many other species thus act. It appears certain that, in those species where this partial double-broodedness occurs, there is a slight hereditary tendency and disposition for a rapid condition of development to be set up under specially favourable conditions, but that this is deferred normally, and goes on slowly throughout the winter months. Now as we get a fair percentage of our normally single-brooded moths forming partial double-broods in hot summers, it is only reasonable to suppose that those species which are normally partially double-brooded, are capable of undergoing more rapid development if subjected to a high temperature, and so ought to be capable of being readily forced, and this we find is the case. One can easily, under artificial conditions, produce imagines of such species at almost any time of the year, but, with regard to those which normally undergo a slow process of development, those which are in the pupal stage for some months (say from September to May or June) it seems impossible, as we have already noticed, at an early stage of the pupal condition to force them, and, in fact, such a process frequently ends in killing them, and it has come to be recognised that some species will not be forced unless they have already become matured up to a certain point; that is, until the pupal tissues have broken down, and the imaginal structures have begun to be differentiated, it seems impossible to force a rapid completion of the imaginal tissue. It is remarkable, however, that, in some species, as we have already noted (*anteà*, p. 36), the imagines are fully formed at an early period of the pupal stage, the scales and ordinary imaginal structures being fully developed. This, we have pointed out, is the case with all the British Tæniocampid species, the imagines of which are fully-formed in the autumn, and may be taken fully-scaled from the pupæ in October, yet these do not normally emerge till March, and, in confinement, one often finds considerable difficulty in getting these species out, even at the proper time. *Lachneis lanestris*, too, which we have stated has been known to exist in the pupal state as long as seven years, cannot be persuaded to take on the imaginal stage by any amount of forcing, except just at that period which is its normal time for emergence, and yet the imago may have been fully formed within the pupa for some weeks previous to the attempt made to force it. Strangely enough, it would appear that all imagines of this species that are to emerge in spring, mature in the autumn, or early winter, preceding the time of their appearance, those that continue in the pupal stage for a lengthened period, remaining apparently in the same undeveloped state as when first formed, only changing a few months before emergence. On these and several other matters connected with the pupal stage of lepidoptera, the collector can make a host of useful observations. [Students requiring to carry on their studies of pupal structure and pupal habits are referred to *Natural History of British Lepidoptera*, vol. ii., pp. 1-100.]

JANUARY, FEBRUARY, AND MARCH.

ERIOCRANIIDES.—At the end of March and in early April, if the season be forward, the females of the birch-feeding Eriocraniids may be observed, sitting on the young birch leaves just protruding from the buds, and cutting little pockets, on the underside of the leaf, in which the eggs are laid. *Eriocrania sangii* lays only one egg in a pocket, *E. purpurella* usually lays three, &c.

ADELIDES.—The flat cases of *Nemophora swammerdammella* may be found in March and April; they are almost polyphagous in their habits and feed indifferently on sloe, chickweed, or almost anything green. They appear to take two years to come to maturity.

ELACHISTIDES.—The mined grass leaves containing larvæ of Elachistids should be broken off as far below the larva as possible ; they should be carried home in a tin, the different kinds placed in well-closed glass tubes, kept in a cool place, when the species may be reared with ease.

The beautiful red-spotted larva of *Elachista cinereopunctella* mines down the leaves of *Carex glauca* in spring, having apparently wintered in the withered top of the leaf ; it is fullfed in March, and almost immediately it quits the mine it fixes itself in the angle of the leaf, and changes to a pupa.

In February and March the larvæ of *Elachista rufocinerea* are to be found abundantly in the leaves of those plants of *Holcus mollis* which grow under hedges and by the sides of ditches; making broad whitish mines, in which there is very little excrement.

In March the fresh long puckered mines of *Elachista gangabella* are to be found in the leaves of *Dactylis glomerata ;* the larvæ move freely from leaf to leaf, being fullfed about the middle of April. The species is very local, and the larvæ have been recorded also from *Holcus mollis*.

The larvæ of *Elachista gleichenella* are to be found in March and April making rather small whitish blotches on the upperside of the leaves of grass and *Carex*, moving from one leaf to another.

The larva of *Elachista bedellella* mines the tips of the leaves of *Avena pratensis* in March and April, when the upper cuticle becomes conspicuously whitish, the tip itself dull pink.

Larvæ of *Laverna propinquella* (*paludicolella*) may be searched for early in March, mining in the leaves of *Epilobium hirsutum*, *E. palustre*, and *E. montanum*. The best way to rear moths is to dig up the plants having mined leaves containing larvæ and plant them in flower-pots or seed-pans.

The larvæ of *Laverna propinquella* are to be obtained generally from February to April, mining the young leaves of *Epilobium hirsutum*; they make their cocoons within the mine.

In early spring sickly-looking buds should be pulled off apple-trees in order to obtain the larvæ of *Chrysoclista* (*Laverna*) *vinolentella*. The larva mines into the bud in autumn, and in the winter is to be found in the alburnum of the bearing spur of the apple.

The pupæ of *Chrysoclista aurifrontella* (*flavicaput*) are to be found in March and April in the last year's wood of hawthorn; a small hole will lead to the discovery of the burrow, at the bottom of which lies the pupa; the hole being situated usually about an inch from a fork, whilst the pupa is just at the fork.

BUCCULATRIGIDES.—The larvæ of *Bucculatrix nigricomella* (*aurimaculella*) are to be found in mid-April, feeding on the leaves of *Chrysanthemum leucanthemum* (oxeye daisy), at first mining the leaves, and later feeding on them externally, and making, when mature, a ribbed whitish cocoon.

ARGYRESTHIIDES.—The shining pale-brown larva of *Cedestis farinatella* is to be found in March, mining the leaves of *Pinus sylvestris*; it mines from the tip of the leaf downwards, leaving the mined portion full of excrement.

The brown larva of *Ocnerostoma piniariella* also mines the leaves of *Pinus sylvestris* in March, working from the tip downwards, but though it leaves the end of the leaf full of excrement, at some distance from the tip it makes a hole and ejects its excrement through it, so that the lower half of the mine is tolerably empty, and appears whitish.

TORTRICIDES.—Early in March the imagines of *Amphysa walkerana* may be found on Scotch and north of England moors. The males fly freely in bright sunshine, but the females are extremely sluggish, and, therefore, require to be diligently searched for.

When the hybernated females of *Peronea ferrugana* are taken in March they are well worth retaining for ova, the moths appearing in July being more given to variation than those occurring in the autumnal emergence.

Roots of *Ajuga reptans* dug in February and March, in localities where *Penthina fuligana* occurs, and planted in seed-pans or shallow boxes will, in due course, yield imagines of this moth.

Last year's stems of *Impatiens noli-me-tangere*, if kept exposed to all weathers, should produce imagines of *Penthina postremana*.

The larvæ of *Paedisca oppressana* are to be obtained from the middle of March in tubes, formed of silk interwoven with particles of their excrement, on the terminal buds of several species of *Populus*. The shoots with attacked buds if stood in damp sand will yield imagines freely.

Heusimene fimbriana first appears about the middle of March, flying in the sunshine amongst oaks, and during dull weather may be obtained by beating.

Heusimene fimbriana flies freely in the bright sunshine during March and April in oak-woods. In dull weather, it may be beaten from oaks, appearing to have a preference for resting in the higher parts of the trees. It is as well when collecting this moth to stand in the centre of a riding, as then one obtains a clear view of the insect against the sky. Under such a condition it is easily seen, which is not the case with a dark background.

Ephippiphora nigricostana can be bred in numbers by collecting, in February and March, dead stems of *Stachys sylvatica* and enclosing them in any suitable receptacle.

Roots and stems of *Centaurea nigra* should in March be gathered for larvæ of *Ephippiphora cirsiana*. The roots may be potted very closely together.

Where the stems of *Artemisia vulgaris* are broken off and the holes in centre covered with silk, larvæ of *Ephippiphora foeneana* may be judged to be present in the roots. The imagines are easily reared from potted roots.

The imagines of *Phloeodes crenana* are to be found in March and April, flying freely on fine sunny mornings from 10 a.m. to 1 p.m., on moorland, amongst heather, stunted oaks, birches, sallows, nut-bushes, and furze-bushes, frequenting mostly the sallows on which the larvæ feed. [Appear again in July.]

Larvæ of *Laspeyresia (Lobesia) servillana* occur in shoots of sallow, betraying their presence in early spring by causing gall-like swellings to arise. The imagines may be reared by placing the tenanted shoots in damp sand. A keen look-out must be kept for the moths, as they are active, and very soon injure themselves.

Steganoptycha pygmaeana should be worked for from the end of March. The males fly round spruce-firs in the sunshine after mid-day. The females, being very sluggish, require to be beaten from the trees, and a sharp look-out must be kept for them, as they invariably fall to the ground, and so require to be intercepted with a net in their descent. Hitherto this species has been recorded only from Cambridge-shire and Norfolk.

Old and prostrate stems of *Eupatorium cannabinum* may be gathered in February and March if a supply of *Eupoecilia rupicola* be required. The stems should be kept out of doors, or any larvæ they contain may otherwise perish.

Stems of *Onobrychis sativa* collected in March, &c., in the Dover and St. Margaret's Bay districts, will supply larvæ of *Grapholitha caecana*.

COLEOPHORIDES.—The larvæ of *Coleophora juncicolella* feed on *Calluna vulgaris* in early spring until May, living on the young shoots, in small cases, which look like small heath-leaves. The case is difficult to detect unless it assumes a direction contrary to the growth of the shoot.

GELECHIIDES.—From December to March the larvæ of *Bryotropha affinis* are to be found feeding on mosses growing on walls, &c.

The pale greenish black-spotted larva of *Gelechia luculella* feeds in the decayed wood of trees (? usually oak-trees) in most of our large woods in March (Stainton). [Meyrick says between joined or spun-together leaves of oak in September.]

The larvæ of *Aristotelia* (*Gelechia*) *brizella* winter in the flower-heads of *Statice armeria*, changing to pupæ the following April (Schmid). [Meyrick notes larvæ as occurring in June-July and again in September-October.]

The larvæ of *Aristotelia* (*Gelechia*) *arundinetella* are to be found from March to May, mining when young, up and down one of the large leaves of *Carex paludosa* growing by river-banks ; leaving one leaf for another as it gets older, and sometimes going to a third, in which it spins its cocoon and changes to a pupa. The imagines appear at the end of June and in July. [Zeller says the imago is to be found in July among *Arundo phragmites* and *Scirpus lacustris*.]

CRAMBIDES.—In February, plants of *Plantago lanceolata* should be examined for larvæ of *Homoeosoma sinuella ;* the larvæ feed in the solid root-stocks, bear considerable resemblance to a Tortricid larva, and in March spin rough, soft, silken cocoons within the cavities mined by the larvæ, and they remain unchanged therein until early June.

In searching for fullfed larvæ, or larvæ already in their puparia, of *Homoeosoma sinuella* in March, it is well to remember that the small stunted root-stocks frequently contain a larva, and the larger roots sometimes have two or three; the affected roots often show portions of the plants killed, whilst fresh vigorous shoots are growing from the side, and these frequently become so luxuriant, that it is difficult to find the old root-stock in which the larvæ or puparia are hidden.

In March, the larvæ of *Nephopteryx genistella* lengthen their hybernating cases and extend, as they increase in size, the silken web, feeding on the young buds and blossoms beneath it. On fine days they may be seen sunning themselves on the outer part of the web, but they retreat into their tunnels at the least alarm.

PYRALIDES.—The larvæ of *Scoparia mercurella* (*frequentella*) are to be found in March and April feeding in galleries made in mosses growing upon tree-trunks, rocks, stone walls, &c., in fact, in all sorts of situations, even on the spreading roots of trees, on the ground, and on the roofs of houses (Barrett).

The larvæ of *Scoparia murana* are to be found from February to May on various mosses—*Hypnum cupressiforme*, *Dicranum scoparium*, *Bryum capillare*, &c.—growing on old walls, dykes, rocks, &c.

In January and February, if the weather be mild, the hybernating larvæ of *Scopula prunalis* are to be found on the underside of the leaves of *Galeobdolon luteum*, *Lamium*, &c., nibbling little channels out of the lower cuticle, causing a change of colour on the upper surface and betraying their situations ; but cold weather at once causes them to become dormant again ; also found on nettle, elm, *Teucrium*, *Stachys*, &c.

In March, the hybernating larva of *Scopula olivalis* feeds rapidly, drawing together the leaves of its foodplants—*Sambucus nigra*, &c.— tightly round itself with a few threads, as it eats portions out of them, and feeds secure from observation until about the middle of April, or the end of the first week of May, according to the season, when it is fullfed.

The hybernating larvæ of *Psammotis* (*Botys*) *hyalinalis* recommence feeding on *Centaurea nigra* in middle March, when they begin to spin short galleries from their hybernacula to the young *Centaurea* leaves just coming out of the earth; they then feed up pretty rapidly, and are fullfed from mid-May to early June, pupating about a fortnight after the cocoons are spun.

GEOMETRIDES.—The hybernating larvæ of *Phorodesma smaragdaria* recommence feeding again in early March, as soon as the *Artemisia maritima* is appearing above the ground. They soon moult, and then cover themselves with pieces of the fresh green foodplant. They are fullfed in mid-June.

The hybernating larvæ of *Acidalia contiguaria* recommence feeding about the end of March, and may be fed in confinement on ling, buds of whitethorn, *Polygonum aviculare*, &c.

The hybernating larvæ of *Acidalia rubricata* commence feeding again in early March, and prefer *Polygonum aviculare* to other foodplants; they feed on until early June, when pupation takes place.

DIMORPHIDES.—The pupæ of *Dimorpha versicolora* must be well watched in February and March, as, in confinement, many of the pupæ leave their cocoons before emergence, and lie exposed for some days before the appearance of the imagines.

PSYCHIDES.—At the end of March the imagines of *Bankesia staintoni* are to be found flying, not uncommonly, in the neighbourhood of spruce fir-trees, on the shore of Southampton Water, opposite Calshot Castle.

ANTHROCERIDES.—In mid-February the leaves of *Centaurea nigra* show small watch-pocket shaped apertures cut in the upper surface, sometimes with minute black atoms of frass protruding, thus denoting the presence of the larvæ of *Rhagades globulariae*.

The hybernating larvæ of *Anthrocera trifolii* are on the move in the first days of March, and should be supplied at once with food, *Lotus corniculatus*.

The hybernating larvæ of *Anthrocera exulans* recommence feeding about the end of February; they often take two years to come to maturity.

As soon as the larvæ of *Anthrocera purpuralis* show a disposition to move in February, after hybernation, a supply of *Thymus serpyllum* should be given them; a frame that receives the sun is a good place in which to grow the thyme for rearing them. The species knows little of frost in its British habitats.

ÆGERIIDES.—The larva of *Trochilium apiformis* bores in the bark and wood of *Populus nigra* a few inches above the soil, and also below it, and its gallery is driven not only into the trunk of the tree, but also through the main roots, sometimes removing to a considerable distance from the trunk; when fullfed the larva comes to the outside again, and spins its cocoon in the winter or very early spring near the exit.

The larva of *Trochilium apiformis* spins its cocoon in winter, and remains therein throughout February, March, April, and May as a larva, pupating in May and June, the imago appearing in June or July.

The larvæ of *Trochilium bembeciformis* are to be found throughout the winter up to June in galleries in the wood of *Salix caprea*, most frequently near the level of the ground; they spin their cocoons in the spring, and pupate in June therein, at the upper end of their galleries. The larva of this species also feeds in poplar.

NOLIDES.—The hybernating larvæ of *Nola centonalis* recommence feeding in early March and continue to do so on flowers and young leaves of *Trifolium procumbens*, *T. pratense*, and *T. repens*, and spin their boat-shaped cocoons in June.

DELTOIDES.—The hybernating larvæ of *Rivula sericealis* usually recommence feeding in the first mild days in early March; they feed openly on *Brachypodium sylvaticum*, and are usually fullfed in early June. The larvæ will also eat *B. pinnatum*, and do not object very much to *Phalaris arundinacea*.

The hybernating larvæ of *Herminia barbalis* are to be obtained in late March and early April, feeding on birch; in confinement they should be supplied with catkins. [Buckler fed them on female birch catkins.]

The larvæ of *Herminia tarsipennalis* are to be found in March and April, feeding, by night, on sallow. They pupate in early May.

NOCTUIDES.—The hybernating females of *Dasypolia templi* lay their eggs in March and April on the old stems or early leaves (underside) of *Heracleum sphondylium*. The larvæ appear in about a month.

In confinement, ♀ s of *Dasypolia templi* lay their eggs freely in late March and April on the early leaves or dry stems of *Heracleum sphondylium*. The young larvæ (hatching towards the end of April) attack the leaves and stem, eating their way into the interior and drinking the sap that flows into their little tunnels; they bore downwards, and by the end of May pass into the roots and practically destroy the plants attacked by the time they are fullfed, in mid-July.

The eggs of *Xylina conformis* are laid in early spring (late February and throughout March) after hybernation, the egg-stage lasting from six to seven weeks. The young larvæ feed up on *Alnus glutinosa*, and are fullfed by mid-June. [Note.—Leaves with the secretion of aphides on them have been found very hurtful to the nearly fullgrown larvæ.] The pupal stage lasts till September and October, when the imagines appear.

The larvæ of *Phlogophora empyrea* are to be found from February-April (after hybernation), feeding on the leaves of *Ranunculus repens* (Buckler says they appear to neglect *R. ficaria*), and forming a cocoon of a rather open network of silk under the upper leaves of this plant.

The young larvæ of *Cirrhoedia xerampelina* burrow into ash-buds, ash-flowers, &c., in March and April; in May the later larvæ feed on leaves; they may be found by day hiding in chinks of bark, &c., in the evening crawling up the trunks, or at night by searching or beating the flower-bearing branches, &c.   The larva assimilates very perfectly with the colour of the crevices of the bark in which it hides.

A hybernated ♀ of *Dasycampa rubiginea*, taken at sallow-bloom on March 21st, 1868, at Babbicombe, enclosed in a glass-topped box about six inches square, fed on plum-jam, laid some fifteen eggs from March 28th onwards.   These hatched between April 19th and 23rd, the larvæ fed freely on plum leaves, less so on sloe and knotgrass, and were full-fed from June 15th-20th; the imagines appeared between September 8th and 20th (Hellins).

The eggs of *Taeniocampa pulverulenta*, laid in March or April, hatch very quickly, and at first the larvæ live in a shelter formed by spinning together several leaves of its various foodplants.   In confinement they do well on hawthorn.

The ova of *Diloba caeruleocephala* go over the winter without hatching, the normal period for the appearance of the larvæ is about mid-March.

By the end of March the bristly young larvæ of *Diloba caeruleo-cephala* are to be found feeding on the forward buds of whitethorn.

LITHOSIIDES. — Hybernating larvæ of *Lithosia griseola* attacked vigorously some slices of turnip, but afterwards preferred, and fed up steadily on *Lichen caninus*, which is probably the natural food (Hellins).

In the early spring the hybernating larvæ of *Miltochrista miniata* were found to nibble slices of turnip, but fed up later on the red waxy tips of *Lichen caninus* (Hellins).

PAPILIONIDES.—Do not forget in late February to keep a sharp eye on the hybernating eggs of *Plebeius aegon*, and have a plant of *Hippocrepis comosa* ready for the larvæ as soon as they emerge.   This emergence is sure to take place either in the last few days of February or the commencement of March.

The hybernating larvæ of *Polyommatus icarus* are sometimes to be found on *Lotus corniculatus*, when searching for cases of *Coleophora discordella* (Jordan).

In March and April the larvæ of *Polyommatus astrarche* feed on the underside of the leaves growing on the young tender shoots of *Helianthemum vulgare*, making marked brown blotches where they feed, and thus betraying their whereabouts ; they are fullfed from the middle to the end of May.

The larvæ of *Pararge megaera* are to be obtained feeding on grasses on the outskirts of woods, by wild hedgesides, &c., in March, the larvæ being fullfed and pupating in early May.

APRIL.

ERIOCRANIIDES.—Many species of the genus *Eriocrania* may be found flying in the midday sun round their respective foodplants, and, on sunless days, may be beaten from them.

The larvæ of the Eriocraniids are most easy things to study, for, by removing all but the mined leaf (containing a larva) from the spray, the latter may be kept in water fresh to the last, and the whole larval history, from beginning to end, be accurately watched (Wood).

At the end of April or beginning of May, the larvæ of *Eriocrania unimaculella*, *E. semipurpurella* and *E. sangii* are to be found feeding in birch leaves before these are fully grown ; all start from the edge of the leaf with a short twisting gallery, that is soon lost in the after-formed blotch.

The full-grown larva of *Eriocrania unimaculella* may be told by its brown head and darker mouth-parts, the posterior points of head forming two black spots ; *E. sangii*, by its black head with brown mouth ; *E. semipurpurella*, by its very pale ochreous-brown head with darker mouth-parts. In addition *E. unimaculella*, is pale-coloured with dark markings on back of head, but no dark markings on prothorax ; *E. semipurpurella*, whitish, with no dark spots on back of head or prothorax ; *E. sangii*, grey, with the black marks on prothorax breaking up.

The mine of *Eriocrania sangii* is browner than those of *E. unimaculella* and *E. semipurpurella*, and shows only a very narrow greenish border at the advancing edge.

The mine of *Eriocrania purpurella* is to be found in young birch leaves, being greener in tint than those of *E. semipurpurella*, *E. sangii* and *E. unimaculella*. It is sometimes solitary, but often 2 or 3 in a leaf, but the larva is pale, of a watery-white colour, the head concolorous, the eyespots black, and the mouth pale brown. Its pallid colour makes it very difficult to see in the mine when full-grown, and this is quite impossible when young.

ADELIDES.—The larvæ of *Nemotois schiffermillerellus*, in their curious saddle-like cases, are now to be found feeding on *Ballota nigra*. After carefully searching the plants it is advisable to examine the ground under them, as the larvæ loosen their very slight hold on the slightest provocation. Should cases not be found at once, search should be continued, for it would not follow that the species was absent from the district, as when it occurs freely it is often confined to one spot.

TINEIDES.—During April the larvæ of *Psychoides verhuellella* are to be found in their burrows amongst the spores on the underside of fronds of *Asplenium ruta-muraria*, *A. trichomanes*, *Scolopendrium vulgare*, and *Ceterach officinarum*.

ELACHISTIDES.—In April, by looking at the bases of the leaves of *Carex glauca* that have been mined by the larvæ of *Elachista cinereopunctella*, the pupæ may be readily collected, their position being in the angle of the leaf. which it has mined as a larva. It frequents those plants that grow in the shelter of bushes. [Those growing among the junipers on Sanderstead Downs are very prolific in this species.]

The larvæ of *Elachista albifrontella* are to be found feeding in the upper part of the leaves of *Holcus mollis* in April.

In April the larvæ of *Elachista biatomella* are to be found making whitish blotches in the leaves of *Carex glauca*, the pupæ being placed in the angle of a lower leaf than that which had been mined. [Larvæ are to be found again in July.]

If, from mid-April to mid-May, the leaves of *Dactylis glomerata* are critically examined, some will be found to have a whitish streak. This is caused by a larva of *Elachista atricomella* or *E. luticomella*. When such a leaf is found it should not be separated from the plant, but the division of the roots producing it should be dug up and placed in a box or flower-pot.

The yellowish larva of *Elachista magnificella* may be found in mid-April and in May, mining in the leaves of *Luzula pilosa*, but owing to its concealed mode of life is very difficult to find, and requires much patience; after hybernation the larva commences at once to make its peculiarly inflated blotch-mine, very different from the narrow gallery of the autumn.

On bright, sunny days, *Heliozela* (*Tinagma*) *sericiella* may often be found flying in quantities at the end of oak twigs.

BUCCULATRIGIDES.—The larvæ of *Bucculatrix cristatella* are to be found in early May, mining the leaves of *Achillea millefolium*, and then eating the leaves halfway through from the outside. [Again in July.]

If leaves of *Chrysanthemum leucanthemum* have small holes through them, most likely larvæ of *Bucculatrix nigricomella* (*aurimaculella*) will still be found on their underside; but, should larvæ be absent, the beautiful white shuttle-like cocoons spun by them may be discovered attached to the plants and grass culms near at hand. The larvæ hang by a thread if the leaves are in the least shaken.

GRACILARIIDES.—In April and May the larvæ of *Gracilaria ononidis* make flat dipterous-looking blotches in the leaves of clover and *Ononis*.

Larvæ of *Gracilaria tringipennella* are to be found in leaves of *Plantago lanceolata* from mid-April to mid-May. They betray their presence by causing the leaves to contract through being mined down the centre. The larvæ do not quit their mines to pupate.

ARGYRESTHIIDES.—*Argyresthia praecocella* occurs towards the end of April amongst *Juniperus communis*. When beaten from its foodplant it is very loth to fly, except on calm and warm days. When such favourable conditions do not prevail, many may be secured by placing a sheet or umbrella under the bushes to be beaten.

The larva of *Argyresthia aurulentella* mines the leaves of juniper at the end of April, it enters from the upper surface and eats out the apical portion of the parenchyma; it then leaves its excrement in the mined leaf, and proceeds to repeat the process in another leaf, it never enters the stem.

TORTRICIDES.—*Spilonota pauperana* may be taken in its very restricted haunts during the latter part of April, and, in backward seasons, until the middle of May. It is not a free flier, and is most readily obtained by beating rose, on the bloom of which its larva feeds. A fairly mild and calm day must be selected when working for the species, as with an east or north wind it is well nigh impossible to induce a moth to move.

Catkins of sallows and aspens, which easily fall when the bushes or trees are shaken, should be collected, and will, in due time, give a varied series of *Grapholitha nisana*. The catkins should be placed in a box or small tub and kept out of doors, but not exposed to rain, or they may decay, and so destroy any larvæ or pupæ they contain.

Coccyx argyrana should be searched for in dull weather from the middle of April to the middle of May, on fences or oak-trunks, but, on sunny afternoons, may be found flying round the tips of oak branches.

If the main terminal shoots of *Pinus sylvestris* be carefully examined, it will be observed that, in some, the central bud is very much shorter than those surrounding it. This dwarfing is caused by the larva of *Retinia turionana*. On breaking off these abortive buds its pupa will be found inside. Needless to say, it is only young trees that can be conveniently worked.

The males of *Stigmonota internana* sometimes fly in late April and May, in little crowds round bushes of *Ulex europaeus* in the bright sunshine, from 11 a.m. to 3 p.m. The females are sluggish, and are best obtained by beating them from the bushes.

COLEOPHORIDES.—In April, the brown larvæ of *Coleophora wockeella* are to be found in elongated cases on the leaves of *Betonica officinalis* (*Stachys betonica*).

The larvæ of *Coleophora vitisella* are to be found in mid-April on *Vaccinium vitis-idaea*, after having hybernated, in dark brown, pistol-shaped cases; the larvæ may also be found in the autumn before hybernation.

PLUTELLIDES.—The terminal shoots of *Hesperis matronalis* are drawn together by larvæ of *Plutella porrectella* about the end of April, often doing considerable injury to garden varieties. Later in the season, pupæ are to be found in their silken hammocks on the under surfaces of the leaves.

The imagines of *Semioscopis steinkellneriana* are to be found flying freely on the wing from daybreak to sunrise in April.

HYPONOMEUTIDES.—*Swammerdamia spiniella* larvæ occur in white silken webs at the junction of twigs of *Prunus communis*.

When very young, in the middle of April, the larvæ of *Hyponomeuta plumbellus* eat the pith of the young shoots of spindle, causing them to droop; they afterwards quit this retreat and feed on the leaves.

GELECHIDES.—Those who are not averse to early rising may now, in April, spend some very profitable hours by searching the moss on old walls before the dew has evaporated. By so doing, the probabilities are that their energy will be rewarded by a good supply of larvæ of *Bryotropha domestica* and *B. affinis*. These larvæ are easily reared if kept in seed-pans on moist patches of their foodplant.

From about the middle of April, *Xenolechia* (*Lita*) *aethiops* is to be obtained in heathy places. Like other black moths that frequent heath and moors, it is always to be found most freely on those parts where the heather has been burned to encourage young growth for the benefit of grouse.

Lita fraternella larvæ occur during the early part of April in the drawn-together leading shoots of *Stellaria uliginosa*, *S. graminea*, and *Cerastium triviale*. Railway-banks and commons are the most likely places for them to occur. After having found one tenanted screwed-up shoot, it is best not to open others, as by so doing the larvæ will not have the chance to wriggle out and escape.

When lying on the sand, in April and May, searching for the sand-tubes of the larvæ of *Lita marmorea* attached to the low trailing stems and leaves of *Cerastium triviale*, the ground on which one has been lying should be examined, for it is often swarming with larvæ which have been disturbed and driven from their tubes.

Butalis incongruella is well out by the middle of April, and should be looked for in heathy places. It is almost useless to work for it except on absolutely calm and bright sunny days. On such occasions (unfortunately rare at this season) it flies abundantly for about two hours before noon.

The larvæ of *Depressaria nanatella* are to be obtained in April mining the upperside of the young leaves of *Carlina vulgaris*, frequently seven or eight occurring on one plant.

CRAMBIDES.—At the end of April, the larvæ of *Crambus inquinatellus* are to be found at the roots of the grasses growing in the localities the species frequents; they live in slight silken galleries spun near the roots on the surface of the soil, apparently hiding by day and feeding only at night. They pupate in May and the imagines emerge in late August and early September.

The larvæ of *Myelois cribrum* are to be obtained in April, feeding in the dried stems of thistle, eating neat circular holes through to enter a fresh stem, or to quit an old one.

PYRALIDES.—The larvæ of *Scoparia resinea* are to be found in April and May, feeding on the lichens and mosses growing on the trunks of ash, apple and elm trees (Barrett).

The larvæ of *Scoparia angustea* are to be found in April, feeding in silken galleries formed through the thick rounded cushions of wall-mosses, such as *Tortula muralis*, *T. revoluta*, &c., eating the solid substance of the moss without attacking the surface.

The larvæ of *Scoparia lineola* are to be found from April to June feeding on the lichens —*Parmelia parietina*, *P. olivacea*, &c.—growing on palings, the branches and twigs of hawthorn, blackthorn, &c., on old fences, stone walls, and rocks.

The larvæ of *Scoparia cembrae* are to be found in April and May feeding on the fleshy roots of *Picris hieracioides*, feeding on the superficial part of the large fleshy root, just beneath the surface of the ground, although sometimes working down to a depth of five or six inches, channelling the surface, and occasionally boring " through " it (Wood).

The puparium of *Scopula olivalis* is spun from mid-April to mid-May, by the larva folding and twisting a leaf up tightly, or joining two leaves together and spinning within the folded leaf, by way of cocoon, a very open-worked web of coarse meshes, in which it changes to a black pupa, with brown segmental divisions, antenna- and leg-cases.

In April, the larvæ of *Scopula prunalis* are to be found on *Galeobdolon luteum*, *Lamium*, *Stachys sylvatica*, *Teucrium scorodonia*, *Lychnis*, marjoram, *Sambucus nigra*, elm, honeysuckle, *Mercurialis perennis*, &c., taking possession of the underside of a leaf, which they turn down, thus making a covered-in abode that they leave to feed on the other part of the leaf not being thus used, piercing this with many holes, and continuing this habit until May, when the larvæ are fullfed.

In April, by collecting the nest-like bunches of twigs growing at the ends of branches on birch-trees, and placing in a large breeding-cage, large numbers of *Pyralis glaucinalis* were bred from mid-June until August. In mid-June larvæ, pupæ, and imagines were all alive in the breeding-cages at the same time (Walsingham).

One of these nests was composed of a mixed mass of birch twigs, decomposed leaves, and earthy matter containing many old cocoons and pupa-cases, small tufts of sheep's wool, &c., in which the larvæ of *Pyralis glaucinalis* constructed silken galleries uniting all the materials of which it was composed. The cocoons appear to be spun from the end of April till well into June (Buckle).

In April, the larvæ of *Hydrocampa stagnata* excavate large irregular perforations through the inner leaves of *Sparganium racemosum*, the ravaged parts, as the plants grow, turn blackish and become exposed to view above the surface of the water, thus affording a sure indication of the larvæ below.

The larvæ of *Stenia punctalis* are to be found in April and May on the dying and decayed leaves of bird's-foot-trefoil, clover, plantain, grass, &c., usually hiding under stones, even when they are wetted with seawater, feeding also, however, on the living leaves of the same plants.

CYMATOPHORIDES.—The eggs of *Asphalia flavicornis* are to be found in April, laid singly in the forks of the twigs of birch.

GEOMETRIDES.—The eggs of *Mesotype virgata* (*lineolata*) are to be obtained in April-May [also July-August]. The larvæ appear in about eight or ten days, feed readily on *Galium verum*, are fullfed in June, and the imagines emerge in July and August.

The eggs of *Carsia imbutata* hatch in April (after having gone through the winter in this stage). The young larvæ feed slowly on *Vaccinium vitis-idaea* and *V. oxycoccus*, spinning up in mid-June, the imagines appearing in July.

ALUCITIDES.—The larvæ of *Alucita pentadactyla* recommence feeding in April, by the 15th are about ¼in. long, and by mid-May are practically fullfed. They are to be found on *Convolvulus* (Porritt).

The larvæ of *Aciptilia tetradactyla* are to be found from mid-April to mid-June on wild thyme. They are difficult to detect and readily drop (Bankes). In April they are, of course, very small, and require much patient search in order to discover them.

The larvæ of *Stenoptilia pterodactyla* (*fuscus*) are well on the move by the end of April, if the season be a warm one, and the small larvæ, having left the winter-quarters into which they mined in the autumn, are now to be found feeding openly on the young flowering-shoots of *Veronica chamaedrys*.

HEPIALIDES.—The fullfed larvæ of *Hepialus humuli* may be dug up in April and May feeding at (and on) the roots of dock (and possibly many other low plants).

In April, the larvæ of *Hepialus lupulinus* are to be found feeding on the roots of *Lamium purpureum* and other herbaceous plants; the larva and pupa are both subterranean. Very fond of roots of lily-of-valley, and often destroys them in gardens.

At the end of April and throughout May, the larva of *Hepialus velleda* makes a cocoon of silk covered with particles of earth, which is generally placed on or near the surface of the ground amongst the loose vegetable soil in the neighbourhood of the foodplant, *Pteris aquilina*.

In April, the larvæ of *Hepialus hectus* leave their underground earthen hybernacula, attack the young shoots of *Pteris aquilina*, making oval excavations, which allow the sap to exude, and are full-fed at the end of May or early in June.

ÆGERIIDES.—The stems of black or red currant containing larvæ of *Ægeria tipuliformis* should be collected towards the end of April. Pruned shoots are the most profitable.

The best date to obtain pupæ of *Ægeria culiciformis* is on, or about, April 26th, when, by the aid of a hard chisel and heavy hammer, they may be readily split out of the two-year-old birch stumps. Keep moist and the imagines will emerge about July 7th (Musham). [Better to saw off 3 or 4 inches of the tops of the stumps.]

The cocoon of *Trochilium apiformis* is constructed in the larval gallery near its exit, and sometimes in the soil near the tree of *Populus nigra* on which the larva has been feeding. April and May are good months in which to search for them. They must be carefully treated and should not be opened, as the fullfed larvæ remain in the cocoons, before changing to pupæ, for a considerable time, sometimes for three or four months.

ANTHROCERIDES.—The hybernating larvæ of *Adscita statices* recommence feeding in early spring, and by the end of April they have attained a fair size; they enjoy the sun, often many feed quite close together, but are difficult to find owing to their feeding so low down on the lower leaves of the sorrel plants.

The fullfed larvæ of *Adscita geryon* are to be found in April and May on the rockrose. *Helianthemum vulgare*: they have been taken in numbers near Hartlepool, &c.

PSYCHIDES.—About the middle of April, larvæ of *Taleporia tubulosa* are to be found wandering over palings, tree-trunks, &c., searching for suitable places to which to affix their cases for pupation. If these are collected and enclosed in a box, the sides of which have been roughened, they will very shortly attach themselves, and, in due season, produce moths.

The made-up cases of *Solenobia inconspicuella* may now be gathered off lichen-covered tree-trunks, fences, walls, &c.

LYMANTRIIDES.—A batch of eggs of *Notolophus antiqua* will provide two or three, or more, larvæ a day for almost a month in April and May; the earliest and latest hatchings of a batch of eggs are usually at least a fortnight apart.

NOTODONTIDES.—In April, breeding-cages in which pupæ of *Petasia nubeculosa* have been reared, should be well watched, as the pupæ sometimes go over two, three, four, and five years before disclosing the imagines.

The young larvæ of *Ptilophora plumigera* emerge from the eggs during April, rarely before the middle, but in backward seasons quite at the end, of the month.

NOCTUIDES.—The, at first, conspicuous yellow eggs of *Taeniocampa opima* are obtained in large numbers, deposited on the old stems of marram-grass, ragwort, houndstongue, etc., on the sandhills of the coast of Cheshire and Lancashire, and many other coast and inland localities. After a few days they change to a leaden tint, and are then inconspicuous; they hatch towards the end of April or in early May.

The eggs of *Taeniocampa gothica* hatch in April and May, the young

APRIL. 53

larvæ feed freely on sallow and willow, pupating towards the end of June (or earlier in the south, later in the north).

The larvæ of *Citria flavago* and *C. fulvago* occur freely in sallow-catkins in April and early May, rarely both species in exactly the same haunts ; they feed well on sallow-leaves in confinement.

The larva of *Citria fulvago* has the ground colour of a reddish- or purplish-brown tint as far as the black spiracles; freckled with dark brown except the segmental divisions, the venter being paler, of a greyish-violet tint (with a bluish-green tinge on the anterior segments); in *C. flavago* the freckling extends only as far as the subdorsal region, and, whilst the larva of *C. fulvago* has a dark brown diamond-shaped mark of thickly packed freckles on the middle of each segment, that on the back of *C. flavago* is of an irregular squarish shape, whilst, in the latter, the subdorsal region forms the upper boundary of a very broad stripe of paler freckling, followed by a much narrower and still paler one, and then a broad one of similar depth of colour to the back, but with more of a violet hue (Buckler).

The hybernating larvæ of *Apamea unanimis* may be commonly found under loose bark on willows growing near damp ditches in April ; place the larvæ in a box with a little earth and moss, and without further care on your part, the perfect insect will appear in June (Greene).

The fullfed larvæ of *Xylophasia hepatica* may be found in April under damp moss on the stumps of trees, &c.; poplar trees are particularly favoured ; the larvæ simply want putting in a box with earth and moss, and require no further attention. The fullfed larvæ of *X. rurea* occur in similar places, and should be treated in a similar manner.

The eggs of *Xylina rhizolitha* are laid in April (after the hybernation of the moth) and hatch in about twelve days, the young larvæ feeding freely on the tender young leaves of oak, at first only eating the soft green tissue, but afterwards gnawing little holes right through the thickness of the leaves. They are fullfed towards the middle of June, and spin earthen cocoons some distance beneath the surface of the ground in which to pupate ; the moths appear about the end of September.

The eggs of *Xylina semibrunnea* are laid in April (after hybernation) on ash twigs ; the young larvæ appear during May and feed on the young ash leaves. They come to maturity very rapidly, becoming full-grown (when they are suffused with brown), and undergoing pupation in a month from the time of hatching. The moths appear in September and October.

The eggs of *Lithomia solidaginis* hatch towards the end of April, the young larvæ feeding up on whitethorn and bilberry, preferring the latter, which is possibly their food in nature.

The eggs of *Polia chi*, laid in August and September, hatch the following April (somewhat irregularly) ; the larvæ feed well on dock, sallow, osier, &c., preferring apparently a change of diet. They are fullfed about the end of May.

The eggs of *Catocala sponsa* hatch in April, just as the oak-buds begin to swell and the blossoms to appear ; the young larvæ will feed readily on the former, but prefer the latter, although after the second moult they feed freely on the leaves.

The eggs of *Catocala promissa* hatch towards the end of April. If the oak-leaves are not yet out it is advisable to find the most advanced oak-buds, especially those containing catkins on which the young larvæ will thrive well ; they feed only after dark, and rest all day stretched out at full length.   The larvæ are fullfed in early June.

In Perthshire, the larvæ of *Cirrhoedia xerampelina* are to be found, the first week in April, at dusk, crawling up from a burnside, at the roots of old stunted ash-trees, but only at those trees that have very prominent flower-buds.

DELTOIDES.—In April, the awakening larvæ of *Herminia derivalis* should be provided with fresh supplies of decaying oak-leaves, of which they will now consume a great quantity, being fullfed in late June and early July ; the larvæ pupate either in the corner of a leaf or between two leaves joined.

The hybernated females of *Hypena rostralis* taken in April or May should be enclosed with the foodplant (hop), for eggs, which will hatch in about 10-12 days from the time of laying.

LITHOSIIDES.—In a state of nature the larvæ of *Miltochrista miniata* have been found feeding upon the lichens that grow on the boles of oak-trees (Hellins).

PAPILIONIDES.—Eggs of *Cyaniris argiolus* are laid in April and May at the footstalks of flowers of holly, or on those of *Rhamnus frangula*, as well as on the young leaves of ivy; they hatch in a few days, and feed up on the flower-buds, the young green berries and tender young leaves, in about four weeks, when they pupate on a leaf or stem of the foodplant.

The larvæ of *Polyommatus bellargus* are to be obtained on *Hippocrepis comosa* in April and May, they pupate about the middle of May and emerge in June.

In late April the nearly fullfed larvæ of *Polyommatus icarus* are to be found on *Lotus corniculatus* and *Ononis arvensis*.

The larva of *Brenthis selene* feeds on *Viola canina*, appears to have an aversion to the sun's rays, reposing either on the undersides of the leaves or on the stems shaded by the leaves, selects always the youngest and tenderest leaves until nearly mature, eating out large portions of them and making its whereabouts conspicuous.

The larvæ of *Brenthis euphrosyne*, approaching full growth in April and early May, are to be found by searching the leaves of *Viola canina* and primrose, where there is much sign of the plants being eaten ; they generally hide and are to be found on the underside of a leaf, but when the sun is shining, love to bask in it, and are very active, retiring, however, as soon as the sun disappears.

The eggs of *Polygonia c-album* are laid in April on stinging-nettle and currant.   The larvæ appear in about eight days and require quite fresh young leaves if they are to be reared successfully in confinement.

The young hybernating larva of *Limenitis sibylla* begins to move in early April, moults almost at once, becomes reddish-brown in colour and spiny, feeds on the fresh bursting honeysuckle buds, and, by the middle of May, has usually assumed a miniature resemblance to the adult larva.

In late April or early May, the young larvæ of *Limenitis sibylla* may be found, and should be fed on young and tender shoots of *Lonicera periclymenum.* They nearly always commence to feed at the top of a shoot, and eat their way downwards, being especially fond of the sun, and always eating greedily when the sun is shining on them.

The larva of *Hipparchia semele* should be swept for in its known habitats, in April and May, by night, when it comes up to feed; it hides by day, often beneath the surface of the ground.

The hybernating larvæ of *Pararge egeria* are to be obtained in early April by the sides of the ridings and paths in woods, feeding on grass (*Dactylis glomerata*); the earliest spin up before the end of the month, and the imagines usually appear in early May.

The larvæ of *Coenonympha typhon* can be obtained on the moors in the early spring, on the beaked rush (*Rhynchospora alba*), on which they feed. They are fullfed in early June.

MAY.

ERIOCRANIIDES.—The larva of *Eriocrania salopiella* begins to mine the birch-leaves about the middle of May, rather more than a fortnight later than the early birch-feeding Eriocraniid larvæ. It commences with a long *Nepticula*-like mine, filled with black frass, making it permanently visible ; next comes a small, *pear-shaped*, dark brown blotch, followed by the usual wide-spreading blotch, which often surrounds and encloses the primary blotch.

The larva of *Eriocrania kaltenbachii* feeds in the earliest expanding nut-leaves (*Corylus avellana*), in the very first days of May. The blotch has very little preliminary gallery, and the fullgrown larva with pale brown head looks rather yellowish in the mine, the colour being partly due to that of the mine, and partly to that of the intestine.

The mines of *E. kaltenbachii* are also to be found on hornbeam, and are most conspicuous towards the end of May or in early June.

ADELIDES.—Often, at the end of May, *Veronica chamaedrys* has its charming flowers made even more attractive by the beautiful *Adela fibulella* resting on them.

If the heads of bloom of *Sisymbrium alliaria* are inspected during the latter part of May, *Adela rufimitrella* will be seen upon them. Later on in the season its larvæ are to be found in the seed-pods of the same plant.

ARGYRESTHIIDES.—At the end of May and beginning of June, in Leigh Woods, near Bristol, the imagines of *Röslerstammia erxlebella* were reported to be found freely on the wing, or *in cop.* upon the leaves of lime. The insect has since been proved to be attached to birch.

ELACHISTIDES.—If we look closely into ,the herbage of meadows among *Achillea*, we may see miniatures of the Coleophorid, *Coleophora murinipennella*, minute pale Tineids, *Bucculatrix cristatella*. While boxing these we catch sight of a small, compact Tineid sitting on a blade of grass, it looks black and white, and, if nimble enough, we may box *Elachista obscurella* ♀ .

As we continue our search almost among the roots of the grass, we notice one or two little white moths crawling on the lowest portion of the grass stems. These will be early specimens of *Elachista cygnipennella*. As we leave the meadow, we notice several pale moths flying along just above the herbage under the hedge ; these will be *Elachista rufocinerea*.

The fullfed larvæ of *Elachista rhynchosporella* are to be found in May mining down from the top of the leaves of *Eleocharis*. Imagines appear in June.

Towards the end of May a critical examination of leaves of *Aira caespitosa* and *Sesleria caerulea* may possibly result in finding larvæ of *Elachista adscitella*. The larvæ mine the grass leaves. The mines not being at all conspicuous are at first difficult to see, but when a few have been found one's eyes readily detect them. Collect the leaves when the larvæ are nearly fullfed, and place them in a jar (with water) and there will be no further trouble, as the larvæ, when fullfed, will quit their mines and pupate on the leaves.

The dull greenish-yellow larvæ of *Elachista triatomea* are to be found in May mining the tips of a fine grass.

The black larva of *Laverna conturbatella* feeds on *Epilobium angustifolium* in May and June, spinning together the terminal leaves of this plant, much as *Halias chlorana* does the terminal leaves of willows (Stainton).

TORTRICIDES.—In the latter half of May, an afternoon can profitably be spent in a meadow, if *Luzula* and *Achillea* be among the "weeds" growing there. While the sun is up we may net a good series of *Dicrorampha plumbagana* and *Glyphipteryx fuscoviridella*, both species flying over the herbage, though the Tortricid has much more dash in its flight than the Glyphipterygid. The latter frequently settling across a bent and "fanning" its wings.

We know that *Pamene rhediella* flies over the hawthorn bushes at noon in the sunshine, in May, and so does *Spuleria* (*Chrysoclista*) *aurifrontella*, about a fortnight later than the Tortricid.

The pupæ of *Argyrolepia maritimana* work up the dead stems of *Eryngium* to the top in May and early June, when the perfect insect makes its escape.

The imagines of *Argyrolepia maritimana* appear at the end of May and beginning of June, resting on the lower leaves of *Eryngium maritimum*, or on the sand under the plant. When disturbed their flight is short, seldom more than three or four yards, alighting on the sand or some neighbouring plant of *Eryngium*.

The females of *Argyrolepia maritimana* deposit their eggs on the top or heart of a plant of *Eryngium* (seldom more than one on a plant). The larvæ hatch in about twelve or fourteen days, and begin to work down the stem to the first joint, where they open a hole to work out their excrement ; they then work down again to the next joint, where they make another hole as at the first joint. Finally they work down the stem to the root, where they change to red-brown pupæ.

Towards the end of May, the larvæ of *Penthina capraeana* are to be found in the spun-together shoots of *Salix capraea*. This species has a decided preference for woods. The larvæ can be "sleeved " on growing sallow and left until they have pupated in the leaves.

If the leaves of *Vaccinium vitis-idaea* and *Arctostaphylos uva-ursi* be carefully examined, some may be found joined together and discoloured; these will contain, according to the season, larvæ or pupæ of *Euchromia mygindana*.

A close inspection of *Scilla nutans* will disclose silken webs amongst its spikes of blossoms; in these webs are feeding larvæ of *Sciaphila sinuana*. This moth is much more generally distributed than is commonly supposed, and, when more thoroughly worked for, will, no doubt, be found to have a very wide range.

About the middle of May, the larvæ of *Bactra furfurana* occur in stems of *Eleocharis palustris*. They should be collected in quantity, as the moths vary very considerably.

Sericoris euphorbiana flies freely in the afternoon sunshine amongst its foodplants, *Euphorbia paralias* and *E. amygdaloides*.

Coccyx ochsenheimeriana flies at the ends of branches of *Pinus cephalonica* and *P. smeathmanni* in the afternoon sunshine. Its small size and dark colour make it difficult to detect. If this moth were more sought after there is no doubt it would prove to be not so local as is the prevailing opinion.

At the end of May, *Phoxopteryx upupana*, in its restricted haunts, flies freely in the afternoon sunshine. It keeps very high up, so that, unless provided with a fifteen- or twenty-foot pole, hardly a specimen will be caught, though numbers may be seen.

About the middle of May, *Phoxopteryx lactana* is to be taken in considerable numbers at rest on stems of *Populus tremula*. Search for them should be made before the sun shines on the tree-trunks, as then they are comparatively restful, but when warmed by the sun's rays they are most annoyingly frisky.

The larvæ of *Exapate congellatella* are to be found in May and early June between united leaves on the terminal twigs of privet. [Wilkinson records the larvæ as having been found between united willow leaves in July, and Bower finds it on sallow.]

COLEOPHORIDES.—As the sun gets low a pale, narrow-winged moth may be seen softly flying from one grass bent to another, often settling on the *Luzula*. This will be *Coleophora murinipennella*. The cases of *Coleophora caespititiella*, are now to be obtained in quantity, on almost any clump of rushes in damp situations. The larvæ will be practically fullfed. A bunch of the seed-heads of the rush should be placed in an inverted glass-bottle, the bottom of which has been cut off, the stems being passed through the neck and standing in a jar with moist sand or water. The living cases can then be placed on the top of the rush-heads, and muslin tied over the bottle. In this way, if the apparatus be placed out-of-doors, abundance of imagines may be obtained. This method is a very convenient and successful one for the breeding of many of our Coleophorids.

In many places, if the plants of *Ballota nigra* growing luxuriantly by the roadsides (even very dusty ones) be examined, one may frequently find the leaves, with large membranous white blotches. Upon looking on their lower surfaces, large, nearly upright, side-flattened bunches of leaf-fragments are seen. These are the cases of the larvæ of *Coleophora lineolea*, and where they are found, are generally in abundance, but always considerably sheltered by a thick hedge.

The larvæ of two species, *Coleophora solitariella* and *C. olivaceella,* are now to be met with on *Stellaria holostea,* that brilliant little hedge-row star, which is so conspicuous at this time of the year. The former is always in much greater numbers than the latter; indeed, to get a respectable number of them is difficult. The cases are attached to the grass-like leaves of the foodplant, and the larvæ make conspicuous blotches. *C. olivaceella* is easily distinguished from that of *C. solitariella* by its darker case and slightly different angle of attachment to the leaf. The species also feed under some sheltering hedge or bush.

On bright afternoons in the middle and end of May the imagines of *Coleophora murinipennella* may be swept in numbers, flying low down in the fields near woods, where the wood-rush (*Luzula*) grows.

On most of our large heaths, assiduous sweeping near the shelter of trees and bushes, will produce a quantity of cases of *Coleophora juncicolella* and *C. pyrrhulipennella.* The former is our smallest species of the genus, and the cases are most difficult to find. The best way is to save all the sweepings in a bag and examine each day to see if any larvæ have crawled out. In the course of a week or ten days, no doubt a number will be obtained. The cases of the latter species are more conspicuous, and may easily be found by searching the sweep-net. To breed these species, one needs to have established plants of heath in pots and cover with muslin. Of course the plants must be kept out-of-doors.

The cases of *Coleophora saturatella* are to be found in May, the larvæ feeding on the leaves of broom, on which they make brownish blotches; the plants may be searched or beaten for the cases, which are large and rubbishy-looking, and the larvæ should not require much more feeding before they pupate.

A visit to the saltings, on which the seawormwood grows in abund-ance, may perchance produce the cases of *Coleophora artemisiella* and *C. maritima.* It is best to place a sweep-net under the base of the stems of the bunches of *Artemisia,* and beat the plants and the basal rubbish into it for examination. This is a tedious process, but other-wise very few larvæ will be obtained. The imagines can be bred by placing the larvæ on the garden *Artemisia,* known as "old man."

The cases of a species which seems known to but very few collec-tors, viz., *Coleophora ardeaepenella,* are now obtainable on birch. They are often mistaken for the cases of *C. ibipennella* among which they are usually found, both species often feeding on the same leaf. The cases of the latter species are almost prostrate, while those of the former are more upright in their attachment.

Young larch plantations should be searched for the presence of *Coleophora laricella.* The needles will be extensively browned by the depredations of the nearly fullfed larvæ if present, and large numbers of the cases may be obtained. As pine is an easy foodplant to keep, there will be little difficulty in breeding the imagines.

Many elms, we have been told in previous "Hints," will produce abundance of *Coleophora fuscedinella.* Among them, a careful search will produce the cases of two other species, the much darker, less bulky case of *C. badiipennella,* and the longer more compressed case of *C. limosipennella.* The case of the last species can never be mistaken for that of either of the others, as its anal opening is two-valved and not three-valved as they are. This species also occurs on birch, but the case is then much more slender and fragile.

If one meets with patches of *Eupatorium cannabinum*, they are worth examination to ascertain if the larvæ of *Coleophora troglydotella* are feeding on the leaves. The plants are found under shelter of the cliffs in some coast localities.

The larvæ of *Coleophora genistae* are to be found early in May on *Genista anglica*, upon the leaves of which it makes white blotches ; the case is formed of little bits of leaves added alternately in front and behind, and is pale greyish-ochreous in colour.

The cases of *Coleophora ibipennella* are to be found on birch in May, the form of the case of this species is very similar to that of *C. anatipennella*, but it is not so well developed behind, and the mouth of the case is cut off so obliquely that the case does not stand up perpendicularly to the surface to which it is attached as in *C. anatipennella*, but lies almost prostrate, the belly of the case resting on the leaf on which the caterpillar is feeding.

The larvæ of *Coleophora chalcogrammella* are to be found feeding on the leaves of *Cerastium arvense* in May ; the imagines appear in June.

HYPERCALLIDES.—The larvæ of the very local *Hypercallia christierniana* are to be obtained towards the end of May in united terminal shoots of *Polygala vulgaris*.

GELECHIIDES.—At the end of May and in June, the young larvæ of *Depressaria badiella* are to be found feeding on the underside of the leaves of *Hypochaeris radicata*, inhabiting fine white silken webs, or between two leaves spun together, and thus concealed, eating away the lower layers of the leaf till a transparent blotch becomes visible and betrays their presence, the brown marks thus formed being conspicuous on the upperside of the plant.

In May, the larvæ of *Depressaria nervosa* are to be found feeding on *Œnanthe crocata*, the pupæ are to be found later in the stems of the same plant.

In early May, the larvæ of *Gelechia domestica* are still to be found living in little silken pouches in moss on old walls.

The larvæ of *Gelechia pinguinella* are to be found feeding under moss on the trunks of poplar trees in May.

The larvæ of *Lita marmorea* are to be found in May, forming a loose web of silk interwoven with particles of sand on the surface of the ground, under plants of *Cerastium triviale ;* it frequents almost all our coast sandhills.

The slender larvæ of *Lita pictella* are to be found in May living in silken tubes formed just under the surface of the ground beneath plants of *Cerastium triviale*.

The green larva of *Sophronia humerella* feeds in May in the terminal shoots, and between spun leaves, of *Artemisia campestris*.

The larvæ of *Butalis senescens* are to be found in May, making little web-galleries amongst moss at the root of thyme.

The larva of *Nothris durdhamellus* (*schmidiellus*) is to be found in May, feeding on *Origanum vulgare*, retreating to the earth when disturbed. It is said to betray its presence by the scarlet leaves, but these certainly frequently remain green.

Œcophora grandis was not uncommon at the end of May and in June on one fence composed of dead and living hazel and birch. The imagines flew only in warm sunshine, from 10 a.m. to 1 p.m. ; when the sun was off the fence they could not be made to fly (Ashworth, *Zoologist*, p. 4814). [This locality has since been destroyed, and the insect has never been found again in Britain.]

By watching, on a sunny day, a fence well advanced in a state of decay, *Œcophora olivierella* will in all probability be seen flying along or running over it.

The cases containing larvæ of *Œcophora flavifrontella* are to be found in May on the trunks of beech-trees ; they are made of an elliptical piece of dead beech-leaf folded lengthwise, and closed by a silken structure. The larvæ remain concealed during the day among the withered leaves, and can be fed on beech-leaves and lichens (Fologne).

The brown larvæ of *Œcophora unitella* are to be found at the beginning of May, feeding on decayed bark of oak, &c.

CRAMBIDES.—In May, localities where *Platytes* (*Crambus*) *cerussellus* occur should be searched for larvæ of this species; they are to be found under stones, feeding on the roots of a short stiff species of grass.

In late May, the larvæ of *Crambus fascelinellus* are to be found on sand-denes (Yarmouth, &c.), in tubular galleries formed of spun-together particles of sand and silk, some five inches or more in length, which are usually extended up through the sand to the plant of *Triticum junceum*, the hinder part being packed with frass.

The larvæ of *Crambus selasellus* are to be found in May and June in green grass-covered tubes or galleries partly attached to stones, among mixed growths of *Poa maritima*, *Spartium stricta* and *Hordeum maritimum*, in marshy places near the sea; in similar situations among other grasses in fenny and marshy districts.

In late May and June, by turning over stones on salterns, sea-banks, &c., that rest on or near small tufts of *Poa maritima* and *P. borreri*, the shortish tubular galleries of *Crambus salinellus* are readily seen, attached to the lower whitish sheaths of the grass towards the roots, being conspicuous, however small, by their covering of fine greenish frass, or frass and fine grains of earth together. Sometimes the gallery is partly spun against the stone itself, when the sudden removal of the latter tears open the gallery and the surprised larva drops out.

In May and June, the larvæ of *Chilo phragmitellus* are sometimes to be obtained in abundance, by pulling at the tall withered previous year's reed-stems along ditchsides; the stems tenanted by larvæ or pupæ break off near the roots. The larvæ feed in the reed-stem, just below the surface of the ground, and frequently, probably usually, under water (Porritt).

The larva of *Chilo phragmitellus*, just before pupation, in May and June, gnaws an oval hole from within, at the side of the reed, covering it carefully with silk and bits of dry reed, so as almost entirely to conceal it ; this serves later as an outlet for the escape of the moth.

The larvæ of *Rhodophaea marmorea* are to be found in May on dwarf sloe-bushes, forming a loosely spun web on the stems, often uniting leaves with the web thus spun, in which they live.

The larvæ of *Nephopteryx roborella* are to be beaten in May from oak; they live under a silken web, but are by no means gregarious, and injure one another if kept too crowded in confinement.

At the end of May, the larvæ of *Euzophera* (*Myelois*) *pinguis* should be sought in the living bark of ash, forming little galleries or chambers somewhat near the surface of the bark, never entering so far as the wood. The presence of particles of frass blocking the point of entrance, and often lodging on projecting points near the opening, betrays the whereabouts of the larvæ. [The larvæ and pupæ may be obtained throughout June, July, and August, some larvæ possibly passing two years in this state.]

The hybernating larvæ of *Ilythia carnella* commence to move about the beginning of May, and spin over the leaflets of *Lotus* to the stem, their presence being shown by the frass pellets held in the newly-spun silk.

The pupæ of *Myelois cribrum* are to be found in May in a net-like cocoon of white threads in the cavity formed by having eaten away the pith in the dried stems of the thistles in which the larvæ have fed.

PYRALIDES.—The larvæ of *Scoparia crataegella* are to be found in tubular webs formed in the mosses growing on tree-trunks, old walls, dykes, &c., in May (Barrett).

The pupæ of *Scoparia murana* are to be found in May (and August) in cavities, lined with silk, hollowed out at the roots of the mosses growing on old walls, rocks, &c.

The imagines of *Scoparia murana* are to be found in late May and early June (also again in August), on moss-covered walls and rocks in moorland and mountainous districts, rarely in the neighbourhood of towns or in the south. It is very quiet and not easily disturbed, and so can be boxed with ease.

The imagines of *Scoparia angustea* are to be found in May (and again in September), sitting sluggishly on mossy walls, and usually allowing themselves to be taken quite easily. At night, in its localities, it is sometimes quite common at light.

In May, the cases of *Cataclysta lemnalis* are to be obtained. They are of an irregular oval form, and somewhat variable in width, made of leaves of *Lemna minor* overlapping each other irregularly, and with a leaf or two hanging down at each end, hiding the openings; they resemble almost exactly an accidental accumulation of duckweed on the surface of the water. The larvæ pupate within the same cases that have served them for dwellings.

In May, the larvæ of *Stenia punctalis* were found at Freshwater, feeding under stones, beneath their silken coverings, on vegetable rubbish composed of grass stems and roots, dead leaves of plants, and withered *Zostera marina*, these pupated in May and June.

The larvæ of *Endotricha flammealis* spin their cocoons on the surface of the ground in May; they form oval silken cocoons covered with particles of earth and dead leaves, or they may be placed in curled decayed leaves of sallow, hazel, and hornbeam.

In May, the larvæ of *Ebulea crocealis* are to be obtained in the growing central shoots of *Inula dysenterica* and *I. conyza.*

The fullfed larvæ of *Botys asinalis* are to be found in May, feeding on *Rubia peregrina:* in some seasons, the larvæ are so abundant in the neighbourhood of Bristol that the conspicuous marks made by them on the madder plants form quite a feature in the locality (Porritt).

At the end of May and in early June, the large larvæ of *Botys verticalis* are to be found spun up in the twisted leaves of stinging-nettle ; they make a large roomy puparium of nettle leaves drawn together ; neither the leaves containing larvæ nor the puparia can be overlooked, so conspicuous are they.

DIMORPHIDES.—The imagines of *Dimorpha versicolora* rarely emerge before early May in Rannoch ; the ♀ s are then to be seen at rest, the ♂ s flying swiftly in the sun. Eggs are laid freely in confinement and usually hatch in about three weeks.

GEOMETRIDES.—The larvæ of *Phorodesma smaragdaria* sometimes feed at dusk, but more often during the morning sunshine, and, at times, when the sun is hot, they eat most voraciously, appearing to be in a very excited state during the whole time the sun is shining upon them (Elisha).

The larvæ of *Cleora glabraria* are to be obtained in May and June in the New Forest, feeding on the points of *Usnea barbata* growing on oaks, sometimes eating the lichen well down for an eighth of an inch or more. The larvæ will starve unless the lichen is kept well moistened with water, as they are quite unable to feed if it gets at all dry.

In early May, the fullfed larvæ of *Boarmia abietaria* are to be beaten from yew. They appear to be very quiet during the day remaining quite still, but are more active at night, which seems to be their natural time of feeding.

The eggs of *Boarmia cinctaria* hatch towards the end of May, the young larvæ eating minute patches of cuticle from the underside of birch leaves, causing transparent specks to appear on the upper surfaces of the leaves, a good sign when searching for them. They are fullfed in early July and want earth in which to pupate.

The eggs of *Tephrosia consonaria* hatch towards the end of May ; the larvæ feed exceedingly well on oak and birch, and are fullfed about the end of July.

In confinement, the ♀ s of *Nyssia lapponaria* require some folded paper, leno, or chinks of some kind into which to push their ovipositors to entice them to lay their eggs. So also do those of *Amphidasys strataria*, *A. betularia*, *Nyssia hispidaria*, *Tephrosia bistortata*, *T. crepuscularia* (*T. biundularia*), *Orrhodia vaccinii*, &c.

The eggs of *Nyssia lapponaria* hatch during May ; the larvæ feed on birch and are fullfed about the end of June.

The larvæ of *Nyssia lapponaria* are to be obtained in May and June, frequently resting in a straight position along the stems of birch, at other times with only the claspers grasping a stem and the head a short distance away from it. The colour varies from pale yellow to dark purplish-brown and putty-colour.

The larvæ of *Nyssia hispidaria* are remarkably rapid feeders, going ahead during May at a great pace and being fullfed by the end of the month.

In May and early June, the larvæ of *Phigalia pedaria* are to be found on hawthorn, oak, birch and elm, and those of *Nyssia hispidaria* on oak. The larvæ of both species have eight pairs of dorsal and eight pairs of subdorsal warts on the abdominal segments. The larvæ, how-

-ever, may be separated, because in *N. hispidaria* the warts, although not uniform in size, do not vary much, the dorsal ones on the 8th abdominal and the subdorsal on the 2nd abdominal being largest, whilst in *P. pedaria* the dorsal and subdorsal warts of the 2nd and 3rd abdominal segments are much larger than the remainder, whilst some ∧-shaped ochreous marks on the same segments in the latter species also help to distinguish it.

In May, watch the eggs of *Ennomos autumnaria*, their dark colour masks any outward sign of the nearness of approaching hatching, which usually takes place between May 15th and May 30th, dependent on the season.

The larvæ of *Oporabia filigrammaria* are to be obtained about the middle of May in Yorkshire, on open heaths, on *Calluna vulgaris;* the imagines appear throughout August.

The imagines of *Hypsipetes ruberata* will fly in May, on nights when scarcely anything else is on the wing, but the species is exceedingly local although sallow may be widely distributed over the neighbourhood where it occurs.

The eggs of *Phibalapteryx lapidata* pass the winter in that stage, hatching in May, the larvæ feeding in confinement on *Clematis*, and becoming fullfed about the end of June (Hellins).

The larvæ of *Mesotype virgata* are to be found in May and June feeding on *Galium verum* (and *G. saxatile*); they are fullfed and pupate in June, most of the imagines emerging in August.

In late May, the larvæ of *Tanagra atrata* are to be obtained from the flowers of *Bunium flexuosum*, one of the Umbellifers. [It is to be noted that though this species has been named *chaerophyllata*, *Chaerophyllum*, *i.e.*, chervil, is *not* its foodplant.]

The eggs of *Eupithecia consignata* are laid towards the end of May, hatch in a few days, the larvæ feeding on apple and being fullfed about the middle of June. The larva is very long and slender and twists the abdominal portion in a very active manner.

The larvæ of *Eupithecia subciliata* are to be beaten from maple blossom during May; the species appears to be very generally distributed in Kent, Surrey, Essex, Suffolk, Norfolk, &c. (Boxhill is a well-known locality); the larvæ are fullfed by the end of the month, the imagines appearing in July and August.

ALUCITIDES.—The larvæ of *Aciptilia baliodactyla* are to be found at the end of May on the top shoots of marjoram, *Origanum vulgare;* they bite practically through the stems near the tops of the plant, causing them to hang down and wither, which is the sign betraying the presence of the larvæ; they eat large holes through the leaves as well as portions out from the edges (Grigg).

The larvæ of *Aciptilia spilodactyla* are to be found feeding on *Marrubium vulgare* in late May and June; they are to be found again in late August and September; a partial second-brood of imagines appears to occur in late September and early October (Buckler).

The larvæ of *Leioptilus tephradactylus* are to be found throughout May, feeding exposed on the leaves of golden-rod, the pupa being attached by the anal segment to a stem or leaf of the foodplant (Porritt).

The larvæ of *Stenoptilia pterodactyla* (*fuscus*) are to be found during May and the first half of June feeding on speedwell in more or less exposed situations—banks, &c. (Porritt). They are generally found half-hidden among the flower-buds.

The larvæ of *Stenoptilia* (*Mimaeseoptilus*) *bipunctidactyla* are to be found in mid-May working up inside the young shoots of *Scabiosa columbaria*, *S. arvensis*, and *S. succisa;* the infested portions of the plants are concealed by the healthy shoots, and the whereabouts of the larvæ are not to be discovered without difficulty (Barrett).

The larvæ of *Marasmarcha phaeodactyla* are to be found readily by searching plants of restharrow (*Ononis*) at the end of May and throughout the first half of June ; they are moderately exposed, and occur mostly towards the upper parts of the plants.

The larvæ of *Oxyptilus heterodactyla* are to be found on *Teucrium scorodonia* during May ; they eat the stem about half-way through, about 1½in. from the bottom of a shoot, causing the part of the plant above to bend down, and soon this withered portion on which the larva feeds is overtopped by the neighbouring plants (Greening).

In the second week of May, the larvæ of *Platyptilia isodactyla* are to be found in marshy places mining the stems of *Senecio aquaticus*, feeding in the thick main stem of the plant, in which each larva hollows out a space in which to assume the pupal state (Barrett).

HEPIALIDES.—At the end of May and early in June, the fullfed larvæ of *Hepialus hectus* spin, near the plants of *Pteris aquilina*, on which they feed, an oblong cocoon, covered with soil, on the surface of the earth under moss or among dried leaves, in which to pupate ; the pupal stage rarely lasts more than a fortnight.

ZEUZERIDES.—The larva of *Phragmatoecia arundinis* lives for two years as an internal feeder in the lower part of the stems of *Arundo phragmites*, *i.e.*, underground, but comes up in May and June when the pupation-period is approaching.

During May, the larvæ of *Phragmatoecia arundinis* are of various sizes, but many quite fullgrown. These pupate within the stems of *Arundo phragmites* in June, the imagines emerging in July.

ÆGERIIDES.—The larvæ of *Ægeria chrysidiformis* are to be found in May and June, feeding in the thickest portion of the roots of *Rumex acetosa*, ejecting heaps of brown frass at both ends of the mine, and spinning a tough brown silken covering over any part of the side which has been eaten quite through, so that an affected stem is easily detected.

The full-grown larvæ and pupæ of *Ægeria cynipiformis* are best collected in late May and early June, the imagines appearing towards the end of the latter month and on into July. Care must be taken not to interfere with the pupæ when they emerge from their cocoons, which they do some little time before the imagines appear.

The pupa of *Ægeria tipuliformis* is to be found in a gallery in the stems of currant-bushes (pruned preferably) the thinnest possible layer of rind alone being left to separate it from the outside ; in the gallery the larva spins a slight silken cocoon, woven with the sawdust-like frass.

In the Gateshead district, the larvæ of *Ægeria formiciformis* is best obtained in decaying stems of *Salix capraea*, more especially in those stems of which only one side has commenced to decay ; large numbers of pupæ are sometimes to be found by removing a piece of bark from such a sallow-stem; the old sallows growing by river-beds are here certainly the most frequented (Harrison). [As we have pointed out in previous notes the species is especially fond of osier-beds in most parts of the country.]

ANTHROCERIDES.—In early May, the mines of *Rhagades globulariae*, in the leaves of *Centaurea nigra*, become very conspicuous blisters, occupying almost the whole of a leaf, and appearing very marked when the frass is not entirely extruded. The fullfed larvæ, at the end of the month, sometimes eat right through the leaves.

In late May and early June, the fullfed larvæ of *Rhagades globulariae* burrow just below the surface of the soil, making oval cocoons covered with grains of earth and grass fibres, and lined with silk. In confinement, great care with the quality of the earth should be taken.

The larva of *Adscita geryon* spins its tough little web-like cocoon in May or June, low down among the stems of *Helianthemum vulgare*; the imago emerges in June or July.

The larvæ, pupæ, and imagines of *Adscita geryon* are sometimes, in favourable seasons, very forward in their transformations. On May 18th, 1864, larvæ, pupæ, and imagines were taken at the same time (Horton).

The fullfed larvæ of *Anthrocera purpuralis* spin up from the middle of May, and the imagines emerge in about a month. The species is earlier in its appearance in Britain than on the continent.

Full-grown larvæ of *Anthrocera trifolii* are to be found in May; they pupate low down near the ground, and are largely restricted to dry situations, the imagines emerging in early June (or even in late May in very early seasons).

The so-called *Anthrocera filipendulae* captured in late May and early June, with a tiny sixth spot (in male), should be put aside for examination as probable *Anthrocera stephensi* (*hippocrepidis*, St.).

NOTODONTIDES.—The ova of *Petasia nubeculosa* hatch from the beginning to the middle of May in normal seasons, and the young larvæ will feed on birch or oak (preferring birch).

In late May, the young larvæ of *Petasia nubeculosa* are to be found on birch eating small holes quite through the leaves, and spinning a few silken threads in order to ensure a safer foothold.

The larvæ of *Petasia nubeculosa* appear to be exceedingly easy to rear on a birch diet, with a little oak given occasionally, provided that, in their later stages, they are well-syringed each morning with a fine-nozzled syringe. This practice holds good for all the larvæ of Notodontids, and effectually does away with all cannibalism (Musham).

The earliest hatched larvæ (May and early June) of *Stauropus fagi*, frequently feed up rapidly, and produce pupæ and imagines in August. The extent of this partial second-brood depends much on the season.

The brown eggs of *Cerura vinula* are readily found in May, in early seasons, on the leaves of poplar, sallow, and willow; they are not unlike the little fungoid buttons found on sallow leaves.

The young larvæ of *Cerura vinula* are easily found on poplar and willow in late May and early June, although their peculiar dark coloration when young makes them very similar to the black curled edges so frequently seen in spring on quite young poplar leaves.

The larvæ of *Lophopteryx carmelita* feed up very rapidly in May and June, only about four weeks elapsing from the time the eggs hatch until the larvæ commence to spin.

In early May, the young larvæ of *Ptilophora plumigera* feed on the buds of maple, their tint exactly matching that of the enveloping sheath of the bud.

As soon as the young leaves of *Acer campestre* unfold, the yellow-green larva of *Ptilophora plumigera* takes up its characteristic position on the underside of a leaf, where it rests in a sort of curved posture, the head bent round on one side until it almost touches the 2nd abdominal segment.

The larvæ of *Ptilophora plumigera* will feed on sycamore as well as on maple. When young, in May, they are often more or less gregarious in habit, two or even three being folded round on the underside of a maple-leaf, so that where one larva is found others are usually near.

The imagines of *Pterostoma palpina* are sometimes attracted freely to light in May and June. [Also in August.]

The whitish-green eggs of *Notodonta ziczac* are to be found on leaves of willow and sallow in May and June by searching. [Also in late July and August.]

The eggs of *Peridea trepida* are laid in May on the leaves or bark of oak; the young larvæ rest with the head and tail free from the leaf as when older.

DELTOIDES.—In May and June, where the leaves of *Brachypodium sylvaticum* and *B. pinnatum* show signs of having been eaten, search should be made for the green larva of *Rivula sericealis*. The feeding of the larva is rather conspicuous on the grass, but it neither attacks the extreme point nor the midrib until after its last moult, when it commences feeding at the top of a leaf of the grass, and eats downward through the midrib from one edge to the other in rather an oblique direction across the full breadth of the leaf. Sometimes two or three leaves are thus eaten, more or less, but when its appetite is nearly satisfied, it eats only about three parts across the truncated top edge so as to leave a portion uncut on one side, generally about ½in. long (Buckler).

When fullfed, the larva, a little below the cut edge described in the last hint, spins a few silk threads, which draw the sides of the leaf a little towards each other, and form an oval-shaped hollow, in part lined with silk, wherein the larva, forming a Pierid-like cincture, changes to a pupa. Before pupation, however, the larva draws down obliquely the uncut portion of the top edge of the leaf over the top of the hollow, thus hiding the pupa, although it is all on the upper surface of the leaf (Buckler).

NOCTUIDES.—The eggs of *Agrotis cinerea*, laid at the end of May or in early June, hatch in about twelve days; the young larvæ should be fed on wild thyme (not on grasses), of which at first they gnaw the under-surface of the leaves. The larvæ are almost fullfed by the time they are ready to hybernate in October.

The larvæ of *Agrotis nigricans* sometimes occur in May in large numbers, doing great damage to clover and cultivated crops, but are more frequently found by ditch-sides, where they are very general feeders on low plants, though they appear to object to grass. In confinement, they will eat clover quite freely. The larvæ are distinguishable by the double white stripe above the feet.

The larvæ of *Agrotis obscura* (*ravida*) are fullfed from the beginning to the end of May, when they are to be found just below the surface of the soil in ground suitable for burrowing, chiefly at the roots of thistles and dandelion. In confinement, they feed freely on the leaves of the latter plant.

The larvæ of *Pachnobia hyperborea* (*alpina*) are obtained in numbers by the professional collectors, in the Rannoch district and in the Shetland Isles, in May and June, where they are said to be taken on *Empetrum nigrum*. Buckler found they would eat bilberry, birch-leaves (a little withered), arbutus, etc.

In May, the larvæ of *Taeniocampa opima* may be reared on sallow, *Rosa spinosissima*, &c., feeding most voraciously for about a month, becoming, as a rule, fullfed towards the middle of June.

The larvæ of *Tethea retusa* are to be found not uncommonly in some seasons by searching in the drawn-together leaves and shoots of various kind of sallows and willows about the end of May and beginning of June. The larva may readily be confounded with that of *Cleoceris viminalis*, but the small and rather flat yellow-green or blackish-brown head, the delicate and thin skin, the occasional dark collar on the pro-thorax, and the loss of all traces of the usual dots, distinguish the larva of *T. retusa* from that of *C. viminalis*, which has a much firmer texture of skin, a thick and more corneous head (pale grey in colour, with the lobes outlined in black), the lines more sharply defined, and two pairs of whitish dots on the back of each segment.

The young larvæ of *Dyschorista suspecta*, in May, are very like the young larvæ of *Orrhodia vaccinii*, but, whilst the former spin leaves together in which to hide, the latter appear to be satisfied with such shelter as curled or touching leaves afford. [The most marked differences in the larger larvæ are the distinctness of the pale dorsal line and the breaking of the subdorsal in *D. suspecta*, and the distinct-ness of the subdorsal and narrow, inconspicuous dorsal line in *O. vaccinii*.]

The larvæ of *Cosmia affinis* are sometimes to be beaten in numbers from elm towards the end of May or in early June ; the fullfed larvæ spin loose cocoons among or under the leaves of their foodplant, and are easily reared.

The larva of *Dicycla oo* should be beaten from the middle of May onwards ; it appears to be somewhat retired in its habits, keeping itself in seclusion among the oak-leaves on which it lives, and becoming fullfed about the end of the month.

The larva of *Tiliacea citrago* feeds on lime in May, drawing two leaves together in which to hide at the moulting-season (Porritt).

The eggs of *Heliodes arbuti* are laid singly at the end of May or in early June in or on the flowers of *Cerastium arvense* and *C. vulgatum*. They are usually hidden in the calyx or corolla of the flower, and the young larvæ, which hatch in a few days, feed, well covered, on the flowers and unripe seeds. After a few days, they eat their way out of the seed-capsules and feed more exposed, clearing out the capsules somewhat after the manner of Dianthœciid larvæ. They are fullfed towards the end of June, and enter the ground for pupation. The imagines emerge the following May.

68 PRACTICAL HINTS FOR THE FIELD LEPIDOPTERIST.

LITHOSIIDES.—Lichens on old stone walls are to be searched for larvæ of *Nudaria mundana* in May and June.

In May, the fullfed larvæ of *Setina irrorella* are to be found feeding on a blackish-brown lichen growing on stones, often, in seaside localities, only just above high-water mark, and, in some cases, mixed with a yellow lichen.

The larvæ of *Setina irrorella* are fond of sunshine, moving slowly over the stones and sunning themselves.

The cocoons of *Setina irrorella* are to be found spun-up amongst the stones and *débris* beneath the foodplant.

Larvæ of *Cybosia mesomella* may be obtained in May feeding on a pale lichen intermixed with the moss growing on the trunks of oak-trees.

The larvæ of *Lithosia muscerda* are fullfed in May and June, and, in the Norfolk Broads, the species affects the sallow-bushes inhabiting the wettest parts, the larvæ most probably feeding on the lichens growing on these bushes throughout the autumn and winter from August to May.

The larvæ of *Lithosia caniola* are fullfed in May and June, and, in confinement, will eat clover (Buckler).

The larvæ of *Lithosia complana* are to be found in May and June ; they feed well on lichens from fir-trees (Buckler).

The larvæ of *Deiopeia pulchella* feed from May to June on *Myosotis palustris*, preferring the flowers and young seeds ; they are fullfed at the end of June and early July, when pupation takes place, the imagines emerging in late July and early August.

HESPERIIDES.—The pale greenish eggs of *Nisoniades tages* are laid on the leaflets of *Lotus corniculatus* from the end of May to the middle of June ; the egg-stage lasts about a fortnight.

The fullfed larvæ of *Pamphila sylvanus* are to be found in early May on *Luzula pilosa*, the edges of a leaf of which are folded over and lined with silk to form a puparium, in which the larva changes to a chrysalis (Buckler).

The larvæ of *Thymelicus actaeon* are to be found in May and June on the sea-slopes from Swanage to Weymouth, feeding on the leaves of *Brachypodium sylvaticum*. In confinement they will eat *Triticum repens* and the allied grasses.

The presence of the larvæ of *Thymelicus actaeon* is best told by the wedge-shaped pieces which they eat out of the side of the blades of *Brachypodium sylvaticum*. When such traces are observed, search for the silk-lined hollow tubes in which they hide, and which are made by spinning the two edges of a leaf together, so as to enclose themselves therein.

PAPILIONIDES.—The larvæ of *Polyommatus astrarche* var. *salmacis* are to be taken in late May and early June on *Helianthemum vulgare* in the northern counties of England. From pupæ formed from larvæ obtained June 3rd, 1877, near Hartlepool, three imagines—apparently *salmacis*, *artaxerxes*, and *astrarche* on the upperside, but more like *salmacis* on the underside, emerged (Buckler).

The larvæ of *Polyommatus* var. *artaxerxes* are to be found on the undersides of the leaves of *Helianthemum vulgare* throughout May, their colour assimilating remarkably well with that of the under-side of the leaves of the foodplant ; the larvæ pupate towards the end of the month in a nearly perpendicular position amongst, and slightly attached to, the stems of the *Helianthemum*, by a few silk threads near the ground.

In May, the eggs of *Nemeobius lucina* may be obtained by enclosing caught females on potted plants of cowslip. A few will be laid on the upper- but most on the under-surface of the leaves. The egg-stage lasts about 17 days.

In early May, the larvæ of *Melitaea athalia* are to be found on *Melampyrum pratense*, *Plantago major*, and *P. lanceolata*, the first-named foodplant being apparently the most preferable.

Captured females of *Brenthis euphrosyne*, taken in late May or early June, lay their eggs freely on the leaves of *Viola canina*. These hatch in a few days, feed well on to hybernation, but they want great care to bring them through to maturity.

The greenish-white bluntly conical eggs of *Pararge megaera* are laid singly on grass blades in late May and early June ; they hatch in about twelve days, the larvæ feeding up, and producing pupæ and imagines in about two months.

The larvæ of *Limenitis sibylla* are to be searched for in May, on honeysuckle, when they may be found with careful work. They are about half-an-inch in length at the commencement, and nearly fullfed towards the end, of the month, or in early June.

The eggs of *Gonepteryx rhamni* are laid singly, generally on a rib on the underside of a leaf of *Rhamnus frangula* or *R. catharticus*, sometimes on the upperside. Although laid singly, two or three eggs may be found on the same leaf, possibly deposited by different females.

In confinement, the larvæ of *Pieris napi* will feed well on horse-radish. They also eat *Nasturtium officinale*, *Barbarea vulgaris*, &c.

The eggs of *Leptosia sinapis* are laid in May on *Vicia cracca* and *Orobus tuberosus*. The larvæ will eat either of these plants in confinement, and pupate in July. Some of the pupæ disclose their imagines in late July or early August as a second brood, others go over the winter until the spring.

JUNE.

ERIOCRANIIDES.—The larva of *Eriocrania sparrmanella* is the latest of the birch-mining Eriocraniids, and does not make its appearance until the middle or end of June. It commences with a long, narrow, *Nepticula*-like gallery, expanding into a rectangular, dark-brown blotch, followed by the usual wide-spreading mine, which sometimes surrounds the primary blotch.

NEPTICULIDES.—The larva of *Nepticula poterii* is to be found early in June, mining the leaves of *Poterium sanguisorba* ; the mine is at first slender and nearly filled up with dark grey excrement, going round the edges of the leaf ; the larva ultimately eats out the central portion of the leaf, when the mine appears almost as a blotch.

ADELIDES.—*Phylloporia bistrigella* flies throughout this month among its foodplant (*Betula alba*). It is to be obtained equally freely either in

the morning or afternoon.   Unless a position is chosen giving a fairly
open and clear space, so that one may see the moths flying against a
light background, very few will be caught or even seen.
A sunny afternoon during the early part of June should produce
Lampronia luzella. The moth flies fast, and just over low growing
vegetation, being most partial to wide ridings in woods. If carried
for only a short time in pillboxes in one's pocket or satchel, this
species is utterly spoiled and generally dies. The only method to
obviate this trouble appears to be by filling a tin with freshly gathered
grass and burying in it the pillboxes containing moths.

ELACHISTIDES.—From the commencement to the third week in
June, *Elachista gangabella* flies in the late afternoon, but not unless
the day is a calm one.   Unlike most of the species of the genus
Elachista, it does not fly amongst its foodplant (*Dactylis glomerata*), but
from four to five feet from the ground.   It frequents the ridings in
woods and hedgerows.
The larvæ of *Elachista trapeziella* are to be found mining the leaves
of *Luzula pilosa* in the middle of June.

ARGYRESTHIIDES.—*Argyresthia abdominalis* occurs amongst *Juniperus
communis* at the end of June.   It is best obtained by beating, but a
sheet or an umbrella should be placed under the bushes, as far more
moths fall to the ground than take wing.
Argyresthia glaucinella occurs during this month amongst oak,
preferring the scrubby pollards usually to be found in well-kept hedgerows.   The moth is most unwilling to fly when beaten from its place
of concealment, and has a very unpleasing knack of falling to the
ground, when, unless one is prepared for the habit, it is more likely to
be taken for anything than a moth.

GLYPHIPTERYGIDES.—In mid-June, if *Sedum acre* be carefully watched
in the sunshine, most probably the beautiful *Glyphipteryx equitella* will
be found flying over or resting upon it.

GRACILARIIDES.—The larvæ of *Coriscium brongniardellum* are excessively abundant in June and July on the oaks between Woking and
Guildford, making blotches in the leaves.

TORTRICIDES.—If, during the early part of this month, the terminal
shoots of *Salix capraea* are carefully examined, some will be found to
have their leaves spun together. These may contain pupæ of *Penthina
capraeana.*
Anyone visiting the coast in the Shields and Hartlepool districts
during the first fortnight in June, should obtain pupæ of *Ephippiphora
grandaevana.*   They are to be found in long silken tubes at the roots
of *Tussilago farfara.*   The best way to secure the pupæ is to thrust
one's fingers well into the sand round the coltsfoot and gently move
them from side to side. This will cause the sand to fall away and
leave the tubes exposed.
Throughout this month, *Stigmonota nitidana* is to be found flying
round oak-trees or sitting on their leaves in the sunshine. This species
is seldom moving before 3 p.m. or after 6.30 p.m., and is confined to
woodland districts.
About the third week in June is a good time to collect pupæ of

Tortrix branderiana. They are to be found in leaves of *Populus tremula*, having a corner turned over and secured at intervals by sundry strands of white silk. It is well to sever these silken bands, or the aspen leaf, by contraction in drying, may press upon and injure the pupa. The pupa, which is jet black, is small in comparison with the moth it produces.

Towards the end of the month, *Peronea shepherdana* larvæ are to be sought in drawn-together terminal shoots of *Spiraea ulmaria*. The species is very local, being strictly confined to fen and marsh lands.

The larvæ of *Paedisca consequana* are common in June at Portland on *Euphorbia portlandica*.

The imagines of *Eupoecilia sodaliana* occur between mid-June and mid-July; sluggish by day, but found sitting on buckthorn-leaves or on grass-stems below the buckthorn-bushes in the early evening. Although very local, it may, where it occurs, sometimes be taken in fairly large numbers. It flies from sunset to almost 9 p.m., being most active from 8.30 p.m. to 9 p.m., after which it is to be found at rest on the leaves of its foodplant, *Rhamnus catharticus*. It is best to stand at this time by the side of a tall buckthorn-bush and net the moths as they fly over and around it, their white colour rendering them easily discernible.

COLEOPHORIDES.—The larvæ of *Coleophora ochrea* are to be found during the latter half of June on the leaves of *Helianthemum vulgare*; the cases are cylindrical, brownish-ochreous, and more than half-an-inch long.

The cases of *Coleophora troglodytella* are to be obtained in numbers in mid-June on the leaves of *Eupatorium cannabinum*, &c.

PLUTELLIDES.—The larvæ of *Plutella dalella* are to be found on *Arabis petraea* at the end of June. The cocoon they make is of open network, similar to that made by the larva of *P. porrectella*.

GELECHIIDES.—In late June and July, the larva of *Depressaria badiella* makes a tunnel or gallery under a plant of *Hypochaeris radicata* on the soil, so that when the rosette of leaves is removed the larva is left behind. When full-grown the larva often eats out the heart of the plant, and bores down far enough into the root to kill it (Fletcher).

The greenish *applana*-like larvæ of *Depressaria capreolella* are to be found in June on the truncate radical leaves of *Pimpinella saxifraga*. The imagines emerge in July.

The young larvæ of *Depressaria alstroemeriana* are to be found in spun-up leaves of *Conium maculatum* in June.

The very scarce *Ptocheuusa* (*Gelechia*) *osseella* may be looked for from the middle to the end of June. It occurs in woods, and on downs and rough pastures, flying, after 4 p.m., low down amongst grass, &c., and is easily passed as a rush-feeding *Coleophora*.

The larvæ of *Gelechia cuneatella* are to be found towards the end of June, feeding on willow.

The greyish-white, black-headed larvæ of *Acanthophila* (*Gelechia*) *alacella* feed at the end of June on a lichen growing on orchard-trees (Grabow).

In June, on coast sandhills (also inland sandhills of old coast lines), the sand-tubes made by the larvæ of *Lita marmorea* at the base of *Stellaria triviale* should be carefully searched for; the closely-spun cocoons are placed within the last larval sand-tube inhabited by a caterpillar.

The larvæ of *Aristotelia* (*Gelechia*) *subdecurtella* are to be found in June, feeding in spun-together shoots of *Lythrum salicaria*.

The cocoons of *Aristotelia* (*Gelechia*) *arundinetella* are to be found in June in the leaves of *Carex paludosa*. They are difficult to collect, for, by the time the fullfed larvæ spin their cocoons, the leaf is generally withered and the cocoons scarcely discernible; the plant generally grows in the water, and the cocoon is only an inch or so above water-mark.

The larvæ of *Anacampsis* (*Gelechia*) *albipalpella* are to be found in mid-June on *Genista anglica;* they draw several leaves together round the stem, and then eat them half through, thus discolouring them and forming conspicuous clusters of yellowish-white leaves.

The larvæ of *Aphanaula* (*Gelechia*) *leucatella* feed between united hawthorn leaves in early and mid-June.

In June, the larvæ of *Butalis fuscoaenea* are to be found feeding on *Helianthemum vulgare*.

CRAMBIDES.—In June, the puparia of *Crambus fascelinellus* are to be obtained on sand-dunes (Norfolk coast, etc.), being spun in the sand near the former opening of the larval gallery, at right angles to it, and in a perpendicular position near the root of a plant of *Triticum junceum*. They are about 1½in. in length, about as thick as a goose-quill, with both ends rounded, the point of junction with the larval gallery about midway in its length, formed of sand-particles, lined with white silk.

In June, the larvæ of *Crambus falsellus*, in different stages of growth, are to be found in silken-lined galleries running through wall-moss (*Barbula muralis*).

In confinement, the eggs of *Crambus pratellus* can be dropped among grass planted in a large flower-pot. The larvæ from these will feed well until the end of September, when they hybernate, recommencing to feed in March, and becoming fullfed in May.

In June and early July, the larvæ of *Chilo mucronellus* are to be found in the stems of a *Carex*, growing in marshy or fenny districts, about an inch above the root, a small, round hole in the stem 2in. or 3in. above the root being the outward sign of the presence of a larva.

The larvæ of *Anerastia lotella* may be obtained in June, when they reside in tubular cases made with grains of sand spun together, irregular in form, and varying in length from 2in. to 3in. The anterior end of a case is narrow, the posterior end connected with a cluster of from three to nine rather rounded and bulb-like terminal pouches stuffed full of frass.

The cases of *Anerastia lotella* lie in a more or less horizontal position, their mouths in connection with the plant stems of *Ammophila arenaria* near the crown of the roots, on which part the larvæ feed; the depth in the sand at which they are found varies from 1 in. to 4 ins., as the surface shifts according to the action of the wind, so that, sometimes, the cases are quite exposed to view, and, at other times, buried deep in the sand blown over them. So great are the ravages made by the larvæ that a plant is frequently so hollowed out as even to be killed (Buckler).

Besides *Ammophila arenaria*, the larvæ of *Anerastia lotella* feed on *Aira canescens*, *Festuca ovina*, and probably *Calamagrostis epigejos* (Zeller). The puparia of *Anerastia lotella* are to be obtained in June and July ; each consists of a dumpy tubular cocoon of sand, smoothly lined with silk, half-an-inch in length, as thick as a goosequill, tapering to a point at one end, abruptly and rather irregularly truncated at the other ; they are spun near the larval tubes, but are not in any way connected with them.

In mid-June, the larvæ of *Homoeosoma senecionis* are to be found mining in stems of *Senecio jacobaea ;* pushing out little heaps of frass, which are agglomerated together by webs (Buckler).

The larvæ of *Pempelia dilutella* are to be found in June, inhabiting silken galleries extending from a plant of *Thymus serpyllum* to a stone or other object on the ground near. The leaves are often drawn together, the gallery passing up some distance into the plant.

In June, the larvæ of *Pempelia dilutella* live in a loose silken pouch or purse at the end of a passage of loose silk, under a spreading plant of *Thymus serpyllum*, coming out at night to feed on the leaves. The loose patches of fine silken threads look as if they might conceal the retreat of a spider, but they are connected beneath with a thicker, dirty-white, loose pouch or passage of silk, in which, if traced far enough, will be found a dull, dark larva. The silken passages of the larvæ were only found where the plants were crowded with the habitations of the yellow ants, and were completely mixed up with the *débris* of the ants' nest, &c. (Barrett).

In the middle of June, and on through July, the webs, spun by the larvæ of *Ilythia carnella* so as to fasten the leaflets of *Lotus* to the stems near them, become more conspicuous, being dense and white, and, as they get older, having many old stems and partly-consumed leaves blotched with white bound up with them.

The cocoons of brown-grey silk, covered with pieces of leaf and frass, spun on the stems of dwarf sloe-bushes in early June, at the end of the looser larval webs running along the twigs, contain the pupæ of *Rhodophaea marmorea*.

The larvæ of *Rhodophaea suavella* are to be obtained in June on stunted sloe and hawthorn bushes, feeding under a whitish web on the underside of the leaves, or in silken galleries along the branches ; the galleries are often quite conspicuous owing to the presence of frass, pieces of dead leaves, &c.

The green, red-brown-lined, larvæ of *Rhodophaea advenella* are to be found in June feeding on the flowers and leaves of hawthorn, and may be occasionally beaten therefrom.

Twisted leaves of oak, spun together by a dense web, the leaves having the green cuticle dissected away and quantities of frass spun up with them, the undersurface of the leaves gnawed away and making a large bundle, contain larvæ of *Rhodophaea consociella* in early June, pupæ in late June.

The larvæ of *Euzophera (Myelois) pinguis* are to be found in June and July in the living bark of ash, frequently pollard trees, never affecting any dead or decayed portions of a tree, nor penetrating to the wood, nor eating far into the bark, however thick, generally less than an inch, and mining more of a chamber than a gallery. A few long black grains of frass always block the original small round hole of entrance, and these, being very characteristic, should always be looked for, when searching a tree, on any projecting bosses, as well as on the spreading foot, upon which they sometimes fall and lodge (Buckler).

PYRALIDES.—The larva of *Scoparia lineolalis* feeds under *Parmelia parietina* which grows upon the rocks at Howth, and on the coast rocks of the Isle of Man. It is fullfed in June, spins a slight web under its food in which to pupate, imago appearing in July (Gregson).

The puparia of *Scoparia cembrae* are to be found in June at the roots of *Picris hieracioides;* they are made of pieces of the root mixed with particles of soil and spun together by silk to the root on which the larvæ have fed.

The imagines of *Scoparia cembrae* are to be found in June and July on the trunks of trees, or on stone walls, with the head upwards and the wings drawn down and partly overlapping in the form of a narrow triangle; usually on the slopes of rough hillsides, rocky quarries, &c.; they fly swiftly off when disturbed to settle at some distance in a similar situation, and must be quickly followed up if they are to be netted.

In June, a walk over the limestone and chalk downs, almost any-where, will result in putting up specimens of the conspicuous *Scoparia dubitalis;* they fly wildly for a short distance, and, once having been disturbed, are very alert, and so should be followed up warily and covered without delay.

Sometimes, where the chalk-downs are edged with woods, as at Cuxton, the trunks of the large beech-trees, oaks, &c., are selected as resting-places for large numbers of the imagines of *Scoparia dubitalis.*

The natural time of flight of *Scoparia dubitalis*, however, is at dusk, when, as it flies quickly over the herbage, it can be captured in large numbers without trouble. Some localities give a fair share of almost pure white aberrations.

In June and July, at early dusk, in suitable marshy or fenny localities, the imagines of *Scoparia pallida* are to be seen in great numbers flying just above the herbage, and falling to the ground if disturbed; they are, however, to be captured with the greatest ease.

The imagines of *Scoparia alpina* are to be found in June and July, on grassy ridges at a considerable elevation on the higher Scotch mountains; they fly up as one walks through the herbage, settling again at a short distance, when their place of rest should be marked and the moths captured.

The imagines of *Scoparia ambigualis* are to be found in great abun-dance in June, on the trunks of trees in woods, throughout the British Islands. During the day they can, if not disturbed, be boxed in great numbers whilst thus at rest. At dusk, the imagines fly freely, and are very puzzling when one is at work for other species.

The imagines of *Scoparia mercurella* are to be found in June on the trunks of trees, and are very active and easily disturbed. [Often abundant on the trunks of the lime-trees in the grounds of the Boys' Naval School at Greenwich, where there is no trace, on the trees, of mosses on which the larvæ could feed. Possibly they live here on ground mosses.]

In June, the full-grown larva of *Ebulea crocealis* draws the edges of two or three leaves of *Inula dysenterica* together, and, in the cavity thus formed, changes to a deep rich brown pupa.

The larvæ of *Scopula lutealis* are to be found at the end of June on the underside of leaves of bramble, wild strawberry, *Plantago lanceolata, Ranunculus*, dock, &c. (Porritt).

In June, the imagines of *Scopula decrepitalis* fly among *Asplenium filix-foemina* and *Lastraea spinula*, on the western shores of Loch Goil and other suitable places.

The flat, scale-like, translucent eggs of *Botys pandalis* are laid about mid-June; the young larvæ hatch towards the end of the month, and can be fed on golden-rod, marjoram, and *Teucrium*, living at first on the underside of the leaves or between leaves, in little silken webs.

The eggs of *Agrotera nemoralis*, laid in early June, hatch in about ten days; the larvæ want placing at once on young hornbeam leaves just unfolded from the bud, or they will die off, being apparently unable to eat the old and mature leaves. When young, the larvæ hide between two ribs on the underside of a young leaf beneath a loosely-spun web made of a few silken threads, and are most difficult to detect.

From the middle to the end of June, after the second moult, larvæ of *Agrotera nemoralis* gnaw little round holes in the leaf, just large enough for them to crawl through on their feeding excursions, and through which they re-enter their little silken abodes for rest and shelter; if touched, they crawl either backwards or forwards like a Tortricid larva (Tugwell).

The larvæ of *Cledeobia angustalis* are to be found in June, feeding in damp localities on a species of moss, supposed to be *Hypnum cupressiforme*; in confinement, the moss must be kept saturated with water. The larvæ live completely hidden in the moss, but their whereabouts are easily seen from the frass thrown out from where the larvæ are feeding.

The eggs of *Spilodes sticticalis* are laid in early June, and the larvæ hatch in a few days. They feed up rapidly on the upperside of the leaves of *Artemisia vulgaris* and allied plants, leaving the veins of the leaf and not touching the underside. They are fullfed in early July, and part of the pupæ produce imagines in late July and August, the other part in late May or June of the following year.

The aquatic larvæ of *Paraponyx stratiotalis* are to be obtained in June and July on the leaves of *Anacharis (Elodea) alsinastrum*. They form retreats of pieces of the food (or some other aquatic plant), which they line with silk.

In June, the fullfed larvæ of *Hydrocampa stagnalis* eat through the leaves of *Sparganium racemosum* near the surface of the water, causing the upper portion to fall and be suspended in the water.

The fullfed aquatic larvæ of *Hydrocampa nymphaealis* are to be found in June or early July in somewhat flat oval cases made of pieces of leaves of their foodplants, *Potamogeton natans, Myosotis caespitosa, Hydrocharis morsusranae, Sparganium simplex*, &c., floating on or near the surface of the water.

In June and early July, the puparia of *Hydrocampa stagnalis* are to be found in brooks, ditches, pools, &c., attached to pieces of *Sparganium*.

ALUCITIDES.—In the first fortnight of June, the larvæ of *Platyptilia ochrodactyla* are to be found in various stages of growth mining in the stems of *Tanacetum vulgare*, the mouth of the mine being generally between the axil of a leaf and the stem, with a few silk threads spun from one to the other, among which the blackish frass gets entangled and becomes conspicuous (Buckler).

In early June, the imagines of *Agdistis bennetii* are to be found flying freely at dusk over the plants of *Statice limonium*, on salt-marshes all along our north-eastern, eastern and southern coasts.

PSYCHIDES.—On bright and still afternoons, *Whittleia retiella* may be found flying over low herbage in salt-marshes. Its chequered markings and rapid flight cause it to be very easily overlooked.

HEPIALIDES.—The eggs of *Hepialus velleda* are scattered on the ground, amongst plants of *Pteris aquilina*, etc., in June.

ZEUZERIDES.—In June, the pupæ of *Phragmatoecia arundinis* are to be found in the stems of *Arundo phragmites*, up and down the hollow stems of which they move actively; the fullfed larva gnaws a thin spot in the stem, and lines it with a thin layer of silk, in preparation for the emergence of the imago.

In the localities where *Phragmatoecia arundinis* occurs, the pupæ may be found protruding from the affected reed-stems for some time before the moth emerges therefrom.

ÆGERIIDES.—The larvæ of *Ægeria ichneumoniformis* are to be obtained in June and July, mining in the main roots of *Lotus corniculatus;* they scrape out a channel along the side of the root, covering the open end of the groove with silk in which frass and *débris* of the plant are entangled, thus maintaining the outline of the root.

The workings of *Ægeria ichneumoniformis* are readily detected, the external covering of the groove being of a pale yellowish (sawdust) tint, forming a strong contrast with the dark grey-brown colour of the rind of the root.

Throughout June, the plants of *Statice armeria* should be collected for the full-grown larvæ and pupæ of *Ægeria musciformis*.

The newly-emerged imagines of *Ægeria tipuliformis* are to be found on the leaves of old red- or black-currant bushes, *in cop.*, between 4 p.m. and 6 p.m., on fine days in June.

In June and early July, the pupæ of *Trochilium apiformis* may be found emerging from the puparia, the escape of the moths taking place some little time afterwards.

ANTHROCERIDES.—The young larvæ of *Adscita statices*, in June and July, burrow into the substance of a sorrel leaf, being, however, never quite hidden, making semitransparent blotches by eating away the under epidermis and clearing out the soft cellular tissue leaving only the transparent upper epidermis.

The young larvæ of *Adscita geryon* leave the eggs in late June or July, burrow into the underside of the flower-buds or leaves of *Helianthemum vulgare*, eating out a little blotch, each one inserting half its body into its burrow ; the upper skin of the leaf is untouched.

The full-grown larvæ of *Anthrocera exulans* are to be found on a great variety of plants in their habitat throughout June and July, burying themselves in the fleshy leaves of some alpine plants, or feeding on the leaves of alpine *Trifolium, Geum, Alchemilla,* or *Silene.*

Larvæ of *Anthrocera palustris* are fullfed in June, and are generally restricted to boggy or marshy places, follow the habit of *A. filipendulae* in pupating fairly high up on grass culms or other tall herbage, the imagines appearing in July (in early seasons they may emerge in late June).

The larvæ of *Anthrocera lonicerae* are fullfed in early June, they feed freely on *Lathyrus pratensis, Lotus corniculatus,* &c.

For pupation, the larvæ of *Anthrocera lonicerae* try to reach as high a point as possible ; the cocoons are readily seen on tall grass culms, and on bushes (sometimes at a height of 5ft. or 6ft.).

The eggs of *Rhagades globulariae* hatch at the end of June and commencement of July ; the young larvæ immediately bore into the leaves of *Centaurea nigra* making a small semitransparent spot between the upper and under epidermis of the leaf, whilst feeding on the soft cellular tissues.

DREPANULIDES.— In June, the spun-together leaves of beech should be carefully examined for pupæ of *Drepana cultraria (unguicula).* [Also in September.] (Greene).

The eggs of *Drepana sicula* are laid in June ; the larvæ hatch in nine or ten days, and must be at once supplied with young leaves of *Tilia parvifolia.*

Where birch is common, examine the leaves that are joined together and you will not infrequently find the pupa of *Platypteryx falcula* between them. [Also in September.] (Greene).

BREPHIDES.—The larvæ of *Brephos notha* are to be obtained in June, feeding on *Populus tremula,* between leaves that have been spun flatly together for concealment. They are fullfed by the end of the month.

GEOMETRIDES.—During late June and early July, the larvæ of *Ennomos tiliaria* are to be found on birch (*Betula alba*), to the twigs of which they cling firmly in a very stiff position, resembling greatly the twigs of the plants on which they rest.

The half- to full-grown larvæ of *Ennomos autumnaria* are reputed to be great cannibals if crowded, and to require plenty of space if they are not to show this unsatisfactory propensity.

In breeding *Angerona prunaria,* give the imagines the chance of pairing more than once. [It is possible that this is a much commoner habit than is generally supposed in very many species, and necessary if infertile eggs are to be avoided.]

The larvæ of *Venilia maculata* can be fed up throughout the summer on woodsage, *Teucrium scorodonia.*

When the larva of *Phorodesma smaragdaria* is fullfed, it is so entirely covered with the dead and brown portions of the foodplant that its detection is most difficult.

The loose cocoons of *Phorodesma smaragdaria* are spun among the foodplant. The larvæ draw together with silken threads the pieces of *Artemisia* attached to their bodies, which they make into oval-shaped puparia fastened to a stem of the foodplant, and in this they change to greyish pupæ with striped wing-cases (Elisha).

About the middle of June, the larva of *Geometra vernaria* rests in a nearly straight position, firmly attached by its anal and ventral claspers to a twig or leaf-stalk of its foodplant, from which it projects with the rigidity of a stick at an angle of about 45°.

Eggs of *Nemoria viridata* produced larvæ June 30th, 1864, the larvæ choosing young leaves of hawthorn for food, and were full-grown about the commencement of August, when they drew some of the leaves of their food together with a few threads to form a puparium, in which they changed to pupæ (Hellins).

The eggs of *Tephrosia crepuscularia* (*biundularia*) hatch in June; the larvæ feed on oak, birch, and sallow, being especially partial to the former.

In some years, in early June, the imagines of *Tephrosia extersaria* (both sexes) visit the sugared trees in numbers. Females should always be kept for eggs, as the imagines are rarely in first-class condition for cabinet.

The eggs of *Tephrosia punctulata* laid in June, hatch in about a fortnight, the larvæ feeding up so rapidly on birch that they are full-fed in a month, and pupate in a loose cocoon formed just below the surface of the ground.

The eggs of *Boarmia consortaria* hatch in June, and the larvæ feed well on oak, birch, and sallow, being fullfed towards the middle or end of August. The larvæ require earth or cocoa-nut fibre in which to pupate. [Larvæ have been known to refuse oak.]

The pupæ of *Boarmia abietaria* are to be found in profusion at roots of fir-trees in Gloucestershire during the last week of June. This time should be strictly adhered to, as the insect sometimes only remains eight days in the pupal state (Greene).

In confinement, in June and July, plenty of damp moss should be kept below the lichens on which the larvæ of *Cleora glabraria* are feeding. Not only does this help to keep the lichen moist, but it forms a very suitable place for the delicate puparia when the larvæ are fullfed.

The eggs of *Zonosoma* (*Ephyra*) *omicronaria* laid in June, hatch in about a fortnight, and the larvæ feed freely on maple and sycamore in confinement, although the former alone appears to be the foodplant in nature.

The eggs of *Phibalapteryx lignata*, laid in June or July, hatch in a few days, and the larvæ feed well and rapidly on *Galium palustre* and *G. saxatile* (*G. mollugo* does not suit them). The larval stage only lasts about three weeks, the moths appearing as a second-brood in August.

The full-grown larvæ of *Eubolia mensuraria* are to be found in mid-June feeding on *Vicia*, and possibly other plants low down near the ground, pupating before the end of the month, the imagines appearing in July.

The bright pale-green larvæ of *Cidaria fulvata*, with grey dorsal and subdorsal lines, yellow spiracular lines, and yellow segmental incisions, are to be found in mid-June on wild-rose, pupating at the end of the month, among the leaves of their foodplant.

The larva of *Thera simulata* is to be found in mid-June on juniper. The head is yellowish-green in colour, the dorsum pale greenish-blue, with a slender, dull grass-green mediodorsal line, and a subdorsal stripe of the same colour, below which is a white stripe, and then a broad stripe of dark green edged along the spiracles with dark brown ; between this and the legs is a pale yellowish stripe ; the legs greenish.

In confinement, the females of *Hypsipetes ruberata* lay, in early June, their eggs on the catkins of sallow, inserting their ovipositors into the catkin, and placing the eggs chiefly on the central stem.

The larvæ of *Lobophora viretata* have been found in June feeding between united leaves at the end of sycamore twigs, collected to feed larvæ of *Ptilophora plumigera* ; the larvæ spun up, and the imagines appeared about mid-August. [See previous " Hints " for usual habits.]

The eggs of *Coremia designata* (*propugnata*) hatch in June, and the larvæ feed up rapidly, in confinement, on young cabbage leaves, being fullfed in about three weeks, the imagines emerging in August (Porritt).

The eggs of *Emmelesia decolorata* are laid in mid-June on the flower-heads of *Lychnis dioica*, the females depositing two or three eggs each time they settle, as they fly from one flower-head to another.

The eggs of *Emmelesia albulata* are deposited in early June on the flower-bracts of yellow-rattle ; they hatch in about a week.

The larvæ of *Emmelesia albulata* are to be found feeding in the green and tender seedpods of yellow-rattle in middle June. The affected seedpods are easily detected, as those containing larvæ look discoloured, although the larvæ themselves are completely hidden within the seedpods. The larvæ feed up very rapidly, being fullgrown by the end of June or in early July, when they go down to the ground for pupation.

Towards the end of June, the eggs of *Eupithecia plumbeolata* are deposited on *Melampyrum pratense*, the young larvæ appearing in about a week, and feeding on the fresh flowers of this plant.

The larvæ of *Eupithecia pygmaeata* feed in June on the petals and anthers of *Stellaria holostea*, and, in confinement, just as readily on petals and stamens of *Cerastium tomentosum* (Crewe).

The larvæ of *Eupithecia irriguata* are to be beaten in early and mid-June from oak. The green tint of the larvæ is almost exactly like that of the oak-leaves at this time. They are usually fullfed about the middle of the month.

The imagines of *Eupithecia extensaria* appear in June, and, in confinement, no difficulty is found in pairing them by enclosing them on a growing potted plant of *Artemisia maritima*. The eggs are placed singly, the moth, after laying three or four eggs, flying off to another sprig, and continually repeating the same performance. The eggs deposited in mid-June hatch at the end of the month, and the larvæ are full-fed from the middle to the end of August (Porritt).

The common garden southernwood, *Artemisia abrotanum*, makes an excellent substitute for *A. maritima*, the larvæ of *Eupithecia extensaria* eating it with evident relish. The larvæ of *Phorodesma smaragdaria* are also easily reared on the same substitute.

The eggs of *Acidalia virgularia* (*incanaria*) are laid in June, hatch in a few days, and the larvæ feed freely in confinement on *Polygonum aviculare;* some pupate in August, the imagines from which emerge as a partial second-brood in September ; other larvæ go over the winter, pupate towards the end of April, the pupæ producing their imagines in June or even July.

The eggs of *Acidalia trigeminata*, laid in June, hatch very quickly; the larvæ feed freely in confinement on *Polygonum aviculare*, and are fullfed in about a month. The pupæ go over the winter, although, in confinement, a fair proportion of imagines usually emerges the same year, in August, as a partial second-brood.

LACHNEIDES.—In early June, the full-grown larvæ of *Poecilocampa populi* are to be found at rest on the branches and stems of oak, frequently exposed to the full sunshine.

DIMORPHIDES.—The fat, fullfed larvæ of *Dimorpha versicolora* should be searched for in June, as they feed up very rapidly; in spite of their large size, they are not always easy to discover.

SPHINGIDES.—The larvæ of *Phryxus livornica* are to be occasionally found on fuchsia, dock, &c., in June and July (progeny of spring immigrants) and again in August and September (progeny of July-August immigrants).

The eggs of *Amorpha populi* are deposited singly, or in pairs, on the upper- and underside of the leaves of the various species of poplar, at almost any height from the ground, from about two or three feet up to the tops of tall trees. They are not at all difficult to find in June.

At the end of June and in early July, the young larvæ of *Amorpha populi* are not at all difficult to find on the underside of the leaves of various species of poplar.

The eggs of *Smerinthus ocellata* are easily found on the leaves of sallow and willow in June and early July.

In June, the young larvæ of *Smerinthus ocellata* may be found resting on the midrib of a sallow, apple, or willow leaf, both sides of which have been eaten away by the larva.

NOTODONTIDES.—The young larvæ of *Stauropus fagi* appear to like the chaffy stipules of beech leaves, although they soon set to work and eat out long pieces down to the midrib. They will also thrive on oak.

The full-grown larvæ of *Cerura bifida* should be searched for on aspen and poplar in August, small larvæ may be obtained in June and July.

Towards mid-June, the larvæ of *Petasia nubeculosa* are to be seen hanging to the birch-sprays, quite motionless, their peculiar attitude of rest. At this time, they eat large pieces out of the leaves and so betray their presence. At the end of the month they are nearly fullfed, are very lethargic, and more often seen lying quite flat along the birch-twigs.

The fullfed larvæ of *Petasia nubeculosa* are to be beaten from birch in their northern haunts, from the middle of June until early July.

In confinement, at the end of June, the fullfed larvæ of *Petasia nubeculosa* should be provided with a liberal supply of cocoanut fibre and leaf-mould, into which they will burrow to a depth of four or five inches for pupation.

The eggs of *Microdonta bicolora* are laid in June on birch, the young larvæ feed up readily in ordinary leno sleeves, and are no trouble whatever till they are fullfed, when they are very liable to die off, unless plenty of room, well sifted earth, dried leaves, &c., be given them, among which to pupate.

The larva of *Ptilophora plumigera* is fullgrown in early June, and should then be searched for, resting in its well-known curved pot-hook posture on the underside of a maple leaf.

The fullfed larva of *Ptilophora plumigera* changes to an uniform semitransparent green colour just before pupation, which takes place from the commencement to the middle of June, in a thin brittle earthen cocoon, placed just below the surface of the ground and near the roots of the foodplant.

The fullfed larvæ of *Lophopteryx carmelita* are to be beaten from birch about the middle of June (more rarely at the end of the month and in early July).

The fullfed larva of *Lophopteryx camelina* is to be found on, or beaten from, oak, alder, and hazel bushes in late June and early July, also on poplar (Buckler).

The pale bluish-green larvæ of *Pterostoma palpina* are to be found, in late June and early July, on poplar, aspen and sallow. [Also in August and early September.]

The lilac-tinted larva of *Notodonta ziczac* with its two dorsal humps, is to be found on sallow and willow in June. [Also in August and September.]

The elongated, shiny, variable-tinted, fullfed larvæ of *Leiocampa dictaea* are to be found in June on poplar and aspen. [Also in late August.]

The greenish-white granulated egg of *Leiocampa dictaeoides* is to be found by searching birch-bushes in June. [Also in August.]

The translucent white, black-headed, young larva of *Leiocampa dictaeoides* is to be found on birch-bushes, in late June and early July, eating the under-surface only of the young birch leaves. [Also in late August.]

The fullfed larvæ of *Drymonia chaonia* are to be beaten from oak from about the middle of June until the middle of July; the pupal stage lasts till the following April or May.

NOLIDES.—The fullfed larvæ of *Nola strigula* are to be obtained in early June feeding on oak, principally on the under cuticle of the leaves.

The boat-shaped cocoons of *Nola strigula* are to be obtained, in late June, on the trunks of oaks, the colour of the cocoons assimilating perfectly in tint with that of the surrounding surface of the bark; very difficult to detect.

The larvæ of *Nola albulalis* are to be found in June feeding on the leaves of *Rubus caesius*. They are readily discovered in the daytime, when they hide under a leaf, the latter simply having to be turned over to expose the hiding larvæ.

DELTOIDES.—In June and July, the larvæ of *Hypena rostralis* are to be found on hop, and, when young, rest along the midrib of a hop-leaf, where they are very inconspicuous, the larvæ closely resembling the leaf in colour.

The eggs of *Hypenodes costaestrigalis* are laid in late June or early July, and the larvæ appear about the middle of the latter month. They feed readily enough on the flowers of *Thymus serpyllum*, want separating as much as possible as they have a bad reputation for cannibalism, and are fullfed about the middle of August.

In confinement, the larvæ of the summer brood of *Herminia tarsipennalis* feed well on knotgrass, *Polygonum aviculare*. The eggs are laid in June, give up their larvæ in July, the latter are fullfed by mid-August, and the imagines appear about mid-September. [The autumnal larvæ hybernate, feed up in spring on sallow, pupate in May, and the imagines appear in June.] (Buckler).

NOCTUIDES.—The comparatively large eggs of *Agrotis exclamationis* hatch after about fourteen days, and the young larvæ can be fed on *Chenopodium*, *Plantago*, &c., but prefer slices of carrot. The larger larvæ, in August and September, can be readily recognised, when compared with the closely allied species, by the difference between the dorsal and ventral ground colours, the former (to the spiracles) being of a warm brown, the latter a pale drab. The large spiracles are also characteristic.

The best time to sugar for *Noctua sobrina*, in Perthshire, is from the third week in June until mid-July. In late years, it may, however, be taken as late as mid-August (Bush).

Towards the end of June, searching the lyme-grass, growing on the Lincolnshire coast, with a light, will result in the capture of *Tapinostola elymi* and *Mamestra albicolon* in the best condition. (The dates given in part i., p. 63, and ii., p. 93, of *Practical Hints* appear to be too late for an average season.) (Musham). [The later dates have been recorded for Scotch localities.]

The eggs of *Hadena genistae* are laid in June, and the young larvæ feed up well on *Polygonum persicaria* and *Alsine media*. They are full-fed in early August, when pupation takes place, the imagines, however, not appearing till the following June.

The eggs of *Hadena suasa* are laid in June, and the young larvæ feed up very rapidly, in July, on plantain, knotgrass, &c., pupating at the end of the month. A few imagines usually emerge in August as a partial second-brood ; the other pupæ go over the winter, the imagines appearing the following June.

The eggs of *Hadena thalassina* are laid in June, and the young larvæ feed up well on knotgrass, pupating in early August. The imagines do not emerge till the following June.

The eggs of *Hyppa rectilinea* are laid in June ; the young larvæ feed up well on sallow until the end of October, when hybernation takes place.

In the third week of June, the larvæ of *Polia nigrocincta* may be found after dark on the coast of the Isle of Man, &c., on *Statice armeria* and *Silene maritima*, but most are to be found on *Plantago maritima*, very few, comparatively, occurring on the two first-named species.

The larvæ of *Dianthoecia caesia* are to be obtained, at the end of June and throughout July and early August, on *Silene maritima* (Isle of Man, &c.) ; in confinement, they feed freely on the seeds of this plant and *S. inflata*. The larva of this species is strikingly distinguished from those of its allies by the diamond-shaped chevrons, composed of dark grey-brown freckles, thickly aggregated together, as well as by the absence of positive outlines in the subdorsal and spiracular regions.

The pale, whitish-green larvæ of *Eremobia ochroleuca* are to be found by searching or sweeping the seedheads of *Dactylis glomerata* from early June until early July.

The eggs of *Aplecta advena*, laid in June, hatch in about a fortnight, the young larvæ feeding freely on *Polygonum aviculare*. Like those of most of its nearest relatives, the larvæ of this species will sometimes feed up very quickly, whilst others do so slowly and hybernate, the former pupating in August and September, the imagines appearing shortly afterwards.

The fullfed larva of *Xylina conformis*, in June, will (in confinement) fold itself up into a leaf, or fasten a leaf loosely to the surface of the soil, and there spin an oval cocoon of close, but semitransparent, whitish silk, closely adhering to surrounding substances (Buckler).

The larva of *Cucullia verbasci* is to be found, in June, feeding on *Verbascum thapsus* and *V. nigrum*. It is rather larger and thicker than that of *C. scrophulariae*. A transverse, equally broad, band of yellow, extending to below the spiracles on either side, is seen on the middle of each segment. [This character is alone sufficient for its identity; and, although the larvæ of this species vary much in colour and size of markings, yet the design remains in all.] Also feeds on *Scrophularia nodosa*, *S. aquatica*, and *Verbascum lychnitis* (Buckler).

The larvæ of *Euperia paleacea* (*fulvago*) are to be taken from early June onwards, by beating birches; they are best obtained in the early morning, when they are more readily dislodged from their hiding-places.

When fullfed, the larvæ of *Euperia paleacea* (*fulvago*) are to be found coiled round in the middle of a birch-leaf, with the head close to the other extremity, resting just as do the larvæ of *Asphalia flavicornis* when full-grown. The fullfed larvæ should be given plenty of suitable *débris* in which to pupate, their cocoons being usually placed just below the surface of the earth.

The larvæ of *Cosmia diffinis* are to be beaten or searched for, on elm, during the early part of June; when at rest, the larvæ lay curved round on the underside of a leaf.

The thin silken cocoons of *Cosmia diffinis* are occasionally to be found in June and July, spun between two or more elm leaves, the pupæ being firmly fixed therein, by the cremaster, to a dense patch of silk spun on one of the leaves forming the chamber. The pupa, like those of its congeners, is covered with the delicate plum-coloured, waxy secretion, so characteristic of certain quite unrelated pupæ.

In early June, the larvæ of *Tiliacea citrago* form their cocoons by drawing closely together several of the growing leaves of lime, and remain inside for two or three weeks before assuming the pupal state.

The larvæ of *Mellinia gilvago* and *M. circellaris* (*ferruginea*) are to be obtained from seeds of wych-elm during the first fortnight of June. That of *M. circellaris* is a trifle the larger, and the series of dark central marks on the back, with their dark wedges, assume together more compact forms of an urn-shape, being attenuated behind, so that a constant character appears in the hinder pair of tubercular dots, being outside the dark urn-shapes (Buckler).

The cocoon of *Dasycampa rubiginea* is spun in June, just below the surface of the ground, moss and pieces of leaves being worked into the upper part of the structure, and bits of earth, &c., below; it, however, maintains its oval form. The pupa is of a dark purplish-brown colour.

The eggs of *Acontia luctuosa*, laid in early June, hatch in nine days, the young larvæ, which are nocturnal feeders, progressing well on the leaves, flowers and seeds of *Convolvulus arvensis*, the larger larvæ lying along, and closely embracing, the stems of the foodplant close to the ground, where it is most difficult to detect them. Many of the imagines appear in August. [A second-brood of larvæ feeds up in September similarly to the early brood.]

The eggs of *Anarta melanopa* hatch in early or mid-June, the young larvæ feeding, in confinement, on tender leaves of *Arbutus unedo*, *Luzula pilosa*, sallow, *Vaccinium vitis-idaea*, and flowers of *Helianthemum vulgare*, although, later, they restricted themselves to *Salix capraea* and *S. acuminata*. They are fullfed in mid-July. It is supposed that *Menziesia caerulea* is the natural foodplant.

The eggs of *Anarta cordigera*, laid in early June, hatch in about twelve days, the young larvæ feeding on leaves of *Luzula pilosa*, *Arbutus unedo* and *A. uvaursi*, but some, reared by Hellins, soon neglected the other foodplants for *A. unedo*, on which they fed up, although *A. uvaursi* is most probably the natural foodplant.

In June, the fullfed larvæ of *Catocala sponsa* spin a loose kind of hammock amongst the oak-leaves, in which they change to purplish-red pupæ covered with a delicate violet bloom.

The fullfed larvæ of *Plusia interrogationis* are to be swept, in early June, from the heather on the moors of the northern parts of the British Islands.

The eggs of *Agrophila trabealis* hatch towards the end of June and in early July, the young larvæ feeding slowly on *Convolvulus arvensis*. [In many localities on the continent we have seen the imagines in great abundance at the end of July and in early August, a good month later than the time of their appearance in England.]

The eggs of *Banksia argentula* are laid in early June and soon hatch, the young larvæ feeding upon *Poa annua*, and becoming full-grown in early August, when they spin their cocoons just beneath the crown of the grass-roots, almost close to the surface of the earth.

The eggs of *Hydrelia unca* are laid towards the end of June, the young larvæ appearing in a week, and feeding well on *Carex sylvatica*, along the blades of which they rest in a stretched-out position. They go underground for pupation.

The eggs of *Euclidia glyphica*, laid in June, hatch towards the end of the month or in early July, the young larvæ feeding up well on the various species of clover, entirely by night, resting at full length by day along the stalks of the foodplant; they are fullfed by early August, when they spin a cocoon on the surface of the ground, the imagines appearing the following June.

The eggs of *Euclidia mi*, laid in June, hatch towards the end of the month; the larvæ feed well on grass and common white clover, have a most remarkable Geometrid appearance, and are fullfed about the middle of September.

LITHOSIIDES.—The eggs of *Eulepia cribrum*, arrayed in the neatest possible manner, are to be found, in June and July, upon little stems of heather in their well-known localities in and near the New Forest.

Young larvæ of *Œonistis quadra* are to be beaten from the lichens on oak-trees in early June; they appear to nibble the oak-leaves, but feed on the lichens, preferring *Lichen caninus*, of which they eat the dark cuticle quite voraciously, not caring for the pale, fleshy substance beneath.

The fullfed larvæ of *Œonistis quadra* may be beaten from oak at Lyndhurst, in late June ; the favourite food is *Parmelia caperata*, which grows in abundance on all the trees where they occur, but they will also eat other species of lichens. They are very restless in confinement, and especially prone to cannibalism (Lockyer).

The cocoon of *Œonistis quadra* is to be found spun up on palings in the neighbourhood of trees covered with lichens at the end of June.

The cocoon of *Lithosia muscerda* is spun under cover of a leaf, piece of bark or lichen ; it is thin, webby, formed of greyish silk, enclosed in a finer and thinner web of white silk.

The thin web-like cocoons of *Lithosia griseola* are spun in June and July, and placed under some protecting cover—stone, piece of wood, &c.

The cocoons of *Lithosia griseola* are to be found, in June, under moss or lichen on moss- or lichen-bearing trees (Greene).

In June, the pupæ of *Lithosia* var. *molybdeola* are to be found enclosed in a very slight web of silk, under cover of a stone or piece of moss.

The larvæ of *Lithosia lurideola* may be fed in confinement on oak ; larvæ have also been taken on buckthorn, dogwood, and clematis.

From early to middle June, the almost fullfed larvæ of *Lithosia helveola* are to be found feeding on a large coarse lichen that grows on the bark of yew-trees.

At the end of June and in early July, the loose cocoons of *Lithosia helveola* are to be found spun up on the underside of loose pieces of bark on yew-trees.

The ova of *Miltochrista miniata* are laid in June and July, and hatch in about ten days ; the larvæ feed slowly through the winter and become fullfed in May.

HESPERIIDES.—The eggs of *Steropes paniscus* are laid in June, and the young larvæ, hatching in about a fortnight, feed satisfactorily on the leaves of *Brachypodium sylvaticum*.

When still small, the larvæ of *Steropes paniscus* make tubular homes of the leaves of *Brachypodium sylvaticum*, leaving an opening at each end, whence they emerge to feed on those parts of the plant near their domiciles.

The young larvæ of *Nisoniades tages* are to be found, at the end of June and in early July, in little hollows formed by drawing together three leaflets of *Lotus corniculatus* ; the two outer ones are drawn close together, and the third one bent over like a curved roof; the structure looks almost exactly like a leaf not quite expanded.

Females of *Pamphila sylvanus* will lay their eggs, in confinement, in late June and early July, on cock's-foot grass, if they be enclosed under a leno cover and placed in the sun.

In early June, the larvæ of *Thymelicus thaumus* (*linea*) are to be swept from the soft grass, *Holcus lanatus*, with the colour of which their tints assimilate remarkably well ; they may also be swept from *Brachypodium sylvaticum*. The larva of this species is often found in one's net when one is working for micro-lepidoptera among the long grass in Chattenden Roughs.

PAPILIONIDES.—The egg of *Lycaena arion* is deposited, in June, among the flowers of *Thymus serpyllum*, being circular in outline, flattened, and covered all over except a central, depressed spot on top, with fine, raised, irregular reticulation, which, in profile, stands out strongly; the colour of the shell is of the blue-green of a hedge-sparrow's egg, the reticulation transparent-white (Hellins).

In mid-June [and mid-August] the young larvæ of *Polyommatus icarus* feed on the leaves of *Lotus corniculatus*, eating into the substance of a leaf either from the upper- or underside, leaving the opposite skin as a white spot, although they sometimes eat the flowers, the petals of which they devour entirely.

In June, by following up a female *Polyommatus bellargus* when on egg-laying intent, one can obtain eggs quite freely by picking the leaves one after the other as she quits the plants on which she has been engaged.

The larvæ of *Polyommatus corydon* are to be found on *Hippocrepis comosa* throughout June. The larva of this species can only be distinguished from that of *P. bellargus* by having the ground-colour of a lighter, brighter green (a green with more yellow in its composition) and the hairs light brown, whilst that of *P. bellargus* has the ground colour deeper green, with the hairs or bristles black (Hellins).

The circular, flattened, greenish-drab eggs of *Polyommatus astrarche* are laid in June in little groups of two, three, or more, on the underside of the leaves of *Helianthemum vulgare*.

The larvæ of *Polyommatus astrarche* are to be found, in late June and July, on the underside of the leaves of *Helianthemum vulgare*. The feeding of the smaller larvæ makes small pale spots on the upper dark-green surface of the leaves, the spots becoming larger and browner, until, at last, almost the whole undersurface of the leaves is entirely eaten, although, with an indefinite supply of food, they rarely remain long enough on one leaf to more than blotch it very markedly before moving to another.

Females of *Cupido minima*, enclosed over a plant of *Anthyllis vulneraria* and allowed plenty of sun, will lay eggs freely among the flowerets.

Although the eggs of *Cupido minima* are placed low down on the calyces of *Anthyllis* flowers, and thus hidden from casual observation, they may be easily detected on careful search (Hellins).

At the end of June and in early July, the larvæ of *Callophrys rubi* can be beaten from broom, *Genista tinctoria*, and many other plants; bramble, after which the species was named, appears to be rarely chosen.

In early June, the eggs of *Nemeobius lucina* can be found fairly readily, in the localities where the species occurs, on the underside of cowslip or primrose leaves. The young larvæ eat little holes in the leaves, but, later, they eat out large pieces of the leaves, and their whereabouts become conspicuous.

The eggs of *Melitaea aurinia* can be obtained freely by enclosing caught females in a leno sleeve over a plant of *Scabiosa succisa* or honeysuckle, the eggs being laid in heaps on the surface of the leaves.

Females of *Brenthis selene*, captured in early June, and enclosed on a plant of *Viola canina*, lay their eggs freely on the upper- and under-surfaces of the leaves, as well as on the stems of the plants.

The young larvæ of *Brenthis selene* usually divide into two sections in this country, one very small part feeding up rapidly and producing a few imagines in August, the others hybernating when about 10mm. long, and going through the winter in this stage.

The fullfed larva of *Argynnis aglaia* feeds on *Viola canina* in June; it is difficult to find, and is best obtained when feeding, as its movements are rapid and may attract attention; when not feeding, it usually hides below the leaves of the plant which it has been eating.

The fullfed larvæ of *Dryas paphia* are to be found, in early June, feeding on the leaves of *Viola canina*, freely exposing themselves (according to Buckler) on the violet plants.

The gregarious larvæ of *Vanessa io* are to be found in considerable-sized companies, in late June and early July, spread out over beds of stinging-nettles by roadsides, behind hedges, or sunny corners on the edges of woods.

The fullfed larva of *Apatura iris* is to be found in June on sallow, eats rapidly, and is easily alarmed, when it draws itself in and is somewhat difficult of detection on its foodplant.

In June (and August), the larvæ of *Pieris napi* may sometimes be found in numbers, feeding on *Nasturtium officinalis* and *Barbarea vulgaris*.

In late June and early July, the larvæ of *Gonepteryx rhamni* are to be found on *Rhamnus frangula* and *R. cartharticus*; stunted bushes in sheltered nooks on the outskirts of a wood, &c., are usually good localities for them.

The globular, greenish-yellow, or greenish-white (when newly-laid) eggs of *Papilio machaon* are to be found, in early June and on through the month (often indeed until August), laid usually on *Peucedanum palustre*, in its local haunts in Cambridge and Norfolk.

In late June and July, on Wicken Fen, stand over a plant of *Peucedanum palustre* and look most carefully if you wish to see the little black larvæ of *Papilio machaon*.

Immigrant females of *Pontia daplidice* lay their eggs occasionally, in June, on *Reseda luteola*, on which the larvæ feed up in July, producing occasional imagines in August or September. [The species is quite unable to go through the winter in our climate.]

In confinement, the larvæ of *Papilio machaon* feed very freely on the leaves of garden carrot, *Angelica sylvestris* and other umbellifers.

The larvæ of *Pieris napi* may be found, in June, on *Hesperis matronalis*, &c. They grow very rapidly, and are fullfed in early July, pupating during that month, and emerging towards the end of July or in early August.

Females of *Colias edusa*, enclosed on a growing plant of clover (*Trifolium repens*) or *Lotus corniculatus*, placed in the sun, and supplied with a little honey and water for food, will lay their eggs pretty freely so long as the weather is bright and sunny. During dull weather the butterflies will not lay.

The larvæ of *Colias edusa*, obtained from eggs laid by immigrant females in June, will feed up well, in confinement, on *Trifolium repens* and *Lotus corniculatus*, pupating in July, the imagines emerging in August.

The fullfed larvæ of *Hipparchia semele* require light soil, peat, or similar material in which to burrow. They hide therein by day, feeding by night, and, when mature, form their puparia just beneath the surface of the ground in a manner altogether different from any other British butterfly.

The imagines of *Erebia epiphron* will lay their eggs in confinement, if placed in a suitable receptacle, with a supply of grasses on which the larvæ will feed, *e.g.*, *Nardus stricta*, *Aira flexuosa*, &c.

The pupæ of *Limenitis sibylla* are to be found in June, suspended to a button of silk spun by the larvæ on the underside of a leaf of honeysuckle. In no case was one found attached to a stem (Barrett), although Buckler says that those he had chose this position.

<hr>

JULY.

TINEIDES.—*Scardia corticella* occurs throughout the month of July on the stems of partly decayed trees. Very careful search must be made to detect the moth, as it sits in the fissures of the bark, where it is easily passed over for an unevenness in the bark, or a piece of lichen.

ADELIDES.—The larva of *Phylloporia bistrigella* mines the leaves of birch in July and August, at first in an excessively slender gallery, which almost invariably commences near the tip of the leaf, runs nearly parallel to, and very near, the midrib to the base of the leaf, then turns off to one side, where, at length, the larva commences making a large blotch, in which it eventually cuts out an elliptical case, and descends to the ground, but does not eat any more. The imago appears the following June.

The larvæ of *Tinagma resplendellum* are to be found, at the end of July, mining the leaves of alder. The mine is commenced along the midrib, then follows a lateral rib for a short distance, and then crosses over from one lateral rib to another, and, in so doing, makes a very slight but visible track. This is the track to be looked for when collecting the larva, for, as soon as it reaches the next lateral rib it mines along it to the midrib, then down the midrib for an inch or more (the track being quite concealed), when it turns round and mines back up the midrib to about the place where it had entered it; then it makes a broad, flat mine in the leaf, and, lastly, cuts out an oval case like that of *Phylloporia bistrigella* and descends to the ground. The escape hole is visible in deserted leaves, as well as the track between the two lateral ribs; if only the track be present, the leaf is still tenanted.

ELACHISTIDES.—In July and early August, the small mining larvæ of *Stephensia* (*Elachista*) *brunnichiella* are to be found making dark-brown blotches in the leaves of *Clinopodium vulgare* (*Calamintha clinopodium*).

The larvæ of *Perittia obscurepunctella* make blotches in the leaves of honeysuckle in July, quitting the leaf when fullfed and changing to a singularly flat pupa.

The larvæ of *Anybia langiella* are to be found mining the leaves of *Circaea lutetiana* in July (Boyd).

Laverna stephensi occurs from the middle to end of July, is very local, but not uncommon in its haunts. It rests on the stems of old oak-trees, usually very low down, often only just clear of the ground.

An especially good spot to search for this moth is in the " bays " at the bases of the trees.

Towards the end of July, the larvæ of *Laverna raschkiella* are to be found mining the leaves of *Epilobium angustifolium*, the imagines of this second-brood appearing in August.

LITHOCOLLETIDES.—The larvæ of *Lithocolletis comparella* are to be found, in July and early August, mining the undersides of the leaves of Lombardy poplar.

BUCCULATRIGIDES.—The larvæ of *Bucculatrix cicadella* are to be found feeding on the leaves of alder in July.

Towards the end of July, the cocoons of *Bucculatrix maritima* are to be found near the tips of blades of grass, rushes, &c., growing among plants of *Aster tripolium*, very rarely indeed on the foodplant, whose leaves show the recent mining done by the larvæ.

GRACILARIIDES.—The larva of *Gracilaria imperialella* is to be obtained in July, mining the leaves of *Orobus niger:* it loosens the lower epidermis of the entire leaf, and eats much of the parenchyma, the leaf slightly curving and becoming quite bladder-like, whilst the lower skin is very white.

TORTRICIDES.—During the first week of July is a good time to work for *Tortrix piceana*. The moth flies at dusk round *Pinus sylvestris,* usually near the tops of the trees, therefore a long-handled net is necessary to secure it. So far this species is only recorded from Hampshire and Surrey.

Tortrix transitana should be sought for about the first fortnight of this month amongst its foodplants (elm, birch, and poplar). It may be found at rest on tree-trunks, or by beating, but, when the latter plan is adopted, the late afternoon will prove most productive. Its time of flight is at dusk, when it flies briskly around tree-tops.

Tortrix viburniana is to be found during July, and is most partial to moors and rough, swampy ground. It flies at dusk, and can be disturbed from low-growing vegetation during the day. It is as well to net and examine a number of specimens, the species, in some localities, being very variable.

Larvæ of *Peronea mixtana* may be obtained during this month in terminal shoots of *Calluna vulgare*.

The larvæ of *Peronea rufana* live, in July and August, on *Myrica gale*, drawing neatly together the terminal leaves and eating out the heart of the shoot. The imagines appear in September (Barrett).

The imagines of *Sciaphila cinctana* flit over the tops of grass on sloping banks by the sides of fields, appearing conspicuously white while on the wing, in early July.

In fenny districts, *Sericoris doubledayana* often occurs in considerable quantities, flying in the late afternoon sunshine.

Ephippiphora tetragonana is to be found throughout this month, and occurs commonly in many woods in the southeast of England. The simplest method to obtain it is to visit some fairly broad riding on a bright afternoon about five o'clock, at which hour its flight commences. The moth appears very small when flying, owing to the white spot on the forewing only then being visible.

If, after four o'clock in the afternoon, a position be taken up amongst beech trees, towards the beginning of July, the imagines of *Carpocapsa grossana* will most probably be observed flying wildly round and over them. A long pole as handle to one's net is necessary to make captures. This insect is very liable to injure itself if carried in a pill-box.

The larva of *Phtheochroa rugosana* is to be taken in July and August feeding in the fruit and shoots of *Bryonia dioica*.

In July, the imagines of *Catoptria caecimaculana* occur fairly freely in the Cuxton district amongst *Centaurea nigra*.

The imagines of *Catoptria citrana* occur very freely among *Achillea millefolium*, in early July, at Tuddenham.

The larvæ of the second-brood of *Eupoecilia atricapitana* feed in the green stems of *Senecio* in July (Thurnall).

At the end of July and through August, the larvæ of *Eupoecilia sodaliana* are to be found burrowing into the berries of *Rhamnus catharticus*. The infested berries turn purple much earlier than the sound ones ; and the larva has a habit of fastening together with silk those berries in its immediate neighbourhood. In confinement, it burrows freely into virgin cork, forming therein a tough, leathery cocoon, in which it passes the winter unchanged.

COLEOPHORIDES.—The larvæ of *Coleophora siccifolia* construct ill-made cases of hawthorn and wild apple in July and August, so ill-made that it long ago received the name of the " clumsy tailor." By keeping the larvæ out-of-doors during the winter and spring, imagines may readily be bred the following June. The larvæ cause peculiar brown blotches, whilst the cases are far larger than is necessary for their homes, and are formed of pieces of mined leaves, the large superfluous piece overlapping, giving them the appearance of dried leaves.

The imagines of *Coleophora lutipennella* may be commonly seen resting on the trunks of oaks early in July.

Coleophora ochrea appears to affect plants of rock-rose growing on sunny slopes. The imagines are to be found freely some years on the chalk slopes in Kent, usually quite up under the shelter of the woods.

Coleophora saturatella is most easily found in early July, after the larvæ have spun up their cases near the tips of the old black sprigs of the broom.

The larvæ of *Coleophora flavaginella* are common, in July and August, among *Suaeda maritima*, at Portland ; the hybernating cases are frequently especially common on this plant and on *Suaeda fruticosa*.

GELECHIIDES.—During this month, many species of micros may be obtained by smoking them out with a pair of bee-bellows, from low and thick-growing herbage. This is pre-eminently the way to secure the Gelechiids frequenting sand-dunes, heath, and rough ground generally. Great care must be exercised or a conflagration may arise.

The larvæ of *Depressaria pulcherimella* are to be found in July feeding on the flowers of *Bunium flexuosum*.

At the beginning of July, the handsome larvæ of *Depressaria nervosa* are to be found in the heads of *Œnanthe crocata*. When fullfed, the healthy larvæ bore into the main stem of the plant, weaving a transverse piece of web above the opening and directly below themselves.

[The ichneumoned larvæ are generally found in the smaller branches of the stem.]

With the larvæ of *Depressaria nervosa*, the lively greenish larvæ of *D. applana* are to be found in the heads of *Œnanthe crocata* in July ; these, however, feed on almost all sorts of *Umbelliferae*.

The pale yellowish-green larvæ of *Gelechia hippophaëella* feed in the shoots of *Hippophaë rhamnoides* in the middle of July. The imagines emerge in August and September (Stainton).

On the trunks of aspen and white poplar, in mid-July, the imagines of *Gelechia nigra* appear in their chosen haunts in companies. About an hour before dusk they run briskly from the crevices, where they lay previously, to other hiding-places on the trunk, moving with a shambling gait, taking flight afterwards to the branches above. If the wind blows they move little, and, if disturbed, usually fly off at once, when they should be smartly snapped up with the net.

In the middle of July, larvæ of *Gelechia velocella* of all'sizes are to be found feeding in tubes at the roots of *Rumex acetosella*, spinning up the stem to some of the leaves (Schmid).

The larvæ of *Aristotelia* (*Gelechia*) *ericinella* make very light gossamer-like webs near the ends of the shoots of heather, in early July, frequently changing to pupæ in the webs thus formed.

If the patches of *Thymus serpyllum* growing on chalk-downs be carefully watched on a bright afternoon towards the end of July, the imagines of *Lita artemisiella* should be found flying amongst them.

By visiting granaries and carefully examining the walls and sacks, large quantities of *Sitotroga cerealella* may often be discovered during July. The moths prefer dark corners, and press themselves into small holes in the brickwork and folds in sacking.

Paltodora cytisella is well out by the middle of July, and is to be obtained amongst *Pteris aquilina*. On mild and calm evenings it sometimes occurs very freely at about eight o'clock.

Early in July, *Butalis senescens* is to be found flying in the afternoon sunshine amongst its foodplant, *Thymus serpyllum*. It has a preference for chalk-downs and railway-banks.

CRAMBIDES.—The imagines of *Anerastia lotella* rest by day, in July, very closely to the ground on stalks of grasses, and are most difficult to disturb ; they fly readily in the early dusk, and are frequently attracted to light.

In early and mid-July, the cocoons of *Rhodophaea suavella* are to be found in greyish silken cocoons on sloe- and hawthorn-bushes, spun in or near the silken galleries used by the larvæ, most frequently found on small, stunted, exposed bushes, and preferring sloe to hawthorn.

The larvæ of *Salebria* (*Pempelia*) *formosa* are to be found on elm from July to September (they will also feed on birch in confinement), spinning together two leaves, or turning down the corner of a leaf for shelter whilst they feed.

The young larvæ of *Salebria* (*Pempelia*) *palumbella* leave the eggs in July, and spin fine webs among the leaves of ling, some of the webs becoming more or less conspicuous in August through particles of frass being scattered about on their surface.

In July, the larvæ of *Pempelia dilutella* spin a number of silken threads into the thyme plant on which they have been feeding, drawing the leaves together, and thus extending the long silken galleries in which they have been feeding, forming their puparia within, and there changing to pupæ.

The eggs of *Nephopteryx genistella* are laid in July and August on young shoots of *Ulex campestris*. As soon as the young larvæ emerge, they spin a thick network of silk round the branches, and, under this, feed until the approach of winter, when each forms for itself a close cocoon or tunnel of silk in which to hybernate (Moncreaff).

The pupæ of *Euzophera* (*Myelois*) *pinguis* are to be found in little mines or chambers in the bark of ash, during July and early August, enclosed in puparia of whitish silk, the head lying very near the entrance of the mine, which is lightly blocked with frass.

The eggs of *Euzophera cinerosella* (*artemisiella*) are laid in July, the young larvæ entering the woody stalks near the roots of old plants of *Artemisia absinthium*, excavating mines or chambers for themselves while feeding on the central substance of the root-stalks much after the manner of some of the larvæ of Ægeriids (Buckler).

In July, the puparia of *Ilythia carnella* are spun among the upper shoots of a plant of *Lotus corniculatus*. A bower-like shelter is formed of detached leaflets and web, any growing stems passing through the chamber being bitten off to prevent growth disarranging it. In this, finally, a dense cocoon is spun, and pupation takes place.

The eggs of *Galleria mellonella* (*cerella*) are laid, in July and early August, on old honeycomb in deserted beehives; the young larvæ feed on the comb, are fullfed in October, when they spin cocoons, in which they remain as larvæ all the winter, pupating therein the following May or June, the moths emerging in July or early August.

PYRALIDES.—The imagines of *Scoparia lineola* are to be found, in July, on old lichen-covered fences, branches of blackthorn, apple-trees, and also on rocks (Isle of Man). They are easily disturbed, and, if beaten out of blackthorn, &c., fly sharply to a more secure hiding-place.

In July and August, the imagines of *Scoparia truncicolella* are to be found resting on the trunks of Scotch fir-trees, &c., in the districts they haunt. The moths rest in the usual vertical posture, the head raised, the wings forming a narrow triangle, and are exceedingly active and alert, flitting off the tree at the approach of a collector, flying generally to the ground, where it is difficult to get them in good condition, and spoiling themselves in a pill-box if confined too long or much shaken.

The imagines of *Scoparia resinea* sit during the day-time on the trunks of trees—ash, elm, apple, &c.—where they are very conspicuous, in July and August (Barrett).

The imagines of *Scoparia ulmella* are to be found, in July and August, resting on the trunks of elms, oak-trees, &c., in woods, especially those on which lichens are common; they are smaller and paler than *S. ambigualis*, are very restless and active, easily startled in the day-time, and are rather confined to the woods of the midland and northern counties, although also recorded from Hants and Berkshire.

The imagines of *Scoparia ambigualis* var. *atomalis*, the mountain and moorland form of the species, are to be found in great abundance in July, resting on rocks, stones, heather, and other bushes that are to be found in their exposed habitat, where trees are practically unknown.

The larvæ of *Scoparia murana* are to be found in July, on mosses

—*Hypnum cupressiforme, Dicranum scoparium, Bryum capillare*, &c.— growing on old walls, rocks, &c.

The larvæ of *Scoparia angustea* are to be found, in July and August, feeding in silken galleries formed through the thick, rounded cushions of wall-mosses, eating out the solid substance of the moss without attacking the surface.

The puparia of *Hydrocampa nymphaealis* are spun in July, about an inch above the level of the water, between two or three stems of aquatic plants, the larvæ leaving their floating cases for the purpose when fullfed. (Réaumur states that they are sometimes spun under water.)

The imagines of *Hydrocampa nymphaealis* are to be disturbed by day from among the tall herbage at the sides of rivers, ponds, ditches, fens, drains, &c., in July and August, flying quickly to a safer hiding-place. At dusk, they are to be found flying freely over the water in these habitats, seeking out the *Potamogeton* and other aquatic plants on which they lay their eggs.

In July and early August, the eggs of *Hydrocampa nymphaealis* can be found laid in a flattish mass on the under-surface of the leaves of *Potamogeton natans; Alisma plantago*, &c.; they hatch in about a fortnight, and the young larvæ mine into the under-surface of the leaf, near the base of the midrib, leaving it three days after to cut out their first tiny cases from the under-cuticle of the leaf. At the end of August, the ponds sometimes contain thousands of tiny cases, not ¼ in. long.

The imagines of *Hydrocampa stagnalis* are to be disturbed from similar hiding-places to the last, and at the same time. They fly, however, much more freely, and, in some cases, occur in the greatest abundance in the late afternoon and early evening, swarming, like animated snowflakes, in some districts, just over the surface of the water and among the water-side herbage.

In July, the eggs of *Hydrocampa stagnalis* are laid on *Sparganium simplex* or *S. racemosum;* the young larvæ, as soon as they are hatched, bore into the stems of the plants and mine the pith in every direction, hybernating therein from about mid-September ; they recommence feeding about early April, when they are to be found devouring the young leaves of *S. simplex*, just below the surface of the water, in their domiciles, made by spinning together parts of two adjacent leaves, and lining the interior with silk ; on *S. racemosum* they still mine in the spring, the edges of the eaten parts turning blackish, and affording a sure indication of the larvæ below.

The imagines of *Cataclysta lemnalis* are to be disturbed by day, in July, from the herbage at the edges of streams, ditches, ponds, large fen-land pools, &c., fluttering only a short distance, in order to find a safer hiding-place. At dusk, however, they are sometimes to be seen in thousands at the sides of their marshy habitats, flying just above the surface of the water ; the females are much less active than the males.

The imagines of *Acentropus niveus* rest during the day, in July and August, upon weeds close to the surface of the water of streams, pools, marsh-ditches, fen-drains, &c., or on floating sticks, &c. The weeds, sticks, and other objects on which they rest can be pulled through the water and the moths immersed without causing them, apparently, any discomfort ; and, if knocked off, they will buzz along the surface of the water until something else be found to which to cling.

The natural time of flight of the imagines of *Acentropus niveus*
appears to be in the evening after dusk, when the males are to be
sometimes seen in great abundance, flying just over the surface of the
water. The more or less apterous females are comparatively rare.

The Anthrocerid-like cocoon of *Cledeobia angustalis* is to be found,
in July, in a moss, supposed to be *Hypnum cupressiforme*. It is formed
of fine white silk, and is spun among the moss in damp localities.

The imagines of *Endotricha flammealis* are to be disturbed, in July,
from among bushes in woods, on heaths (especially among furze-
bushes), in gardens, &c., flying rapidly to seek another hiding-place
during the daytime. Their natural time of flight is at dusk.

The eggs of *Stenopteryx hybridalis*, laid in July or August, hatch
in from two to three weeks, and the larvæ feed up rapidly, in confinement, on *Polygonum aviculare*. When fullfed, at the end of September,
they form rather tough, white, glossy, silken cocoons, amongst leaves of
the foodplant, in which they pupate, the imagines either appearing in
a few days, or the pupæ going over the winter until the following
summer (Porritt).

In July, the fairly well-grown, glossy, translucent-looking larvæ
of *Agrotera nemoralis* feed between two united leaves of hornbeam, and
continue this habit until fullfed towards the end of the month.

From mid-July to the end of the month, the fullfed larva of
Agrotera nemoralis forms for itself a puparium from a leaf of hornbeam. It cuts out a portion, which it neatly folds over, and fastens
the edges carefully together all along their length, so that, in shape, it
is something like a turnover tart; this it lines with silk, making it,
doubtless, a secure and watertight abode in which to pass the winter.

The eggs of *Pionea stramentalis* are laid at the end of July, and
hatch in early August, the young larvæ feeding on various cruciferous
plants, e.g., *Barbarea vulgaris*, *Sinapis arvensis*, *Cardamine amara*, &c.

The eggs of *Botys hyalinalis* are laid in late July or early August,
hatching about the middle of the latter month, when the larvæ should
be supplied with fresh leaves of *Centaurea nigra*, or, much better,
sleeved on a growing plant.

In July, the larvæ of *Botys pandalis* inhabit little cases cut from
the leaves of *Teucrium*, *Solidago*, *Origanum*, or any other of their food-
plants, or, if available, of more permanent material, such as beech-
leaves, bramble-leaves, &c.

The larvæ of *Botys fuscalis* are to be obtained in July, feeding on
the flowers and seed-capsules of *Melampyrum cristatum* and *M. pratense*,
inhabiting webs or galleries formed with silk, uniting the flowers and
capsules to the stalk, and much covered in parts with frass.

In July, the young larvæ of *Botys asinalis* are to be found feeding
on the flowers and young seeds of *Rubia peregrina*, and, later on, the
leaves, from which they eat out the thick substance from below,
leaving the upper skin quite perfect in large transparent blotches,
whilst they themselves live in little silken hammocks, from which
they emerge in order to feed (Hellins).

The eggs of *Aglossa cuprealis*, laid in July and August, hatch in
about ten or twelve days, the larvæ forming (like those of *A.
pinguinalis*) silken galleries, about three inches long, with pieces of
straw intermixed, among and under rubbish accumulated on a barn-
floor, in stables, and similar situations; they live in these all the
winter and spring, and form their cocoons and pupate in mid-June.

DREPANULIDES.—The young larvæ of *Drepana harpagula* (*sicula*) must be given the tender leaves of *Tilia parvifolia*; they will not eat those of *T. europaea*.

GEOMETRIDES.—The variable larvæ of *Ennomos fuscantaria* are to be found in July, on ash, resembling greatly, when at rest, small twigs of the branches of the tree on which they rest.

In the middle of July, the newly-emerged imagines of *Phorodesma smaragdaria* generally appear in the morning, but do not seem to pair until very early the following morning (*i.e.*, after midnight), they then remain *in cop.* during the whole of that day, separating towards evening, the females beginning to oviposit during the next night.

The eggs of *Phorodesma smaragdaria* are laid, in July, on the stems and leaves of *Artemisia maritima*, generally near the top of the shoots. When first laid they are of a light yellowish colour, changing, in about a fortnight, to dark greyish, soon after which the young larvæ emerge.

Boarmia abietaria occurs in the New Forest, on larch trunks, in July. It is a rather difficult species to see, has a habit of flying off suddenly as one approaches the trunk, and dropping sharply to the ground and resting there. Fanning the trunks for it is a good dodge (Bayne).

In confinement, the females of *Boarmia abietaria*, when sleeved on any of the foodplants, neglect, as a rule, the branch, and, pushing the ovipositor through the holes in the muslin, leno, &c., lay their eggs on the outside of the material. Many allied species have the same peculiar habit.

The eggs of *Boarmia abietaria*, laid in July, hatch in a fortnight, and the larvæ feed well but slowly on yew, not disdaining, however, birch, oak, beech, and fir, until about the middle of October, when they commence hybernation.

The eggs of *Boarmia roboraria*, laid in early or mid-July, hatch in about a fortnight, and the larvæ feed slowly on oak-leaves or birch-leaves as long as they are obtainable. They then hybernate, but, through the whole hybernating period, should be supplied with tender oak- or birch-twigs (better sleeved thereon), as the larvæ nibble at the bark all the winter, and, unless well supplied, will nibble right through the twigs. They moult in early March, and at once commence to feed very eagerly, as the birch-buds have now begun to swell; the larvæ are fullfed in early June.

The larvæ of *Tephrosia extersaria* (of two forms—pale pea-green and purplish-brown) are to be taken throughout July and August. They feed well, in confinement, on birch, oak, sallow, and willow, preferring the two last-named.

The sallow-feeding larvæ of *Zonosoma* (*Ephyra*) *orbicularia* are usually fullfed in late July or early August, when they pupate in the well-known characteristic butterfly fashion. It is well to remember that some of these pupæ will give up their imagines in about a fortnight, whilst others stand over until the spring.

The eggs of *Acidalia ochrata* are laid in early July, and hatch in about twelve days; the larvæ feed on flowers of *Galium verum*, *Lotus corniculatus* (especially if somewhat withered), *Solidago virgaurea*, *Crepis virens*, *Picris hieracioides*, *Apargia autumnalis*, &c. They lie up for hybernation in early September, but a few larvæ will usually feed on in confinement, and pupate, when the imagines emerge in the winter.

The eggs of *Acidalia rubricata*, laid in late July, hatch in early August, in about nine days, feed freely on *Polygonum aviculare* in confinement, also on *Lotus corniculatus*, *Medicago lupulina*, and *Trifolium minus*; they feed until the end of October, when hybernation takes place.

The eggs of *Acidalia dimidiata* (*scutulata*), laid in July, hatch in about five or six days; in confinement, the larvæ will eat withered leaves of dandelion, and, in spring, have been seen to eat sliced turnip.

The eggs of *Acidalia bisetata*, laid in late July, hatch towards the end of the month or in early August, and feed freely on *Polygonum aviculare* (Hellins says also on withered bramble-leaves); they go into hybernation towards the end of September, and feed up the following spring, being fullfed in May.

In early July, the imagines of *Acidalia rusticata* are to be found in their chosen localities, sitting throughout the day with outspread wings on leaves of elm, *Parietaria*, &c., at the foot of hedges. When one has been detected, many can generally be obtained. We have always found that searching the *Parietaria* is the best way to obtain them in numbers.

The larvæ of *Acidalia rusticata* feed freely in confinement on *Polygonum aviculare* (tender leaves of ivy, lilac, and withered bramble-leaves are also mentioned). The eggs hatch in July, the larvæ hybernate from October to March, and are fullfed about the beginning of June.

The eggs of *Acidalia dilutaria* (*holosericata*) are laid in July, and hatch in a few days; the young larvæ, feeding on the leaves of *Helianthemum vulgare*, prefer to congregate three or four together, near the bottom of a shoot, which they strip, for some distance, of its bark or skin, and then feed on the withered leaves at the tip of the shoot as it hangs down. They hybernate from October to February, and are fullfed in May.

The eggs of *Acidalia interjectaria* hatch in July, and the larvæ feed freely in confinement on *Polygonum aviculare*, withered dandelion leaves, &c.; they hybernate from September to March, and are fullfed towards the end of April or in early May. (Hellins also adds that the larvæ will eat scarlet pimpernel and the leaves of almost anything so long as they are withered.)

The bright green larvæ of *Eupisteria heparata* are to be obtained on alder in July, and are easily known by the broad velvety dorsal band, intersected by a pale greenish-yellow mediodorsal line; when fullfed the larvæ become dark green tinged with purple, and with purple segmental incisions.

The eggs of *Hyria auroraria*, laid in July, hatch in about a fortnight; the young larvæ feed on *Polygonum aviculare*, growing very slowly, and reach about half-an-inch in length in October; they then rest until February, when they recommence feeding, and are fullfed in June.

The females of *Asthena blomeri* lay freely in confinement. The eggs hatch in July and the young larvæ are, owing to their vagrant habits, most difficult to deal with. The only way to keep them with their food is to get a quite tight-fitting tin box, cover with muslin and then force on the lid, so that the larvæ cannot possibly get away from their food. If the box be not opened for some days, *i.e.*, until they have made considerable growth, they will generally be found all right and sufficiently large to manipulate with success.

The fertile eggs of *Asthena sylvata*, laid in July, do not change colour as they mature, but remain pale yellowish-white until the larvæ are hatched. It is, therefore, necessary to watch carefully for the appearance of the latter, as there is nothing to give warning of their approaching exit, and, being very delicate, they will soon die if not at once supplied with food (Hellins).

The eggs of *Asthena sylvata* hatch about mid-July; the young larvæ feed on alder, preferring the tender open leaves and avoiding the sticky leaf-buds; they grow rapidly and are fullfed from about mid-August until early September.

The larvæ of *Asthena candidata* are to be found in July and August on birch. They remind one of those of *Cidaria sagittata* in their hunched posture, and, in colouring, those of *Venusia cambricaria*.

The larva of *Emmelesia decolorata* bores into the capsule of *Lychnis dioica*, feeding on the seeds, and remaining therein until the end of July, when it quits it to pupate in the earth.

At the end of July and commencement of August, the fullfed larvæ of *Emmelesia affinitata* are to be found feeding in the seed-capsules of *Lychnis diurna*. They usually pupate on the surface of the ground.

The larvæ of *Eupithecia plumbeolata* are to be found in July, feeding on the flowers of *Melampyrum pratense*; when quite small they feed on the stamens alone, but afterwards on the whole corolla-tube.

The larva of *Hypsipetes ruberata* best likes the underside of a sallow-leaf, spinning thereon a white silken covering for itself, on the outside of which it sticks the down from the leaf, and so makes its abode very inconspicuous. Gregarious when young, it becomes more solitary as it gets older, and, when nearly fullfed, generally draws together two leaves and lives between them, although sometimes it makes its home by folding over a part of a leaf. It feeds by night on the neighbouring leaves, not on the part spun over, and is fullfed early in September.

The larvæ of *Chesias obliquaria* are to be obtained continuously on *Spartium scoparium* from July to September; they pupate throughout the late summer and autumn, but no imagines appear until May and June of the following year.

The eggs of *Lithostege griseata* are laid on *Sisymbrium sophia*, on which the larvæ are to be found at Tuddenham. *Erysimum cheiranthoides*, however, proves a good substitute foodplant.

Eggs of *Cidaria populata*, laid in July on bilberry, remain in the egg-stage till early April, when the larvæ appear, and these, by the commencement of June, are fullfed, forming their slight puparia by drawing together a few leaves with silken threads.

The eggs of *Cidaria sagittata* are laid in little groups of four or five together on the seed-vessels of *Thalictrum aquilegifolium* and *T. flavum*. The larvæ appear about the beginning of August, and have a habit of biting half through the stalks of their foodplant and feeding on the leaves which they have thus caused to become partly withered. They feed through August into September, and, although not strictly gregarious, are often found to the number of a dozen or more on one plant, their presence being easily detected from the habit of feeding mentioned above (Hellins).

The eggs of *Cidaria picata* are laid in July, on *Stellaria (Alsine) media*, the larvæ appearing in early August. These are peculiar in their habits, lying stretched, when at rest, along a stem of the foodplant, but, when disturbed, they assume the form of an Ionic volute. They are fullfed in early September, when they make a slight cocoon on the surface of the earth.

The eggs of *Coremia quadrifasciaria* are laid in July ; the larvæ feed on slowly through the winter on *Galium mollugo*, being fullfed in the following April or May.

The larvæ of *Lobophora viretata* have been found in mid-July, living in a slight web amongst flower-buds of *Ligustrum vulgare*, eating out the interior of the flower-buds, portions of the leaves, and the rind of the flower-stalks ; they spin up in the middle of the month, appearing as imagines about the middle of August.

At the end of July, in the Perth district, *Anticlea sinuata* may be found ovipositing where there is an abundance of *Galium verum*, generally upon dry southern slopes, and may easily be mistaken at first sight for *Melanippe montanata* (Bush).

The larvæ of *Thera simulata* are fullfed towards the end of July, and may then be beaten from juniper, on which they feed.

The imagines of *Eupithecia tenuiata* are to be collected from the catkins of large sallows in early July ; the species is best bred, the captured specimens usually being more or less faded in colour.

The fullfed larvæ of *Eupithecia lariciata* are to be beaten during July from larch and spruce-fir. There are two chief varieties—(1) bright grass-green, with dark green mediodorsal line ; (2) yellowish-red, with the mediodorsal line brownish-olive.

The larvæ of *Eupithecia pulchellata* feed in July on the stamens and unripe seed-capsules of the common foxglove, spinning the lips of the flowers together, a habit that makes them somewhat easily detected.

HEPIALIDES.—From the commencement to the middle of July, the very young larvæ of *Hepialus velleda* burrow into the earth by the side of the underground rhizomes of *Pteris aquilina*, eating the surface of the stems on their way to the root-stock.

The fullgrown larvæ of *Hepialus sylvinus* are to be found feeding in the roots of dock throughout July.

In late July, the cocoons of *Hepialus sylvinus* are to be found near the surface of the earth, at or near the tops of the mines excavated in the roots of dock by the larvæ. The pupæ, like those of all Hepialids, emerge some hours before the imagines appear in August and September.

ÆGERIIDES.—The larva of *Ægeria musciformis* spins a silken tube, covered with frass, through the solid materials of a tuft of *Statice armeria*, leaving a projection of an inch or more out of the tuft ; in this, the larva pupates, and, through it, the pupa works its way before the emergence of the imago.

The larval and pupal tubes of *Ægeria musciformis* sometimes stand out perpendicularly from a tuft of *Statice armeria*, at other times, they are almost horizontal, and may be at any angle between these positions.

The imagines of *Ægeria musciformis* always jump backwards when one attempts to box them. The habit usually leaves the collector without a victim.

The dark purple-brown eggs of *Ægeria chrysidiformis* are laid on *Rumex acetosa* in July and August, the larvæ burrowing into the root-stocks in which they feed.

In July, the pupæ of *Ægeria ichneumoniformis* are to be found in the larval-tubes at the sides of the roots of affected plants of *Lotus corniculatus*.

ANTHROCERIDES.—During July (and August and September) clear blotches, observed on the surface of the leaves of *Centaurea nigra*, indicate the feeding of the larvæ of *Rhagades globulariae;* such spots are usually vacated, but the larvæ are to be found quite near in a fresh leaf. By September, the large blister-like mines are conspicuous.

The fullgrown larvæ of *Anthrocera exulans* climb upon the taller herbage in their neighbourhood, and spin their thin swollen cocoons, often one upon the other, until a collection of three to six is found massed together.

The pupæ of *Anthrocera exulans* are exceedingly soft and delicate, and, when collected in July or August, must be treated with great care.

The young larvæ of *Anthrocera purpuralis* feed freely on *Thymus serpyllum* in July.

SPHINGIDES.—The young larvæ of *Hyloicus pinastri* leave the egg throughout July, and feed on *Pinus;* the fullfed larvæ may be beaten at the end of August and throughout September.

The fullfed larvæ of *Mimas tiliae* are best obtained in late July and early August, more generally on elm than any other foodplant ; as they usually feed high up they are more frequently beaten than found by searching. On lime-trees the larvæ are sometimes lower down and more easily seen.

The fullfed larvæ of *Smerinthus ocellata* are usually to be found from the middle of July until well into September, on willow and sallow bushes almost everywhere.

Throughout July, the underside of the leaves of plants of *Scabiosa succisa*, especially those growing near woods, should be carefully searched for larvæ of *Hemaris tityus*.

NOLIDES.—Larvæ of *Nola centonalis*, hatching in July and August, and kept in a warm room with an abundance of food—*Trifolium*, *Lotus* and *Medicago*—will feed up in the late autumn, pupate in December, and the imagines will emerge in due course.

The boat-shaped or rather fusiform cocoons of *Nola albulalis* are spun on the stems of *Rubus caesius*, or near the base of the stems of bushes, etc., growing among the dewberry, etc., on which the larvæ have been feeding.

NOTODONTIDES.—At the end of July, keep a sharp eye on your breeding-cages in the south of England for the imagines of many Notodontids, which, though single-brooded in Scotland and northern England, are double-brooded in the south of England.

The larva of *Stauropus fagi* should be very carefully treated during its moulting periods ; the operation is a long and often troublesome one, the actual moult usually taking place during the night.

The fullfed larvæ of *Cerura vinula* are easily found on sallow, willow, and poplar, in July and August ; they prefer to rest on the top of the leaves near the midrib.

The fullfed larvæ of *Lophopteryx cucullina* are to be obtained on maple in late July and throughout August. [They also feed on sycamore.]

The tough, dirty-grey, silken cocoon, covered with fine earth, of *Lophopteryx carmelita*, is to be found at the base of birch-trees in July and August.

The fine, gauzy, silken cocoon, covered with fine earth, of *Lophopteryx camelina*, is to be found at the roots of oak, alder, and hazel-bushes in late July. [Also in October and November.]

The weak cocoon of *Pterostoma palpina*, formed of greyish silk, with a small quantity of earth intermixed therewith, is to be found near the roots of sallow and aspen, and, more rarely, poplar, from about the beginning to the middle of July. [Also in late August and September.]

The imagines of *Pterostoma palpina* are sometimes attracted freely to light in late July and August. [Also in May and early June.]

The whitish-green eggs of *Notodonta ziczac* are to be found on leaves of willow and sallow in July and August. [Also in May and early June.]

The reputed British species, *Notodonta tritophus*, appears in July, the young larvæ leaving the eggs in early August and feeding throughout this month and September on poplar.

The puparia of *Microdonta bicolora* are made at the end of July, usually between two or three birch-leaves spun together, if the larvæ be left on the tree.

The fullfed purplish larvæ of *Leiocampa dictaeoides* are to be found on birch in July. [Again towards the end of September.]

The large grey silken cocoon of *Leiocampa dictaea*, covered with loose earth, is to be found at the roots of aspen and poplar in early July. [Again in September.]

The fullfed larvæ of *Peridea trepida* are to be beaten from oak from the commencement to the middle of July ; the oblique lateral stripes make the larva a very fine-looking creature.

The larvæ of *Drymonia chaonia* may be taken, towards the end of July, about half-grown, on oak. The best way to obtain them is to stand beneath the tree and look up, and they can then be seen resting along the midrib of a leaf. This will happen even after the branches have been beaten thoroughly, for the larvæ have the most tenacious grip of any that I know (Bush).

The cocoons of *Asteroscopus sphinx* (*cassinea*) are to be found at roots of elm in July, in Gloucestershire (Greene). [The fullfed larvæ can often be beaten freely from elm in late May and early June in the Kentish woods.]

LYMANTRIIDES.—The cocoons of *Lymantria monacha* are to be found at the end of July, in the crevices of the bark of the trunk of oak-trees (Greene).

DELTOIDES.—The eggs of *Herminia cribralis* hatch in July, and the young larvæ feed on sallow, *Carex sylvatica*, and *Luzula pilosa*, preferring the latter ; the larvæ hybernate at the roots of the food-plant from October to April, and are fullfed early in June.

The eggs of *Herminia derivalis* are laid in July, and the young larvæ usually hatch in early August. The latter feed on fallen leaves of sallow and oak which must not be allowed to get dry, and, in November, hybernate in corners of the leaves, formed by turning down the edges or by joining one leaf against another by a few silken threads.

The eggs of *Rivula sericealis* are laid from mid-July to the end of the month on *Brachypodium sylvaticum;* usually side by side in considerable-sized groups.

The young larvæ of *Rivula sericealis* usually leave the eggs towards the end of July or in early August, and feed freely on *Brachypodium sylvaticum* until they hybernate in early October.

NOCTUIDES.—When fullfed, in July and August, the larva of *Dasypolia templi* leaves the foodplant and makes up in the earth. Larvæ fed on wild parsnip, produce larger and brighter coloured imagines than those fed upon cow-parsnip (Gregson).

The best method of rearing Agrotid larvæ is to fill large flower-pots with sea-sand, the food being laid upon the surface and supplied frequently ; this has been proved repeatedly to be better than keeping the larvæ on growing plants, unless the latter are specially healthy and well attended to.

The eggs of *Agrotis corticea*, laid in July, usually hatch before the end of the month, and feed up pretty rapidly on *Chenopodium, Polygonum,* clover, and are of almost full size by November, when they hybernate, but commence feeding again in March, and do not pupate until the end of April or early May.

It may be added that the young larvæ of *Agrotis corticea* feed exposed in July, and make holes in the leaves of knotgrass, clover, etc., but, by August, should be treated as other Agrotids, *i.e.,* placed in large pots partly filled with sea-sand, into which they burrow by day ; dock, mullein, and hollyhock, are also eaten freely, but slices of carrot are, as usual, preferred. They are chiefly night-feeders, but will feed by day in dull weather. The larva is more rugose than that of *A. segetum,* but the back and sides are similarly coloured with brownish-grey.

The eggs of *Agrotis lunigera* hatch in July ; the young larvæ prefer withered leaves of dandelion (which they perforate with small round holes), *Polygonum aviculare* and *Plantago major* to almost any other of their known foodplants. They should be kept in large flower-pots three parts filled with sea-sand, on which the food should be placed, as they insist on burrowing. They feed up rapidly in confinement, and are frequently almost fullfed by the end of October. They, however, continue to eat sparingly through the winter, and should be supplied with slices of carrot throughout January and February.

The ova of *Agrotis ashworthii* hatch towards the end of July ; the larvæ can be fed well on knotgrass, sallow, and dock, but, whilst some will feed up, pupate, and produce imagines in November, others, kept under precisely similar conditions, insist on hybernating. These should, as soon as it is seen that they will not feed up, be allowed to hybernate, and they will wait until spring before going on. Any attempt to force such, usually results in killing them.

The captured ♀ s of *Pachnobia hyperborea* (*alpina*), taken in July, are not averse to laying their eggs within glass-topped or glass-bottomed boxes. [We suspect in any pill-boxes.]

The eggs of *Noctua ditrapezium* hatch in July ; the young larvæ feed freely on dock and sallow, and, through the winter, can be satisfied with bramble and dock, eating during the milder, and hybernating in the colder, periods ; in the spring, hawthorn should be added to their bill of fare ; the larvæ become fullfed in May.

The eggs of the Scotch forms of *Noctua festiva* are obtainable in July and August, and the young larvæ feed freely on dock, knotgrass, bramble, &c. ; they are not difficult to rear, and occasionally, in confinement, feed right ahead and produce moths in December and January, although their habit is to hybernate more or less from the end of October till February.

The eggs of *Noctua umbrosa*, laid in July and early August, hatch in about ten days. The young larvæ feed readily on dock, bramble, plantain, periwinkle, strawberry, &c., throughout the winter, if at all mild, and, towards the end of February, become quite active and hungry. Before this time, they have usually split up into "forwards" and "laggards" in confinement, the former pupating in February and March, the latter in May. The larva (except when quite young) is very like that of *N. xanthographa*, and can easily be mistaken for that species.

The young larvæ of *Aplecta herbida* will feed on dock, *Plantago major*, strawberry, &c., and, in confinement, can readily be made to forego their usual hybernation, and reach the pupation-period by the end of November or early December.

The eggs of *Aplecta tincta*, laid in July, hatch in about ten days, the young larvæ feeding freely on *Polygonum aviculare*, and less so on *Plantago lanceolata* and birch.

The young larvæ of *Hadena adusta* are easily reared on lettuce, knotgrass, hawthorn, sallow, &c., being fullfed at the end of September or in early October, when they hybernate, awakening and forming their cocoons in early March, when pupation takes place. [The fullfed larvæ, passing the winter in this manner, are most difficult to hybernate safely, being frequently attacked with mildew if kept too moist, whilst they die off if kept too dry.]

The fullfed larvæ of *Polia flavicincta* are to be obtained in July, and feed on plants as widely different as mint and apricot.

The pupæ of *Agriopis aprilina* occur in the utmost profusion at the roots of oak in July and August; one has merely to turn up the earth and break it, and they will tumble out of their brittle cocoons in plenty (Greene).

The larvæ of *Dianthoecia albimacula* are to be obtained from mid-July, feeding on the unripe seeds of *Silene nutans*. In confinement, they feed well on this plant, eating out the contents of the capsules, also on those of *Silene inflata*, *S. maritima*, and *Lychnis dioica*. They can be distinguished from the larvæ of the closely-allied species by the absence of the slanting streaks or chevrons which they so generally have.

In July, the young larvæ of *Dianthoecia barrettii* eat out little sinuous channels in the leaves of *Silene maritima*, surrounding themselves with frass, working, in a few days, into the stems at the axils of the leaves or into the root into which they bore, throwing out heaps of minute, pale, cream-coloured frass that adhere to the aperture. The larvæ are fullfed about mid-September.

In confinement, the larvæ of *Dianthoecia irregularis* will feed on the seedheads of *Lychnis floscuculi*. They are best swept, when small, from the seedheads of *Silene otites* in July, or searched for by day when larger at the roots of this plant, just below the surface of the ground, in August.

The eggs of *Cucullia umbratica* are laid in July. They hatch in five or six days, and the young larvæ feed well on *Sonchus* (sow-thistle); they show a great aversion to light, hide under the lower leaves by day, and ascend at night to feed on the upper leaves and flowers, being fullfed in August, when they spin a loose, silken hammock, among the leaves and flowers of the foodplant, in which to pupate.

The larva of *Cucullia scrophulariae* is to be found in early July feeding on flowers and seed-vessels of *Scrophularia nodosa* and *S. aquatica*. It can be distinguished from that of *C. verbasci* by the bright yellow dorsal mark, for, whether little or much intersected by black, it is distinctly seen to be a blunt-pointed triangle of yellow, close to the beginning of each segment, pointing forward, its trans-verse base being longer than the sides, and placed on rather less than the first half of each segment (Buckler).

The eggs of *Plusia pulchrina* are laid in June or early July, hatch-ing quickly, the young larvæ feeding on *Lamium*, cow-parsley and honeysuckle till October, when hybernation takes place, the larvæ beginning to feed again in March and being fullfed by the end of April and May, when they spin flimsy, light, whitish, silken cocoons among the foodplant.

The eggs of *Plusia iota* are laid in early July and hatch in a few days, the young larvæ feeding on *Lamium album* and *L. purpureum*. They feed slowly till October, when they hybernate, commencing to move about again towards the end of January, and occasionally nibbling *L. album*, honeysuckle, &c., till the end of March, when they moult and feed up rapidly till the middle of May, when they are fullfed, and spin their cocoons between the stalk and leaves of their foodplant.

The eggs of *Plusia bractea* are laid towards the end of July, and hatch in early August ; the larvæ feed well on stinging-nettle, *Stachys sylvatica, Lamium purpureum*, &c., hybernating till March, when they can be fed up on *Lamium album*, pupating towards the end of May or in June. [Groundsel is too relaxing a food, and large numbers frequently die off before hybernation when it is given.] (Buckler).

The larvæ of *Heliothis dipsacea* are to be found in July and August, feeding on the blossoms of purple clover (Hallett-Todd), the seed-vessels of *Silene otites* (Harwood), green seedpods of toadflax (Walsing-ham), *Crepis virens* (Cole), *Ononis arvensis* (Harwood), *Hieracium, Linaria* (Buckler), &c. They require sand in which to pupate.

The larvæ of *Anarta myrtilli* should be taken in July and early August ; those collected after the middle of August are usually badly ichneumoned (Brady).

If eggs are required, captured females of *Catocala sponsa* should be confined in a roomy leno cage, in which oak-twigs have been placed, and fed carefully with moistened sugar. They will lay their eggs on the oak-twigs and also on the leno of the cage.

The eggs of *Phytometra viridaria* (*aenea*) laid in July, hatch in a few days; the young larvæ feed freely on milkwort, *Polygala vul-garis*, and are fullfed in September, when they spin closely woven, grey, silken cocoons, in which to pupate. [Eggs laid in May and June, produce larvæ almost at once, imagines appearing in July.]

LITHOSIIDES.—In early July, swamps are to be worked for *Nudaria*

104 PRACTICAL HINTS FOR THE FIELD LEPIDOPTERIST.

senex; the moths fly gently over the low herbage, resting on the stems of grass and reeds. Their favourite time of flight is at dusk, afterwards they are best obtained at rest by searching with a lantern.

Larvæ of *Nudaria senex* feed from July to June on decayed sallow and bramble leaves, on the young growth of *Hypnum sericeum, Weissia cirrata* and *Lichen caninus* (Hellins).

The young larvæ of *Cybosia mesomella* may be fed from July onwards on sallow-leaves, although this is probably not their natural food (Hellins).

The larvæ of *Miltochrista miniata* will feed in confinement on sallow-leaves that have begun to decay, also on withered oak-leaves and various species of lichen.

The ova of *Lithosia* var. *stramineola (flava)* are laid in July and August, the larvæ soon appearing, and feeding throughout the winter on *Lichen caninus*, becoming fullfed in May and June.

The imagines of *Lithosia muscerda* are on the wing in July and August from early dusk till darkness sets in, when they disappear until midnight, after which they have another short flight, and probably there is a third flight in the morning dusk (Barrett).

HESPERIIDES.—In July, the larvæ of *Nisoniades tages*, no longer able to hide within the little caves formed of the leaflets of *Lotus corniculatus*, which they use when young, make larger ones, but their feeding soon exposes their bodies partly to view. They repeatedly change their habitations, always, however, by night, and are most retired in their habits. They are fullfed at the end of the month, when they spin silken hybernacula, in which they remain invisible, not pupating until the following April or May.

The young larvæ of *Syrichthus malvae (alveolus)* are to be obtained, in July, on *Potentilla fragariastrum* and *Rubus fruticosus;* the larvæ that are on bramble seem to be found chiefly on stunted bushes with small leaves, the large juicy leaves of strong bushes apparently offering no temptation to the female.

The larvæ of *Syrichthus malvae* appear to choose the upperside of a leaf, and, resting along the midrib, spin several silken threads overhead for a covering, feeding therein by eating away the upper part of the leaf. When a larva has cleared this and made a blotch of considerable extent, it repeats the work on another leaf. The larger larvæ pull down a second leaf over the first, fastening the edges with silk, and thus form a hollow in which they live, coming out therefrom occasionally to feed on the surrounding leaves.

In July, the young larvæ of *Pamphila sylvanus* feed on cock's-footgrass, couch-grass, &c., resting in the middle of a blade and fastening its edges across with five or six distinct little ropes of white silk (Hellins).

The young larvæ of *Thymelicus thaumas* leave the eggs in late July or early August, and spin little silken web coverings for themselves and little silken ropes across the blades of grass ; but, although they feed until November before hybernation, they are not more than about 2mm.-3mm. in length, almost the whole of the growth being done in the spring.

The puparium of *Thymelicus actaeon* is spun by the larva in a retired corner between some leaves of *Brachypodium sylvaticum*, of which it forms a spacious habitation by spinning, in the open parts, a thin wall of whitish silk web, with large and very irregular meshes, the resting-place being thick, and covered with whitish silk, but even more thick where the tail of the larva is to rest. In four or five days the larva changes to a pupa (Zeller).

PAPILIONIDES.—The eggs of *Plebeius aegon* are laid in July, but do not hatch until the early part of the following March ; the young larvæ can then be fed on *Ornithopus perpusillus*. The larvæ feed up very rapidly (for a Lycænid species), and are fullfed in mid-June.

In the early part of July, collect the flowerheads of *Anthyllis vulneraria* for larvæ of *Cupido minima*. They eat little holes through the calyx and corolla so as to get into the flowers, when they feed on the miniature seed-vessels, leaving a floweret when cleared and entering another. As they get older, their bodies cannot be wholly contained in the corolla, and they may be then seen with the fronts of their bodies thrust into the flower, the hinder part hanging out, but still difficult to distinguish among the dense inflorescence of the flowerhead.

In confinement, the larvæ of *Cupido minima* are fullfed before the end of July ; they then take up a position as if for pupation, but remain quite still and immovable until the following May, when pupation takes place.

In July, the little larvæ of *Polyommatus icarus* make small, pale, transparent blotches on the leaflets of *Lotus corniculatus*, *Ornithopus perpusillus*, &c., the paleness being due to the eating away of the soft parts of the leaf, and leaving only the transparent skin. [Also common again in September.]

In late July, the now nearly fullfed larvæ of *Polyommatus icarus* are to be found on *Lotus corniculatus* and *Ononis arvensis*.

The eggs of *Argynnis adippe* are laid in July and August on the leaves of *Viola canina*, generally on the underside or on the stems. They change colour very rapidly as the embryos mature, but the larvæ do not appear till late February or early March the following year.

When fullfed, the larva of *Argynnis aglaia* will spin together several of the large leaves of its foodplant into a hollow, tent-like enclosure, and, in this, suspend itself before changing to a chrysalis.

The eggs of *Dryas paphia* are laid in July and August, the egg-stage lasting a fortnight, the larvæ feeding very little (or not at all) on *Viola canina*, before hybernation, being only about 3mm. long in spring (March), when they recommence to feed.

The larvæ of *Pyrameis atalanta* are to be found in July and August in little chambers, formed by drawing together the leaves of *Urtica dioica* and *Parietaria officinalis ;* they generally hang up and pupate within these larval chambers.

In July, the larvæ of *Pyrameis cardui* fasten together the leaves of *Onopordon acanthium*, and other thistles, with a few tough silken threads, eating out the thick, fleshy parts of the enclosed leaves. They generally hang up and pupate within these larval chambers. [Larvæ also feed on *Echium vulgare*, *Malva*, &c.]

The eggs of *Apatura iris* are laid in July and early August on sallow-leaves, and the egg-stage lasts about eight days ; the young larvæ are not difficult to rear in leno sleeves on a healthy sallow-bush.

The eggs of *Melampias epiphron* are laid in early July, and hatch before the end of the month ; the larvæ, in confinement, feed on *Aira praecox* and *A. caespitosa*, growing to the length of about half-an-inch before winter; they then hybernate until the end of February, when their food should be carefully attended to.

The females of *Coenonympha tiphon* will lay their eggs in confinement, if placed in a suitable receptacle, and exposed to the sun, with a supply of their foodplant, the beaked rush (*Rhynchospora alba*) ; this should be potted, and the young larvæ will feed thereon until their hybernating stage with little trouble, provided care be taken that the active little fellows do not escape.

The larvæ of *Euchloë cardamines* are to be found, in July, on many cruciferous plants, of which *Hesperis matronalis*, *Sinapis arvensis*, *Cardamine pratensis*, *Sisymbrium officinale*, *Alliaria officinalis*, *Turritis glabra*, and, in gardens, garden-rocket and horseradish are the best known.

The eggs of *Leptosia sinapis* are laid by the second-brood females in July or early August. The larvæ will feed up on *Vicia cracca* or *Orobus tuberosus*, and are generally fullfed in early September, the pupæ going through the winter, the imagines emerging the next spring.

In July, the young larva of *Papilio machaon* may be searched for with every prospect of success, the black larva, with its white saddle, being very easily found when once the eye of the searcher is in. Until then, it is most difficult to detect, although many may be on the plant under examination. The habit of repose, with the neck arched something like a Sphingid, is very striking.

AUGUST.

NEPTICULIDES.—The autumn-collected mines (and contained larvæ) of *Nepticula angulifasciella* should be kept out-of-doors all the winter if success be desired ; the imagines do not appear until the following July.

OCHSENHEIMERIIDES.—A visit to some grassy spot, from noon to 2 p.m., on a sunny day about the middle of August, is likely to disclose small moths flying amongst the vegetation. These are very likely to prove the local *Ochsenheimeria birdella*, as so it is to be procured.

LYONETIIDES.—Imagines of *Lyonetia padifoliella* are to be bred in August, from broad mines in sloe. These are not at all narrow like those of *L. clerckella* (Frey).

GRACILARIIDES.—Towards the end of August, a careful examination of the leaves of *Artemisia vulgaris*, growing on hedge-banks, will most probably result in some being found to have an inflated bladder-like appearance. This is the work of the larva of *Gracilaria omissella*. If the larva be of a rich crimson colour, it is fullfed and about to quit its mine to form its silken cocoon.

The larvæ of *Gracilaria phasianipennella* are to be found, in August and September, in cones on the leaves of *Polygonum hydropiper*, *P. persicaria*, and *Rumex acetosella*.   [Zeller also found the larvæ on *Rumex obtusifolius*.]

The fullfed larvæ of *Gracilaria imperialella* leave the bladders they form in the leaves of *Orobus niger*, and spin pale ochreous cocoons outside the leaves, but attached to the white loosenèd skin of the undersides, so that they are hardly perceptible.

The pupæ of *Coriscium cuculipennellum* are to be found in August, concealed in very symmetrical cones formed at, and of, the end of privet leaves. When withered, the cones become distorted. Very local ; only a bush or two out of dozens surrounding them will contain pupæ.

Throughout the month of August, the larvæ of *Phyllocnistis suffusella* may be found mining the leaves of various poplars ; not confining themselves to British species. The mines are very likely to be passed by, as they have a very strong resemblance to the slimy track left by a slug or snail in crawling over a leaf.

The larvæ of *Phyllocnistis suffusella* mine indifferently the upper- or underside of the leaves of *Populus tremula*, &c. The mined leaves show no distinct track, as the larvæ drink only the juice by which the skin of the leaf is fastened to the parenchyma, but, when the leaves are viewed at an angle, they appear shining, and reflect prismatic colours as though a snail had crawled across them.

GLYPHIPTERYGIDES.—If the seed-heads of *Dactylis glomerata* be gathered about the middle of August, they will often yield a good supply of larvæ of *Glyphipteryx fischeriella*.

The fullfed larvæ of *Glyphipteryx fischeriella* are to be found from mid-August to the end of the month, feeding on the seeds of *Dactylis glomerata* and other grasses, boring into the grass-flowers and making holes at the sides ; you can, at the time of collecting, see little or no trace of the larvæ. Having obtained some of the seed-heads, put them into a glass and look at them in a day or two, when you will soon observe where the larvæ are at work.

TORTRICIDES.—Larvæ of *Penthina dimidiana* are to be found, during August, in the spun-together leaves of terminal shoots of *Myrica gale*.

If the patches of *Armeria vulgaris*, growing on salt-marshes, be carefully watched on a bright afternoon in August, *Sericoris littoralis* will, in all probability, be seen flying over them in numbers.

Towards the end of this month, the larvæ of *Phtheochroa rugosana* are to be found feeding in the fruits of *Bryonia dioica*, which they often attach to the stems with silk. As these larvæ do not pupate until the spring it is expedient to keep them in a cool, and not too dry, place.

From the middle of the month, *Semasia spiniana* may be captured flying over whitethorn in the bright sunshine. Its time of flight commences about midday and lasts well into the afternoon. By stooping down by a close-cut hawthorn hedge, so as to get a clear view of the moths against the sky as they fly along, large numbers are sometimes to be obtained.

About the close of August, a careful examination of the leaves of various species of *Salix* may result in some being found drawn together at the edges, so as to form a pod-like chamber. These leaves should contain the active larvæ of *Phoxopteryx biarcuana*.

COLEOPHORIDES.—The larvæ of *Coleophora wilkinsoni* mine the leaves of birch in August and September. The case is of a dark brown colour, with a rounded projection towards the middle on its underside.

The elongate, soft, grey-green cases of *Coleophora artemisiella* are to be obtained in August, the larvæ feeding at this time on the leaves of *Artemisia maritima*.

The cases of *Coleophora frischella* (*melilotella*) are to be obtained in early August, the larvæ feeding on the seeds of *Melilotus officinalis*. The case is made of the seed-husk; at first, only a single seed is used, then two are clumsily attached together, and, ultimately, they are so blended as to form a symmetrically cylindrical case.

GELECHIIDES.—The larvæ of *Enicostoma lobella* occur from the middle to the end of August. They are found on the underside of leaves of *Prunus communis* and its cultivated varieties. The larvæ spin silk on the under-surface of a leaf, which causes it to contract, but not to a very marked extent. Unless great care is exercised in gathering the tenanted leaves, their occupiers will be found to have abandoned their homes.

Psoricoptera gibbosella occurs towards the end of August, and is best found by searching oak-stems. It sits tightly pressed into a crevice in the bark, and, when the stems are lichen-covered, it is well-nigh imperceptible. Under such conditions, it may be dislodged by gently blowing on the stems.

Larvæ of *Teleia scriptella* are to be found, at the end of August, in leaves of *Acer campestre* having a corner turned over. The larvæ occur on the lower shoots of their foodplant, and appear to have a liking for those in close-cut hedges.

From 6 p.m. to 7.15 p.m., on calm and mild evenings throughout August, the imagines of *Lita maculiferella* may be found flying along whitethorn hedges. This species is local, but generally abundant where it occurs.

ALUCITIDES.—In mid-August, the larvæ of the second-brood of *Platyptilia isodactyla* mine the buds and stems of *Senecio aquaticus*. They appear to enter the larger branches at the axil of a leaf, frequently devouring the tender side-shoots, boring down the interior, and feeding on the pith till full-grown. In every case, a round hole is left for the extrusion of excrement, which leads to a knowledge of the whereabouts of the hidden larvæ.

At the beginning of August, the larvæ of *Amblyptilia acanthodactyla* are to be found feeding on restharrow and many other plants.

CRAMBIDES.—The young larvæ of *Gymnancyla canella* are to be found, towards the end of August, mining within the stems of *Salsola kali*, generally in the side-shoots. In September, however, they change their habit, going outside, attacking the unripe seeds, and, from the cavity thus made, proceed to burrow into the main stem, where they feed in concealment until nearly fully grown. At this time, the entrance-hole of the burrow is protected by a few silk threads, into which the sand is often blown, rendering the burrows conspicuous and pointing out the position of the larvæ.

The young larvæ of *Ilythia carnella* are to be found, in early August, on *Lotus corniculatus*. They spin the stem and leaves together with white silk, the frass pellets being somewhat conspicuous thereon.

In August and September, cones of spruce fir, which show extruded particles of light, fawn-coloured frass, adhering by a few silk threads to the scales of the cone, should be carefully collected for larvæ of *Nephopteryx abietella*.

Several larvæ of *Nephopteryx abietella* will often be found in a single cone, but each takes care not to encroach on the particular mines belonging to the others.

The cones containing the larvæ of *Nephopteryx abietella* are best kept on their side, and not on end. The larvæ do not appear difficult to rear. The only time when they want a little management is just after the last, or penultimate, moult. Should they, at these times, be out of the burrow, it is useless to put them on an ordinary cone, as they will not make any attempt to eat into it, but wander about and ultimately die. The best plan is to get a dry cone and break off some of the scales so as to leave a rough surface ; as soon as a larva is put upon this, all tendency to wander vanishes ; it soon sets about spinning a hiding-place, making it very secure, and taking plenty of time over it, for it is sometimes as much as 36 hours before it runs out a little covered way to the fresh cone that has been placed by the side of the other (Wood).

The larva of *Nephopteryx abietella* seems to be very impatient of exposure, more particularly when left feeble after the process of moulting, for it has been only at this particular time that the care of putting it on a rough cone has been required ; at an earlier age it is able to creep under a scale, and a few threads will then complete its concealment (Wood).

The larvæ of *Homoeosoma nebulella* are to be found, in August, feeding on the flowerheads of thistle (*Carduus*).

At the end of August, the larvæ of *Homoeosoma senecionis* feed in the flowers of ragwort, drawing together the clusters of flowers with silken webs (Porritt).

In August and September, the larvæ of *Homoeosoma nimbella* are to be found feeding on the seeds of ragwort ; they spin little webs, in which they hybernate, pupating in the spring (Buckler).

In August, the larvæ of *Homoeosoma nimbella* var. *saxicola* are to be found commonly, feeding in wild chamomile flowers on the rocks near Douglas, Isle of Man, &c. (Porritt).

The larvæ of *Homoeosoma binaevella* are to be found in mid-August in the flower- and seedheads of *Carduus lanceolatus*, eating the young seeds and excavating large cavities in the solid substance at the bases of the flowerheads, in which cavities they live. When fullfed, the larvæ leave the head, and spin tough, brownish cocoons among rubbish, in which (like those of the allied species) they remain unchanged through the winter and spring (Barrett).

PYRALIDES.—The larvæ of *Herbula cespitalis* can be found in August, under a web, spun quite at the base of a leaf of *Plantago lanceolata*, beneath which they rest in companies, feeding taking place chiefly by night.

In early August, the eggs of *Ennychia nigrata* (*anguinalis*) are laid on the leaves and flower-bracts of *Origanum vulgare, Mentha arvensis,* or *Thymus serpyllum* ; the young larvæ hatch about the middle of the month, and feed on any of these plants equally well.

The young larvæ of *Aglossa pinguinalis* form slender tubes among barn-rubbish, on which they feed ; they live in these until the following May, when pupation takes place.

The eggs of *Endotricha flammealis*, laid in July, hatch in about ten days or a fortnight, the young larvæ feeding on flowers or leaves of *Lotus major*, and, possibly, many other plants (we beat the imagines freely from furze). The larvæ lie on the underside of a leaf, or in a flower, beneath a gossamer-like web, and appear to feed only by night. *Agrimonia*, hazel, sallow, hornbeam, also appear to be suitable foodplants.

The imagines of *Stenia punctalis* are to be found freely in August in many places; they will lay eggs in confinement, and the young larvæ, hatching in September, will feed freely on flowers and leaves of *Lotus corniculatus*, making awnings (not tubes) of very sticky silk. They prefer, however, decaying leaves, and hybernate among such satisfactorily, becoming fullfed in May.

In confinement, eggs of *Ebulea verbascalis* hatch in August. The young larvæ should be placed on *Teucrium*, under the leaves of which they form a silken tubular gallery, eating small holes through the leaves; later, they turn down a part of a leaf, or join two leaves together with a few silken threads in which to hide.

In August, the larvæ of *Pionea stramentalis* are to be found living, more or less gregariously, on leaves of various *Cruciferae*, apparently preferring *Barbarea vulgaris*; they are fullfed in September, and form cocoons in which to pupate under leaves on the surface of the ground.

The work of the larvæ of *Pionea stramentalis* is very conspicuous in August, the larvæ at first eating little pits and channels from the cuticle, and so causing transparent blotches on the leaf, but later eating holes completely through the leaf.

The young larvæ of *Lemiodes pulveralis* live on the undersides of leaves of *Mentha hirsuta*, in mid-August, eating out tiny hollows from the under-surface, where little heaps of frass disclose their whereabouts.

In August, the larvæ of *Scopula decrepitalis* are to be found in slight webs, spun under leaves of *Lastraea spinulosa*. A stronger web is spun during the moulting season. The fullfed larvæ make strong silken puparia, in which they remain until the spring before changing to pupæ.

In late August and early September, search should be made for cylindrical webs of greyish glistening silk, spun within the seedheads of *Daucus carota*, which are then closed up together, so that a good indication of the presence of larvæ of *Spilodes palealis* is afforded by the closeness of the mass. Each web is just sufficiently large and long enough to contain a larva, the outer stalks of the umbel being brought up together by a few single outlying threads. The larvæ are full-fed towards the end of September, and then make puparia under the surface of the earth, sometimes remaining therein for more than twelve months without changing to the pupal stage.

In mid-August, the young larvæ of *Botys hyalinalis* are to be found on the leaves of *Centaurea nigra*, spinning little webs along the side of the midrib towards the stem, eating away the underside of the leaves, and leaving the upperside untouched, thus making little transparent blotches, which indicate where they are feeding, although their translucent colour makes them most difficult to discover.

The fullfed larvæ of *Botys asinalis* are to be found in early August, making large conspicuous blotches in the leaves of *Rubia peregrina*, whilst they inhabit silken galleries as a sort of hiding-place. They spin their puparia during the first fortnight of the month, when they pupate, and the imagines of the second-brood emerge therefrom about the end of the month.

In August and September, the larvæ of *Botys lancealis* spin webs among the leaves of *Eupatorium cannabinum*. The webs are generally placed under the end of a leaf which they fold down, and, in this, they live, coming out at intervals to feed on the neighbouring leaves.

CYMATOPHORIDES.—The pupæ of *Asphalia ridens* should be sought as early as possible, *viz.*, middle to end of August. Detached oaks, growing in meadows, of a dry, loamy soil, seem the best, the situation evidently preferred being the corners filled with dry rubbish and little stunted brambles. Insert the trowel well into the earth, six or seven inches from the angle, and turn up the sod, bramble and all, if possible. To find the pupa after this is done is a work both of time and pain ; it will not do, in this case, to tap the sod. First, carefully examine the dead leaves, for the larvæ frequently spin up in them ; you must then tear the roots asunder as gently as possible. The cocoon is very weak, composed of little bits of stick, dried leaves, &c., and requires delicate handling. Indeed, the whole affair demands an elaborate manipulation (Greene).

DREPANULIDES.—The eggs of *Drepana hamula* are laid in August on the very edge of the leaves of oak. They hatch in a few days, and the larvæ feed up well on oak in confinement.

GEOMETRIDES.—In August and September, the young larvæ of *Phorodesma smaragdaria* are to be found on *Artemisia maritima* on the Kent and Essex salt-marshes. The smallest are like little balls of white wool, owing to the body being covered with the mealy portions of the *Artemisia*. As they increase in size, the pieces of foodplant used are increased in length, and the larvæ resemble more closely portions of broken and dying leaf.

The hybernating larvæ of *Gnophos obscurata* are best reared in con- finement on strawberry, as it is a food that can be usually fairly easily obtained ; the larvæ nibble all through the winter, and generally die unless supplied with something that they can use for this purpose.

The eggs of *Dasydia obfuscata*, usually laid in early August, hatch in about three weeks. It is to be noted that, in Perthshire at least, the purple bell-heather (*Erica tetralix*) is undoubtedly its foodplant, although nearly all authorities give other plants (Bush).

The larvæ of *Dasydia obfuscata* (which leave the eggs in August and early September) can be fed readily in confinement on *Calluna vulgaris* and *Polygonum aviculare*.

The larvæ of *Amphidasys betularia* are known to feed on a variety of foodplants, of which, however, broom is not often recorded.

The larvæ of *Biston hirtaria*, *Selenia tetralunaria*, and *S. bilunaria* are found commonly on birches in Sutherland, and they are not hard to see when you know how to look for them ; they sit, during the day, motionless among the clusters of dead twigs, only moving at night to the fresh twigs (Chamberlain).

The imagines of *Oporabia filigrammaria* are to be obtained, through- out August, on the open heaths among *Calluna vulgaris* in our northern counties, often somewhat later in Scotch localities.

The hybernating larvæ of *Acidalia virgularia* (*incanaria*) should be given withered leaves of dandelion, on which they will feed on all mild days through the winter, and will continue to live on this food very well until pupation in May.

The larvæ of *Asthena blomeri* are to be beaten, in August and September, from wych-elm. They should be obtained as early as possible, as the late ones appear to be, in some years, dreadfully infested with parasites. Some larvæ can be taken even in October.

If search be made for the larvæ of *Asthena blomeri* in August and September, it is well to look for leaves of the wych-elm that have little holes eaten through them, the ribs having been avoided. When quite young, the larvæ eat only the under-surface of the leaf, and keep to the underside for shelter throughout their whole existence.

In August, the feeding larva of *Asthena sylvata* loves to rest along the midrib at the back of an alder-leaf, with the head held up, the thoracic segments kept close to the leaf, the 2nd-5th abdominals raised in an arch, or sometimes a loop, and the 6th-10th abdominals again pressed close to the leaf.

The puparia of *Asthena sylvata* are spun on the surface of the soil in late August and early September, and are formed of small particles of earth, leaves, &c., fastened together with a tough, although not hard, lining of pale silk.

The eggs of *Emmelesia blandiata* are laid in August on the under-side of the leaves or among the open flowers of *Euphrasia officinalis*. The larvæ bore into the unripe seed-vessels and feed on their contents, leaving those first attacked for fresh ones when they have cleared out the former. They are fullfed in September, going down to the earth for pupation, forming compact little cocoons. The fullfed larvæ, with their heads buried in the seed-capsules and the greater portion of their bodies resting outside, assimilate most perfectly to the stems and leaves, so that even when their presence is known their detection is most difficult.

Eggs of *Emmelesia taeniata*, laid in confinement, hatch during the second week of August. They feed on garden nasturtium, making round holes through the leaves (Hodgkinson). Later, Hodgkinson suggested that the larvæ after hybernation fed on the fruits of many of the mosses, more particularly on a species of *Bryum* growing in wet places.

The eggs of *Camptogramma fluviata* hatch in a few days. At large, no doubt, the larva is polyphagous, and has been found on *Senecio vulgaris*, *Polygonum persicaria*, *Agrimonia eupatorium*, and other low plants, its movements being quiet and sluggish.

The larvæ of *Camptogramma fluviata* are very easy to rear in confinement, being quite tame and domestic, only the temperature must be warm enough ; they feed rapidly on groundsel (a food that grows everywhere), and spin up contentedly, ninety-nine out of every hundred producing imagines, which pair in confinement without difficulty and become continuously-brooded, up to six or seven broods, under suitable conditions (Hellins).

When fullfed, the larva of *Camptogramma fluviata* retires into any cover it can find at hand, and, either just below the surface of the soil or amongst moss or dried bits of its food, constructs a perfect, but thin and weak, cocoon of silk, drawing in enough particles of dust, etc., to give it an oval form.

The larvæ of *Cidaria reticulata* thrive, in August and September, on the seed-vessels of *Impatiens noli-me-tangere*, which they apparently prefer to any other parts of the plant, although they will also eat the leaves if they are in good condition. They rest perfectly quiescent on the stem of the plant all day, looking rather shorter and stouter than when they wake up at sunset and feed, which they continue to do at intervals throughout the night. During this time the larvæ are very active and lively. Pupation takes place in late September and early October.

The larvæ of *Mesotype virgata* are to be found on *Galium verum*, in August and September, being fullfed by the end of the month, when pupation takes place, the imagines not appearing until the following April-May (sometimes April-June).

The larvæ of *Eupithecia pimpinellata* are to be found, feeding on the flowers and seeds of *Pimpinella saxifraga* and *P. magna*, apparently preferring the latter, throughout August and September, choosing generally those plants growing on hedgesides and banks.

In August, gather the plants of *Campanula trachelium*, the nettle-leaved bell-flower. Knock them against the sides of an umbrella, and the young larvæ of *Eupithecia campanulata* will soon be seen crawling on the bottom and sides; the larvæ feed upon the unripe seeds and seed-capsules of this plant, and live till nearly fullgrown, either in the dry corolla-tube, or just at the crown of the capsule. In confinement, they will feed upon garden species of *Campanula*, and are fullfed at the end of August or beginning of September.

If one, however, knows himself to be in the locality for *Eupithecia campanulata*, he need not beat out the larvæ from the flowers of *Campanula trachelium*, but the heads can be collected, and placed in a strong linen bag until home is reached, when they can be transferred to a large bandbox, and the larvæ collected therefrom and placed on fresh food, as they leave the withered plants on which they were first found.

The larvæ of *Eupithecia extensaria* are to be found, in August, on *Artemisia maritima*. The species seems to be gregarious or excessively local in its habits, frequenting sheltered clumps of the foodplant, but not extending its range very far, although the *Artemisia* is very plentiful on the coast of northern Norfolk (Barrett).

In August, the larva of *Eupithecia extensaria* spins a tough, oval, brown, silken cocoon among the *débris* on the surface of the ground; sometimes also among the leaves or stalks of the foodplant (Porritt).

In August and September, the larvæ of *Collix sparsata* feed on *Lysimachia vulgaris;* the fullfed larvæ spin up between the commencement and middle of September.

HEPIALIDES.—The young larvæ of *Hepialus hectus* are to be found, burrowing in the lower part of the stem and root of *Pteris aquilina*, from August throughout the winter, following spring and summer, taking two years to come to maturity.

ANTHROCERIDES.—The larvæ of *Adscita statices*, in August, (September and October), eat completely through the whole substance of the young leaves of sorrel, remaining, however, mostly buried in the leaf, and only exposing themselves in the hot sunshine.

The young larvæ of *Anthrocera exulans* emerge from the egg in

August, and feed very freely on *Lotus corniculatus*; they hybernate very small, fixing themselves about the middle of September. They should be placed with pieces of absolutely dry cork, in a dry, but not warm, place, and will then usually hybernate well.

SPHINGIDES.—In August and September, search the dwarf sallows and willow-bushes by the sides of the ditches of coast sandhills for larvæ of *Smerinthus ocellata*, which are often common in such localities.

The fullfed larvæ of *Amorpha populi* are to be found on poplar-trees in August and September. In our experience, they get much higher on the trees than do those of *Smerinthus ocellata*, which prefer dwarf bushes.

The larvæ of *Celerio gallii* are to be found in August and September, feeding on *Galium verum* and *Fuchsia*, in those years in which migrants reach us in July.

The larvæ of *Agrius convolvuli* are usually found on the plants of *Convolvulus arvensis*, growing in potato-fields, in August and September. The larvæ are very lethargic, rest quite still in the sunshine on a stem of the foodplant, and clear the leaves completely from a plant in their order on the stem before moving to another plant.

NOLIDES.—The eggs of *Nola centonalis* hatch in August, and the young larvæ feed freely on the leaves and flowers of *Trifolium procumbens* and *Lotus corniculatus*, preferring the former. They will also eat the flowers and leaves of *T. minus*, *T. pratense*, and *Medicago lupulina*.

NOTODONTIDES.—In August, the young larvæ of certain species of Notodontids may be readily obtained; the eggs of many species are laid on the underside of the leaves of birch, and, by turning over the long twigs at the ends of the lower branches, the little larvæ are easily seen, and large numbers may be taken, e.g., *Lophopteryx camelina*, *Notodonta dromedarius*, *Leiocampa dictaeoides*, and possibly *Lophopteryx carmelita*. [Other larvæ to be found at the same time are those of *Demas coryli*, *Acronicta leporina*, *Selenia tetralunaria*, *S. bilunaria*, *Biston hirtaria*, and *Cidaria miata*.] (Chamberlain).

The cocoons of *Cerura vinula* are to be found on the bark of poplar, sallow and willow trees from August onwards; they form a somewhat swollen oval on the surface, and are very similar to the bark of the tree, particles of which are woven into the surface, in colour and appearance.

The pale bluish-green larvæ of *Pterostoma palpina* are readily found, in August and early September, on sallow, aspen, and poplar. [Also in late June and early July.]

The weak cocoon of *Pterostoma palpina*, formed of greyish silk, thinly coated with earth, is to be found near the roots of sallow and aspen in late August and September; also more rarely near poplar.

The fullfed variable larvæ of *Notodonta dromedarius* are to be found on, or beaten from, birch and alder throughout August; the four dorsal humps are very characteristic. [Searching is better than beating.]

The handsome lilac-tinted larva of *Notodonta ziczac*, with its two characteristic dorsal humps, is to be found on sallow in August and September. [Also in June.]

The somewhat elongated, shiny, fullfed larvæ of *Leiocampa dictaea* are to be beaten from, or found on, poplar and aspen, towards the end of August. [Also in June.]

The fullfed larvæ of *Drymonia trimacula* (*dodonaea*), which are superficially like those of *Lophopteryx carmelita*, are to be beaten from oak in August and September. They are usually four to six weeks (or more) behind those of *D. chaonia* in feeding up.

The cocoons of *Drymonia chaonia* are much like those of *D. dodonaea*, being tough, regular in outline, and covered evenly with bits of earth ; they are to be found in the fine dry soil in the angles of the roots of oak-trees.

The cocoons of *Drymonia chaonia* are to be found at the roots of oaks in August and September, the larvæ of this species feeding up at least a month earlier than those of *D. trimacula* (*dodonaea*). They are sometimes attached to the trunk, but more usually are to be found among the dry friable sods collected in the corners, or even in the corners themselves without any sod, but, in all cases, great caution is necessary in obtaining them ; it is a good plan, when a sod has been pulled out, to gently feel the trunk for any cocoons which may adhere to it (Greene).

DELTOIDES.—In August, where the leaves of *Brachypodium sylvaticum* and *B. pinnatum* are observed to have little pieces eaten out of the sides, quite on the edge, search should be made for the pale greenish larvæ of *Rivula sericealis*.

NOCTUIDES.—Eggs of *Agrotis puta*, laid in August, hatch towards the end of the month or in September, and the larvæ can be reared easily on lettuce, dandelion, knotgrass, and slices of the root of carrot ; they much prefer the lettuce. They are mostly fullfed by the end of December or in early January.

The eggs of *Agrotis cursoria* are laid in August and early September, and hatch in a few days. The larvæ should be treated as are those of other Agrotids, but, owing to their hatching so late, they feed more continuously during the winter, on *Arenaria*, *Viola*, *Triticum*, &c., but are not fullfed until towards the end of June, when they will burrow deep into the sand for pupation, the imagines emerging in due course.

The eggs of *Noctua sobrina* hatch towards the end of August ; the young larvæ feed freely on heather and birch, and do not altogether disdain grass. The larvæ hybernate (in the nibbling stage) from the end of November to March, and are fullfed in June.

The eggs of *Pachnobia hyperborea* (*alpina*) hatch in early August; the young larvæ feed freely on bilberry, riddling the leaves with small holes, and, as they get older, skeletonise the bilberry leaves. (Buckler found great trouble in getting the larvæ through the winter, so that possibly one should pot up some bilberry plants on which to place them, if their natural food, *Empetrum nigrum*, be not available. In the Alps of central Europe the species occurs apparently where whortleberry is abundant.)

The eggs of *Aplecta occulta* are laid in August, hatch in a rather short time, and, in confinement, the larvæ will sometimes feed up very rapidly, being fullfed by the end of October or in November, and, occasionally, as early as the end of September, pupate at once and produce imagines, if carefully looked after, even in the middle of winter. Heather, bramble, sallow (leaves, buds, and catkins), knotgrass, birch, bilberry, dock (*Rumex pulcher* and *R. crispus*), &c., are given as some of the food-plants.

The pupæ of the Tæniocampids are to be found, from August through the winter, at the roots of various trees, and may be discovered by simply turning the soil over carefully and passing it through the fingers, or by shaking carefully any turf that may be dug from near the roots.

The apparently almost shapeless eggs of *Dyschorista suspecta*, laid in August and September, hybernate as eggs throughout the winter, hatching towards the end of April or in early May; the young larvæ feed well on birch, and come to maturity very rapidly, pupating by the end of May, or in early June.

The pupæ of *Cirrhoedia xerampelina* are to be sought at the roots of ash-trees of good growth; those on the borders of streams and damp ditches will be found most productive. They are contained in a hard egg-shaped cocoon. Turn up the loose dry earth, rubbish or moss about or adhering to that side of the tree which faces the stream; crumble it very carefully with the hand, and should you see something resembling a cocoon, of a dark muddy colour, take it up and try whether you have obtained a prize, but, in this trying lies the danger. Though hard, the cocoon is extremely brittle, and almost the slightest pressure crushes it; the best way, therefore, when you think you have a cocoon, is to pare one end with a penknife as gently as possible, and if, after scraping it in this manner, you find it is a cocoon, the chances are that you have found *C. xerampelina*, and may congratulate yourself. You may look for it as early as the beginning of August, certainly not later than the first week of September (Greene).

During the last fortnight of August, the rich ochreous-brown aberration of *Cirrhoedia xerampelina* occurs on the trunks of the ash-trees growing near Douglas; the best way to collect them is to go round and examine the trunks from 4 p.m. to dusk, when they will be seen extending their wings, newly-emerged from the pupæ. About one-third of the captures made here belong to the aberration (Gregson).

Near Wilsden, *Lithomia solidaginis* occurs in some years abundantly; at rest, this species has a most remarkable resemblance to the excrement of grouse—the male particularly so. It folds its wings round the body, clasps a stone with its legs, and raises its body to an angle of about 30°. Its markings, colour, shape, and mode of attachment make the imitation almost perfect (Butterfield).

A larva of *Cucullia gnaphalii*, found at Tilgate, end of August, was feeding on the small lanceolate leaves of *Solidago virgaurea* just below the flowers; the larva does not appear to eat the flowers (Buckler).

The pupa of *Catocala nupta* occurs not infrequently under loose bark on willows in August; it never appears to enter the earth for pupation (Greene).

LYMANTRIIDES.—In Sutherland, the favourite foodplant of *Demas coryli* is birch (not hazel); the larvæ are very plentiful on stunted birch-trees or bushes, growing near the sea-coast, exposed to the full force of the cold northerly winds (Chamberlain).

LITHOSIIDES.—The eggs of *Lithosia griseola* hatch in August, and the young larvæ will feed on withered leaves, especially delighting to riddle decaying sallow leaves full of holes; they also eat a little clover, knotgrass, and various lichens and mosses (Hellins).

In August, the larvæ of *Lithosia sororcula* (*aureola*) are to be found feeding upon the lichens attached to oak (Buckler).

HESPERIIDES.—In August, the larvæ of *Steropes palaemon* (*paniscus*) feed within long cylindrical tubes made of the leaves of *Brachypodium sylvaticum*, quickly, however, eating their domiciles and forming fresh ones; they first eat the lower part of the leaf below the tube all but the midrib, then devour the top of the leaf above the tubular part, and, lastly, the tube itself, until, by degrees, it becomes too short to shelter them, when they desert it and cut through the midrib, causing the tubular remains to fall away, after which they select a fresh leaf for the construction of another tube as above (Buckler).

The eggs of *Pamphila comma*, laid in August, do not hatch until the following March. As most of the " skippers " hybernate as larvæ, care must be taken to look after the eggs of this species and not throw them away with the idea that they are infertile.

The pupa of *Syrichthus malvae* (*alveolus*) is to be found, in August, in a hollow between two leaves of *Potentilla fragariastrum* and *Rubus fruticosus*, the ends protected by a loose, pale yellow webbing of silk.

PAPILIONIDES.—Eggs of *Cyaniris argiolus* are laid in August, beneath the flower-heads of the umbels of ivy. The larvæ feed on tender young ivy-leaves and -flowers, and pupate in early September.

The circular, flattened, greenish-drab eggs of *Polyommatus astrarche* are laid in August and September, in little groups of two, three, or more, on the underside of the leaves of *Helianthemum vulgare*.

Captured females of *Colias hyale* will lay their eggs on *Trifolium repens*, *Medicago lupulina*, *M. sativa*, &c., in August. They hatch in about a fortnight, and the young larvæ feed up slowly until October or November, when they will hybernate until March, provided they are not exposed to a really low temperature, are kept quite clear of any decaying leaves, and have a perfectly dry spot to rest on. They must be supplied with food very early in the spring, and should be given as much sun and air as possible, but not exposed to a low temperature.

Females of *Colias edusa*, enclosed on a growing plant of clover (*Trifolium repens*) or *Lotus corniculatus*, placed in the sun, and supplied with a little honey and water for food, will lay their eggs pretty freely so long as the weather is bright and sunny. During dull weather the butterflies will not lay.

The larvæ of *Colias edusa*, obtained from eggs laid by females captured in August and September, will feed up well in confinement on *Trifolium repens* and *Lotus corniculatus*. These will try to feed up the same year, and must be carefully nurtured. They might, indoors, be induced to partially hybernate until early March, but then would have to be kept perfectly free from damp and well away from any decaying leaves of their foodplant.

The females of *Epinephele tithonus* will lay eggs fairly freely in confinement in August. The larvæ appear in about three weeks, and feed well on *Poa annua*, *Dactylis glomerata*, and other common grasses; they hybernate when exceedingly small, but nibble in winter when the weather is mild, not feeding very much, however, till mid-March. They are usually fullfed about mid-June, and turn into pale drab pupæ, which are suspended by the tail.

Females of *Erebia aethiops* will lay their eggs quite freely, in August, if supplied with their foodplant in a suitable receptacle and placed in the light and sun. They are glued to stems of various species of grass (*Aira praecox*, *A. caespitosa*, &c.), and are large and conspicuous. The young larvæ appear in about three weeks.

The eggs of *Melanargia galathea* hatch in August, and the larvæ feed well, in confinement, on almost all common garden grasses—*Dactylis glomerata* has been noted as a specially favoured one. The larvæ hybernate from about the end of October, feeding occasionally when the weather is mild, going ahead more rapidly in March and April, and being fullfed in June.

The larvæ of *Hipparchia semele* can be reared in confinement on *Triticum repens*, *Aira praecox*, and many other grasses; they are very sluggish, hide low down among the foodplant, nibble slowly most of the winter, feed only at night, and often bore under the ground, if at all suitable, by day. They are fullfed about mid-June, when pupation takes place.

The fullfed larvæ of *Papilio machaon* are to be found in August, usually resting in a perfectly vertical position on a stem of the food-plant, or on a plant near; although such a large conspicuous cater-pillar, when separated from its food, it is not at all easy to see when surrounded by the herbage of the fen districts in which alone it lives in Britain.

SEPTEMBER.

ADELIDES.—The larvæ of *Nemotois scabiosellus* are to be found in *Scabiosa arvensis*, from September to May, the young larvæ use one of the husks of the seeds (on which it feeds) as a case; it is very difficult to find these seed-feeding larvæ, which are to all intents and purposes invisible; an inhabited seed cannot at first be distinguished from an uninhabited; one must wait patiently to detect some movement. The larger larvæ construct cases from pieces of dried leaves.

ELACHISTIDES.—The long whitish cocoons of *Mompha* (*Laverna*) *decorella* are to be collected in September and October, within the rounded gall-like swellings (about the size of a pea), made in the stems of *Epilobium montanum* and *E. palustre*, several galls often being on the same stem; some white web protrudes through the upperside of each gall in a tubular form, and from this the imago escapes.

The larva of *Limnoecia phragmitella* feeds in the heads of *Typha* from September to May, secreting itself in the woolly down, where it is about as easy to find as a needle in a bundle of hay; the down itself hangs out in the spring in large conspicuous masses.

CHRYSOCORIDIDES.—In September, the larvæ of *Chrysocorys festa-liella* are to be found feeding either on the upper- or underside of bramble-leaves, eating the leaves half through or making conspicuous blotches, which are very evident even when one is not specially look-ing for them.

LITHOCOLLETIDES.—From mid-September to mid-October, leaves of all trees and low-growing plants should be most carefully examined for various Lithocolletid and Nepticulid mines and blotches.

In late September, the mines containing larvæ or pupæ of *Lithocolletis bremiella* are to be found in leaves of *Vicia sepium* growing at the edge of woods. The imagines appear throughout the early part of October. [Larvæ of an earlier brood are to be found similarly in July.]

The larvæ of *Lithocolletis stettinensis* are sometimes very abundant in autumn, on alders growing in meadows, &c. The larvæ are very partial to the terminal leaf of the twig, frequently four or five larvæ being in one leaf.

At the end of September, pod-like excrescences on plants of *Polygonum aviculare* contain the larvæ of *Asychna aeratella*. The larvæ, towards the end of May, eat small openings near the end of their habitations, and, through them, eject some frass, a certain sign that the larvæ are about to change into pupæ.

TORTRICIDES.—Towards the end of September, the pupæ—and in late seasons larvæ—of *Peronea logiana* may be found in " pockets " in screwed-up leaves of *Viburnum lantana*. Unless the pupæ are removed from the leaves, crippled imagines are likely to be bred.

The larvæ of *Paedisca consequana* are common in September and October on *Euphorbia portlandica*.

During the last fortnight of September, larvæ of *Phoxopteryx derasana* are to be obtained on *Rhamnus frangula*. These larvæ draw together the sides of a leaf, causing it to strongly resemble the seed-vessel of a leguminous plant. The larvæ must be kept out-of-doors during the winter, as they do not pupate until the spring, and, if kept indoors, will in all probability die.

Phoxopteryx upupana larvæ occur on *Betula alba* during September. A good account of their manner of feeding will be found in the *Ent. Mo. Mag.*, vol. xxvi., p. 192. The treatment of the larvæ is similar to that of those of *P. derasana*.

At the end of September, imagines of *Paedisca ophthalmicana* occur on the trunks of various species of poplar. It is best to work for them in the morning, and, if possible, to select a dull day ; after mid-day, or in bright weather, they give one little chance to " box " them, being very skittish, and having an unpleasant habit of flying off, so as to place the trunk between themselves and their would-be captor.

The last week in September is the best time to collect larvæ of *Catoptria albersana*, which are to be found in folded leaves of *Lonicera periclymenum*.

By gathering seed-heads of *Lactuca virosa* from the middle to the end of September, the larvæ of *Catoptria conterminana* may be secured. Place the seed-heads in a flower-pot half filled with light soil, and the larvæ will readily pupate in it.

A bag full of flower-heads of *Artemisia maritima*, collected from a salt-marsh, will generally result in a quantity of *candidulana* (*wimmerana*) being reared.

COLEOPHORIDES.—The larvæ of *Coleophora virgaureae* are to be found from September to November, on the seeds of golden-rod ; they inhabit small and cylindrical cases, with some of the loose filaments of the seeddown woven into them and fastened only by the anterior end. The imagines appear in August.

The cylindrical, ochreous-brown cases of *Coleophora olivaceella* are to be found from September to May, on, or near, *Stellaria holostea*, on which the larvæ feed. The cases of *Coleophora bicolorella* are to be found from September to May on alder ; they are something of the shape of the case of *C. viminetella*, but are much stouter and stumpier than the latter, and conspicuously of two colours. [Found much more commonly on hazel.]

GELECHIIDES.—At the end of September, the lively, yellowish-grey larvæ of *Ypsolophus fasciellus* are to be found feeding on sloe, doubling up the leaves and leaving an opening at each end, through which they quickly escape when alarmed. Best obtained by beating.

The larvæ of *Dasystoma salicella* are to be found, in September and October, on *Potentilla anserina* or silverweed. They are troublesome larvæ to feed, eating very little at a time, growing very slowly, and wandering about the breeding-glasses when not eating. They appear to have nothing to do with *Salix*.

CRAMBIDES.—On sandy patches near the shore, where *Salsola kali* grows, the plants should be searched, in September, for larvæ of *Gymnancyla canella*. After a storm of wind and rain, their little webs, which would ordinarily escape observation, now become the certain means of finding them, for the wind blows the sand into the webs, and this remains there to show the whereabouts of the larvæ.

The imaginés of *Ephestia semirufa* lay their eggs in September. The young larvæ can be fed up on nut-kernels ; they pupate in May, and the imagines appear in June. Although the larvæ take kindly to nuts, it is possible that they live on almost any animal or vegetable refuse. Possibly the ♀ parent of those, here referred to, fed on the refuse materials in an old ivy-bush, quite near where it was captured (Wood).

In late September, the young larvæ of *Nephopteryx angustella* are to be found entirely hidden in the red berries of spindle, but may be detected by a circular hole in the side of a berry, through which the frass is extruded.

In mid-September, the larvæ of *Homoeosoma sinuella* are to be found mining in the root-stocks of *Plantago lanceolata ;* the affected roots should be planted in a pot, and kept growing through the winter and spring, the larvæ spinning up at various times during the spring.

In mid-September, the fullfed larvæ of *Cryptoblabes bistriga* are to be found within more or less skeletonised leaves of oak, eating holes through the substance between the veins, always keeping the sides of the leaf folded to within a quarter-of-an-inch of each other by means of a quantity of lightly-spun web, the upper surface being the one generally folded together (Buckler).

In late September, the larvæ of *Pempelia hostilis* are to be found feeding between two leaves of *Populus tremula*, spun together with silk, to which large quantities of frass are adhering. When hunting for larvæ, look for dead or dying aspen leaves spun to a living green one ; the silken tubes or " nests " of the larvæ are in the dead leaves, usually two or three larvæ living together, each in its own gallery, in or between the leaves.

Small fragments of rotten-wood (touch-wood) are best placed in the pots in which you are breeding *Nephopteryx abietella*. In these, the larvæ will bore in September or early October, forming their hybernating cocoons therein, and, later, their pupal cocoons in which, in the spring, they pass on to the pupal stage ; the imagines do not emerge till June. Sometimes the hybernacula are occupied by larvæ that start feeding again in the spring, and take two seasons to come to maturity.

PYRALIDES.—Just above the roots of *Plantago lanceolata* and *P. major*, will be observed galleries of web, in which, about mid-September, will be found fullfed larvæ, or the snow-white tough cocoons, of *Herbula cespitalis*.

The larvæ of *Pyrausta punicealis* are to be found in September, feeding on *Nepeta cataria*, the flower-heads of which they are covering with a number of confused silken threads, not regular galleries. They continue to feed until mid-October, when they spin very tough cocoons in which the larvæ pass the winter, pupating the following spring and early summer (Hellins).

In September and October, the larvæ of *Ennychia nigrata* (*anguinalis*) are to be found feeding on *Thymus serpyllum* and *Origanum vulgare*; at first, they eat only the under cuticle, and thus make small transparent blotches, later they eat small holes quite through the leaf-substance, until, when nearly fullgrown, they consume whole leaves of medium size.

In mid-September, the creamy-coloured (green mediodorsal line) larvæ of *Ennychia octomaculata* are to be found in slight whitish webs, on the underside of the lowest leaves of *Solidago virgaurea*, eating away large portions from them, and thus rendering their own whereabouts conspicuous. [Hofmann says that the larvæ skeletonise the underside of the leaves of *Bellidiastrum michelii*.]

In mid-September, the larvæ of *Endotricha flammealis* are to be found in the grey-brown webs that they spin, among the leaves of hazel, sallow, *Lotus*, &c., the dwellings being divided into from three to five chambers, in one of which they rest, with the tail curled either across the prothorax or head.

The larvæ of *Scopula ferrugalis* are to be found in September, on *Eupatorium cannabinum*, *Stachys palustris*, strawberry, *Arctium minus*, etc.; they hide themselves by drawing together, with white silk, a part of a leaf, or by folding under a part of one edge ; later, they partially join two leaves together, or lie in a very slight and open web made of a few fine threads, which, spun on the under surface of a leaf, create and retain the hollows the larvæ design to dwell in, and where they find secure footing, stretched out on the threads (Buckler).

The young larva of *Scopula olivalis* resides, in the autumn, in the twisted top of a leaf, or under a part of the turned-down edge, or between two leaves partly spun together with white silk, of *Sambucus nigra*, *Galeobdolon luteum*, *Stachys sylvatica*, *Mercurialis perennis*, *Urtica dioica*, *Humulus lupulus*, &c. (Buckler).

The larvæ of *Ebulea sambucalis* are not to be found on the tall bushes with stiff leaves, but on the young growth, of *Sambucus nigra*, a foot or two high. Each larva lives under a whitish silken web, spun on the undersurface of a leaf, and causing a narrow fold, which, though slight, is perceptible even on the upper surface. When the leaf is turned up, the larva is to be seen lying in the hollow, covered with this semitransparent screen of silk, open at each end, and from which, at night, it emerges to feed on other parts of the leaf (Buckler).

The larvæ of *Nascia cilialis* are to be found, about the middle of September and in mid-October, in *Carex riparia*. They are most difficult to obtain, as they crawl deep down into the herbage, where they hide by day, so that they cannot be beaten into the tray, and have to be taken after 5.30 p.m. or 6 p.m., when they appear to come up for the evening's meal.

These larvæ (*Nascia cilialis*) eat large pieces out of the sides of the leaves of *Carex riparia*, a plant known in Wicken Fen as "lisp"; when traces of this feeding are found, the spot should be worked with a lantern at dusk, for a bright yellow larva with an olive-green mediodorsal line, and bright red-purple subdorsal line.

The larva of *Nascia cilialis*, fullfed in September and October, spins up in an old reed-stem or similar hiding-place; it does not pupate until the following June, and is only about a fortnight in the pupal stage.

The larvæ of *Stenia punctalis* feed from September to May, under stones, on vegetable rubbish composed of grass-stems, roots, dead leaves of plants, etc. The species occurs at Freshwater, the east side of Chesil Beach railway, Portland, etc. It is usually very abundant locally, hundreds occurring in its chosen haunts.

From September onwards, searching the plants of *Potamogeton natans*, *Alisma plantago*, *Nymphaea alba*, &c., growing near the margin of ponds, will almost certainly show cases cut by the larvæ of *Hydrocampa nymphaealis* from these and other plants.

In early September, the young larvæ of *Botys asinalis* may still sometimes be collected on *Rubia peregrina*, blotching the leaves conspicuously where they have been feeding, the upper skin that is left untouched forming very noticeable pale areaes. These larvæ hybernate, and are fullfed the following May.

The larvæ of *Botys terrealis* are to be collected in September, on *Solidago virgaurea*; they are usually most abundantly found on plants denuded of flowers and generally shabby, although they feed on the blossoms.

The larvæ of *Scoparia truncicolella* are to be found from September until June, feeding in silken tunnels, bored under or through mosses, growing on stones or the earth, in moorland districts, especially in the neighbourhood of pinewoods or plantations of Scots fir.

In September, the larvæ of *Ebulea verbascalis* are to be found on *Teucrium scorodonia*, making very marked holes in the leaves; they can be easily beaten from the plants, or dislodged by shaking them. They are usually fullfed before the end of the month.

DREPANULIDES.—In September, where birch is common, the spun-together leaves should be examined for the pupa of *Platypteryx* (*falcataria* (*falcula*) (Greene).

In September, the joined-together leaves of beech should be carefully examined for pupæ of *Drepana cultraria* (*unguicula*) (Greene).

GEOMETRIDES.—The larvæ of *Zonosoma punctaria* are to be obtained in September on oak. They rest in a very peculiar position, the food being grasped by the claspers, and the whole remaining portion of the body turned sideways against the foodplant, which gives them a very ludicrous appearance (Porritt).

The larvæ of *Abraxas ulmata* are frequently to be beaten freely, in September, from wych-elm. They must be supplied very frequently with fresh food or the moths bred will be very much undersized. The fullfed larvæ of *Hypsipetes ruberata* remain on their foodplant for five or six weeks before going down for pupation. This they generally do about the middle or end of October, spinning a tough cocoon, mixed with earth and rubbish, on the surface of the ground. [The moths emerge the following May.]

The larvæ of *Hypsipetes trifasciata (impluviata)* are to be found in early September, in curled-up leaves of alder, feeding and living altogether in concealment, either by uniting the leaves, after the manner of the Cymatophorids, or by curling one side of a leaf over the other.

The larva of *Lobophora viretata* has been found in early September, living in a thin and transparent open-meshed web spun round a small umbel of blossom-buds of *Hedera helix ;* several of the buds were eaten out, and a few grains of frass clinging to the web ; it fed up on ivy-buds, went down into the earth for pupation on September 21st, and appeared as an imago in early May the following year.

The larvæ of *Emmelesia alchemillata* are to be found about the middle and end of September, feeding on the flowers of hempnettle, *Galeopsis tetrahit.*

The young larva of *Emmelesia unifasciata* lives within the unripe seedpods of *Bartsia odontites*, which it enters by a hole in the side, remaining hidden until all the seeds are consumed, the frass at the entrance alone showing its whereabouts. After the last moult, it no longer hides itself, and seems to have no difficulty with the ripening capsules and seeds, still making a hole as before in the side, and inserting its head and front segments as far as it finds it necessary to get at the seeds, all the while holding on with its prolegs to the stem outside. It is fullfed in October, and goes just under the surface of the ground for pupation (Hellins).

The young larvæ of *Eupithecia succenturiata* are to be found, in late September and in October, feeding on the leaves of *Artemisia vulgaris*, eating away the upper green skin, so that the leaves appear very soon to be covered with white traces, where, in little patches, the green part has been removed. They are fullfed towards the end of October.

In September, spruce-fir cones should be collected for the larvæ of *Eupithecia togata ;* many of those thrown down, in the localities for this species, will have a copious quantity of fresh frass protruding; in some, as many as six or seven larvæ will be found ; they feed between the scales of the cone upon the ripe seeds at the base. When fullfed, they quit the cones, and spin slight cocoons on the surface of the earth.

The larvæ of *Eupithecia jasioneata* are to be found in mid-September, feeding in the seed-heads of *Jasione montana.* [The plant has been mistaken in the field for *Scabiosa succisa*.] The best known localities are on the cliffs in the neighbourhood of Lynton, in North Devon.

The long slender tapering pupa of *Eupithecia fraxinata* is to be found enclosed in a cocoon under moss, on the trunks of ash-trees in September and October.

In rearing *Larentia viridaria (pectinataria)* from June ova, always keep a sharp eye on the pupæ in September, as odd emergences frequently take place in confinement during this month. [The same is true of many other species.]

HEPIALIDES.—The fourteen or fifteen month-old larvæ of *Hepialus velleda* are to be found in September and October, channelling out the rhizomes of *Pteris aquilina*, in some places leaving nothing but the rind. Sometimes, however, a larva will feed from the outside.

COCHLIDIDES.—The larvæ of *Heterogenea cruciata* are to be found on beech in September and October. The species is widely distributed in beech-woods, but difficult to find.

ANTHROCERIDES.—The larvæ of *Anthrocera purpuralis* should be removed from the foodplant in early September, placed with a quantity of small pieces of dried cork in a tin box, among which they will hybernate successfully, not stirring until February.

XYLOPODIDES.—*Xylopoda pariana* is out from mid-September, and, in its restricted haunts, is often common. Like *X. fabriciana* it is fond of sitting (in the sunshine) on flowers of *Compositae*. During sunless days it may be obtained by beating thatches into a net or an umbrella.

SPHINGIDES.—The pupæ of *Amorpha populi* are usually to be found very near the surface of the ground, generally without any cocoon, and comparatively near where the poplar trees enter the ground; September is the best month in which to find them.

The pupæ of *Smerinthus ocellata* are to be found about two inches below the soil in the neighbourhood of sallow and willow trees and bushes ; a hollow earthen cocoon without silk is made by the larva in which to change.

The pupæ of *Celerio gallii*, resulting from autumnal larvæ, should be kept protected in September and October and then forced. The imagines are rarely perfected in Britain in unforced pupæ.

The larvæ of *Agrius convolvuli* form large, dome-shaped, friable cocoons, in which they change to pupæ ; the latter are very delicate and usually have to be forced in order to obtain the imagines in this country.

DELTOIDES.—The fullfed pinkish-grey larvæ of *Herminia grisealis* are to be found in September, feeding on oak. Each spins a whitish-grey silken web about the middle of the month, in the centre of the uppersurface of an oakleaf, the sides being slightly drawn together to form a hollow in the middle of the leaf, and, in this, the larva changes to a deep chestnut-brown pupa, which gives up its imago about early or mid-June the following year.

NOTODONTIDES.—In September, the larvæ of *Cerura furcula* are full-fed, and may be found on sallow and osier ; the larvæ are very similar to the distorted and discoloured willow-leaves that have been attacked by gall-insects, and are, no doubt, advantaged by the resemblance.

The cocoons of *Cerura furcula* are to be found under the bark and on the trunks of willow in September and October (Greene).

The cocoons of *Cerura bifida* are to be found on the trunks and under the bark of poplars, in September and October, at distances varying from one to three feet from the ground, or on the bark of

aspen, and are somewhat difficult to detect unless placed in a crack
(Greene) ; often about a yard or so from the ground on the trunk of a
poplar, the larva will gnaw out a hollow and mix the gnawed bark
with silk, until the surface is exceedingly similar both in contour
and colour with its surroundings (Hellins).

The fullfed larvæ of *Cerura bicuspis* are to be beaten throughout
September from birch. Tilgate Forest is one of the best known
localities.

The cocoons of *Cerura vinula* are to be found on the trunks of
poplars and willows near the ground, in September, &c. (Greene).

In September, the reputed British species, *Notodonta trilophus*,
forms its puparium by spinning together two poplar leaves, between
which it pupates.

The pupæ of *Notodonta dromedarius* are to be found, at the roots
of alders and birch, in September and October ; the imagines appear
the following May and June.

The cocoons of *Peridea (Notodonta) trepida* are to be found at the
roots of oak, in September, preferably at those trees growing in a sandy
soil ; the larva does not seem to be partial in its selection of corners, as
do those of most Notodontids (Greene).

The purple-coloured, fullfed larvæ of *Leiocampa dictaeoides* are to
be found on, or beaten from, birch, from the middle to the end of
September. [Also towards the end of July.]

The large, grey, silken, earth-covered cocoons of *Leiocampa dictaea*
(larger than those of *Pterostoma (Ptilodontis) palpina*) are to be found at
the roots of aspen and poplar bushes in September. [Also in July.]

The cocoons of *Pterostoma palpina* are to be found in September,
occasionally at the roots of poplars, but much more frequently at those
of willows, especially if they be on the banks of ditches, streams, &c.;
when in such situations, that side of the trunk which faces the stream
is often clothed with grassy sods of loose, dry, friable earth ; shake the
sods well, and the cocoon, which is greyish and of weak consistency,
will generally be found among the dry roots. The cocoon is easily
distinguishable from that of *Leiocampa (Pheosia) dictaea*, being much
smaller and not so much mixed with earth (Greene).

The fullfed variably-tinted larvæ of *Lophopteryx camelina* are to be
beaten from oak, alder, birch and hazel, in September and October ;
they pupate near the surface of the ground.

The gauzy, silken cocoon of *Lophopteryx cucullina* is spun under
loose earth or leaves at the foot of maple-trees, and may be found
from September onwards, until the following June.

The cocoons of *Drymonia trimacula (dodonaea)* are to be found at
the roots of oaks in September and October, often mixed with those
of *D. chaonia* (which species, however, pupates a full month earlier),
the larvæ of *dodonaea* being fullfed about August 25th. They are
sometimes attached to the tree, but more usually are to be found
among the dry friable sods collected in the corners, or even in the
corners themselves without any sod, but, in all cases, great caution is
necessary in obtaining them ; it is a good plan, when a sod has been
pulled out, to gently feel the trunk for any cocoons which may adhere
to it (Greene).

The tough cocoons of *Drymonia trimacula (dodonaea)* are well
covered with fine earth, and, as noted above, are to be found at the

roots of oak-trees, as are those of *D. chaonia*, in late September and October; the pupa is less stout than that of *D. chaonia*, but more glossy.

The cocoons of *Drymonia chaonia* can be distinguished from those of *D. dodonaea* by the brownish-golden colour of the silk lining, and by their larger size and greater strength (Harwood).

NOCTUIDES.—The eggs of *Triphaena subsequa*, laid in September, hatch in about three weeks; the larvæ can be fed, in confinement, on *Potentilla reptans*, *Ranunculus acris*, *R. repens*, and riband-grass; they nibble throughout the winter and feed up in the spring, usually pupating in May.

The eggs of *Noctua dahlii* hatch in September; the young larvæ feed freely on dock (preferring *Rumex crispus* and *R. pulcher*), hybernate when quite small, and commence to feed again in April, rarely pupating before the end of June. In confinement, the larvæ act much as do those of *Peridroma saucia*, *i.e.*, some will feed up rapidly and produce moths in December and January of the same year, others more slowly and produce moths in March, April and May, few remaining until August (the normal time of appearance in nature).

The last fortnight of September to the first fortnight of October is the best time for *Peridroma saucia* and *Aporophyla australis*. Both species occur freely at Deal, Freshwater, Portland, &c. (Rogers).

The eggs of *Anchocelis rufina* are laid in rows, in September and October, on the bark of oak-twigs, elm-twigs, &c., to the colour of which they soon assimilate. They remain as eggs all the winter, becoming much darker towards the beginning of April, and hatch towards the middle of the month. The young looping larvæ feed well on elm-buds when young, but prefer oak to elm later, becoming fullfed in early June, when pupation takes place. (In confinement the larva of this species is somewhat of a cannibal.)

The eggs of *Anchocelis lunosa*, laid in September and October, hatch very quickly; the larvæ, however, feed slowly on grass throughout the winter, and become fullfed in April. (A great deal of variation, from green to olive-brown, occurs in the colour of the larvæ.)

The almost shapeless, shiny eggs of *Anchocelis litura*, laid in September and October, in cracks and crannies, remain as eggs all the winter, hatching in early April, the larvæ then taking from six to eight weeks to come to maturity; they feed freely on bramble, rose, and other allied plants.

The pupæ of *Habrostola triplasia* and *H. urticae* are both to be found, though not commonly, under moss on ash-trees, near nettles, throughout the autumn (Greene).

The young larvæ of *Plusia interrogationis* are to be swept in autumn, when very small, from heather growing on the moorlands. They hybernate from the end of October, commencing to feed again in March.

LITHOSIIDES.—Larvæ of *Lithosia* var. *molybdeola* feed throughout the winter; some, kept in 1867-1868, ate various lichens from trees or banks, wall-moss, withered sallow and oak-leaves, slices of carrot and turnip, knotgrass, &c. They spun up in May, and the imagines emerged in early July (Hellins).

The larvæ of *Lithosia sororcula* (*aureola*) are fullfed in September, when they spin up among the lichens growing on oak-trees, and hyber-nate as pupæ, the imagines emerging the following May and June.

PAPILIONIDES.—The young larvæ of *Polyommatus icarus* make little, pale, transparent blotches on the leaflets of *Lotus corniculatus, Ornitho-pus perpusillus,* &c., the paleness being due to the eating away of the soft parts of the leaf and leaving only the transparent skin ; com-monest in July and September.

The larvæ of *Pararge egeria* can be reared, in confinement, on *Dactylis glomerata,* &c. They appear to nibble throughout the winter, and to pupate as soon as there is any mild weather in the spring, often by the commencement of April.

OCTOBER.

NEPTICULIDES.—The amber-coloured larva (with green dorsal vessel) of *Nepticula betulicola* makes a small contorted gallery (of which the commencement is filled with brown excrement, the latter half having the greenish-grey excrement only in the central portion), in leaves of birch, in October.

The larva of *Nepticula myrtillella* mines the leaves of *Vaccinium myrtillus* in October and November ; the mine is rather broad and considerably contorted, and, when fullfed, the larva leaves it for pupation.

ADELIDES.—The larvæ of *Nemotois scabiosellus* are to be secured at this season feeding on the seeds of *Scabiosa arvensis.* The species is almost entirely confined to downs.

ELACHISTIDES.—The larvæ of *Elachista gangabella* make long puck-ered Lithocolletiform mines in the leaves of *Dactylis glomerata* in the autumn, and pass the winter inside the withered leaves without eating.

LITHOCOLLETIDES.—The larvæ of *Lithocolletis stettinensis* are to be found in October, mining the upperside of the leaves of alder.

GLYPHIPTERYGIDES.—The larvæ of *Glyphipteryx haworthana* can be collected during the winter in the prostrate heads of the cotton-grass (*Eriophorum vaginatum*).

The curious colourless transparent larvæ of *Rösslerstammia erxlebella* are to be found on birch in early October. They make tubular cocoons of very white silk, each furnished with a hinged lid.

TORTRICIDES.—In October, old stunted birch-trees should be searched in Perthshire, &c., for the imagines of *Leptogramma scotana,* which can be seen from a good distance in the localities where it occurs.

The larvæ of *Stigmonota weirana* are to be found from the com-mencement of October, between united leaves of *Fagus sylvatica.* The joined leaves are easily seen by standing under the trees and looking up through their branches.

If discoloured rose-hips be gathered during the first fortnight of October, and enclosed in a box with pieces of virgin cork, a goodly supply of *Stigmonota roseticolana* will most likely be the result. When the larvæ have formed their cocoons in the cork they must be put out-of-doors.

During October, a visit to a pond or stream, in which *Alisma plantago* is growing, will often result in larvæ of *Eupoecilia alismana* being found in the stems of the plants. Tenanted stems should be tied into a bundle, and left exposed to the weather until the following May.

In October, the larvæ of *Phoxopteryx (Anchylopera) mitterpacheriana* are to be found in oak-leaves. the upper surface of each leaf containing a larva, being folded together longitudinally, and the edges of the leaf spun closely together from one end to the other; the larvæ feed on the green cuticle of the surface enclosed between the veins of the leaf.

COLEOPHORIDES.—About the middle of October, larvæ of *Coleophora fuscocuprella* are making numerous small blotches on the undersurfaces of leaves of *Corylus avellana*. These larvæ should be wintered in the open, as pupation does not take place until the following spring. It is useless to collect larvæ found feeding in the early summer, as these, without exception, contain parasites.

The larvæ of *Coleophora salinella* may be collected at the beginning of October, on the seeds of *Atriplex portulacoides*.

In mid-October, the larvæ of *Coleophora binotapennella* are to be found boring the shoots and stems of *Salicornia*, throwing out much ochreous-white frass, and using a piece of the bored stem for a portable case. For pupation they quit their cases and enter the ground.

PLUTELLIDES.—Larvæ of *Swammerdammia griseocapitella* occur about mid-October on *Betula alba*. The larvæ spin silken pads on the uppersurfaces of leaves, causing them to slightly contract. The best time to find them is in the early morning, as then the dew is resting on the silken pads, which are made most conspicuous.

GELECHIIDES.—The strange-looking sluggish larvæ of *Dasystoma (Cheimophila) salicella* are to be found, in October, on *(Potentilla) anserina*; they are exceedingly slow growers, and want considerable patience in breeding them.

CRAMBIDES.—In October and November, the sand about the roots and stems of *Salsola kali* should be passed through the fingers for the ovate puparia of *Gymnancyla canella*, composed of grains of sand, spun together and smoothly lined inside with silk. The imagines will not emerge until the end of July and beginning of August of the following year.

In October, the presence of the larvæ of *Homoeosoma sinuella* is only slightly indicated by the drooping heart-leaves of *Plantago lanceolata*, as the larvæ, feeding on the mined substance of the root-stock, appear, otherwise, to affect the plant little.

The larvæ of *Nephopteryx angustella* are to be collected from the red seed-berries of spindle, in October; they are often plentiful where the berries are abundant; the larvæ grow very rapidly, and want rotten wood in which to spin up in confinement.

Larvæ of *Cryptoblabes bistriga* should be searched for on oak, from the middle to the end of October; at the same time, one is sure to find the small hybernating larvæ of *Rhodophaea consociella*.

PYRALIDES.—Leaves of *Thymus serpyllum* and *Origanum vulgare* are often seen, in October, drawn partly round the stem so as to appear to be a natural result of plant growth, in these, and beneath the fine whitish silk—spun under the leaves—are to be found the larvæ of *Ennychia nigrata (anguinalis)*. The whereabouts of a larva can only be guessed by noticing pieces absent from the neighbouring leaves. The larvæ spin their puparia just after the middle of the month.

Flower-heads of *Nepeta cataria*, covered with silken threads, should be collected in October, for in these will be found the winter puparia of *Pyrausta punicealis*, in which the larvæ remain unchanged until the following spring (Hellins).

The young larvæ of *Cataclysta lemnata* are to be obtained in October and November, and may be kept satisfactorily in a vessel of water with a supply of *Lemna minor* floating therein. Their cases are open at both ends, and they emerge with equal ease at either end, being sometimes wholly, sometimes only partially, submerged. They shut themselves up in their cases for hybernation from early December to March.

In October, the young larvæ of *Scopula prunalis* are to be found in a silken spinning, under the turned-down edge of a leaf of *Galeobdolon luteum*, *Lamium*, &c.

In October, the green, black-spotted larva of *Scopula olivalis* spins an opaque, white, silken, oval, cocoon-like hybernaculum, firmly and closely attached to part of the under-surface of a leaf of *Sambucus nigra*, *Galeobdolon luteum*, *Stachys sylvatica*, *Mercurialis perennis*, *Urtica dioica*, *Humulus lupulus*, &c., the edge of the leaf being so turned down, as to hide the hybernaculum completely (Buckler).

In October, the larva of *Scopula ferrugalis* spins its puparium among its foodplants, *Eupatorium cannabinum*, *Stachys palustris*, &c., cutting out a portion of a leaf, which it draws around itself, and spinning a silken cocoon, in which it pupates; the imago may appear in three weeks, three months, or eleven months after pupation, so uncertain is the length of the pupal period.

At the end of October, the young larvæ of *Ebulea crocealis* may be found spun-up in their hybernacula, under the turned-down tips of the leaves of *Inula dysenterica* and *I. conyza*.

In October, the larvæ of *Endotricha flammealis* leave the trees or plants on which they feed during the autumn, and live on the decaying leaves of sallow, hazel, and hornbeam, gathered about the bases of the trees, maintaining this habit until May, when they spin oval cocoons on the surface of the ground, each one using a decayed leaf in which to form its puparium.

In confinement, the pupæ of *Botys pandalis* should be watched throughout October and November, as late autumnal imagines occasionally emerge therefrom.

CYMATOPHORIDES.—The cocoons of *Cymatophora or* are found very rarely under moss and dry rubbish on and about poplars in October (Greene).

The cocoons of *Cymatophora ocularis* are to be found at the roots of various poplars in early October. The black and stout pupa is enclosed in a very delicate open net-work of a rusty-brown colour, which is extremely difficult to find ; the larva generally spins on the surface of spreading moss, or barely beneath it—sometimes between leaves—and should be sought as soon as possible after the change, certainly not later than the first week in October (Greene).

The larvæ of *Thyatira batis* and *Gonophora derasa* feed by night on blackberry, as a rule under trees, during October.

GEOMETRIDES.—By hanging a lamp on a tree in woods and standing by with a net, *Himera pennaria*, *Ennomos erosaria*, *Oporabia dilutata*, &c., may be taken in suitable localities ; as also may many Noctuids when sugar fails (Bush).

Search old stunted birch-trees in Perthshire (and elsewhere) during this month for imagines of *Cidaria miata* and *Leptogramma scotana*, the latter can be seen from a good distance.

In Perthshire, you may still, in October, sweep *Calluna vulgaris* for the larvæ of *Scodiona belgiaria*, *Anarta myrtilli*, &c. [Also again in the spring.]

You may also, in October, in Perthshire, beat larch for larvæ of *Eupithecia lariciata* and *Odontopera bidentata*. Often pays well to beat at night (Bush).

The pupæ of *Eurymene dolabraria* occur in plenty in October, under moss on beech-trees, in Bucks ; they occur also, but much more sparingly, on oak ; the larvæ enter the moss at the first convenient place, and, therefore, in tearing it off (which should be done with the hand, not with the trowel), great care should be taken in loosening the edge of the moss, for there the pupæ are almost invariably found (Greene).

Like so many other species, the larvæ of *Boarmia abietaria*, though commencing to hybernate in mid-October, will, if sleeved on their foodplant, continue to nibble during the winter and until the end of March, when they moult and feed up very rapidly, being fullfed before the end of April and mid-May.

The larvæ of *Hypsipetes impluviata* are fullfed by mid-October, remaining from that time until the beginning of December in their hiding-places in leaves, &c., when they become pupæ, the perfect insects appearing towards the end of the following May (Buckler).

Eggs of *Chesias spartiata*, laid in October, hatch in March, the larvæ feeding on *Spartium scoparium*, on which they may be obtained throughout April and May ; they are fullfed in June, the imagines appearing in September and October.

The larvæ of *Eupithecia innotata* are to be beaten in October from *Artemisia vulgaris* and *A. maritima*.

HEPIALIDES.—In October, the larvæ of *Hepialus velleda* leave the stems of *Pteris aquilina*, and are to be dug up, when hybernating in the ground, near the roots of the bracken on which they have been feeding.

The one-year old larva of *Hepialus hectus* leaves the root of *Pteris aquilina*, in which it has fed for some fourteen months, in October, and makes an earthen cocoon in the ground, a few inches from the root, in which it remains dormant until the following spring.

COCHLIDIDES.—The peculiar little cocoons of *Heterogenea cruciata* are to be obtained in October, spun on the surface of the leaves of beech. [Only to be found by keen searching.]

ÆGERIIDES.—The usual habitat of the larvæ of *Ægeria cyniformis* is at the open edge of the bark at the top of the stump of a felled oak, where it has been cut across, during the first year or two after felling, but, in older and decaying stumps, the larvæ in the second and third years burrow under the thick bark lower down, and may be obtained from October to May.

ANTHROCERIDES.—The larvæ of *Rhagades globulariae* leave their mines in October, and hybernate on a silken pad, spun up on the underside of a leaf of *Centaurea nigra*.

A dry box, half filled with small dry pieces of cork, makes an excellent hybernating place for larvæ of *Anthrocera trifolii*; they rarely feed after October, but commence to do so again about the commencement of March. Sometimes the individuals of whole broods take two years to come to maturity.

LACHNEIDES.—*Macrothylacia rubi* larvæ may be obtained in abundance during this month by searching in the early evening on downs, by woodsides, &c.; these may be kept successfully by putting a good-sized turf of grass in a box and placing the larvæ thereon; cover with muslin and stand out-of-doors.

SPHINGIDES.—The pupæ of *Hyloicus pinastri* are to be found from October onwards, beneath the surface of the ground near the trees on which the larvæ have fed.

I have found the following method of rearing *Manduca atropos* work successfully:—Feed the larvæ in large flower-pots half-filled with light mould, in which they can pupate; after they have been underground ten days, take out the pupæ, place in pots partly filled with mould and sand, and well-drained. Keep in a warm room, well saturate with water once a week; place damp moss over them every third day to keep the moisture uniform. In this way imagines will sometimes emerge after a pupal period of only three weeks, but a month is nearer the average time (Rogers).

LYMANTRIIDES.—Beat birch-trees during early October, preferably at night, for fullfed larvæ of *Demas coryli*.

The cocoons of *Demas coryli* are generally very plentiful in Bucks, under moss on beech-trees, at the roots, and not on the trunk, in October, &c. (Greene).

NOTODONTIDES.—The larva of *Clostera curtula* forms its cocoon in October, between two leaves of its foodplant, which it joins firmly together, and remains there until the leaves fall off. The larvæ appear to be more abundant on shrubby poplars (Greene).

The cocoons of *Phalera bucephala* are to be found at the roots of various trees in October (Greene).

The cocoon of *Cerura furcula* is made in a neat hollow, gnawed out of the surface of a depression in the bark of a sallow tree, and the materials thus obtained are strongly cemented until the cocoon is quite hard and level with the rest of the bark. The cocoons should be obtained in October.

The larvæ of *Cerura bicuspis* may be beaten from birch and alder as late as early October.

The fine gauzy silken cocoon of *Lophopteryx camelina*, covered with fine earth, is to be found at the roots of hazel, birch, oak, and alder trees, and bushes, in October and November.   [Also in July.]

The cocoons of *Lophopteryx camelina* are to be found in October, &c., very commonly under moss on various trees, beech, elm, &c. They are weak, and the pupæ (unlike those of other Notodontids), terminate in a single point or spike (Greene).

The cocoon of *Lophopteryx cucullina* is to be found under moss on maples, or on other trees near maples, in October (Greene).

In October and November, a sharp look-out should be kept on the breeding-cages for the imagines of *Ptilophora plumigera*, which are due to emerge by the middle or end of the first-named month.

The cocoons of *Notodonta dromedarius* are to be found at the roots of alder in October (Greene).

The cocoons of *Notodonta ziczac* are to be found at the roots of poplar in October (Greene).

NOCTUIDES.—It is profitable during October to sweep, at night, railway-banks and edges of woods where mixed plants abound, for Noctuid larvæ, as here many uncommon species may be obtained; they may be kept through the winter by planting a variety of low plants in a large flower-pot and covering with muslin; most hybernating larvæ will feed upon these until spring, when their own special foodplants may be found.

Look with lamp at night at blooms of ragwort for *Hydroecia lucens*, *Agrotis obelisca*, and other Noctuids, which are rather later in Perthshire than in the south of England (Bush).

Sugar, north side of woods for preference, for *Agriopis aprilina*, *Agrotis obelisca*, *Epunda nigra*, *Orrhodia vaccinii*, *O. erythrocephala*, *Scopelosoma satellitia*, *Noctua glareosa*, and many other species.

The cocoons of *Triaena tridens* are to be found under the bark of hawthorns, in October, &c. (Greene).

The cocoons of *Triaena psi* are to be found commonly under the bark of various trees in October, &c. (Greene).

The cocoons of *Cuspidia megacephala* are by no means uncommon under loose bark on poplars, and occasionally on willows, in October, &c.   They are not very easy to get at, as the larvæ enter the smallest chinks.   Break off every bit of loose bark with the point of a trowel, and the pupa-case, which, with the pupa, closely resembles that of *T. psi*, will be found firmly glued to the surface; the cocoon being formed of decayed wood.

The cocoons of *Craniophora ligustri* are abundant under moss on ash-trees in October.   The moss must be very carefully torn off; the pupa-case, which is black and very tough, but not hard, will, in most cases, be found adhering to the moss; but if there be no moss, examine the trunk.   There are often long perpendicular slits in the bark of ash-trees, and these are favourite hybernacula for *C. ligustri*.   If both loose bark and moss are wanting, go to another tree (Greene).

The eggs of *Hadena protea*, laid in October and November, hatch the following year in late April or early May, and the larvæ feeding rapidly on oak are fullfed before mid-June.

In October, the young larvæ of *Aplecta occulta* can be swept from heather (in Scotland), although, in confinement (and no doubt in nature), they will eat the leaves of a variety of other plants. They are fullfed in May.

The eggs of *Epunda nigra*, laid in October, hatch in about five weeks, and the larvæ, when young, prefer grass as food, but later become somewhat general feeders on low plants, being found in spring on *Galium*, &c., whilst, in confinement, they show a preference for hawthorn. The larvæ are fullfed from mid-May to mid-June.

The eggs of *Epunda lutulenta*, laid in October, hatch in November (about six weeks after they have been laid), and the young larvæ feed freely, in confinement, on *Poa annua*, *Potentilla fragariastrum*. Flowers of wildmint, leaves of scabious, heather, &c., are also recorded foodplants, although it appears to be in reality a veritable grass-feeder.

Sugar in the late autumn has some difficulties; it must be put on well before dusk, for the moths fly early. In mid-October, (the 14th) I put it on at 5.30 p.m. Directly the patches (almost 100) are finished, I light my lamp, and go round furiously until 6.45 p.m., when the flight is virtually over. I have often tried, at this time of year, going round later on, but I always find the moths sitting about, "chewing the cud," and not upon the sugar (Burrows).

ARCTIIDES.—The cocoons of *Spilosoma mendica* are to be found in October, &c., under moss on trees bordering damp ditches in Gloucestershire (Greene).

Pupæ of *Euchelia jacobaeae* are to be found in countless profusion under loose bark on wych-elms, in suitable places in Ireland, the larvæ having crawled up the trees from their foodplant (*Senecio*) for the purpose of pupation (Greene).

LITHOSIIDES.—In September and October, the larvæ of *Gnophria rubricollis* are sometimes abundant in beech-woods, feeding on the tree-lichens; found them swarming once on a lichen-covered park paling (Buckler).

HESPERIIDES.—In October, the larva of *Steropes palaemon* (*paniscus*) draws a leaf of *Brachypodium sylvaticum* into tubular form around itself; lining the inside carefully with white silk, and this forms the hyberniculum in which the larva spends the winter.

PAPILIONIDES.—Small larvæ of *Polyommatus bellargus* are to be found, in October (and July), on the underside of the leaves of *Hippocrepis comosa*, eating out the undersurface for a small space, but leaving the upper skin untouched, which then turns white; these little white dots or spots show, therefore, where the larvæ are at work; they feed slowly throughout the winter, and the blotches are much larger by early February. In March, the leaflets are eaten from the edge, and often demolished entirely.

The larvæ of *Erebia aethiops* commence to hybernate in October, when exceedingly small, hiding in the thickest parts of the tufts of grass, with which they may be supplied. They commence to feed again very early in the spring, as soon as the grass commences to grow, and are fullfed in May and June.

NOVEMBER AND DECEMBER.

ELACHISTIDES.—The larvæ of *Elachista adscitella* are to be found from November to April, making very white mines in the leaves of *Sesleria caerulea*.

The larvæ of *Elachista taeniatella* are to be found in November, forming brownish mines, something like those of *E. megerlella*, in the leaves of a rather coarse grass, *Arrhenatherum avenaceum*, in which they hybernate, fullfed, pupating in the spring in the mines; the imagines appear in April.

GELECHIIDES.—The fat whitish larvæ of *Parasia metzneriella* are to be found in November, feeding below the seeds of *Centaurea nigra*.

The hybernating larvæ of *Pleurota bicostella* are to be found in November, under webs on the stems of *Erica cinerea*.

CRAMBIDES.—Puparia of *Homoeosoma sinuella* may be found in hollow cavities in the root-stocks of *Plantago lanceolata*, from November to June. Although there is such a vast difference in the time that the larvæ are fullfed and spin their puparia, they rarely pupate until May.

The larvæ of *Chilo phragmitellus* are to be obtained in the winter, in the stems of *Arundo phragmites*. They remain in the old stems, and appear to find sufficient food within them for their needs.

The little companies of larvæ of *Nephopteryx genistella* are very abundant in some districts (*e.g.*, Portsmouth) in winter and spring, and, as summer advances, form conspicuous objects on the furze bushes. The larvæ cannot be dislodged by beating, and it is only by pulling their nests to pieces that their presence can be detected.

The larvæ of *Euzophera cinerosella* (*artemisiella*), variable in size, are to be found in November, December, and on throughout the spring, feeding on the central substance of the root-stocks of *Artemisia absynthium*, mining chambers in the lower parts of the stems and the roots whilst thus engaged.

PYRALIDES.—The larvæ of *Aglossa pinguinalis*, form their tubes usually between an oat-bin and wall, into the cracks between which, husks, chaff, oats, and particles of straw find their way, the larvæ being almost always on the floor; or, otherwise, they may be found in cracks near the rack and manger, where chaffy rubbish fills the chinks, which then form suitable places for their tubes.

The larvæ of *Pyralis farinalis* live in long tubes of dusty flour, spun together rather toughly, and placed under projecting ledges in flour-mills, or in stables among the waste food above or under mangers, oat-bins, &c. They may be found throughout the year, and are said to take two years to come to maturity.

In November, the young larvæ of *Scopula prunalis*, spin up in white silken hybernacula, securely attached beneath the edges of the leaves of *Galeobdolon luteum*, *Lamium*, &c.

If breeding *Botys pandalis*, do not forget that the imagines often emerge in November.

GEOMETRIDES.—Search on oak-trees for *Hybernia defoliaria* in November; the females may be found at night upon the trunks, about a foot from the ground; males come readily to light.

Look under birch and beech-trees at night, in November, for females of *Cheimatobia boreata*; they will be found among the leaves upon the ground ; many males flying about denote their presence.

Hybernia aurantiaria may usually be found at the same time and place as *Cheimatobia boreata*.

Hybernia aurantiaria pair between 9 p.m. and 9.30 p.m., and remain paired for an hour or so only.

Himera pennaria may yet be taken (in November), coming very readily to light in certain Perthshire localities and elsewhere (Bush).

The eggs of *Himera pennaria* are to be found from December to April, laid in rows parallel with the direction of the twig, being firmly attached to the twig and to each other by a shining pale red cement. They are laid on oak, ash, sallow, &c.

COSSIDES.—The fullfed larvæ of *Cossus cossus* are to be dug up in the winter in earthen cocoons, which they make in October, and in which they remain coiled up, without pupating until the following May or June, when, without leaving their puparia, they change to pupæ.

ÆGERIIDES.—The larvæ of *Ægeria cynipiformis* may be obtained freely in November and December, but are less likely to be brought through safely than if collected in the spring.

LACHNEIDES.—Search amongst the short growth upon trunks of oak and beech, and many other deciduous trees, for *Poecilocampa populi* : this is an insect that likes, if possible, to get under a leaf still hanging on the tree.

The imagines of *Poecilocampa populi* emerge most frequently between 3 p.m. and 4.30 p.m., and commence flying about almost as soon as their wings are dry, that is at early dusk. They soon damage themselves if not attended to before they begin to move.

Poecilocampa populi pair from 9 p.m. to 9.30 p.m., and separate about 11 p.m.

SPHINGIDES.—The pupæ of *Mimas tiliae* are frequently found behind loose bark of elm-trees; chinks and hollows where a branch has been broken off are favourite pupation-places.

DELTOIDES.—In November, hybernating larvæ of *Herminia derivalis* were placed, with a plentiful supply of fallen oak-leaves, into a calico bag, which, after being sewn up, was tied to a branch of a tree three or four feet from the ground, being brought indoors and placed in the window of a cool room during frost. The larvæ remained there, alive, well, and healthy, until mid-April, when they were just leaving their hybernacula (Buckler).

NOTODONTIDES.—*Asteroscopus sphinx* pair after 12.30 a.m., and remain *in copulâ* till dusk.

Pupæ of *Petasia nubeculosa*, kept out-of-doors all the winter and exposed to the weather usually emerge the spring after pupation; those kept indoors often go over two or more winters.

In November, the eggs of *Ptilophora plumigera* are laid singly, or in little groups of two or three together, on the young brown shoots of maple, to which they assimilate well. The larvæ appear in mid-April, and feed on sycamore as well as maple (Buckler).

The cocoons of *Notodonta dromedarius* are oval, and made of tough, shining, yellowish-brown silk, covered with earth; the pupæ are of a deep rich red-brown, without a cremastral spike; the cocoons are to be found at the roots of alder and birch throughout the autumn, winter, and early spring.

NOCTUIDES.—Sweeping, on banks and under hedge-bottoms, may be carried on with advantage upon any warm nights, for Noctuid larvæ in November (Bush).

Sugar park (or other) palings for *Calocampa vetusta, C. exoleta,* and *Orrhodia vaccinii,* which may be taken in the best of condition in November.

The young larvæ of *Noctua rubi* may be found in November, will feed on heather, dock, &c., hybernate from the end of November until February, eating, however, a little dock, lettuce, grass, &c., on warm days, feed up more rapidly after mid-February, pupating in April, the imagines appearing (as a first brood) in May and June. [The larvæ of the second-brood feed up in June and July.]

Imagines of *Xylina semibrunnea*, taken in the autumn or winter (November-March), should be sleeved on an ash-tree: they possibly pair during the winter, and live until the end of May, if properly fed; they lay their eggs somewhat sparingly on ash-twigs.

HESPERIIDES.—In November, the hybernating larvæ of *Pamphila sylvanus*, about 12mm. long, are to be found in their long, silken, narrow, tough, close-fitting hybernacula, formed by spinning together the edges of the grass-blades, the opaque webs being not much bigger than the larvæ; in confinement, riband-grass forms a useful substitute for the finer grasses; they commence feeding again in March, and are fullfed about the end of May (Hellins).

PAPILIONIDES.—The hybernaculum, in which the larva of *Limenitis sibylla* passes the winter, may be placed three or four buds down from the tip of a twig, shooting out from the main stalk of a large honeysuckle-bine; it is made of a honeysuckle-leaf, which has been first partly bitten through near its axil, and then securely fixed by its two edges, for about half its length, to the twig from which it grows, and across which its edges are firmly bound together with a spinning of strong silk; just at the point where the leaf meets the underside of the twig there is a circular aperture, apparently designed for the egress of the larva in spring; as the leaf withers the hybernaculum becoms puckered, and little more than half-an-inch in length, and has the appearance of a small shrivelled leaf clinging to the dry stem, and would thus easily escape ordinary observation (Buckler).

EXPLANATION OF PLATES.

PLATE I.—EGGS OF LEPIDOPTERA.

Fig. 1. Eggs of *Polygonia c-album*.
Fig. 2. ,, ,, *Sphinx ligustri*.
Fig. 3. ,, ,, *Anthrocera filipendulae*.
Fig. 4. ,, ,, *Hepialus humuli*.
Fig. 5. ,, ,, *Œonistis quadra*.
Fig. 6. ,, ,, *Poecilocampa populi*.

All magnified 20 diameters (except fig. 2, magnified 10 diameters).

PLATE II.—EGGS OF LEPIDOPTERA.

Fig. 1. Eggs of *Dasychira pudibunda*.
Fig. 2. ,, ,, *Notodonta ziczac*.
Fig. 3. ,, ,, *Triphaena pronuba*.
Fig. 4. ,, ,, *Hypena rostralis*.
Fig. 5. ,, ,, *Boarmia abietaria*.
Fig. 6. ,, ,, *Coleophora laricella*.

All magnified 20 diameters.

PLATE III.—EGGS OF LYCÆNID BUTTERFLIES.

Fig. 1. Eggs of *Chrysophanus phlaeas*, on *Rumex*, July 8th, 1900.
Fig. 2. ,, ,, *Polyommatus corydon*, on *Lotus* ?, August 12th, 1900.
Fig. 3. ,, ,. *Polyommatus bellargus*, on *Hippocrepis*, June 11th, 1900.
Fig. 4. ,, ,, *Nomiades semiargus*, on red clover, July 31st, 1900.
Fig. 5. ,, ,, *Plebeius aegon*, on *Erica cinerea*, July 17th, 1900.
Fig. 6. ,, ,, *Polyommatus icarus*, on *Ononis*, June 13th, 1900.
Fig. 7. ,, ,, *Cupido minima*, on *Anthyllis vulneraria*, June 27th, 1900.
Fig. 8. ,, ,, *Cyaniris argiolus*, on holly, May 29th, 1900.
Fig. 9. ,, ,, *Callophrys rubi*, on *Rhamnus catharticus*, June 11th, 1900.

All of same magnification, and hence showing relative size.

PLATE IV.—DOLLMAN'S BREEDING-CAGE.

Fig. 1. Wire gauze meat-safe with tin back, top and bottom.
Fig. 2. Framework of stout wire.
Fig. 3. Tin baking-dish.
Fig. 4. Frame covered on front and sides.
Fig. 5. Apparatus in position.

PLATE V.—DIAGRAMMATIC REPRESENTATION OF ARRANGEMENT OF LARVAL TUBERCLES AND TUBERCULAR SETÆ.

Fig. 1. (a) Tubercular setæ on typical thoracic segment numbered according to Dyar's original system of notation, *i.e.*, under the assumption that the thoracic tubercles were not homologous with those of the abdomen, *viz.*, i, ii, iia, iib, iii, iv, v, and vi. (b) Tubercular setæ on typical thoracic segment; numbered in accordance with their homology with those of the abdominal segments, *viz.*, i, ii, iii(=iia), iv(=iib), subprimary=iii, v(=iv), vi(=v), vii(=vi). [By following these numbers the comparison with the original system of numbering is readily learned.] (c) Typical abdominal segment, numbered according to Dyar's system of notation, and now in general use, *viz.*, i (anterior trapezoidal), ii (posterior trapezoidal), iii (supraspiracular), No. 0 (subprimary), sp. (spiracle), iv (postspiracular), v (subspiracular), vi (flange tubercle), vii (marginal).

Fig. 2. (a) Thoracic segment of adult larva of *Endrosis fenestrella*. (c) Abdominal segment of same species. [Typical of Œcophorid group. Note v subspiracular and iv postspiracular.]

Fig. 3. (a) Thoracic segment of adult larva of *Eurrhypara urticalis*. (c) Abdominal segment of same species. [Typical of Pyraustid and Crambid group. Note iv+v subspiracular on common plate.]

Fig. 4. (a) Thoracic segment of larva in first instar of *Triphaena comes*. (c) Abdominal segment of same larva. Note absence of vi, by some considered a subprimary tubercle. [The plates are much accentuated to make them prominent.]

Fig. 5. (a) Thoracic segment of adult larva of *Triphaena comes*. (c) Abdominal segment of same larva. The development of vi, and increase of setæ on vii on abdominal segments are to be noted. [Figs. 4 and 5 are typical of Noctuid group, although plates are not always so developed in 1st stage (fig. 4), and the subprimary tubercle vi is present in 1st instar of some

138 PRACTICAL HINTS FOR THE FIELD LEPIDOPTERIST.

species. Note v subspiracular and iv postspiracular. In this figure (5) the plates (bases) are more normally shown than in fig. 4.]

Fig. 6. (a) Thoracic segment of larva of *Peridea trepida* in second instar. [Similarity of arrangement of tubercles with those of fig. 5 to be noted.] (c) Abdominal segment of same larva. [Typical of Notodontid group. Note v subspiracular and iv postspiracular but low down.]

Fig. 7. (a) Thoracic segment of larva of *Phragmatobia fuliginosa* in first instar. (c) Abdominal segment of same larva. [Nearly typical of Arctiid group. Note double hair on iii, which is normal on abdominal segments, but abnormal on the thoracic in this group (compare with pl. vi, fig. 6, which is normal); also absence of vi in first instar as in some Noctuids. Plates much accentuated to bring them into prominence.]

Fig. 8. (a) Thoracic segment of larva of *Hyles euphorbiae* in first instar. (c) Abdominal segment of same larva. [Typical of Eumorphid section of Sphingid group. Some Sphingids have i and ii on the meso- and metathorax approximate or on same plate. Note v as prespiracular and iv as subspiracular ; also absence of vi.]

PLATE VI.—DIAGRAMMATIC REPRESENTATION OF ARRANGEMENT OF LARVAL TUBERCLES AND TUBERCULAR SETÆ.

Fig. 1. (a) Mesothorax of adult larva of *Hepialus humuli* showing thoracic setæ. [Note iii and iv on one plate, v and vi on another.] (b) Metathorax of same larva. [Note especially tubercle x belonging possibly to a lost segment, traces of which are also noticeable in Zeuzerids (fig. 3) and in some Noctuids. Also iii and iv still on one plate, v and vi now separated.] (c) First abdominal segment of same larva. (d) Second abdominal segment of same larva. [Typical of Hepialid group. Note the modification of iv and v in the abdominal segments. Also observe that iii is very puzzling, and looks homologous with iii and iv of thoracic segments, which, however, the authorities will not allow.]

Fig. 2. (a) Thoracic segment of larva of *Cossus cossus* (*ligniperda*) in first instar. (c) Abdominal segment of same larva. [Typical of Cossid group. Note iv + v on common plate beneath spiracle in abdominal segment. Also that vi is absent, as in pl. v., figs. 4 and 7.]

Fig. 3. (a) Mesothoracic segment of larva of *Zeuzera pyrina* (*aesculi*) in first instar. (b) Metathoracic segment of same larva. (c) Abdominal segment of same. [Typical of Zeuzerid group. Note trace of lost segment (as in fig. 1), also iv + v on common plate beneath spiracle on abdominal segments. Also similar arrangement to fig. 2c, except that vi is present in this (fig. 3).]

Fig. 4. (a) Mesothoracic (also agrees with metathoracic) segment of larva of *Stenoptilia bipunctidactyla* in last instar, showing primary tubercular setæ (*i.e.*, omitting secondary setæ). [The tubercle marked v should be v + vi; that marked vi is a subprimary, and should be above v + vi in position.] (c) 1st (also agrees with 2nd) abdominal segment of same larva. (d) 3rd (also 4th) abdominal segment of same larva. [Typical of Alucitid group. Note iv + v on same plate beneath spiracle. Also variation of vi and vii on the different segments.]

Fig. 5. (a) Mesothoracic (also agrees with metathoracic) segment of larva of *Eugonia polychloros* in first instar. (c) 1st (also 2nd) abdominal segment of same larva. (d) 3rd (also 4th) abdominal segment of same larva. [Typical of Vanessid group. Hairs spiculated. Note iv postspiracular, v subspiracular. Compare with pl. v., figs. 4, 5 and 6.]

Fig. 6. (a) Thoracic segment of larva of *Arctia fasciata* in first instar. (c) Abdominal segment of same larva. [Typical of Arctiid group. Note double-hairs of iii in abdominal, single in thoracic, segments, also iv postspiracular and v subspiracular. Compare with pl. v., fig. 7, belonging to same group, with which it practically agrees, except in having single, instead of double, hairs on iii on meso- and metathorax. The single hair on thoracic segments is normal. Also compare with pl. v., figs. 4 and 5, the Noctuid type.]

Fig. 7. (a and c) Same segments of same larva in third instar. [Compare carefully with fig. 6, and notice modification. The hairs are much underdrawn. If done in proportion they would hide segment.]

NOTE.—In studying these plates it is to be carefully remembered that in nearly all cases the setæ (hairs) and basal plates are exaggerated, being larger and more distinct than is normal. Fig. 7 on pl. vi is an exception to this rule as regards the size of the hairs.

The minute prespiracular points and the similar anterior marginal one, on dorsal area in front of ii, are much exaggerated in relation to the other hairs and plates. This remark also applies to the subprimary hairs in some of the diagrams.

PLATE VII.—STRUCTURE OF LEPIDOPTEROUS PUPA-INCOMPLETA AND PUPA-OBTECTA.

Fig. 1.   Ventral view of ♂ pupa of *Diplodomá herminata* :—

1. Frontal headpiece.	13. Forewings.
2. Scape of antenna.	14. Tarsi of 3rd legs.
3. Clavola (flagellum) of antenna.	15. 3rd abdominal segment (free).
4. Labrum.	16. 4th ,, ,, ,,
5. Mandibles.	17. 5th ,, ,, ,,
6. Glazed eye.	18. 6th ,, ,, ,,
7. Maxilla (proboscis).	19. 7th ,, ,, ,, (♂)
8. Maxillary palpi.	20. 8th ,, ,, (fixed).
9. Labial palpi.	21. 9th ,, ,, with ♂
10. Base (coxa and femur) of 1st legs.	tubercles.
11. Tibia (and tarsi) of 1st legs.	22. 10th abdominal segment, with anal
12. Tibia (and 12a tarsi) of 2nd legs.	scar.

Fig. 2.   Dorsal view of ♂ pupa of *Diplodoma herminata* :—

1. Frontal headpiece.	27. Metathorax.
2. Scape of antenna and part of clavola.	28. Base of hindwings.
23. Dorsal headpiece.	29. 1st abdominal segment (fixed).
24. Prothorax (with dorsal suture).	30. 2nd abdominal segment (fixed).
25. Mesothorax (with dorsal suture indicated).	Tubercles i, ii. iii on 4th abdominal segment noted.
26. 1st spiracle.	

Fig. 3.   Headparts of ♂ pupa of *Diplodoma herminata* (from an empty pupa-case, so that 1 and 2 are folded over at top, and the posterior portion is seen through the front translucent portion, *x* being the true posterior (or upper) margin) :—

1. Frontal headpiece.	6a. Eye.
2. Scape of antenna.	7. Maxilla (proboscis).
4. Labrum.	8. Maxillary palpi.
5. Mandibles.	9. Labial palpi.
6. Glazed eye.	x. The true upper margin.

Fig. 4.   Ventral view of ♀ pupa of *Manduca atropos* :—

1. Ventral headpiece (no dorsal headpiece).	17. 5th abdominal segment (free).
	18. 6th ,, ,, (free).
3. Flagellum of antenna.	19. 7th ,, ,, (fixed).
4. Small scar indicating labrum.	20. 8th ,, ,,
5. Small scar indicating mandibles.	21. 9th ,, ,,
6. Glazed eye.	22. 10th ,,
6a. Eye.	31. Abdominal flanges above spiracles
7. Maxillæ (no trace of maxillary or labial palpi).	on 5th, 6th, and 7th abdominal segments.
10. Base of 1st legs (? femur).	32. Corrugated anal spike ending in
11. 1st legs.	two points.
12. 2nd legs.	35. ⎫ Female organs on 8th and 9th
13. Forewings.	36. ⎭ abdominal segments.
14. Bulge of forewings (indicating 3rd tarsi beneath).	37. Scar of anus on 10th abdominal segment.
16. 4th abdominal segment (fixed).	

Fig. 5. Dorsal view of ♀ pupa of *Manduca atropos* :—

1. Frontal headpiece.
2. Scape of antenna.
3. Flagellum of antenna.
6. Glazed eye.
15. 3rd abdominal segment.
16., 17., 18., 19., 20., 21., 22. As above in fig. 4.
24. Prothorax.
25. Mesothorax.
26. Prothoracic spiracle and cover.
27. Metathorax with callosities.
28. Base and hind-margin of hind-wings.
29. 1st abdominal segment (no spiracle exposed).
30. 2nd abdominal segment.
32. Anal spike ending in two points.
33. Scar of horn on 8th abdominal.
38. Mesothoracic bulgings.

Fig. 6. Lateral view of ♀ pupa of *Manduca atropos* :—

1. Frontal (ventral) headpiece.
2. Scape of antenna.
3. Flagellum of antenna.
4. Small scar indicating labrum.
6. Glazed eye.
6*a*. Eye.
7. Maxillæ (no trace of maxillary or labial palpi).
10. Base of 1st leg (? femur).
11. 1st leg.
12. 2nd leg.
24. Prothorax.
25. Mesothorax.
26. 1st spiracle and cover.
27. Metathorax with callosities.
28. Base and inner-margin of hind-wing.
29. 1st abdominal segment (no spiracle exposed).
30. 2nd abdominal segment (spiracle exposed).
32. Corrugated anal spike ending in two points.
34. Scar of spiracle on 8th abdominal (abd. segments 2-7 have actual and exposed spiracles).

ARRANGEMENT OF PLATES.

Plate I.	To be placed opposite title-page as	FRONTISPIECE.
Plate II.	,, ,, ,, ,,	PAGE 4
Plate III.	,, ,, ,, ,,	,, 16
Plate IV.	,, ,, ,, ,,	,, 26
Plate V.	,, ,, ,, ,,	,, 22
Plate VI.	,, ,, ,, ,,	,, 24
Plate VII.	,, ,, ,, ,,	,, 34

Index to Parts I, II and III.

General Index.

156

158   INDEX.

THE ENTOMOLOGIST'S LIBRARY
Books written by J. W. TUTT, F.E.S.

[These books are written by an Entomologist for Entomologists. Up-to-date information ; up-to-date synonymy. Entomology treated on lines of modern science.]

The Natural History of the British Lepidoptera.
(A Text-book for Students and Collectors.)

Four volumes. Price £1 each volume, net. Vols. I-IV, £3 7s. 6d. Demy 8vo., thick, strongly bound in cloth.

Volume I contains 560 pp. + vi pp. Volume II, 584 pp. + viii pp. Volume III, 558 pp. + xi pp. Volume IV, 535 pp. + xvii pp.

The most important work ever offered to lepidopterists. The British fauna is merely taken as the groundwork for the thorough revision of each superfamily treated, and the work thus becomes of first importance to all lepidopterists in the world—systematists, biologists, synonymists, phenologists, &c. This important work puts all others of the kind into the shade. It deserves our full attention and recognition, and the opportunity for its study is not to be missed by any student of European lepidoptera to whom it is no less valuable than the Briton (*Berl. Ent. Zeits.*, December 1902).

Monograph of the British Pterophorina.
BY J. W. TUTT, F.E.S.

(Demy 8vo., 161 pp., bound in Cloth.)

A few copies having unexpectedly come to hand, will be sold as long as they last at 5/- per copy.

This book contains an introductory chapter on " Collecting," " Killing " and " Setting " the Pterophorina, a table giving details of each species—Times of appearance of larva, of imago, food-plants, mode of pupation, and a complete account (so far as is known) of every British species, under the headings of " Synonymy," " Imago," " Variation," " Ovum," " Larva," " Food-plants," " Pupa," " Habitat," and " Distribution." It is much the most complete and trustworthy account of this interesting group of Lepidoptera that has ever been published.

Melanism and Melanochroism in British Lepidoptera.
(Demy 8vo., bound in Cloth. Price 5/-.)

Deals exhaustively with all the views brought forward by scientists to account for the forms of melanism and melanochroism ; contains full data respecting the distribution of melanic forms in Britain, and theories to account for their origin ; the special value of " natural selection," " environment," " heredity," " disease," " temperature," &c., in particular cases. Lord Walsingham, in his Presidential address to the Fellows of the Entomological Society of London, says, " An especially interesting line of enquiry as connected with the use and value of colour in insects is that which has been followed up in Mr. TUTT's series of papers on ' Melanism and Melanochroism.' "

The Migration and Dispersal of Insects.
By J. W. TUTT, F.E.S.

Demy 8vo., 132 pp. Price Five Shillings net.

This book, the only one published on this interesting subject, is of first importance to all students of the geographical distribution of animals, and contains the following chapters :—

1. General Considerations. 2. Coccids and Aphides. 3. Orthoptera. 4. Odonata. 5. Lepidoptera. 6. Coleoptera. 7. Diptera. 8. Social Insects—Hymenoptera, Termites. 9. Final considerations.

Only a small number of copies have been printed. It is trusted that all entomologists will, besides supporting the book themselves, recommend it to any libraries in which they are interested or with which they are connected.

The British Noctuæ and their Varieties.
(Complete in volumes. Price 7s. per vol., 28s. per set).

These four volumes comprise the most complete text-book ever issued on the NOCTUIDES. The work contains critical notes on the synonymy, the original type descriptions (or descriptions of the original figures) of every British species, the type descriptions of all known varieties of each British species, tabulated diagnosis and short descriptions of the various phases of variation of the more polymorphic species; all the data known concerning the rare and reputed British species. Complete notes on the lines of development of the general variation observed in the various families and genera. The geographical range of the various species and their varieties, as well as special notes by lepidopterists who have paid particular attention to certain species.

Each volume has an extended introduction. That to Vol. I deals with "General variation and its causes"—with a detailed account of the action of natural selection in producing melanism, albinism, etc. That to Vol. II deals with "The evolution and genetic sequence of insect colours," the most complete review of the subject published. That to Vol. III deals with "Secondary Sexual Characters in Lepidoptera," explaining so far as is known, a consideration of the organs (and their functions) included in the term. That to Vol. IV deals with "The classification of the Noctuæ," with a comparison of the Nearctic and Palæarctic Noctuides.

The first subscription list comprised some 200 of our leading British lepidopterists, and up to the present time some 550 complete sets of the work have been sold. The treatise is invaluable to all working collectors who want the latest information on this group, and contains large quantities of material collected from foreign magazines and the works of old British authors, arranged in connection with each species, and not to be found in any other published work.

Bombycine Moths of North America.
By A. S. PACKARD, Ph.D., Hon. F.E.S., etc.
Price £2 15s. (*new copy*).

Large quarto Volume. Forty-seven full-size chromo-lithograph plates of larvæ, imagines, neuration, etc., of the Notodontid moths ; 10 maps, with letterpress and indexes.

This book, issued by the Smithsonian Institution, is not on sale in the ordinary way, and is only obtainable when occasional copies come into the market. The few copies sold at present have been priced at from £3 3s. to £3 15s. in the sale catalogues. The number of references to this work in the general chapters of Vols. I and II of Mr. Tutt's large work *A Natural History of the British Lepidoptera* suggests its great importance to scientific lepidopterists.

Also for Sale.
The following Rare and Important Entomological Works and Pamphlets.

	s.	d.
The Aleyrodids of California, by F. E. Bemis, 8vo., with 11 full-page plates of structural details. New. Bound in cloth. Gold-lettered on back	5	0
Revision of American Siphonaptera (with a complete bibliography of the group), by Carl F. Baker, with 17 full-page plates of structural details. New. Bound in Cloth. Gold lettered	7	6
Beetles of the district of Columbia, by Henry Uhle, comprising a complete list of 2975 species and detailed ecological notes. New. Bound in cloth. Gold lettered	4	6
The Blepharoceridæ of North America, by Vernon L. Kellogg, with 5 large full-sized plates (including Head-structures, Neuration, Larval structure, Pupal structure, Imaginal structure, Genitalia, and Mouthparts). New. Bound in cloth. Gold lettered	5	0
Catalogue of the Coccidæ of the World, by Maria E. Fernald, M.A., 1st edition, 8vo., 360 pp. Well bound in half-roan. Gold lettered. New. Original cover bound inside. (This edition went out of print in 6 months and is no longer obtainable)	12	6
The Phasmidæ of the United States, by Andrew Nelson Caudell, 8vo., with 4 full-page plates. Well-bound in cloth. Gold-lettered. New. Original cover bound inside	4	6
Monograph of the insects of the order Thysanoptera inhabiting North America, by Warren Elmer Hinds, with 11 full-page plates of details. 8vo. Well bound in cloth. Gold lettered. New. Original covers bound inside	7	6
Genealogic Study of Dragonfly Wing Venation, by James G. Needham, with 24 beautiful full-page plates of excellently reproduced photographic details, and a large number of woodcuts. 8vo. Well bound in Cloth. Gold lettered. New. Original cover bound inside. The most important work ever issued on the subject	10	0

	s.	d.
Coccidæ of Japan, by Shinkai Inokichi Kuwana, with 7 full-page plates of beautifully reproduced details drawn by the author. Large 8vo. New. Bound in cloth and gold lettered. Letterpress in English	7	6
Butterflies of Switzerland and Alps of Central Europe, by George Wheeler, M.A., 8vo. Strongly bound, 5s. 0d. ; interleaved	6	0
Proceedings of the South London Entomological Society for 1895	2	0
,, ,, ,, ,, ,, ,, ,, ,, 1896	2	6
,, ,, ,, ,, ,, ,, ,, ,, 1897	2	0
,, ,, ,, ,, ,, ,, ,, ,, 1899	2	6
,, ,, ,, ,, ,, ,, ,, ,, 1900	2	6
,, ,, ,, ,, ,, ,, ,, ,, 1901	2	6
,, ,, ,, ,, ,, ,, ,, ,, 1902	2	0
Transactions of the City of London Entomological Society for 1896	2	0
,, ,, ,, ,, ,, ,, ,, ,, 1897	2	0
,, ,, ,, ,, ,, ,, ,, ,, 1898	2	0
,, ,, ,, ,, ,, ,, ,, ,, 1899	2	0
,, ,, ,, ,, ,, ,, ,, ,, 1903	2	0
The South-Eastern Naturalist (being the *Transactions* of the South-Eastern Union of Scientific Societies) for 1899, 1900, each	2	0
The South-Eastern Naturalist (being the *Transactions* of the South-Eastern Union of Scientific Societies) for 1901, 1902, 1903, 1904, each	2	6
Monograph of Peronea cristana and its aberrations (with beautiful chromo-lithograph plates), by J. A. Clark, F.E.S.	2	0
Notes on Hydrids of Tephrosia bistortata and Tephrosia crepuscularia, by J. W. Tutt, F.E.S.	1	0
Some results of recent experiments in hybridising Tephrosia bistortata and T. crepuscularia, by J. W. Tutt, F.E.S.	2	0
The drinking habits of Butterflies and Moths, by J. W. Tutt, F.E.S.	1	6
The Lasiocampids, by J. W. Tutt, F.E.S.	1	0
Some considerations of Natural Genera and incidental reference to the nature of Species, by J. W. Tutt, F.E.S.	1	6
Some considerations of the nature and origin of species, by J, W. Tutt, F.E.S.	1	0
The Scientific aspect of Entomology (1) by J. W. Tutt, F.E.S.	1	0
,, ,, ,, ,, ,, (2) by J. W. Tutt, F.E.S.	1	0
A gregarious butterfly—Erebia nerine, by J. W. Tutt, F.E.S.	1	0
The nature of Metamorphosis, by J. W. Tutt, F.E.S.	1	0
Notes on the Zygænidæ, by J. W. Tutt, F.E.S.	1	0
Random Recollections of Woodland, Fen and Hill (1st edition) by, J. W. Tutt, F.E.S.	3	0
Stray notes on the Noctuæ, by J. W. Tutt, F.E.S.	1	0
Presidential Address to the Entomological Society of London for 1896 by Prof. R. Meldola, F.R.S.	1	6
Presidential Address to the Entomological Society of London for 1897 by Prof. R. Meldola, F.R.S.	1	0
Presidential Address to the Entomological Society of London for 1899 by R. Trimen, F.E.S.	1	0
Presidential Address to the Entomological Society of London for 1900 by G. H. Verrall, F.E.S.	1	0
Presidential Address to the Entomological Society of London for 1902 by Rev. Canon Fowler, M.A.	1	0
Distribution of European Erebias, by J. W. Tutt, F.E.S.	0	6
Chortodes morrisii, Morris = C. bondii, Knaggs, by J. W. Tutt, F.E.S.	1	0
A lutescent aberration of Epinephele tithonus, by J. W. Tutt, F.E.S.	0	6
The Variation of Papilio machaon, by W. Farren, F.E.S.	0	6
Collecting Noctuidæ by Lake Erie, by Prof. A. R. Grote, M.A.	1	0
Catalogue of the Palæarctic Dimorphides, Bombycides, Brahmæides and Attacides, by J. W. Tutt, F.E.S.	0	3
Catalogue of the Palæarctic Lachneides, by J. W. Tutt, F.E.S.	0	6
Philosophical Aspects of Entomology, by J. W. Tutt., F.E.S.	0	6
Nomenclature, Phylogeny and Synonymy, by J. W. Tutt, F.E.S.	0	6
Variation and Natural Selection as factors in species formation, by J. W. Tutt, F.E.S.	1	0
Correlation of the results arrived at in recent Papers on the Classification of Lepidoptera, by J. W. Tutt, F.E.S.	2	0
Catalogue of the Palæarctic Psychides, by J. W. Tutt, F.E.S.	0	3

Entomologists seeking books are invited to send their lists of desiderata.

Orders for the above to **A. H. 41, Wisteria Road, Lewisham, S.E.**